Alternative Strategies for
Desert
Development and
Management

Related Titles in the Environmental Sciences and Applications Series

Biswas CLIMATIC FLUCTUATIONS AND AGRICULTURAL AND WATER RESOURCES PLANNING
Biswas DESERTIFICATION STUDIES
Hinckley RENEWABLE RESOURCES IN OUR FUTURE
Holy EROSION AND ENVIRONMENT
United Nations DESERTIFICATION: Its Causes and Consequences

Pergamon Titles of Related Interest

Biswas/Biswas ENVIRONMENT AND MAN
Biswas WATER AND ENVIRONMENT
Samaha et al WATER MANAGEMENT FOR ARID LANDS

Related Journals*

ATMOSPHERIC ENVIRONMENT
CONSERVATION AND RECYCLING
THE ENVIRONMENTAL PROFESSIONAL
NUCLEAR & CHEMICAL WASTE MANAGEMENT
WATER SUPPLY AND MANAGEMENT

*Free specimen copy available upon request.

Alternative Strategies for
Desert
Development and
Management

Proceedings of an International
Conference held in Sacramento, CA,
May 31-June 10, 1977

United Nations Institute for Training and Research

Water
volume 3

Volume 3 in the Environmental Sciences and Applications Series
Series Editors: Margaret R. Biswas
                 Asit K. Biswas

Pergamon Press   New York Oxford Toronto Sydney Paris Frankfurt

Pergamon Press Offices:

U.S.A.	Pergamon Press Inc., Maxwell House, Fairview Park, Elmsford, New York 10523, U.S.A.
U.K.	Pergamon Press Ltd., Headington Hill Hall, Oxford OX3 OBW, England
CANADA	Pergamon Press Canada Ltd., Suite 104, 150 Consumers Road, Willowdale, Ontario M2J 1P9, Canada
AUSTRALIA	Pergamon Press (Aust.) Pty. Ltd., P.O. Box 544, Potts Point, NSW 2011, Australia
FRANCE	Pergamon Press SARL, 24 rue des Ecoles, 75240 Paris, Cedex 05, France
FEDERAL REPUBLIC OF GERMANY	Pergamon Press GmbH, Hammerweg 6 6242 Kronberg/Taunus, Federal Republic of Germany

Library of Congress Cataloging in Publication Data
Main entry under title:

Alternative strategies for desert development.

(Environmental sciences and applications ; v. 3)
Contents: v. 1. Energy and minerals -- v. 2. Agriculture -- v. 3. Water -- [etc.]
Proceedings of the Conference on Alternative Strate-
gies for Desert Development and Management.
1. Arid regions--Congresses. 2. Arid regions agri-
culture--Congresses. 3. Deserts--Congresses. I. United
Nations Institute for Training and Research. II. Confer-
ence on Alternative Strategies for Desert Development
and Management (1977 : Sacramento, Calif.) III. Series.
GB611.A44   1982   333.73   81-23433
ISBN 0-08-022401-6 (set)   AACR2
ISBN 0-08-022402-4 (Vol.1)
ISBN 0-08-022403-2 (Vol.2)
ISBN 0-08-022404-0 (Vol.3)
ISBN 0-08-022405-9 (Vol.4)

Printed in the United States of America

CONTENTS

Conference Papers

Contents

VOLUME III: WATER

Water Policy and Planning

Contents

Water Conservation and Technology

Case Studies

Contents

Contents

Alternative Strategies for
**Desert
Development** and
Management

THE LAW OF UNDERGROUND WATER IN ARID COUNTRIES

P. A. Towner
Chief Counsel, Department of Water Resources
The Resources Agency, State of California

In this era of limits, the arid regions of the world have assumed increasing importance because of their possibilities for development. The limiting resource in
such regions is almost invariably water, with ground water representing an important, often predominant, and sometimes exclusive source.

This paper analyzes the ground water laws of several arid areas in at attempt to
indicate the spectrum of possible legal systems. The spectrum ranges from community control to private ownership. The observation is made that any system of
ground water law can be a "correct" one - as long as it maximizes the goal of effective management of the resource.

Methodologically, the paper considers - in order - the ground water laws of Iran,
Israel, Mexico, and the Southwestern United States.

Iran and Israel represent examples of comprehensive, community-ownership approaches to all water resources. The goal of providing for a system which maximizes the
use of both surface and subsurface waters appears to be realizable under these two
systems.

Mexico's is introduced as an example of hybridization, one which includes private
and community ownership notions. The legal framework for effective ground water
management is provided by broad governmental regulatory powers.

In the Southwestern United States, there has developed a variety of legal approaches to ground water. Considered in turn are the ground water laws of Texas, Arizona, California and Utah. The selections represent samples of the major variations
of Southwestern ground water law - absolute private ownership, reasonable use,
correlative rights, and appropriation.

In the Southwest, Texas is the best example of the common law doctrine of absolute
ownership - the right of an owner of overlying land to extract underlying ground
water almost without limit. However, even in Texas the strict common law notion
of pure private ownership has yielded somewhat and well-drilling is regulated in
certain critical areas.

Arizona and California represent two major modifications of the common law rule of absolute ownership. Their ground water laws generally limit the overlying land-owner's right to extract ground water to amounts necessary for "reasonable" bene-ficial use. California's system, with greater emphasis upon community ownership, labels its doctrine as "correlative rights". Recent California legal developments which will hopefully allow more efficient management of the ground water resource - including its use in conjunction with surface water - are highlighted.

An analysis of Utah's ground water law leads to the conclusion that it represents an example of the doctrine of "appropriation". In this jurisdiction, the State exercises significant and comprehensive control over both surface and subsurface water.

In summary, any system of ground water law can be a good one if it applies modern hydrology to achieve the highest and best water use.

We hear much these days of spaceship earth and of the limits of the spaceship and our need to live within them. The many people of the world who are living in arid environments have always been confronted with the most critical limit of all: wa-ter. Subsurface water - ground water - is of great importance in these arid re-gions.

To manage ground water effectively, two things are needed - knowledge of its phy-sical characteristics and appropriate legal guidelines.

Today I will discuss the ground water laws of several arid regions. These regions are the Middle East, with particular reference to Iran and Israel, Mexico and the Southwestern United States.

MIDDLE EAST

Few peoples of arid areas have records of longer experience with ground water than those who inhabit the Middle East. Before the Prophet Mohammed there was no sys-tematic water law and the region's various Bedouin tribes and more settled areas had different customs which embodied both individual and community ownership prin-ciples. Intertribal and interareal conflict over water was common.

In his teachings Mohammed proclaimed a spirit of community. The Prophet consider-ed all Moslems to have common ownership of water, pasture, and fire; and at least to some believers, he forbade the sale of water (1).

Now, however, the water laws and customs of many Moslem countries embody signifi-cant principles of private ownership, including marketability. Accordingly, water laws in the Middle East vary considerably. They may be owned privately or collec-tively - and if collectively then by either the tribe, city, or state. They may be subject to rigid state control or they may be sold, subject to various restric-tions. A system very different than that in the West is the concept of water mar-kets. A United Nations study makes the following description of these markets:

> "In some places, water markets are held at certain spots
> and at set times (at sunrise or dusk, at the beginning or
> end of a rotation period or, exceptionally, at a compulsory
> convocation or voluntary meeting of co-owners). Sale,
> purchase or rental of water rights can take place by public
> auction (dellal), depending on demand, season and price, or
> between private individuals in the presence of administrative
> or religious authorities or of the water distribution offi-
> cial. Prices vary with the amount of water, the require-

ments of the persons concerned, the season, the time of
distribution, the quality of water, etc. (2)".

The same publication also notes the tendency, readily observable in all arid environments, that as water becomes scarcer in such regions, the land proportionately
becomes of secondary importance to the water on which its fertility depends. Similar water markets, perhaps reflecting Moslem influence, exist in arid portions
of Spain (3).

Iran

One type of legal system which has developed in the Middle East recognizes only a
community interest in water. It is in Iran, perhaps, that this legal system has
reached its most extreme development.

Iran, of course, has a long history of highly centralized government stretching
back to the early Persian empires. It is also an extremely arid country, with 90
percent of its territory receiving only low and intermittent rainfall. Finally,
it is a country which has always seemed to demonstrate a certain uniqueness in water matters - an example of which is the development of an ingenious system of
deep wells and water delivery, still effective after almost 7,000 years.

It is not surprising, therefore, that Iran should represent a prime example of a
country which has attempted to implement a coherent national water development
plan by nationalizing it. The 1968 Water Nationalization Act provides that all
water, whether surface or underground, is part of the national wealth and belongs
to the public.

As far as ground water is concerned, what this has meant is that any drilling or
use of the resource, other than in minor amounts, is subject to permit. The permits are issued by the Iranian Ministry of Water and Power. They are narrowly pro-
scribed, are limited to use on a particular parcel of land, and are transferable
only upon approval of the Ministry. The Ministry also undertakes surveys and studies of all water resources (4).

Israel

Israel has also adopted a comprehensive solution to the problem based on community
ownership of the resource. The roots of Israel's water law reach back almost
5,000 years. The Talmud adopted the principle of community ownership of water and
detailed water use regulation. In present times the Water Law of 1959 represents
the current comprehensive statement of legal rights to the water resources of Israel. Like Iran, this law provides for a unified public control of all waters
with no individual ownership.

Once again a rigid permit system is prescribed and, in Israel, is administered by
the Water Commission. The permits are issued after the Commission considers the
hydrology of the region, the effect on existing permits, the degree of beneficial
use, and the needs of the community and the nation. As with Iran, stiff fines and
penalties are imposed upon those who either waste or misuse water. Water is generally metered and the Commission is highly aware of the benefits accruing from
conjunctive usage of surface and ground water. The Commission exercises continuing jurisdiction over its permittees and it has the power, among others, to institute rationing. One especially interesting legal rule allows a permittee who
saves water by his conservation efforts to retain that incremental water for his
own beneficial use. Hydrological knowledge is well-developed in the region -
which greatly facilitates effective administration of the Israelis' water management system (5).

MEXICO

The Constitution of Mexico, adopted in 1917, provides that: "The land and water within the national territory belongs to the nation which has had an has the right to transfer their control to private parties (6)".

In implementing this constitutional provision, the Mexican government has recognized the rights of property owners to the underlying ground water, but the Mexican Water Resources Ministry exercises broad regulatory powers to oversee its extraction and use. Accordingly, wells and new drilling are carefully recorded throughout the country, particularly when a region has been declared a "prohibited zone". The declaration of such a zone requires installation of metering equipment to aid in the effective management of the aquifer. Such management by the Ministry has assumed increasing importance in recent years because of the acceptance of the concept of conjunctive management of surface and ground water, now recognized as crucial in the region because of high evaporation losses. As a result, Mexican ground water law represents a type of hybrid law, one which includes notions of both private and community ownership.

SOUTHWESTERN UNITED STATES

This region's chronic water problems have, in less than 150 years, generated a surprisingly diverse collection of ground water laws.

I will restrict this discussion to percolating ground water - water which flows or trickles through the soil or alluvium to the ground water table (7). Most states do, in fact, presume ground water to be percolating and tend to treat the nonpercolating varieties according to surface water principles.

Four major legal theories have evolved in this region: (1) the common law rule of absolute ownership ("rule of capture"), (2) the reasonable use modification of the common law rule, (3) the correlative rights doctrine, (4) the rule of prior appropriation. All four of these theories fit somewhere upon a continuum which stretches from private ownership to community ownership and roughly in the order given. I will give one example of each.

Common Law Rule (Texas)

Our first example of Southwestern ground water law is the common law rule of absolute ownership. Historically, the doctrine was introduced into the English Common Law tradition by the 1843 case of Acton v. Blundell (8). The court based its decision on two major grounds: First, there was the privileged nature of private property inherent in the English common law - a right which gave the property owner title to everything from the center of the earth to the heavens. Second, the court pointed out that the ways of ground water were too mysterious and unknowable and, hence, dictated a measure of judicial self-restraint. Given the advances in knowledge about the behavior of ground water, most consider this basis for the rule is no longer valid.

Although many Southwestern states at one time applied the rule, Texas is the only one which applies it today.

A key Texas case had this to say in summarizing the doctrine:

> "An owner of land had a legal right to take all the water
> he could capture under the land that was needed by him for
> his use, even though the use had no connection with the
> use of land as land and required the removal of the water

from the premises where the well was located (9)".

The landowner also has the right to sell this ground water. In fact, the landowner's right to extract ground water can only be limited in Texas if he is acting with malice or is willfully wasting water - limitations which, though minimal, do represent notions of community control.

In 1949, the Texas legislature departed somewhat from the common law rule when it authorized creation of Underground Water Conservation Districts. These districts may require permits for the drilling of wells in certain areas and also subject such permits to conditions which are designed to prevent waste. Among other things, proper well spacing is a prime consideration. These districts seem to provide a useful service in assuring that particularly delicate ground water basins are subject to at least some water management principles, albeit at the cost of departing somewhat from pure private ownership. Thus, though it is true that the Texas framework probably continues to represent the most authentic embodiment by an arid state of the common law rule, even it has departed in favor of limited community control.

Reasonable Use (Arizona)

The Southwestern United States has, in fact, experienced a general retreat from strict concepts of private ownership of underlying ground water. One avenue of retreat resulted in modification of the original common law doctrine in favor of "reasonable use" - our second example of a major Southwestern ground water scheme. The most notable adaptation of this relatively widespread principle probably occurs today in Arizona. In this state, the rule generally limits the overlying landowner's right to ground water to an amount necessary for "reasonable" beneficial use. Thus, the legal problems arise for an Arizona property owner whenever his usage is judged to be "unreasonable" or where he attempts to export (or sell) ground water to the detriment of other overlying property owners. Arizona further erodes the unfettered private property tradition by statutorily requiring a permit for the drilling of wells in critical ground water areas. In general, however, the community interest in the ground water is protected by court decisions rather than by statutory enactment. In doing this the Arizona courts must continually apply the intrinsically vague test of "reasonableness" to the circumstances of individual cases. I say this not by way of criticism but only to emphasize that as a result of such a system, ground water law is developed in an ad hoc manner by judges rather than by legislatures.

Correlative Rights (California)

In California our departure from strictly common law principles resulted in a unique hybridization of community and private property concepts, eventually culminating in the correlative rights doctrine. This doctrine, the third of our Southwestern examples, was introduced by the 1903 case of Katz v. Walkinshaw (10).

In a nutshell, the doctrine recognizes the rights of all overlying property owners to share equally and reasonably in the underlying ground water. Only if that water is excess to the needs of the owners can there by legally permissible use by nonoverlying users. If there is not enough water to satisfy all reasonable needs of the overlying properties, then apportionment is to be made.

The 1949 case of City of Pasadena v. City of Alhambra (11) established the principle of mutual prescription. In prorating the ground water from an overdrafted basin, a court will, under this doctrine, order allocations to be made on the basis of a property owner's use of water during the five-year period which preceded court action. The doctrine encouraged the extravagant uses of water in order to insure larger court-ordered allocations.

The doctrine was modified in 1975 by the decision in the case of City of Los Ange-les v. City of San Fernando (12). In this case and that of Niles Sand and Gravel Co. v. Alameda County Water District (13), another recent case of importance, the courts have recognized the paramount public interest in the use of ground water and storage space in ground water basins. It has been observed that:

> "Between them, the two decisions have judicially established
> four public rights that, taken together, represent the general
> underground storage right and that of critical importance if
> California is to realize the full potential of its ground
> water basins:

> "1. The right to store water in a natural underground basin
>   without compensating overlying landowners;

> "2. The right to protect the stored water from expropriation
>   by others and from inequitable operational burdens;

> "3. The right to recapture the stored water when it is needed;
>   and

> "4. The public's priority to store water underground when
>   there is a shortage of underground storage space (14)".

But these cases are only small victories in the battle for sound legal principles; and the present drought has only served to intensify our concern as to the capability of our sometimes patchwork water law to permit an effective program of water management. It is likely that California's ground water law, like that of so many other jurisdictions, will continue to handicap such management until surface and ground waters are consistently treated as aspects of the same resource - a view strongly endorsed by the National Water Commission (15).

Then, too, much of our law is judge-made, based upon terms like "correlative rights" and "reasonable use", rather than being parts of a coherent and consistent legislative legal framework. The California Legislature has become increasingly aware of the importance of ground water - as evidenced, for example, by the 1961 enactment of the Porter-Dolwig Ground Water Basin Protection Law (California Water Code Sections 12920, et seq.) which permits the State a role in planning for ground water basins. Still, there remains a need here in California for a more holistic legal approach to ground water management.

In this connection, our State's Director of Water Resources, Ronald B. Robie, had this to say:

> "In my opinion, the acceleration of groundwater problems
> and general absence of solutions since(1962) ... have
> demonstrated beyond any doubt that 'ad hoc' solutions are
> not satisfactory. I find it curious that although regula-
> tion of surface waters is properly a responsibility of the
> State, groundwater regulation is somehow viewed as a
> 'local' concern ... (T)he lack of adequate state authority
> in allocating water resources encourages poor local decisions,
> since the restrictions and conditions imposed on surface
> diversions are often avoided by use of groundwater supplies.
> The result is uncoordinated administration of interrelated
> resources (16)".

Appropriation (Utah)

Turning our attention from California and its correlative rights doctrine, we now consider the last major type of Southwestern ground water law. This is the principle of prior appropriation and it represents the current trend of the law in the whole region (Kansas, Nevada, New Mexico, North Dakota, Oklahoma, Oregon, South Dakota, Washington, and Wyoming are all "subscribers"). It is also the principle that, potentially, at least, embodies the largest amount of community control. It should not be surprising that the Southwest stands in the foreground as far as adopting such a principle since it is nothing more than the ground water analog to the water rights permit system initiated over a hundred years ago in this region to replace the principle of riparian rights.

Utah is a good example because of the comprehensiveness of its solution and because Utah has adopted and rejected two of the rules which we have just considered - absolute ownership and correlative rights.

It was in 1935, when the State declared all waters to be public property, that ground water became subject to appropriation. The statutory scheme which was adopted prescribes that applications to appropriate water are made to the State Engineer. A permit is granted only if it will not interfere with more beneficial use or be to the detriment of public welfare. In an important Utah Supreme Court case, the State's goal of maximum beneficial use of its ground water resources was underscored:

> "It will be seen that the result will much better serve
> the group (all users and society) by putting to beneficial
> use the greatest amount of available water, and ultimately
> also for each individual therein, than would any ruthless
> insistence upon individual rights which simply results in
> competitive drilling of deeper and deeper wells (17)".

The State Engineer's jurisdiction is quite comprehensive, since his permit authority encompasses both surface and subsurface water. The result is a good example of a legal framework which encourages effective management of the State's water resources.

CONCLUSION

In conclusion, I want to emphasize my purpose here today - to display to you the spectrum of possible legal solutions to the ground water problems which people in various areas of the world have grappled with over the years. As we have seen, some legal systems are rooted deeply in the past with only minor modifications having been made to accommodate current technological developments and cirumstances. Other systems are better designed to cope with modern day demands. It is up to us to work to ensure that the ground water laws which prevail in our own arid regions are the type which will foster rather than inhibit the most effective use of our valuable water resources.

REFERENCES

(1) Irrigation and Drainage Paper #20/1. (1973) In Water Laws in Moslem Countries, D. A. Caponera, ed. Food and Agriculture Organization of the U.N., Rome, 12-13.

(2) Ibid, 30.

(3) Ibid, 28-29.

(4) Ibid, 74-96.

(5) Legislative Study No. 10.   (1975) Water Laws in Selected European Countries,
     Food and Agriculture Organization of the U.N., Rome, 86-116.

(6) Framji, K. and I. Majahan.   (1969) Irrigation and Drainage in the World,
     Vol. 2.   International Commission on Irrigation and Drainage, New Delhi,
     702.

(7) State of California, Dept. of Water Resources, Sacramento, California,
     California's Ground Water, Bulletin, 118, 4 (1975).

(8) Acton v. Blundell, 12 M & W 324, 152 Eng. Rep. 1223 (Ex. 1843).

(9) Corpus Christi v. Pleasanton, 154 Tex. 289, 276 S.W. 2d 798 (1955).

(10) Katz v. Walkinshaw, 141 Cal. 116, 70 P. 663 (1902-3).

(11) City of Pasadena v. City of Alhambra, 33 Cal. 2d 908, 207 P.2d 17 (1949).

(12) City of Los Angeles v. City of San Fernando, 14 Cal. 3d 199, 537 P.2d 1250
     (1975).

(13) Niles Sand and Gravel Co. v. Alameda County Water District, 37 Cal.App. 3d
     924, 112 Cal. Rptr. 846 (1974), cert. den. 419 U.S. 869 (1975).

  4) V. Gleason, Water Projects Go Underground, 5 Ecology Law Quarterly, 625,
     667-68.

(15) National Water Commission. (1973) Water Policies for the Future, Arlington,
     Virginia, 233.

(16) Robie, R. B. (1973) Proc. Ninth Biennial Conference on Ground water, Water
     Resources Center, U. of California, Davis, California, 146-147.

(17) Wayman v. Murray City Corporation, 23 Utah 2d 97, 458 P.2d 861 (1969) p.103.

EXPERIENCE TRANSFER: ARIZONA TO THE SAHEL, OR, PASSING
ON THE STAIRWAY

Patricia Paylore, Assistant Director,
Office of Arid Lands Studies, University of Arizona, Tuscon

When UNITAR agreed, with somewhat less than enthusiasm, to allow me to speak here
under the aegis of the Sierra Club International, its representative said my proposal for a topic seemed "a little strong on views and weak on facts," so he urged
me to give you the facts, just the facts, ma'am.

Easier said than done. What is a fact? One man's nectar, another man's poison.
For instance, just one month ago the first major revision to Arizona's groundwater
code in 25 years became law, and its immediate impact was to maintain temporarily
the status quo in continued pumping rights for the City of Tucson and those neighboring mines drawing on a critical underground aquifer common to both. Now Tucson
is already the largest municipality in the United States to live entirely on pumped groundwater, so our Governor praised the Arizona Legislature for providing "this
major step toward finding lasting, long-term solutions to Arizona's water problems,"
but the Tucson City Attorney's Office called the legislation "a mishmash of compro-
mises." What facts are to be derived from these opposing views of the same action?
The fact is that a new 25-member study commission has also been appointed to draft
a comprehensive revision of our groundwater code by 1979, and if the Legislature
fails to enact a new code by September 1981, over four years away, the new study
commission's draft recommendations will become law.

Do these bare facts tell you how to transfer this experience to the advantage of
any other water-short area in a similar environmental dilemma? Does the fact that
this legislation permits Tucson to double its pumping in an adjacent groundwater
basin take account of the fact that of the nearly five million acre feet of pumped
groundwater used annually in Arizona, statistics consistently put the recharge to
Arizona's groundwater reservoirs at 1.8 million acre feet, thus compensating for
only about one-third of current withdrawals? There has been a dramatic rapid post-
war increase in groundwater pumping with a concomitant lowering of the water table,
especially in critical groundwater areas. Because present patterns of demand require amounts of water significantly in excess of surface delivery capabilities,
the resulting supplementary use of groundwater must provide some 60 percent of the
total water used. To change Arizona's archaic groundwater laws to conform to these
hydrologic realities - facts, that is - will require the kind of legislative cour-
age that we have yet to see, to overcome the entrenched special interests that still
dominate our law-making bodies. Legislative intransigence, as demonstrated last

month, has passed the buck on to another commission to struggle with, so that those who opted for this solution can go home and get re-elected because it won't be their fault but rather that of the commission, when, in a few years, the cost of pumping, exacerbated by cost increases in fuel to run the pumps, forces water rates up, and agricultural lands purchased and retired to provide water in critical groundwater areas of industrial and municipal needs become modern dust bowls.

Compounding the chaotic groundwater situation in Arizona are Indian reservation water resources that several tribes are now attempting to control, against all the evidence of history, custom, habit, practice. The 1908 Winters doctrine will surely be invoked increasingly, and court cases will be filed and heard and appealed and finally determined, it is certain, by the U. S. Supreme Court itself.

Meanwhile, the water tables are being lowered, intra-governmental bodies quarrel about highest use, millions are spent on studies, reviews, consultations, surveys, canvasses; elected officials are recalled and their replacements are denounced by the very people who put the anti-incumbent coalition together. And the water? Are we worried that the same fate that has overtaken our California neighbors will be visited upon us? No, we are not. The Mayor of Tucson says comfortingly we are not about to run out of water.

Facts? Nobody knows how much groundwater there remains in the Tucson basin, how much deeper we shall have to pump, what its quality will be, what the rate of recharge will be, or how long it will last.

Perhaps we might have better luck with facts about Arizona's surface water. Fossil water, groundwater, is definitely a non renewable resource whose depletion can contribute significantly to desertification, even though there may be arguments that surface water is not, or that recharge makes even groundwater renewable. These may be no more than semantic arguments, but the frightening reality is that for the time being there is no guaranteed way of appreciably increasing runoff to surface supplies other than by evaporation control, elimination of phreatophytes, cutting down our forests and converting to grassland, as has been proposed by some scientists; increasing storage devices such as catchments, reservoirs, and lakes; or cloud seeding to increase precipitation. This gives us plenty of options, you say. We are indeed working on all these choices, and have been - long enough to know what the problems are:

> 1) evaporation control efforts on large bodies of water such as Lake Powell, or Lake Mead with its 550 miles of shoreline, are difficult, expensive, and unproved, and the total evaporation losses from our two million acre feet of other surface waters is on the order of 95 percent. Arizona is not alone in this statistic, for it joins with 16 other western states by contributing its share of the 17 million acre feet lost annually by evaporation from bodies of surface waters in the region - more, incidentally, than the total allocation of Colorado River streamflow under the Compact

> 2) elimination of phreatophytes as a method for increasing stream-flow has its proponents, but it has its dogged opponents, too, and they are in a standoff position currently. The tradeoffs - increased streamflow at the expense of wildlife habitat, for instance - are typical of those that represent the dilemma we must resolve in Arizona is to survive as a desirable place to live

> 3) the sacrifice of forest resources to increase runoff from high watersheds that provide surface water for the arid lands below is presently a controversial suggestion, meeting vehement opposition,

and as yet theoretical except for isolated experimental plots

4) additional storage devices are feasible from an engineering point of view, but evaporation is ever present, and on small catchments where control is successful, savings are insignificant compared with needs

5) weather modification, still experimental, is also fraught with legal complications that have spawned a whole new class of lawyers

Yet among these options, there must be opportunities for modifications, combinations, new devices, and technologies that will enable us to capture our surface waters, impound them, conserve them. Many factors enter into the equation: the ambient temperature, for instance; humidity, barometric pressure, wind, solar radiation, electrical conductivity, and water temperature all make of evaporation control a complex and frustrating issue, adding its own dimension to the overall need to construct an efficient water management plan to satisfy our needs to exist here in ever-increasing numbers. Just let us not be lulled into complacency by the statistic of 80 million acre feet received in an average year in Arizona from rain and snow, for only two million of those 80 are captured for man's use.

And what of that sacred cow the Central Arizona Project? Will it not relieve us forever of the bugaboo of threatened desertification when our groundwater is gone and our other surface waters used beyond their limits? Or is it likely to become a "stop-gap measure" for reformed groundwater use as President Carter urged in tentatively agreeing to remove it from his list of condemned water projects? As was pointed out in a recent unpopular-in-Arizona television special, it now has a life and a thrust of its own, heedless of statistics, reality, cost, past, or future. Its costs, the amount and impact of such water as may be delivered, depend on assessment of variables most people do not even know about - so I cannot give you facts about the Central Arizona Project, either. I can only go on asking uncomfortable questions.

The legal allocation of the Colorado River's water supply is, admittedly and unfortunately, based on a short historical record which included some unprecedented wet years. Whereas a total of some 16 million acre feet have been allocated in perpetuity to the upper and lower basins, plus what is owed Mexico, long-range studies show the River's average flow to be somewhere between 11.5 and 13.5 million acre feet. If all legally-entitled rights to the Colorado's water were drawn off by all potential users, there simply would not be enough water for everyone's entitlement. While the amount of the deficiency depends on the authority being cited, even the most sanguine of projections are dependent upon an assumption that the upper basin states will not be developing their projects as quickly as originally anticipated. This assumption could well be reversed, depending on the water needs and implementation of oil shale recovery projects in Colorado, strip coal mining in Montana, thousand-mile slurry from Wyoming to Arkansas, power generating plants in the Four Corners, so our short-term solution to Arizona's water shortage, to bring water from the Colorado River via the Central Arizona Project, is predicated on a supply that may not be dependable unless that supply in turn is increased by the import of additional water by out-of-basin transfers, such as from the Columbia River. Robbing Peter to pay Paul, it's called. Will this happen? Should we be confident enough that it will to stake our future and perhaps our very lives on this particular set of cirumstances? It is a chancy path to follow.

The scientific and technical steps we could be taking to bring our soil and water resources into equilibrium with the demands made upon them are on a collision course with economic and political realities. And the fall guy in this scenario is our state itself.

Arizona has made many mistakes, and I detail some of them here in sorrow, even anguish, to help other areas with comparable environmental problems take heed. How many of us, and you, are asking what the prevailing attitudes are that influence decisions relating to the management of natural resources, in Arizona or in the Atacama? Does one need take precedence over another, in Arizona or in Mauritania? Are physical needs more urgent than emotional, in Arizona or anywhere in the Sahel? To develop, or not to develop, in Arizona or in Sinkiang? Are we recklessly wasteful in our use of resources, in Arizona or in Nigeria? In Arizona are such uses needed simply to sustain us in the lifestyles to which we have become accustomed? Have we in Arizona been anesthetized by our abundant society?

Well, if you think I am asking more questions than I am giving answers, your are right. And I have more for you:

Whence cometh the water for the artificial lakes: Mohave, Imperial, Roosevelt, Apache, Lyman, San Carlos, Powell, Mead? How satisfy the fisherman who expects to find these artificial lakes plentifully stocked regularly and artificially with hatchery-reared fish? especially when in dry times the runoff doesn't fill those lakes and the fish already stocked lie gasping in the mud that lines the lowered shore? How accommodate the increasing number of tourists from all over the world who long to ride the Colorado on powered rafts, thinking they are emulating John Wesley Powell and seeing it as he did over a hundred years ago? especially when in dry times not enough water is released from upper dams to float those rafts? In this dry summer of 1977 the 22 commercial river runners who make a living hauling those eager tourists down the River are crying bitter tears because the Bureau of Reclamation won't release the water earlier than planned, so they are faced with the unexpected expense of lifting their customers out by helicopters because the rafts are stranded in low water. Whose interests are at stake in this absurd scenario? Is it for this reason we built those dams and impounded all that water? You tell me the facts, even though I know already what you are going to say!

So how do we manage the spatial distribution so that increased user-intensity associated with all our recreational facilities based on non-renewable resources will not exceed the environmental carrying capacity? When does the breakover point become the breakdown point, in Arizona or in Kenya?

Just as Tanzania is not necessarily only boating on Lake Victoria, or South Africa is not only the best surfing in the world along its Indian Ocean coast, so Arizona is not necessarily only Arizona Highways, nor is Arizona necessarily only Hollywood cowboys singing around the dudes' campfire, or colorfully romantic Indians weaving rugs against a backdrop of Monument Valley, or great lakes cascading over their artificial spillways while the water skiers cavort around the shimmering surface with happy license, or rugged businessmen in their high-heeled specially tooled boots and made-in-Japan bolo ties threading their important ways among the canyons of our proud skyscrapers. Yet this is the perception that much of the desert and non-desert world has of us.

But as our ancient forebears in this arid land learned centuries ago, they could not survive the prolonged droughts that drove them from the Valley of the Hohokam, nor is it likely that the millions more of us, knowing better in the bottoms of our little black hearts but committed seemingly inexorably to a way of life that cannot be sustained, will survive in our own time and place.

Perhaps we can express this dilemma and its lesson for desert pupolations in the developing world as the impending intersection of two lines, where Arizonans' recognition of the inevitable necessity for a scaling down of our way of life meets a developing desert country's "rising expectations," or, in a more homely statement, where we pass on the stairway. We shall have initiated this event by our unwillingness to take off our blinders and see the future clearly, exporting instead technologies not suitably modified to meet differing environmental, cultural, or economic situations, technologies based on the linear process of growth that characterizes our more familiar Arizona world, technologies that feed the hopes of those other strangers to us for that better life we shall be giving up, technologies that hold out a promise for the undeliverable.

Water harvesting, trickle irrigation, solar energy, windpower, are only a few commonly understood and recognized areas in which there is hope for a restoration of the world's expanding degraded lands. But technology transferred without regard for the experience that could articulate the problems it is supposed to solve through a consideration of alternatives and adaptations for use in situations far removed from those for which the technology was originally devised, is technology - and experience - not only wasted, if not doomed to failure, but worse, employed in the service of badly-conceived projects that actually damage the situation sometimes beyond repair. Tradeoffs between degrees of disaster do not justify the exploitation of precariously balanced needs. "Yes, that was a bad thing to happen, but here's a good thing that came of it." Or, "Yes, the High Dam increased the incidence of schistosomiasis and kills marine life off the mouth of the Nile, but it also irrigates much new land and provides crops for hungry people."

I am trying not to be ambivalent.

I acknowledge that technology-experience transfer, even on its simplest level, can be enormously expensive, fraught with uncertainties, undercut by irrational local decisions.

I acknowledge that recipients of technological assistance often fail to define their own goals and responsibilities, fail to encourage indigenous competence at the field level, fail to recognize the necessity of an infrastructure that can persist through changes of government.

But I also acknowledge that donors can be arrogant, self-serving, so thinly oriented that their representatives believe they can force the recipient into a framework of activities utterly out of keeping with project goals.

So, we are all guilty. But while sackcloth and ashes may satisfy our lazy belief that such penance will absolve us from any real responsibility for our technological failures, this self flagellation does not seem to help us do more than make value judgments, usually arrived at in absentia, far from the scene of the problem, by people who are not hungry, who will never have to look upon the periphery of that desertified borehole, whose wife will never have to walk five kilometers tomorrow morning to gather enough roots to cook the evening meal.

How is it possible to put into motion an upward chain of events, based on field experience and understanding of cultural constraints? Must decisions always be predicated on a one-direction process, downward? Who makes the policy decisions that give impetus to a filtering-down-only direction from which there is no recourse?

Perhaps our preoccupation with the question of life on Mars should be recognized for what it really is: the ultimate in desert technology transfer. But to what end and for what purpose? Is it not actually an expression of our longing to believe that we are not alone in the universe? our cry for an answer to the fearful question: Are we unique? the frightened comfort that if we only try hard enough and spend enough money, there will be a chance to start all over again - somewhere? Whether we are willing to invest in our own despoiled planet the resources we have already invested so compulsively in this technological imperative is moot, one of those abstractions whose consideration can be put off until tomorrow.

Jeremy Swift's "technological arrogance" prevails today, however, not tomorrow, and to a global society nourished on science fiction, it is much more fun to "roll back Mar's desert" on television than face the uncomfortable nagging seemingly unsolvable problems of desertification on Earth.

But we must try.

And so, if in my old-fashioned elderly passionate way, I have provoked you, it is because I have tried to provoke you.

And that is a fact.

DEVELOPMENT OF THE POTENTIALITIES OF LARGE RIVERS
CUTTING ACROSS DESERT ZONES

A. Bouchardeau, Technical Adviser
Center for National Resources, Energy and Transports

ABSTRACT

There are in several desert areas in the world perennial rivers with a heavy flow,
fed by basins located in heavy rainfall areas.

In crossing desert zones, these rivers acquire a specific morphology; their flora
and fauna also change. So does the way man accommodates himself to his natural
environment.

While the water supplies of these rivers is mainly used for irrigation of crops,
we can visualize more diversified uses in the future. The adoption of agreements
and the setting up of international institutions is as a·rule an essential prerequisite to the development of the potentialities of these rivers as well as the
coordination of technical methods of exploitation over the entire basin.

It is recommended that studies be undertaken on the fields in which and the
methods through which uses of large rivers in desert zones could be better diversified and made more efficient.

RESUME

On trouve dans plusieurs régions désertiques du monde des rivières de débit
important et permanent alimentées par des bassins situées dans des régions de
fortes précipitations.

A la traversée de la zone désertique ces rivières acquièrent des caractères
morphologiques particuliers; leur flore et leur faune sont également spéciales
ainsi que le mode d'adaptation de l'homme au milieu naturel.

Bien que les eaux de ces rivières soient surtout utilisées actuellement pour la
culture irriguée, on peut pressentir dans l'avenir un usage plus diversifié.
L'établissement d'accords et d'institutions internationales est généralement un

préliminaire indispensable à la mise en valeur du potentiel de ces rivières tout autant que la co-ordination des moyens techniques d'exploitation à l'échelle de l'ensemble du bassin.

Il est recommandé de rechercher dans quels domaines et par quels moyens l'utilisation des grands cours l'eau en zone désertique pourrait-être mieux diversifiée et rentabilisée.

1. L'exploitation des énormes volumes d'eau transportés par certaines rivières au sein de régions désertiques a frappé bien des imaginations, soulevé bien des espoirs et suscité bien des efforts enthousiastes. Mais dans ce domaine il existe entre le rêve et le succès tout un monde de réalités, à vrai dire très diverses, à prendre en compte.

C'est en Afrique que l'on trouve l'exemple le plus frappant de ce type de rivières, le Nil dont le bassin versant de 2,980,000 km^2 et le cours de 6,500 km sont situés pour les 2/3 en zone subdésertique ou désertique qu'il approvisionne en moyenne de quelques 90 milliards m^3/an. Le Niger long de 4,190 km fait une large boucle en zone désertique, où il amène quelques 50 milliards m^3/an, dont une grande partie est perdue par évaporation. De même dans le même Sahel africain le Sénégal (30 milliards m^3/an) longe les confins sud du Sahara; le complexe Chari-Logone alimente le lac Tchad en zone subdésertique (bassin versant 600,000 km^2, volume annuel: 40 milliards m^3.

En Afrique du sud l'Orange (2,090 km, 829,000 km^2, et 90 milliards m^3 annuellement) traverse le désert du Kalahari.

Au moyen Orient l'Euphrate (275,000 km^2), 22 milliards m^3 annuels, et le Tigre (54,600 km^2, 40 milliards m^3 annuels) irriguent – de toute antiquité – un vaste delta en zone aride.

En Asie l'Indus (2,900 km, 958,000 km^2, plus de 300 milliards m^3 transportés actuellement) alimente dans le désert de Thar au Pakistan le réseau d'irrigation le plus étendu du monde entier. Le Tarim s'écoule dans le désert de Takla-Manan. L'Ouang Ho en Chine, d'une longueur de 4,350 km, drainant 715,000 km^2 transporte 47 milliards m^3 dans le désert d'Ordoz. L'Amu Darya 326,860 km^2, 70 milliards m^3 annuels arrose le désert du Turkestan; l'Helmand en Afghanistan (275,000 km, 22 milliards m^3 annuels) le désert iranien.

En Amérique du sud il faut citer le Colorado, 46,900 km^2 plus de 4 milliards m^3/an; le rio Negro, 150,000 km^2, 40 milliards m^3/an, le Senguerr-Chubut 55,000 km^2, 3 milliards m^3/an, et le Chico-Santa Cruz qui traversent le désert de Patagonie avant de se jeter dans l'ocean Atlantique.

En Amérique du Nord le Colorado 2,330 km, 641,000 km^2, environ 16 milliards m^3/an, et le Rio Grande 3,030 km, 471,000 km^2 environ 5 milliards m^3/an traversent le désert nord-américain.

2. Nous avons énuméré quelques grandes rivières typiques, ayant des débits permanents importants en toutes saisons. Ce ne sont pas les seuls cours d'eau existant en zone désertique. En fait il y a peu de régions au monde, même dans les déserts les plus arides, qui soient totalement dépourvues d'eau de manière permanente. Les quelques précipitations qui se forment en zone désertique sont suffisantes pour provoquer des ruissellements qui lorsque le relief est suffisant s'écoulent dans des cours d'eau temporaires aboutissant généralement à des mares dans les zones de piedmont. Nous ne considérons pas dans cette note l'aménagement de ces oueds qui comporte des solutions d'utilisation et d'aménagement différentes de celles des grands cours d'eau permanents.

3. Les déserts ne sont pas tous situés sous les tropiques. Toutefois les caractères morphologiques des lits et des plaines des cours d'eau traversant des régions désertiques, que ce soit au sud du Sahara, en Arabie ou en Amérique du Nord, sont semblables. En effet ils résultent de la diminution des débits et de la puissance de transport des rivières privées d'affluents et d'apports importants, alors que leurs eaux sont soumises à une intense évaporation et subissent des pertes par infiltration. Il en résulte un dépôt progressif des matériaux charriés, amenant l'élargissement des lits, la formation de méandres, l'érosion continuelle des berges et la séparation en plusieurs bras. Enfin le lit principal encaissé entre des bourrelets de berge élevés domine généralement les plaines avoisinantes. En période de crue les eaux débordent ces bourrelets et inondent les plaines avoisinantes. Ce processus aboutit à une morphologie semblable à celle que l'on trouve dans les deltas. En l'absence d'intervention humaine un équilibre relativement stable s'établit conditionnant la profondeur de l'eau dans les plaines d'inondation en fonction de la végétation qui s'y forme, la largeur et la profondeur des lits, et le régime hydrologique.

Les crues s'étendent d'autant plus loin, et subissent des pertes d'autant plus fortes par évaporation qu'elles sont elles-mêmes plus importantes. Les débits sont ainsi régularisés naturellement après un parcours plus ou moins bref du cours d'eau en zone désertique.

Il est important de noter que ce processus ne joue qu'à l'état naturel, et cesse dès que les berges du lit principal sont endiguées.

Ces caractères morphologiques sont favorables à l'épanouissement d'une flore caractérisée en zone tropicale par un peuplement de mytragynes le long des berges, de graminées et d'accacias dans les zones inondables. Le développement de la faune piscicole et avicole est exceptionnel en raison de l'inondation des plaines en saison des crues sur des superficies considérables de profondeur constante, créant les zones de frai du poisson. De nombreux ruminants – antilopes, troupeaux de bovins de pasteurs nomades paissent dans ces plaines en période de moyennes et basses eaux du fleuve. Il ne faut pas oublier les moustiques qui pullulent et servent d'alimentation à une multitude d'oiseaux. Dans le Sahel africain l'homme s'est adapté au milieu naturel sans modifier beaucoup son état sauvage. Les diguettes, construites sommairement par les habitants sont assez fragiles pour céder lors des crues importantes, et ne gênent donc pas le processus de régularisation naturelle des débits du fleuve.

Les secteurs des cours d'eau conservés dans leur état naturel, ou aménagés sommairement par des moyens précaires sont de plus en plus réduits. Même dans les régions désertiques la pression démographique est assez forte pour justifier une exploitation du fleuve qui, apparemment tire un meilleur parti des ressources en terres et en eau. Les caractères morphologiques indiqués ci-dessus: plaines étendues, dominées par le cours principal du fleuve sont en principe propices aux aménagements de périmètres irrigués par gravité. Le contrôle absolu de l'eau dans les périmètres irrigués est un dogme fondamental des ingénieurs chargés des aménagements agricoles le long des grands cours d'eau.

Cette doctrine conduit à la création d'un système hydraulique remplaçant celui existant naturellement et comportant des barrages de régularisation, des endiguements, des réseaux de canaux. Ces systèmes sont conçus pour des cultures déterminées et n'ont généralement pas le souci de résoudre les problèmes posés par la conservation de l'environnement convenable et par l'adaption de l'homme à ce milieu artificiel. En raison des multiples échecs essuyés dans l'application brutale de telles méthodes la conception étriquée et purement technique des aménagements est peu à peu remise en question. On doit toutefois se rendre compte de l'échelle des problèmes à résoudre et de l'étendue des zones aménagées ou en cours d'aménagement pour l'irrigation systématique en zone désertique.

D'après Wynne Thorne (Arid Lands in Transition) l'étendue des terres arables en climats désertiques et subdésertiques du monde entier comprend dans la zone des steppes 38.0 millions d'hectares, dans la zone tempérée semi tropicale 44.4 millions et dans la zone tropicale 28.9 millions, soit un total de plus de 111 millions d'hectares dans l'ensemble du monde. La proportion irrigable est probablement de l'ordre de 20% de ce total, soit 20 millions d'hectares.

Un des problèmes très importants posés par l'irrigation systématique en milieu désertique est le phénomène de salinisation progressive des sols lorsqu'ils ne sont pas drainés convenablement et irrigués avec de l'eau de mauvaise qualité. Les remèdes à la salinisation sont en principe connus mais ne sont pas en général appliqués correctement. La salinisation peut amener une destruction irréversible des sols rendus impropres à toutes cultures.

Contrairement à ce que l'on penserait l'eau n'est généralement pas utilisée avec toute l'économie désirable en zone désertique même si son coût est très élevé. Les cultures ne sont pas toujours les plus adaptées aux exigences de cette économie.

Les avantages d'installer de grands périmètres d'irrigation en zone désertique le long des rivières permanentes lorsque la pression démographique ne l'exige pas, ou lorsque les sols ne sont pas de qualité exceptionnelle, sont très aléatoires. Ceci semble être démontré par l'exemple de l'extrême nord de la boucle du Niger où il n'existait qu'une population sédentaire limitée. Dans un tel cas on peut se demander s'il n'est pas préférable de limiter au maximum les pertes d'eau par évaporation qui se produisent dans les plaines inondées à la traversée de la zone désertique et employer cette eau avec plus de profit dans une zone moins aride. C'est ce qui est tenté pour le Nil à la traversée des marécages du Sudd au Soudan. L'approfondissement du lit principal du Nil Blanc par des dragages appropriés et l'endiguement conjugué des berges, maintenant à l'étude, accroîtra considérablement les débits disponibles en aval des marécages.

On notera que la régularisation des débits par des barrages réservoirs construits en tête des affluents limite également les débordements dans les plaines inondées, et les pertes par évaporation qui s'ensuivent.

Le coût élevé de l'eau en zone désertique justifierait le long des rivières, des cultures diversifiées de plantes qui se prêtent particulièrement au climat et assurent un meilleur revenu que les plantes industrielles irriguées classiques (riz, coton, etc...).

On doit également se demander si la production des périmètres irrigués compense effectivement la production piscicole réduite du fait de la diminution de l'étendue des zones de frai. La modification des conditions naturelles peut s'avérer dans certains cas plus nuisible qu'utile même sur le plan économique.

Certains pensent que l'exploitation systématique des nappes phréatiques alluviales par pompage, et le rabattement de la nappe à un niveau inférieur à ce qu'il est dans les périmètres d'irrigation classiques apporterait la solution à bien des problèmes: meilleure conservation du milieu naturel en limitant la superficie endiguée des périmètres aménagés au minimum nécessaire, diminution de la salinité par un meilleur drainage, et peut-être réduction des coûts d'aménagement en évitant la construction de barrages réservoirs et réseaux de canaux primaires.

Dans l'immédiat, et surtout dans les pays en voie de développement les seules utilisations importantes envisagées de l'eau des grands cours d'eau sont, en plus de l'irrigation des cultures, l'approvisionnement en eau des habitants, directement dans la rivière, ou par des puits alimentés par les infiltrations dans le lit. Mais il pourrait en être autrement.

Dans les pays industriels la tendance est d'utiliser de plus en plus d'eau à d'autres fins que l'agriculture. Il est intéressant d'examiner de ce point de vue le cas de l'Arizona, aux Etats Unis, en majeure partie de climat désertique et cependant très prospère.

La raison primordiale de l'afflux des populations nouvelles en Arizona qui atteint maintenant 2,500,000 habitants et devrait atteindre 14,000,000 en l'an 2,000 paraît être psychologique: attrait d'un climat où le soleil brille la journée 80% du temps et où les pluies n'atteignent que 300 mm par an.

L'eau est fournie par des nappes jusqu'à présent abondantes, mais dont le niveau décroît rapidement, et par plusieurs rivières dont le Colorado. Les problèmes posés par l'irrigation des terres, commencée il y a cent ans sont de plus en plus délicats du fait de l'épuisement des nappes et de l'augmentation de la salinité.

Des expériences de pluie artificielle ont été faites, qui n'ont donné, jusqu'à présent que des raisons d'espérer. Cependant si l'agriculture utilise 92% de l'eau consommée en Arizona elle ne contribue que pour 7% du revenu des habitants de l'état. L'eau est vendue beaucoup plus cher pour les usages non agricoles, (de \$1 à \$250 pour 1,000 m^3 pour les usages miniers, \$4 pour 1,000 m^3 pour les usages domestiques, \$25 pour 1,000 m^3 pour élever du bétail) que pour les usages agricoles (\$1 à \$2.5 pour 1,000 m^3). Les revenus provenant de la vente de l'eau pour les loisirs sont ainsi supérieurs à ceux provenant de la vente de l'eau pour l'agriculture. L'eau utilisée dans l'industrie est soigneusement recyclée. Ainsi les problèmes agricoles ne semblent pas être un obstacle à la prospérité économique globale. Si l'Arizona est un exemple permettant de prendre des options sur l'avenir assez rapidement la plus rentable utilisation des ressources en eau des fleuves en zone désertique pourrait être de développer certaines formes de peuplement et de tourisme, de créer des complexes industriels,les déserts constituant apparemment une zone de peuplement très acceptable.

Un dernier point qu'il convient d'examiner dans l'exploitation des rivières coulant en zone désertique est celui des accords internationaux. Par définition ces rivières s'alimentent loin de la zone désertique du pays intéressé et il y a donc des chances que leur bassin versant et leur cours supérieur soient à cheval sur plusieurs pays. C'est le cas du Nil, du Niger, du Colorado. Il serait hasardeux d'investir en zone désertique sur des projets coûteux sans s'assurer que les pays situés en amont sont disposés à laisser une partie des eaux disponible dans le futur. Un tel accord existe pour l'ensemble des pays du bassin du Nil dont le Soudan et l'Egypte sont les seuls bénéficiaires. Une commission existe pour le Niger, chargée de résoudre ces problèmes.

A un degré encore plus élevé que pour les bassins intéressant les autres zones climatiques, l'aménagement des bassins en zone désertique impose donc des moyens techniques coordonnés et des institutions appropriées.

En conclusion, des études et des recherches sont recommandées pour améliorer l'exploitation du potentiel des grandes rivières traversant des zones désertiques. Pour cela on devra tenir compte des projets des Nations Unies qui ont étudiés certains cas particuliers (Sénégal, Niger, etc.) et de l'expérience acquise dans les pays industriels.

Bien qu'il ne soit pas possible d'appliquer ce dernier modèle aux pays en voie de développement, qui ne disposent d'ailleurs pas des capitaux nécessaires pour le réaliser, des idées neuves sont certainement à examiner et à retenir pour les aménagements futurs de la zone tropicale.

ARIDITY, WATER SYSTEMS AND SOCIETY

K. K. Prah
Senior Lecturer, Department of Sociology
University of Botswana & Swaziland, Gaborone, Botswana

"With a force of 3,000 men, I made the road a river, the desert a meadow" (An officer of Dynasty XI Egypt - 2000 BC - on an expedition through the Wadi Hammamat).

The aim of this short paper is not to present theoretical constructions or hypotheses of any definite sort on the subject of the paper. Neither is it an endeavor to answer major questions either of long or short standing. The effort presented here is towards posing a few questions of interest, based on a one and only fundamental presumption - i.e. that, there is a direct relation between human civilization, settlements, population concentrations, social organization, and water tenure or hydraulic systems, particularly under conditions of aridity or semi-aridity.

If we concede the argument that the ecosystems variously described as deserts in the world, vary considerably with respect to their habitability for flora and fauna, so that in Africa, the Kalahari is naturally more habitable than large sections of the Sahara; and further, that some savannah regions are almost structurally akin to regions described as deserts by others, it seems that a rigid definition of deserts is conceptually a difficult if not impossible task. Perhaps we would be pushed to describing deserts as areas with relatively high degrees of aridity (the qualification relatively is important).

Desert conditions, and the extent to which such ecological conditions are consciously made to support human population concentrations, seem to have been historically manifested by societies with relatively sophisticated hydraulic systems. These systems for making arid and semi-arid regions support life have varied greatly in history. They include cisterns, wells and water holes, canals, aqueducts, water troughs, dams, dykes, tanks, sub-terranean conduits, water-lifting systems such as the Shaduf, the saqiya, etc.

As Drower (1) has argued from classical times in regions of scanty rainfall and aridity, the origins of irrigation systems are bound up with the emergence of agriculture. Both irrigation systems and agriculture also seem to be bound up with Neolithic man's emergence as a being of sedentary corporate social existence. The development of irrigation systems obviously raised the potential for social production above subsistence requirements. This development tended to release the social potential for the development of pottery, animal domestication, etc. Irriga-

tion systems have not only been created to serve production purposes. It seems
that irrigation systems can and have in the past from ancient times to the present
period been used for warlike purposes, both as protective devices and as targets
of war. The Assyrians and Neo-Babylonian kings of the early 6th century B.C. were
particularly adept at using irrigation systems for protective purposes. In more
recent times (1939-45 war) the German forces attempted with some success to breach
Dutch dykes in their war effort.

The quality of these hydraulic systems has, of course, depended on the general development of the forces of production in a given society. So that pre-industrial
or pre-capitalist societies which have attempted to tame and make habitable desert
or semi-desert conditions have achieved this by relatively unsophisticated technological innovations while industrial societies have approached the same problem
with greater quantitative and qualitative technological efficiency. However,there
have been pre-industrial societies whose efforts in this direction have been more
efficient than many present day industrializing societies of the third world. Also,
within the same historical period often different societies with basically similar
relations of production have displayed dissimilar levels of development in the
force of production in such a way that differences in their hydraulic technologies
and water tenure systems have shown wide differences in efficiency and have thus
also been able to support different population densities. What have been the specific or general conditions for these differences?

Some may argue that under favourable ecological conditions, there is little stimulus towards the development of more effective and advanced technology to irrigate
a given area. But then, a favourable ecological system has, as it were, a ceiling
on its possibility to accommodate population density. So that, once a certain
degree of population concentration is achieved, the problem of developing or creating irrigation systems which can provide the basis for sustaining a growing population demand must create stimulus or a rationale for introducing commensurate innovative measures. Is the relationship between the quality of irrigation systems
and population density thus a dialectical one?

Probably no study focussed on this problem area has been as erudite and substan-
tive without necessarily being essentially correct as Karl Wittfogel's (2), Oriental Despotism. He estimates that desert-like areas of more or less full aridity
and steppe-like areas of semi-aridity taken together cover almost three-fifths of
the world. Indeed, these arid regions are the ultimate testing grounds for novel
techniques for agricultural development and water control. Wittfogel suggests the
hypothesis that in the development of hydraulic agriculture as opposed to rainfall
agriculture, cultural diffusion may have been significant, even for such widely
spread areas as ancient Mesopotamia, India, and the western zone of South America.

Further, in each case, the presence or absence of adjacent humid regions influence
the pattern of growth and development. Thus in Egypt, hunters, gatherers, and fi-
shermen practised agriculture as a secondary occupation on the natural flood banks
of the Nile, before water-controlled farming became the dominant system of production. For China and Meso-America, the possibility of diffusion has been suggested (from Inner or South Asia and South America respectively). But here also
questions remain to be answered, for the stimulating base did not originate in a
more humid area. In ancient China apparently the semi-arid north related dominant-
ly and innovatively to the Yangtze basin states, and in India, the arid and semiarid regions of the north formed a more viable center from which innovative rip-
ples radiated.

The difficulty with diffusionist explanation is that they are often evidentially
flimsy and tenuous. Probably one less debatable case is that of the Achaemenid
kings of Persia (circa 650-331 B.C.), who built underground water systems else-

where in the arid Near East;   the underground reservoirs of the Dakhla and Kharga oases have been probably built under their instructions.   Also the system of vaulted conduits which were fairly common in the Sahara from Tripolitan to Morocco which are often regarded to be of Roman origin may well have been due to the Persians (3).

Certain characteristics seem to lie at the foundation of societies in which although naturally arid or semi-arid conditions may prevail, artificial water control systems are developed to support relatively high population densities:

    a)   Complex division of labour, and

    b)   Intensive cultivation.

These characteristics require generally intricately structured bureaucratic structures which can ensure the maintenance, construction, and reconstruction of the water-agrarian systems;   but whether these state structures also require peculiarly despotic and centralized leadership or not, has been an often debated question, which probably cannot be answered if other social structural peculiarities are not taken into consideration.   In any case, it is doubtful if centralized or decentralized leadership can affect the fundamental relations of production in any essential fashion.   If the institution of bureaucracy is understood as an apparatus or instrument of the state for maintaining class rule, then Wittfogel's contention that "under agrodespotic conditions the managerial bureaucracy was the ruling class" (4) is certainly wrong.

Wittfogel's study based itself mainly on Asiatic society under pre-capitalism, with all the specificities which the Asiatic mode of production implies, both in and out of Asia.   It should also be borne in mind that it would be pedantic, even in Asia, to expect identical similarity in the various pre-capitalist Asiatic social formations which existed.

As Goody has explained, in most of pre-colonial Africa, "the lack of the wheel had another consequence for agriculture, since it limited the possibilities of water control.   In the drier regions of the Eurasian continent the wheel has played a dominant part in raising water from wells to irrigate the land.   Simple irrigation there is in Africa, as almost everywhere agriculture is practised.   Some of the inhabitants of the settlement of Birifu (LoWiili) in northern Ghana channel the water from a permanent spring to run among their fields, and thus get two crops a year in place of one.   The Sonjo of Tanzania practise more developed water control. Rice growing in the Western Sudan, and it should be remembered that some of the rice used here (Oryza glaberima) was domesticated independently of Asian rice in the Senegal-Mali region, demands yet more positive measures.

There are other means of water control that do not involve the wheel, that is, using various techniques of temporary storage.   Methods of this kind did of course exist.   Everywhere there was some improvement of natural pools.   In Gonja and in neighbouring areas of northern Ghana there are many ancient cisterns hollowed out of the laterite;   in the famous market town of Salaga, the city of 1,000 wells, these are cylindrical in form and do not seem wholly dependent upon surface water. But these storage systems are very different from the village tank of south-east Asia;   while there is no lack of water in Africa, the problem of its distribution is enormous.   And in terms of agriculture what is lacking, apart from the Shaduf of the Saharan fringe, which used the lever principle, is any mechanical device for drawing water, such as is used in the Middle East and even in the Saharan oasis (5)".   There is a possible relationship between the relative lack of population concentrations in pre-colonial Africa (compared to Eurasia) and the low development of productive forces;   which is the cause and which the consequence is debatable. But obviously population increase itself is a definite plus factor in the development of productive force; indeed population itself is a force of production.

The significant population concentrations on the African continent have in the past been in the main river basin. As one moves towards the desert areas, population thins out, and the expansion and encroachment of deserts like the Sahara, have feuded to push human groups into more humid areas. The threat of desert expansion in African history does not seem to have stimulated significant technological innovations to make the desert more habitable. What is the explanation for this?

These questions and others refer to history, but since the past and the present indeed form a continuous process, their relevance is also for the present, and recent successes of States with as divergent political economies as the Chinese and Israelis in taming arid regions put these old questions in a modern framework.

REFERENCES

(1) Drower, M. S. (1967) Water Supply, Irrigation, and Agriculture. In A History of Technology, C. Singer, E. J. Holmyard, A. R. Hall and T. I. Williams, eds., Oxford, I, 520.

(2) Wittfogel, K. (1967) Oriental Despotism, Yale.

(3) Drower, M. S., ibid, 534.

(4) Wittfogel, K., ibid, 4.

(5) Goody, I. (1971) Technology, Tradition and the State in Africa, Oxford, 27.

Criteria for Long-Term Water Planning

Michael D. Bradley, Ph.D., Associate Professor

ABSTRACT

Long-term water planning is not just a matter of engineering or economics; it is a matter of institutional development. Traditional water planning is based upon static institutional doctrines, such as supply engineering, economic efficiency, and social stability, that are unresponsive to future needs. Long-term water planning is a new institutional role, requiring new methods and skills. Institutional changeover to long-term water planning will involve organizational development, increased professional flexibility, and conscious societal learning.

INTRODUCTION

Once futures analysis was the preserve of dreamers, doomsters, and science fiction writers; however, over the past thirty years, an interest in serious, long-term or long-range planning has grown. Methodologies for systematic analysis were developed during World War II and found extensive application in the following years, particularly in defense systems and private industry. Since then governments at all levels have demanded better planning as a condition for project development and public investment. And, the field of "futurism" has grown in popularity and interest. Prominent scholars in many countries now formally study the future as an important intellectual exercise, complete with its own commissions, institutes, professional organizations, journals, and university courses. Futures planning and analysis is increasingly recognized as a vital component of overall societal policy and guidance (Tugwell, 1973).

Unfortunately, futures planning has seldom been a profound concern of water resources planners. On the whole, water planners have assumed the future will be much like an extended present, with little change in human needs, social values, policies, goals or technologies. A simplifying assumption, this perspective of short-term stability simplifies the planning task, but ignores reality (Sewell, 1977). It hides the future in a gold-dust haze, and gives planners and decision-makers an inadequate intellectual access to reality. Recent experience shows that changes in values, needs and goals are occurring with greater rapidity than before and when they are not anticipated, plans fail (Vlachos, 1973; Hendricks, 1975). Mounting opposition to poorly planned water projects illustrate the point with extreme clarity; water resources planning can no longer respond to the needs of a turbulent and disruptive social future with conceptual models and methodologies more appropriate for the unlimited expansionist ethic of an era now past.

This article will examine the developing criteria for long-term water planning from two perspectives. First, it will describe the philosophy and methodologies of long-term planning. Knowing present shortcomings in water planning will lead to a fuller appreciation of need for long-range water resources planning. Second, it will analyze the institutional requirements for long-range planning. Implementing long-term planning is the basic challenge for water policy-makers, and implementation comes about only by a process of institutional modification and development. Designing institutional arrangements for long-term water planning is the vital first step to assure better organizational efforts in the future in both the developed and developing arid regions of the world.

THE CHANGEOVER TO LONG-TERM PLANNING

The process of water resources planning occurs in the present, but its real object is the future. Water development is future oriented because of the time scale of the impacts of projects. Dams, irrigation canals, and water treatment facilities are expected to serve useful purposes for long periods of time after construction and are economically justified by rates of return and capital expenditure of fifty years or longer. But the social net worth of a project is more than its interest rate or its repayment period. Water projects provide public services and benefits that are more than the sum of money expended (Mishan, 1974). Water projects also serve rapidly changing social goals, and how well these continue to be served is more than a matter of economics. Aspirations and material wants change rapidly under the influence of modernization and economic growth often leaving previously felt "needs" behind. Technological methods also change, making the previously thought impossible a commonplace event. And institutions change in time, collecting another long-term toll upon present decisions and plans. Long-term water planning needs to deal effectively with this entire range of relationships, instead of only the technical or economic sub-set.

Traditional Doctrines of Water Planning

Three basic doctrines have guided the traditional development of water resources by public water agencies in the developed nations: the doctrine of supply engineering, physical structural orientation, and the rationality of short-term stability. Each doctrine explains a method of perceiving and understanding the complexities of water use, and each developed in conjunction with the other. Fundamental perceptions help organize information and expectations in the process of water planning and help explain why certain functions are considered legitimate by water resources planners.

The first doctrine is supply engineering. Basically this doctrine assumes that water resources development is only a matter of supply and that more water solves most water problems especially scarcities, shortages or inefficient allocations. (Hirshleifer, et al., 1960). The perception of increased supply as a solution to most water problems is an engineering doctrine; in fact, most engineering schools and civil engineering curricula contain specific courses on water supply engineering. Many texts used often contain some derivative of that basic idea as a title. The supply doctrine is an explanation of the missions and goals of many construction agencies at the national level of government, for example, the Army Corps of Engineers, the Bureau of Reclamation, the Soil Conservation Service, and the Tennessee Valley Authority.

The second traditional doctrine of water resources planning is the physical or structures orenntation. The physical structures doctrine is a complementary part of the engineering supply doctrine because water is sometimes physically scarce, as in an arid or semi-arid region, but also, it is sometimes physically overabundant, as in a flash-flood. Therefore, water engineering and planning must deal with more than just shortages. Water abundance in the hydrologic cycle can lead to damaging floods. Both problems can be "solved" by structures such as dams and levees, and by storing floodwaters in reservoirs to provide supply or low-flow augmentation during the low-flow seasons. The technological opportunities for expanding a single-purpose project into a multi-purpose project are extensive. Benefits can range from flood protection low-flow augmentation, recreational opportunities, irrigation development, and increased fish yields to and including the increased value of shorelands for homesites as other purposes (Ackerman and Löf, 1959). Multiple purpose benefits can then be assigned to beneficiaries in the benefit-cost appraisal, making the ratio highly favorable and also making construction even more likely (Haverman, 1965; Krutilla and Eckstein, 1958; Steiner, 1959).

The third traditional doctrine of water planning is the primacy of short-run considerations in the project selection process. This dilemma is more apparent than real; projects may have long-range consequences, but they are usually funded for short-term benefits. Projects are usually designed to attain capacity quickly and to contribute to economic development as soon as possible. Widespread problems have recently been found in the method of financing international water resources projects, many attributable to poor project planning. Among the more common are the unrealistic goal and purpose definition given the available time for project completion, the failure to explicitly define both immediate and long-range development goals, with immediate objectives linked to long-term outcomes, and the failure to accurately assess the demand for present or future demand of a project's goods and services (Rondinelli, 1977). The pressures for short-term payoffs are most intense in those poorer countries where the long-range perspective seems an unaffordable luxury (Goulet, 1976).

Difficulties with Traditional Doctrines

At least these major difficulties can be seen with the traditional doctrines of water planning; inadequate resources development, loss of

alternatives and inability to capture secondary benefits. It is becoming
increasingly clear that traditional doctrines of water resources planning
are simply inadequate to provide efficient and effective water resources
planning and development. The supply orientation to water resources and
the doctrines of engineering or economic efficiency are not the only purposes or goals available to water resources planners; they may be the
most easily analyzed goals and purposes, but they are not the only
acceptable ones. In fact, evidence is mounting that the spillover effects
and secondary consequences in the form of environmental and social impacts
are making traditional water supply projects increasingly unattractive
(Milton, 1971). A limited perspective can be served by limited and
"rational" methodologies, but that "better" projects will emerge from
plans subjected to detailed technical and economic analysis; that optimal
choices will come from appraisals using elegant rates of return, shadow
prices, and benefit-cost analysis; that project implementation will be
more efficient if planned and controlled by PERT, CPM and systems analysis, are assumptions now being challenged by contrary evidence (Wildavsky,
1969; Schick, 1973, Hoos, 1972).

A second difficulty with traditional water planning is the assumption that alternatives are somehow given by the physical and economic
boundaries of the project site, and that project planning will eventually
consider and select the most appropriate and effective alternative among
several different projects. Implicit is the assumption that alternatives
must be cast aside until the one best stands revealed for all to see. But
alternatives selection is neither so rational nor so benign. Choosing
among a narrowing range of alternatives is a process of social choice that
can lead to political conflict or can impose the will of an expert or
bureaucrat upon the recipients of a project's benefits. The emerging
considerations of multiple use, participation by affected public in
decision-making and alternatives illumination all serve as symbols of
desirable political goals in democratic society. What is important is
that a pet project of one particular agency or planner may not provide
from the standpoint of the entire nation or the entire international water
resource system, the maximum or optimal net gain (Mishan, 1969). Only
better alternatives generation and explication can encourage the interests
of a broader public than an agencies' clientele, and can avoid the prema
ture loss of alternatives due to short-term political pressures. Alternatives selection is an important component of long-term water planning,
and recent experiences will need to be extrapolated as guidelines for
the future in a changing world (Wolman, 1977).

The third difficulty again follows from the traditional method of
selecting and limiting project alternatives. Alternatives tend to be based
upon geographical locations of technological interventions almost exclusively in traditional planning. This is not surprising considering the
importance of supply engineering in the process; however, the situation
leads to an unfortunate analytical inflexibility when considering other
types of alternatives. The pure water supply orientation of most water
planning in the present dominates the minds of traditional planners,
restricting from view the smaller or more appropriately scaled interventions that might be cost-efficient and inexpensive. The "appropriate
technology" movement in the less-developed world is a reaction against the
hardware-oriented interventions of large-scale technocratic planning
(Schumaker, 1975). Many appropriate alternatives for water planning in
arid and semi-arid regions continue to be overshadowed by traditional
doctrines and methods of water planning; two good examples are waste water
reuse and individual conservation efforts (Barr and Pingery, 1977).

Short-term water planning has traditionally been an allocative process,
but an innovative planning process is more appropriate to the conditions of

the future. Allocative planning is concerned with the distribution of
scarce resources among completing ends, and tries in all respects to be
comprehensive, balanced, quantitative, and economically and technically
rational. Those attributes are not incorrect: they are inadequate.
Allocative planning assumes a stable future of agreed upon goals and
objectives and plans toward that perception. The traits of allocative
planning are deeply ingrained in the myths and doctrines of water planning agencies, although centralized allocative planning has not lived
up to its initial promises and has remained relatively ineffective. The
desire to be comprehensive has produced the illusion of an omnipotent
intelligence, the method of system-wide balances had led to an overemphasis
on stability, quantitative modeling has encouraged the neglect of the
actual conditions governing policy and program implementation, and the
claim to functional rationality has made water planners insensitive to the
value implications of their task (Freidmann, 1973).

Innovative planning has fared much better. This is an approach to
institutional development that is expected to produce a limited, but significant change in the structure of an existing planning agency. Instead
of being bound by traditional goals and methodologies, innovative planning
is based upon designing institutional arrangements to change the direction
of the stream of ongoing events or to change the direction of its flow.
Three salient aspects are apparent: a predominant concern with institutional change, a basic orientation towards action, and a special emphasis
upon the implementation of future-responsive plans (Freidmann, 1973).
Innovative water planning requires the development of methodologies for
long-term planning, the integration of long-term considerations into
project analysis, and the institutional development needed for adequate
long-term water planning.

METHODS OF LONG-TERM WATER PLANNING

The first requirement for more innovative water planning is the continual development of methodologies to assist in directing programs, policies, and organizational change. Several likely methodologies have
recently been developed; these include delphi, cross-impact analysis,
computer simulation models, trend extrapolation, scenarios, and trend
impact analysis. A general description of each method will summarize the
uses of each.

The Delphi Method

This technique is used to gather information and judgments from experts in a particular field or specialty.(Linstone and Turoff, 1975). A
group of respondents is sent a questionnaire posing quantitative and
qualitative questions about the future problem. Answers are collected
and a second questionnaire is sent, containing the first round questions
and statistical summary of the answer. The experts then rank and evaluate
the answers, seeking a consensus or a convergence of answers. The Delphi
method has many variations, including personal interviews, computer-
assisted interviews, varying the number of questionnaires, varying the

feedback to experts, and allowing the experts to introduce new questions
for discussion. Delphi is not a direct scientific tool or a planning
methodology; rather, it is a way to incorporate expert judgment and wis-
dom about the future into the planning and decision-making process (Coates,
1975).

Cross-Impact Analysis

This technique predicts the impact an event can have on the liklihood
of another event occurring, focusing on the relationship if any, between
two individual events (Kane, 1972). The technique quantifies cross-
impacts and is used with other futures research methods. For example, a
technique called "probabilistic system dynamics" combines cross-impact with
systems dynamics to describe not only the impact of a single event upon
other events, but also the impact of events upon all elements of a simulation model (Stove , 1975). Cross-impact analysis has been applied by the
U.S. Army Corps of Engineers to study the effect of external resources
upon internal water developments in the South Platt River Basin (USACOE,
1975).

Scenarios

A scenario is a hypothetical sequence of events constructed for
focusing attention on causal processes and decision points. Scenarios of
the future can be found in all lengths, topics, and formats. If well
structured, they allow a policy-maker to prove consequences, using the
human talents of integration and imagination; if poorly or hastily done,
they are useless objects of scorn and limited vision (Zentner, 1975).
Scenarios are of two important types: environmental scenarios and organizational scenarios. Environmental scenarios describe the future universe
for a policy-maker. The factors may be economic, such as the rate of inflation and energy costs, or political, such as the evolution of laws and
regulations. Organizational scenarios deal with institutional structure,
missions, components, and so on. Scenarios are increasingly used for long-
term planning and their relevance and credibility are under careful review.
Plausable scenarios only indicate what could happen, not predict what will
happen (Elgin, et al., 1975).

Trend Extrapolation

This method projects trend lines, a historical record of events, into
the future, and lists the values of some factor or variable against speci-
ified periods of time, past and future. The time intervals may vary
according to the policy-maker's needs, and the variable's values may increase or decrease with time. These changes are the trend (Gordon, 1972).
This method is regularly used by business analysts for marketing forecasting, but has proven of little help to water resources planners. The
demands for water goods and services are usually too complex to be appropriately bounded by the trend line under analysis.

Trend Impact Analysis

This method is more sophisticated trend extrapolation, using probabilistic concepts to analyze unexpected events likely to change the historic paths of a trend. This can aid planners by identifying the most and
least important events, and by indicating what policy options are likely
to be the most effective (Rosove, 1973). The method has had only limited
use in public policy, and remains unproven.

Computer Simulation Modeling

The method of computer simulation modeling uses two basic approaches:
a system and a mathematical model. Linking a systems approach to a mathematical model results in a detailed quantification of the relationships
among the important variables in a system. Approximation procedures can
reduce the number of factors and can explicitly formulate a planner's
implicit mental model of a system. Computer simulation modeling is
increasingly applied to the physical environment, social systems, military
situations, and economic analysis; but the results are mixed and incon-
clusive. The method simplifies reality, often to the point of distortion.
It lends itself well to experimentation, but it also is the source of
error. It forces risk assessment, without a concomitant acceptance of personal or organizational responsibility for the incidence of the risk. Many
analysts are devoted to developing this methodology, and its uses are
likely to increase as long-term planning continues to develop (Barton,
1970; Enshoff and Sisson, 1970; Meadows et al., 1974). This development
needs countervailing balance against the fascination with computers and
for the primacy of human reason and values (Weizenbau, 1976).
But the development of methodologies for long-term water planning is
only a first step. Those methods must then be integrated into project
planning on a day-to-day basis. Recent legislation has recognized the need
for a long-term prespective and has incorporated those concerns into public
planning and policy making. The most famous effort is the National Environmental Policy Act of 1969, which required an environmental impact statement to discuss, in addition to physical impacts from a project and possible project alternatives, the relationship between local short-term
uses of man's environment and the maintenance and enhancement of long-
term productivity, as well as any irreversible and irretrievable commitments of resources involved in proposal implementation. So far, NEPA has
been procedurally if not substantively successful, but its worth as a
long-term planning tool seems limited (Bradley, 1976). Other recent legislation attempts to incorporate a long-term perspective into resource
programs; for example, forest planning and energy development. The
improved assessment of future demands and alternative supplies is easier
to theoretically discuss than to implement as a practical matter. Many
argue for more policy review of local or regional projects, but weak
methodologies and an uncertain future make such prescriptions generally
unhelpful. In order to seriously approach long-term planning, institutional development for that purpose is necessary.

INSTITUTIONAL REQUIREMENTS FOR LONG-TERM PLANNING

Long-term water planning must be designed into present institutions to increase society's ability to respond to an increasingly disruptive and turbulent future. Future water policy will differ from past policy because societal change occurs rapidly, and the pace of change is accelerating. Planning that provides for past interests, with outcomes legitimized by inappropriate methods, will fail. Institutions must face the future, and plan water resources in ways that are responsible to uncertainty, turbulence, and disruption.

Societal complexity means increased responsibility for the public sector: for more public goods, for more public management, and for more planning of common property resources, such as water. The conditions of future uncertainty, turmoil and scarcity are undermining traditional planning efforts. Not long ago systems science, bigger and more sophisticated computers, management information science, and planning-programming-budget systems were to overcome environmental problems and efficiently allocate resources. Not long ago planning themes stressed consensus, comprehensiveness, and rational order; today the problems are diversity, conflict, and tension. Water planners need to plan for a crucial common property, but have few guidelines to the most likely future. Adequate planning requires more than just scientific information, more than just market economics, and more than just technical or engineering efficiency. What more is still unclear.

As yet, no adequate theory of long-term planning is available. A realistic theory will require overcoming individual and institutional resistance to change. It will require institutional development for allocating common property waters, not just market transactions or engineering structures. It will require planning for the biosphere's limits and processes, not just technological interventions (Vickers, 1968).

But anticipating options, boundaries, and choices is a new, untested role for many water planners, requiring not only new skills and knowledge, but also personal incentives for changing to long-term planning (Michael, 1973). Unfortunately, the present weighs heavily in water planning. Methodologies are oriented toward supply engineering, focused upon quantified indicators, the multi-objective, multi-purpose approach, and the easier parts of the planning process (Water Resources Council, 1972; Utah State University, 1974). Multivalue choices are still left to chance, and are still outside the formal planning process; however, these are the most fundamental issues of all, and ignoring them will lead to less desirable future outcomes.

Long-term water planning must be more carefully concerned with the problems and opportunities of information feedback. Feedback is crucial for organizational selfevaluation and development, for it contains information on the direction of social needs as well as clues about why to change (Kaufman, 1975). In the past, water planners developed elaborate mechanisms to filter feedback, to adjust it to prevailing agency doctrine, and to ignore it, if possible. Much of the current interest in public participation is a reaction to this unfortunate situation; to the assumption that publics can be "educated" to an agencies goals or doctrines, and harmlessly co-opted into a ritualistic adversary process (Mazmanian and Lee, 1975). This tactic is more appropriate to a stable organizational environment than to a turbulent social future (Selznick, 1968). Active responses to unwelcome feedback include denial, flight and uncertainty. What is needed is to channel these responses into a learning system whereby water planning is reiterated to respond and adopt to social and environmental uncertainties.

Water planning as a future-oriented learning system will not be accomplished easily. It will require new professional roles by those with value-linkages important for agency plans. New roles will also develop for those with interpersonal as well as technical skills, especially for boundary-spanners that translate changing social needs to a water agency's technical personnel. Also important are organizational development roles to help a water institution restructure its capability to deal with turbulence and to link knowledge to implementation. But by far the most important new role is for those water planners who can combine social values and technical alternatives into a vision, a controlling synthesis of where the future is and should be (Morison, 1974). Vickers calls this an "appreciation" (Vickers, 1973). Whatever its name, it deals with acceptable, future water planning outcomes. An appreciation of the need for institutional changes is necessary for both the future study and practice of long-term water planning.

References

Barr, James L. and David E. Pingery (1977), Rational Water Pricing in the Tucson Basin, *Arizona Review*, 25, 10, 1-12.

Barton, Richard E. (1970), *A Primer on Simulation and Gaming*, Prentice-Hall, Englewood Cliffs.

Bradley, Michael D. (1976), Institutional and Policy Aspects of Instream Flow Needs, *Instream Flow Needs*, 1, ed. by J. F. Orsborn and C. H. Allman, American Fisheries Society, Bethesda.

Bradley, Michael D. et al. (1976), *Environmental Impact Statements in Water Resources Planning and Decision Making*, Tucson, Arizona: University of Arizona, College of Earth Sciences, Department of Hydrology and Water Resources.

Coats, Joseph F. (1975), In Defense of Delphi, *Technological Forecasting and Social Change*, 7, 2, 193-194.

Elgin, Duane S., David C. MacMichael, and Peter Schwartz (1975), *Alternative Futures for Environmental Policy Planning: 1975 - 2000*, Palo Alto, California: Stanford Research Institute for the Environmental Protection Agency.

Enshoff, James R. and R. L. Sissor (1970), *Design and Use of Computer Simulation Models*, MacMillan, New York.

Friedmann, John (1973), *Retracking America: A Theory of Transactive Planning*, Anchor Press, New York.

Gordon, Theodore J. (1972), The Current Methods of Futures Research, in *The Futurists*, A. Toffler, ed., Random House, New York.

Goulet, Denis (1976), On the Ehhics of Development Planning, *Studies in Comparative International Development*, 11, 1, 25–43.

Haverman, Robert H. (1965), *Water Resource Investment in the Public Interest*, Vanderbilt Press, Nashville.

Hendricks, David W. (1975), *Environmental Design for Public Projects*, Water Resources Publications, Ft. Collins.

Hirsleifer, Jack, James C. DeHaven, and Jeroma W. Milliman (1960), *Water Supply*, University of Chicago Press, Chicago.

Hoos, Ida R. (1972)*Systems Analysis in Public Policy: A Critique*, University of California Press, Berkeley.

Kane, Julius (1972), A Primer for a New Cross-Impact Language – KSIM, *Technological Forecasting and Social Change*, 4, 2, 129–142.

Kaufman, Herbert (1975), The Natural History of Human Organizations, *Administration and Society*, 7, 2, 131–149.

Krutilla, John V. and Otto Eckstein (1965), *Multiple Purpose River Development*, Johns Hopkins Press, Baltimore.

Linstone, Harold A. and Murray Turoff (eds.) (1975), *The Delphi Method: Techniques and Application*, Addison-Wesley, Reading.

Mazmanian, Daniel A., and Lee Mardecai (1975), Tradition Be Damned: The Army Corps of Engineers is Changing, *Public Administration Review*, 35, 2, 166–172.

Michael, Donald N. (1973), *On Learning to Plan - And Planning to Learn*, Jossey-Bass, San Francisco.

Milton, J and T. Farvar (1971), *The Careless Technology*, Natural History Press, New York.

Mishan, E. (1974), *Economics for Social Decisions*, Praeger, New York.

Mishan, E. (1969), *Technology and Growth: The Price We Pay*, Praeger, New York.

Morison, Elting E. (1974), *From Know-How to Nowhere: The Development of American Technology*, Basic Books, New York.

Rondinelli, Dennis A. (1976), International Assistance Policy and Development Project Administration: The Impact of Impervious Rationality, *International Organization*, 30, 4, 573–605.

Rosove, Perry E. (1973), *A Trend Impact Matrix for Societal Impact Assess-
    ment*, Los Angeles, California: University of Southern California,
    Graduate School of Business Administration, Center for Futures
    Research.

Schick, Allen (1973), A Death in the Bureaucracy: The Demise of Federal
    PPB, *Public Administration Review*, 33, 2, 146-156.

Selznick, Philip (1968), *The TVA and the Grass Roots*, University of Calif-
    ornia Press, Berkeley.

Sewell, Derrick, W. R. (1977), The Changing Content of Water Resources
    Planning: The Next 25 Years, *Natural Resources Journal*, 16, 4,
    791-805.

Steiner, P. O. (1959), Choosing Among Alternating Public Investments in
    the Water Resources Field, *American Economic Review*, 48, 4,
    893-916.

Stover, John (1975), *Probabilistic Systems Dynamics*, The Futures Group,
    Glastonbury.

Tugwell, Franklin (1973), *The Search for Alternatives: Public Policy And
    the Study of the Future*, Winthrop Publishers, Cambridge.

U.S. Army Corps of Engineers (1975), *Uncertainties Associated with Water
    and Related Land Resources Planning*, Menlo Park, California: Center
    for the Study of Social Policy for the U.S. Army Corps of Engineers,
    Napa District.

Utah State University (1974), *Water Resources Planning Social Goals, and
    Indicators: Methodological Development and Empirical Test*, Utah
    Water Research Laboratory, Logan.

Vickers, Sir Geoffrey (1973), *Making Institutions Work*, Wiley and Sons,
    New York.

Vickers, Sir Geoffrey (1968), The Multi-Valued Choice, *Value Systems and
    Social Process*, Basic Books, New York.

Vlachos, Evans (1973), *Transfer of Water Resources Knowledge*,Water Re-
    sources Publications, Ft. Collins.

Water Resources Council (1973), Principals and Standards for Planning
    Water and Related Land Resources, *Federal Register*, 38, Part II,
    24777-24869:

Weizenbaum, Joseph (1976), *Computer Power and Human Reason: From Judgment
    to Calculation*, Freeman, San Francisco.

Wildansky, Aaron (1969), Rescuing Policy Analysis from PPBS, *Public Admin-
    istration Review*, 29, 2, 189-202.

Wolman, M. Gordon (1977), Selecting Alternatives in Water Resources Plan-
    ning and the Politics of Agendas, *Natural Resources Journal*, 16, 4,
    773-789.

Zentner, Rene D. (1975), Scenarios in Forecasting, *Chemical and Engineering
    News*, 53, 22-34.

AN APPROPRIATE MULTIDISCIPLINARY METHODOLOGY FOR WATER
RESOURCES PLANNING AND DECISION MAKING

D.J. Percious, Research Associate, M.E. Norvelle, Research
Associate, N.G. Wright, Research Associate, K.E. Foster,
Associate Director, D.A. Mouat, Director, ARSP
Office of Arid Lands Studies,
University of Arizona, Tucson, Arizona

ABSTRACT

The paper draws upon the literature for examples and inferences that point to some
requirements or imperatives that underly the general problem of water resources
planning and development along with a discussion of their importance. The cost-
effectiveness methodology is suggested as an appropriate general methodology for
water resources planning since the methodology is a comprehensive and systematic
approach to resource development problems in general, acts as an effective framework for interfacing multidisciplinary team efforts, and fulfills some of the requirements for technology transfer. It is pointed out that the methodology is able
to incorporate all the stated imperatives. The use of the methodology is illustrated with an agricultural land and water resources development study for the San
Carlos Apache Indian Tribe in southeastern Arizona.

INTRODUCTION

The purpose of this paper is to describe briefly a general methodology appropriate
for water resource planning and development problems. The methodology can also
function as a technology transfer mechanism and as a framework for interfacing efforts of multidisciplinary teams. The methodology is called cost-effectiveness
(CE) and has enjoyed some success in comparing water resources alternative systems.
This paper briefly draws upon the literature for examples and inferences that point
to some requirements or imperatives that underly the general problem of water resources planning and development along with a discussion of their importance. The
use of the methodology is illustrated with a case study involving alternative land
and water resources problem for the San Carlos Apache Indian Tribe of southeastern
Arizona. The application of CE is discussed generally since the inclusion of specifics would detract from the thrust of the paper, that is, its importance as a
comprehensive and systematic methodology for resource development problems in general, as an effective framework for multidisciplinary teams, and as a technology
transfer mechanism in developing situations.

WATER RESOURCES DEVELOPMENT IN ARID LANDS

Water resources development problems in arid lands are, in general, similar to
those in humid regions except they are exacerbated by the extreme variability in
the timing and quantity of usable water available for development. In some cases,
the quality of usable water during times of peak demand may not be at its optimum
level, for example in agricultural applications; thus adding an additional dimension to the overall problem. In addition, existing institutional and cultural
constraints may not encourage conservation practices in utilizing scarce water resources. For example, an irrigator in the Salt River Valley in Central Arizona
used water saving techniques in his alloted water to realize salvaged water that
he later used to bring in new agricultural not included in his original appropriation. The court ruled against the user, requiring the release of his salvaged water as return flow (Holub, 1972).

The six categories of uncertainty, natural, parametric, model, economic, technological, and strategic, already inherent in water resources decision making in humid
regions, are heightened in importance for water resources decisions in arid environments.

Resources are developed to fulfill societal needs and such as, resourse development problems are multiobjective or multicriterion in nature. Approaches to development require multidisciplinary teams to ensure that the diverse criteria be adequately accounted for, since these criteria, by the nature of the problem, cut
across several disciplines. Multidisciplinary approaches are especially necessary
in arid areas where water scarcity presents formidable problems in technical development and is a sensitive social issue as well.

Water resources development problems in general have enjoyed mixed successes.Nar-
row confinement to careful engineering investigations, construction, and operation
of flood control projects have not decreased flood damages since the significance
of alternatives such as flood warning, flood proofing, land use regulations and
insurance have been obscured or ignored (White, 1973). Lack of multidisciplinary
team approaches in the Lower Mekong and Sri Lanka have resulted in development
projects with a narrow range of alternatives (Hewapathirane and White, 1973). It
would seem that irrigation projects in arid and developing countries have also enjoyed mixed success. Wiener, 1973, comments that in regions where there is sufficient motivation on the farmer level, an effective set of rural and regional institutions, and where there is sufficient development agency capacity, then irrigation projects have been highly successful as in northwestern and northeastern
Mexico; but other areas in Mexico not having entrepreneural talent have a rather
unsatisfactory record.

An examination of the literature seems to indicate that there are at least five
resource development imperatives that should condition approaches to resource development in general and are salient for arid development countries in particular.

FIVE RESOURCE DEVELOPMENT IMPERATIVES

An appropriate framework for water resources planning and development should include, 1) a sufficiently clear statement of goals and/or objectives, 2) the opportunity for a multidisciplinary team approach and a methodology for interfacing
their efforst, 3) the treatment of the uncertainty categories, 4) a methodology
that includes the quantifiable, and equally important, unquantifiable aspects of
development, and 5) an appropriate mechanism for technology transfer. Each of
these imperatives are addressed briefly in the following paragraphs.

Goals/Objectives

A proper set of goals and objectives condition the planning effort since they define the "path" or direction of the planning and development effort, and as such,
act as the backdrop against which development success can be measured. An increased awareness of social needs and goals have resulted in dramatic changes, in recent years, in objective formulation in water resource planning and management
(Biswas, 1973). The adoption in the U.S. of the Principles & Standards for Planning, Water and Related Land Resources in 1973, and the Federal Water Pollution
Control Act Amendments of 1973 has had a distinct impact in water utilization and
reflects an increased concern for environmental consequences stemming from water
resources development. A more detailed discussion of these issues are given in
Viesman and Stork, 1973; and in Duckstein, 1975. It seems essential then, that
social goals be well specified or articulated before concise objectives can be
specified. The problem is, however, that the societal values on which goals are
based are transitory (Dalkey, et al., 1970). Rapid technological and sociopoliti-
cal changes in societal values can be expected, a result of disproportionate increases in uncertainties associated with the above (Wiener, 1973). The goal setting process is thus made even more complicated than has heretofore been the case
pointing out the need for flexibility in water development systems, shorter planning periods, and frequent review of the planning process.

It is characteristic then of resources development problems that they seek to fulfill societal needs and hence should be flexible enough to meet a changing value
structure. Given that societal goals are well specified then an appropriate objective set can be formulated. Multidisciplinary efforts can help this process.

Multidisciplinary Team Approach

It seems consistent with the multiobjective-multicriterion nature of water resource
development problems that a multidisciplinary team approach would best be able to
assist in the formulation of appropriate objectives and help specify the diversity
of criteria by which performance is measured. In addition, the team can interact
on a technical level in the inventory and analytical phases, or on an institutional and cultural level as well. For example, geographers, hydrologists, and agricultural economists can interact under the umbrella of applied remote sensing in
the inventory phase to specify or make inferences regarding potential agricultural
land, land use status, vegetative patterns, as they related to geologic structure
and water resources occurrence, and the interrelationship between drainage patterns, geologic structure, and hydrology.

Team interaction can play an important role in analyzing what Wiener 1973, refers
to as the farmer's psychological space or his motivation to accept innovation and
change, the rural and regional institutional structure, and capacity for development assistance.

Finally, team assistance is indispensable in the treatment of uncertainty categories and in accomplishing technology transfer.

Uncertainties

We choose to separate goals, which are general and refer to broad societal needs,
from objectives that are posed as strategy statements to achieve goals. Objectives need to reflect technical and societal considerations and therefore, are
subject to considerable uncertainty.

Six categories of uncertainty inherent in resources development have been enume-
rated by Kisiel and Duckstein, 1972; Bogardi, 1975; and Duckstein, 1975, as fol-

lows:

Natural Uncertainty: The natural processes of rainfall, droughts, and stream runoff always introduce uncertainty in decisions.

Model Uncertainty: It is impossible to choose a model that represents actual hydrological phenomena.

Parametric (Sample) Uncertainty: The finiteness of available sample data to estimate model parameters introduces an element of uncertainty. This is particularly true for developing countries where information on natural processes is lacking.

Economic Uncertainty: Inadequate knowledge of economic factors, particularly costs and losses introduces considerable uncertainty and is typical of most situations.

Technological Uncertainty: In this world of rapid technological advances, it is difficult to forecast system responses. Introduction of highly developed technologies in a developing situation may cause structural changes that may, or may not, be desired, and may even increase economic and strategic uncertainties.

Strategic Uncertainty: It is not usually known at decision time what the future prevailing institutions and priorities will be. If project implementation lead times are large, then this category is heightened. Seeking flexibility in system design is a way of accounting for strategic uncertainty.

As was pointed out by Szidarovsky, et al. 1976, water resources decision makers are often called upon to make difficult decisions for complex engineering works with fairly adequate physical data but poor or inaccurate economic data. These authors and others, for example, Davis, et al., 1972, and Kazanowski, 1972, have investigated ways of coping with economic and parametric uncertainty.

In his address to the First International Conference on the Transfer of Water Resource Knowledge, 1973, Gilbert F. White discusses the importance of uncertainty in water resources decisions and points to the apparent paradox in water resources investigations today. He aptly points out that each advance in technological development is accompanied by increasing uncertainty to the extent that new alternatives are exposed and new situations arise in which human judgement must be exercised. Uncertainties in the decision process need to be specified either qualitatively or in quantifiable probabilistic terms whenever possible. In his final statements, White points out the need for full documentation so that the essential paths of analysis can be identified to trace the technical and social adjustments in the investigation at hand or for subsequent inquiry. This statement speaks directly to transfer of technology mechanisms and the mandate for inclusion of quantifiable and unquantifiable criteria.

Measurable Criteria

In any water resources development problems, criteria are called for that in some way relates project or system performance to specified goals/objectives. Traditionally, criteria are chosen to relate directly to monetary units, i.e., the tangible or quantifiable effects of development; however, there has been an increasing emphasis in recent years on the importance of including the tangible or unquantifiable aspects of development, indeed raising them to a level of equal importance, see for example the U.S. Water Resources Council's Principles & Standards for Planning Water & Related Land Resources of 1973; Popovich, et al.,

1973: Kazanowski, 1972; and White, 1973. We have already alluded to the importance of adopting a broad outlook in water resources planning, including quantifiable and qualitative criteria, that should preclude selection of a narrow set of
alternatives and/or the pitfalls of the sole criterion fallacy, Kazanowski, 1968,
that lumps all criteria into a single unit, e.g., dollars, or the fallacy that
everything of importance can be quantified. A more detailed discussion of the fallacies inherent in water resources planning can be found in Kazanowski, 1968, and
in Chaemsaithong, et al., 1974.

Technology Transfer

The words "technology transfer" are pregnant with meaning. The phrase embraces
the simple documentation of rationale and assumptions underlying an investigation
to the higher order functions of water resources system conceptualization, information retrieval systems, and assistance in decision making and policy formulation. It is obvious that no one mechanism or framework can fulfill all the rami-
fications of technology transfer, and it is not within the scope of this paper to
discuss the many efforts that are in progress in this area. Particularly notable
are the United Nations efforts and the International Programs at many Universities
which are discussed in the First International Conference on Transfer of Water Resources Knowledge, held in Colorado in 1973.

It is important to recognize that water resources decision making is a process and
therefore a framework or methodology that can be used successively on water resources problems would seem desirable. A general and systematic methodology that
could be used for water resources problems and incorporate the imperatives requires the means to interface multidisciplinary team efforts, the means to incorporate uncertainty categories so that their significance is accounted for at decision time, and the means to function on the technology transfer level. In addition, it would be desirable that the functioning of the methodology was predicated
on a specified set of goals/objectives to properly condition the entire planning
and development effort.

The Laboratory for Native Development, Systems Analysis and Applied Technology
(NADSAT) of the Office of Arid Lands, University of Arizona, is presently employing a general methodology that incorporates these imperatives in dealing with water and land resources problems of soutwestern Indian tribes and is discussed in
the following paragraphs.

PROPOSED METHODOLOGY FOR WATER RESOURCES PLANNING & DEVELOPMENT

It is suggested that the cost-effectiveness methodology (CE) is an appropriate
methodology and framework to approach natural resources development problems in
general and water resources development problems in particular.

It is our feeling that the methodology exhibits many of the desirable attributes
that a comprehensive methodology should have and has the ability to function as a
broad transfer of technology mechanism. The methodology also provides a systematic framework to interface multidisciplinary efforts and allows consideration of
all uncertainty categories at decision time.

COST-EFFECTIVENESS METHODOLOGY

The cost-effectiveness methodology was first exposed, in its standardized form, by
Kazanowski, 1968, and consists of the following ten steps:

> 1) Define the desired goals, objectives, or purposes that
>   alternative systems are to meet or fulfill.

2) Identify the requirements essential for the attainment of the desired objectives.
3) Establish system evaluation criteria as measures of effectiveness (MOE) that relate system capabilities to requirements.
4) Develop alternative system concepts for accomplishing the objectives.
5) Determine capabilities of alternative systems in terms of MOE.
6) Generate system versus criteria array (evaluation matrix).
7) Analyze merits of alternative systems.
8) Select fixed cost or fixed-effectiveness approach.
9) Perform sensitivity analyses.
10) Document the rationale, assumptions, and analyses underlying the previous steps.

The methodology has been used to compare water resources systems in the Mekong Basin by Chaemsaithong, Duckstein, and Kisiel, 1974; in analyzing waste water reuse schemes by Ko and Duckstein, 1972; in the analysis of large scale water development alternatives in the Hungarian Great Plain by David and Duckstein, 1976; and for evaluation procedures for water quality control analysis in river basins by Duckstein and Kisiel, 1972. They also pointed out in the same paper, that the U. S. Water Resources Council guidelines for water resources development falls squarely within the cost-effectiveness methodology.

Bokhari, 1975, used CE to accomplish ex post evaluation of river basin development in Pakistan and Popovich, et al., 1973, used the methodology to evaluate various solid waste disposal alternatives. Kisiel and Duckstein, 1972, employed the methodology to evaluate the economics of hydrologic modelling.

Kazanowski, 1972, emphasized the major sources of uncertainty in the use of CE as being in the determination of the goals/objectives, the criteria, and the uncertainty surrounding cost estimates of alternatives. The importance of this statement is discussed briefly in the previous section. Examples of handling the economic uncertainty of cost estimates can be found in Szidarovszky, et al., 1976; and in Kazanowski, 1972.

The CE methodology is a general approach to resource development problems that seeks to find significant differences in the costs or requirements of alternatives that are formulated to attain desired objectives (Duckstein, 1975). The methodology, as is illustrated in the above ten steps is a systematic framework that leads analysts and decision makers alike throughout the process of conceptualization, the formulation of alternatives, and analysis of a resource development problem. Opportunities for the interfacing of multidisciplinary team expertise are present in all the steps. Decision makers are offered the latitude of decisions based either on a fixed-cost (budgeting constraints or a fixed-effectiveness standpoint. Sensitivity analysis on the steps provides an element of feedback for both analysts and decision makers, the latter being able to view how perturbations in system elements can affect alternative outcomes relative to their fixed-cost or fixed-effectiveness decision basis. Finally, by virtue of step 10, the methodology is an invaluable tool to examine the outcomes of implemented decisions in an ex post fashion, in the light of new data or decision making imperatives at some future time. As such, the methodology is akin to an adaptive control process on resource development, whereby the methodology can be used successively, whether for project updating or new ventures, incorporating the significance of past mistakes at each juncture.

We recognize that the CE methodology is not a panacea, but we can't help but agree with Kazanowski, 1972, that is has "intellectual and visceral appeal". The methodology excludes some of the fallacies inherent in other approaches, e.g., benefit-cost approach, again the reader is referred to Chaemsaithong, et al., 1974, and Kazanowski, 1968.

The structure of the methodology lends itself to communication to decison makers, irregardless of their levels of sophistication, of the multifaceted nature of resources development problems, the analytical procedures used in alternative performance evaluation, and the realization that there are two sides to the decision coin that demand consideration in the decision process. In addition, the methodology can be left with the recipient to be used successively in future planning, and implementation efforts. The foregoing indicates that the CE methodology is a promising candidate in satisfying the natural resources technology transfer function. The use of CE in an agricultural land and water resources development problems is outlined in the following section.

AGRICULTURAL LAND AND WATER RESOURCES DEVELOPMENT PROBLEM FOR THE SAN CARLOS APACHE INDIAN RESERVATION

The function of NADSAT is discussed elsewhere in this conference, see Norvelle, et al., 1977; however, suffice to say that NADSAT's role is assistance and technology transfer and therefore has adopted a passive, rather than an active role in decision making. The emphasis is on alternative formulation and performance analysis, relative to their costs and effectiveness in attaining tribal objectives. Imbedded within this commitment is the obligation of technology transfer to assist the tribes in attaining an increased level of self determination. Current technology is to be applied to Indian Resource development problems by NADSAT and transferred to the tribes so that they can deal with similar problems in the future with a greatly reduced level of reliance upon outside assistance.

The primary function of this technology transfer commitment is to provide tribal decision makers and planning personnel with an appreciation for the non-Indian method of resource problem conceptualization, the factors to be considered in the analysis, and the formation of alternatives that a multidisciplinary analytical team perceives.

The cost-effectiveness methodology has been adopted for this study not only because it provides a systematic methodology for problem conceptualization and subsequent analysis, but it also is a coherent framework to interface various disciplines in a multicriterion problem and affords a mechanism for technology transfer.

Physical Setting

The Gila River portion of the reservation lies in the southeastern portion of Arizona and is part of the Basin and Range Physiographic Province. The Gila River Valley is part of a northwest trending structural trough, bounded by volcanic mountains on the north and granitic and gneissic mountains on the south. Ground water occurs in a deep, basin-fill aquifer of low permeability. under semi-confined conditions, and in a shallow alluvial aquifer occupying the terrace and floodplain alluvium. The drainage area of the Gila, upstream of the reservation gage, is 11,470 sq. miles. Rainfall averages over 13 inches per year with precipitation dominated by winter frontal and summer cyclonic storms. Some 390 acres are currently being irrigated with pumped river water, and shallow ground water during the low river flow periods. Crops raised are alfafa, barley, and grain sorghum. Water quality of the Gila varies inversely with discharge exhibiting a very high sodium hazard during low flow periods. Ground water is pumped for irrigation from the alluvium; but is of poor quality presenting a high to very high sodium hazard to its use for irrigation. Numerous thermal springs discharge to the surface from the basin fill in close proximity to the river on the north, contributing to the salinity problem.

Cultural Setting

Indian culture, with few exceptions, is essentially non-manipulative in its perception of the natural environment. Historically, the Indian has viewed himself as not being engaged in a contest with nature, as has typified non-Indian society; however, it has become increasingly obvious to Indian societies that the economic development of their natural resources can yield considerable benefits. Development of their natural resources, consonant with the Indian desire of self-determination, can not only provide sizeable social and economic advantages to individual tribes; but contribute to the gross national product as well. As was pointed out by Anderson, 1976, three conditions necessary for economic development are, 1) jurisdictional control over tribal natural resources, 2) availability of capital, and 3) management assistance. Hence, there is a need for strengthening tribal resource ownership and sovereignty for if Indians lack control over their resources, they lack control over their economy.

COST-EFFECTIVENESS APPLICATION

The discussion of the applied methodology is in a step form, similar to that originally suggested by Kazanowski, 1968.

The ordering of the steps is not of utmost importance since each problem may present significant differences. The order that we are utilizing in this problem is 1, 2, 4, 5, 3, 6, 7, 8, 9 and 10. The salient difference is the precedence of step 4 over step 3, to avoid possible bias towards formulating MOE to conform to prior specified alternative systems.

Objectives

   Agricultural Production. Maximize irrigated agriculture production from the available land resources along the Gila River.

   Water Resources. Increase the utilization and economic development of the valley's water resources.

   Resource Utilization. The use of natural, social, economic resources to implement and operate alternative systems should be minimized but may be conditioned by tribal and non-tribal institutional and cultural imperatives.

   Salinity. The salinity impact from resource development should be minimized.

   Flood Protection. Flood protection should be provided to lessen deterioration of the natural resources of the valley.

   Flexibility. Alternative system development should be flexible enough to meet the needs of the future and avoid foreclosures of future options open to the Tribe.

Since the objective set conditions the entire process, it is important that the objectives reflect as much as possible, the "true" desires of the tribal decision makers. The input from team and tribe can complement one another so that important omissions may be neglected or that preference hierarchies can be ascertained. This is a difficult process in any resource development problem and may require several consultations to ensure their appropriateness.

System Specifications

These should be specific as completely as possible and follow in one-to-one order
with the objectives.

Agricultural Production. Available agricultural land along the Gila
should be maximized. Yield increases of 20% could be realized in the present 390
acres with improvements in water deliveries and farm management practices. Agricultural lands are categorized as presently irrigated; having high, medium, and
low potential; and improved pasture. Categories are based on location, accessibility, soil characteristics, and potential for inundation from specified flood
return periods. The agricultural land distribution lands itself well to investment phasing and demonstration projects.

Potential crops are wheat, barley, grain sorghum, cotton, jojoba, and bermuda grass
for improved pasture. Farm budgets specify break-even production levels. Unit
cost-production/acre curves will assist in defining management requirements and
target yields. Production periods are a function of time; years for jojoba, and
annually for the remaining crops.

Water Resources. The San Carlos Apache Tribe has a decreed allottment
(Globe Equity No. 59) to 6000 AF/year from the natural flow of the Gila for reclamation and irrigation of 1000 acres of land. The amount of withdrawals are
further constrained as not exceeding 12.5 cfs at any one time or 1/80 cfs per irrigated acre. Use of the shallow ground water system should be minimized due to
the water quality problems. This constraint can be released with potential of
quality improvement under a given alternative. Different contingency pumping
schedules can arise from different alternatives. Monthly crop water demands are
determined from published consumptive use of data for the various crops (Halder-
man, 1973). Different farm management specifications can alter total crop deliveries, and diversion requirements are determined from various efficiency levels.

Resource Utilization. The resource categories are water, land, manpower, energy, capital, and environmental resources.

Water Losses: Represent a valuable resource and should be minimized.
Losses are determined by specifying water use and conveyance efficiencies for
sprinkler and flood irrigation, and leaching requirements.

Land Resources: Are specified through the results of a basin-wide
soil survey and flood inundation potential determined from the results of the Army Corps of Engineers HEC-2 program for water surface profiles.

Manpower Requirements: Occur both in the implementation and the operational stage of development. The potential for manpower utilization may be considerable and should be maximized to increase personal income, social well being,
and other benefits.

Energy Consumption: Should be minimized. The prospects of additional
energy from the existing source is highly uncertain and additional power from any
source will be expensive and uncertain in the light of the present energy situation. The current average monthly consumption for 13 existing irrigation wells is
1,336 Kwh with a peak demand of 100 Kwh, assuming a peaking load factor of 1.1.
Estimated irrigation horsepower is 165, based on total power use per year of
208,416 Kwh at 1650 Kwh/horsepower (Loftin & Associates, 1977). Improvements in
the existing pumping system can lessen the annual energy consumption.

Capital Resources: May be broken down into categories to assist decision makers (DM) and lending institutions in financial feasibility analyses. Categories are land development costs, equipment costs, production costs, and the cost of irrigation facilities. These costs may be tempered by considerable uncertainty. In general, one would tend to minimize initial capital investment.

Environmental Resources; Should be protected from deterioration as much as possible. It is likely that wildlife population may be enhanced by relatively large scale farming in the valley since the available land will be in parcels spread through the valley, affording population's cover and food supplies for wildlife.

Salinity. Regression analyses of mean daily discharge with electrical conductivity and the sodium adsorption ratio (SAR) allows specification of river discharge thresholds relative to irrigation suitability. Mean daily discharge thresholds as a function of SAR are as follows: a) 38 cfs, SAR = 11.3, high sodium hazard, b) 70 cfs, SAR = 8.7, medium sodium hazard, and c) 125 cfs SAR = 5.2, exhibit a low sodium hazard.

Ground water quality is specified as it bears on its use for irrigation purposes or its impact on river and/or alluvial ground water aquifer quality. Ground water issuing from 10 thermal, limestone springs north of the river, indicated total dissolved solids (TDS) from 1,357 to 4,037 ppm and a SAR range of 7.86 - 23.37. Analyses of irrigation well water from the alluvial aquifer exhibited a TDS range from 2,107 to 3,409 and SAR's from 7.02 to 12.46. These waters fall either in the C4-S4 or C4-S3 class.

Analyses from stock wells that penetrate the basin fill aquifer indicate low TDS, low SAR's, and irrigation classifications from C2-S1 to C3-S1.

Return flows should be minimized to avoid quality problems for downstream agricultural areas. Pesticide residue levels 15-20 miles downstream at San Carlos Reservoir is already near FDA standards (Arizona Water Quality Council Hearings, 1976). Return flows should subscribe to Federal Standards. State standards specify only boron and turbidity limits for agricultural purposes.

Flood Protection. Annual peak flood of the Gila River for 46 years of record were analyzed and found to distribute as a Log Pearson III pdf. This hypothesis was not rejected by the K-S test. The basic descriptive statistics for the record are: mean of 10,400 cfs, standard deviation of 13,700 cfs, skew of 3.27, and a coefficient of variation of 1.32 for the unlogged data. For the period of record, 12 events of flows greater than 12,000 cfs occurred, and ranged from 12,800 cfs in 1937 to 80,000 cfs in 1973. Computed flood events corresponding to the 2, 5, 10, 25, 50 and 100 year return periods will be processed through the HEC-2 Water Surface Profile Program (Army Corps of Engineers, 1973) to obtain flood inundation levels. These levels will be validated by comparison with available aerial photography of major flood events. This information will yield estimates of flood protected measures for irrigation facilities, crop protection, and assist in crop selections for flood prone areas. For example, jojoba can suffer no flood damages over its 12 year maturation and production period requiring that jojoba cropping be protected or there will be a certain risk associated with raising jojoba inside the 100 year flood level.

Flexibility. Flexibility refers, in a somewhat general manner, to the ability of alternative systems to integrate with other Tribal needs, the robustness of alternatives to uncertainties, the susceptibility of alternative systems to development staging, and the ability of alternatives to function in various operational modes.

Improved pasture acreage could be used to help maintain the tribal breeding stock for the Tribal Cattle Association; however, the management requirements will be high. The grazing of 150 to 200 head per 100 acre of improved pasture is optimum. Management requirements dictate partitioning a 100 acre parcel of improved pasture into four 25 acre portions. The grazing herd is rotated weekly on each 25 acre portion, consequently each 25 acre portion is grazed for one week followed by a 3 week rest period.

There are presently 295 residential and commercial power consumers in the valley, exclusive of irrigation requirements. Due to the poor water quality of the shallow ground water aquifer, water for domestic uses is obtained from a shallow, but apparently highly productive, alluvial aquifer in Goodwin Wash, a tributary to the Gila River. The lower end of the tributary is outside the eastern border for the Reservation, requiring piping the water to storage tanks for community distribution. The demand for municipal water supplies can be expected to increase as a result of agricultural expansion and other factors; consequently the ability of a given alternative system to help meet this demand would be a desirable feature.

The use of poor quality water for irrigation purposes poses rather high level management requirements and any capability of specific alternatives to mitigate the quality problem would be desirable. Further, alternative systems that are amenable to investment and development staging would be beneficial from a training and demonstration project standpoint, with the investment benefits notwithstanding.

The set of specifications should be defined as precisely as possible since they identify the specifications or requirements that formulated alternatives are to fulfill to approach objective attainment. The word approach is used since it is probable that no particular alternative will satisfy all the desired objectives exactly. Uncertainties in the specifications can be encoded into probability distribution functions, if possible, and be used later in the sensitivity analysis.

System Evaluation Criteria

The criteria or measures of effectiveness (MOE) are used to evaluate how well a given system performs with respect to meeting the specifications, and hence the objectives. Systems responses may be quantitative such as crop production, system water deliveries, volume of ground water pumped, and the level of flood protection required; or may be expressed qualitatively for system flexibility, tribal employment potential, and potential for soil and water quality. The criteria list should be complete but not exhaustive and should follow the specifications sequence. The following is only a partial list for brevity.

     Agricultural Production. The MOE for agricultural production is tons/acre or tons/acre-year, for a rate criteria.

     Water Resources. Use is source specific and to quantity used, a) surface water in AF/year, and b) ground water pumped in AF/year.

     Resource Utilization. Land MOE is acres in production per year. Capital MOE may be in terms of land development costs, rate-of-return, and total investment. Energy MOE may be expressed as Kwh/year. Water losses MOE are similar to water use in AF/year. Manpower potential may be expressed in terms of high, medium, and low potential for employment.

Kazanowski, 1972, pointed out that a complete listing of MOE is critical since it not only relates directly to objective attainment, but plays an important role in the next step, because with a fixed cost approach, decisions among alternatives are based on the evaluation criteria or MOE.

Select Fixed Cost or Fixed Effectiveness

This step is of considerable importance in the decision making process and is not a trivial matter. The "true" preference hierarchy of the Tribe's goals and objectives may be imbedded in the selection process and will become more obvious once the final selection is made and implementation begins. Goals are high order statements and specific objectives may be difficult to vocalize, especially regarding natural resources development. Ill specified objectives are recognized as an important source of uncertainty in multiobjective-multicriterion problems, as pointed out previously.

In the fixed-cost approach, a definite budgetary expenditure is given, and the alternatives are compared in terms of levels of effectiveness. On the other hand, the fixed effectiveness approach requires that the alternatives be compared through various implementation costs. Analysts cannot dictate which approach should be taken since each decision approach will really be based on value judgments of the decision makers. The role of cost uncertainties, inherent in any problem, takes on a special significance in this step. Kazanowski, 1972, emphasized this point when he considered costs as a major uncertainty source.

It should be added that despite scarce financial resources, a fixed-effectiveness approach is still feasible because there may be a viable effectiveness level due to the availability of funds from outside sources. In addition, it is illuminating to comment on the phrase "it is cost-effective" that can often be heard conversationally or in the literature. As discussed above, the two words are not correlative but speak to two different approaches in decision making. The desire for maximum effectiveness for the least cost is fallacious since minimum cost (zero) corresponds to minimum effectiveness (zero).

Determine Alternative Systems

Various systems formulated to achieve specified objectives should be distinct so that their costs and measures of effectiveness (MOE) can be explicitly related to each. Extremal alternatives will tend to broaden a decision makers satisfaction space within which he operates, whereas marginal alternatives (not too distinct) will narrow this space so that trade-offs may be difficult to evaluate. This should be tempered with practicality and a foreknowledge of what institutional and cultural frame the DM is situated in. For the purposes of discussion, only four alternatives are discussed below in general terms.

System I. Considers maximizing planting of bermuda grass and jojoba on the available agricultural land to take advantage of the salt tolerant aspects of jojoba and to examine the potential of improved pasture. Jojoba is indigenous to the soutwest and thrives on lesser amounts of water, in addition, its long maturation period, economic future, and production capacity/acre can create spillover effects for farm management training, personal income, and tribal welfare by strengthening the existing Apache Tribal Jojoba Enterprise. Raising jojoba commercially is a no risk decision, requiring considerable flood protection.

System II. Continue irrigation of crops in the conventional manner using river pumping plants and ground water wells to augment river supplies during low flow periods having poor ground water quality would severely hamper agricultural production. This alternative would entail improvement of existing facilities, including flood proofing and refurbishing of poor productive irrigation wells; and the installation of new systems.

System III. Entails fixed river pumping plants at various points, with flood protection, that can function as System II and also lift river water up

to gravity distribution points for distant fields or to spreading basins with small
retention structures built on highly permeable alluvial fan areas. The structures
can also function as a retardant for tributary flood waters. Recharge from these
spreading areas can promote the flushing of poor quality ground water in the shallow alluvium by steepening the gradient towards the river. Environmental effects
may be both good and bad.

System IV. Entails developing the alluvial ground water aquifer in
Goodwin, Wash to provide high quality water to the Gila River agricultural area
near the eastern border. Conservative estimates indicate that 8 AF per day or some
2000 gpm could be made available to help mitigate the water quality problem and
promote increased agricultural yields. Depending on the operations of the overall
irrigation system, the use of unlined canals could promote recharging of high quality waters to the shallow Gila ground water system. System IV can easily function as an adjunct to either of the two previous systems.

All of the systems are amenable to investment phasing and/or demonstration projects with spillover benefits in training and strengthening lines of credit.

Determine Alternative System Capabilities

This step analyzes the capabilities of alternative systems in terms of various MOE.
The methods employed can range from simulation processes, stochastic hydrology, to
pilot plant studies, analogues, and economic analysis. The various uncertainty
categories can be incorporated here depending on models chosen to evaluate system
performances.

Generate System Vs. Criteria Array

For this step, an array is constructed that assembles all the system capabilities,
as they relate to the specifications through their measures of effectiveness (MOE).
This allows the decision makers (DM) an alternative by alternative comparison regarding their individual cost requirements, water deliveries, energy consumption
or requirements, manpower potential, management requirements, etc. The usefulness
of the array tableau is heightened by the inclusion of quantitative and qualitative MOE's.

Analyze the Merits of Alternatives

The merits of the alternative systems require ranking in order of importance and
is largely a subjective process. Ranking is the responsibility of the Tribe;
however, assistance can be offered in examining the problem with specific cost and
effectiveness levels to illustrate the process and perhaps generate an appreciation of the non-trivality of the matter. It is also important to appreciate that
all the information in the array is assumed correct at this juncture, and that the
ranking should not be determined by arbitrary weighting.

There are various methods that can be employed here to assist in the ranking process if serious conflicts arise between alternatives. For example, the multicriterion algorithm ELECTRE described by Roy, 1971 and used by David and Duckstein,
1976, in an analysis of water resources alternatives in Hungary is based on an out-
ranking or preference ordering of alternatives. Another is multi-attribute utility that has been employed in city airport development by Keeney, 1973; and in water pricing utility by Duckstein and Kisiel, 1971. The latter method involves extensive interaction between decision makers and analysts to assess and utilize
multi-attribute utility functions.

Perform Sensitivity Analysis

Sensitivity analyses are performed on all the previous steps to introduce a feed-back element into the procedure. Perturbations are introduced to the various MOE's specifications, parameters, approach, etc. to observe changes in the system criteria array. The effect of various uncertainty categories can be assessed in this manner.

Documentation

The essential value of the standardized CE methodology lies in its ability to order one thinking regarding water resources planning and development, both on the part of the analysts and the decision makers. The organized approach is not only valuable in evaluating alternative courses of action, prior to implementation, and expected results; but also holds consierable value in looking back on actions taken and evaluating the results in the light of current information. The methodology can be used successfully for project planning, and subsequent reevaluation before new projects or updating is required, thereby retaining control over development and retaining flexibility. To accomplish the foregoing, it is essential that the assumptions, rationale, and analytical methods employed are fully documented.

This step is an important part of the technology transfer function since the essence of the evaluation effort is contained here, i.e., the conceptualization of the development problem and the assumption underlying the approach and analytical methods are discussed in this step. Further, the transmittal of uncertainties also occurs here insofar as they relate to the assumptions and analytical methods. The content of this step is essential for project updating, future reevaluation, or for future planning efforts. It is retained by recipient decision makers as a valuable adjunct to their entire planning and decision process.

CONCLUSIONS

The following concluding points may be drawn from the above discussion:

1) The CE methodology is an appropriate general method to approach natural resource development problems.

2) CE can avoid some of the common fallacies inherent in other approaches, e.g., sole criterion as in benefit-cost, the trade offs are determined in the fixed-cost or fixed-effectiveness approach. Both quantitative or qualitative criteria are easily included.

3) The methodology offers a systematic approach for analysis of resource development problems, with a built-in feedback element to avoid omission of important factors.

4) The methodology is able to account for all categories of uncertainty, e.g., natural, model, parameter, economic, strategic and technological.

5) CE offers a means of ex post evaluation of decisions, can incorporate new data for subsequent planning and implementation, and allows decision makers to retrace their steps through the entire decision process. The planning and decision process is strengthened by familiarization and in learing from past implementation results.

6) The methodology is essentially a "soft" appro?. to water resources planning and development; however "hard" analytical methods can easily be included depending on the dictates of the problem at hand.

7) CE offers a logical and ordered framework wherein the efforts of multidisciplinary teams can be interfaced throughout the entire problem.

8) Cost-effectiveness methodology provides an important mechanism for technology transfer in that the conceptualization, factors, and alternative formulation and analysis, are communicated to decision makers in an orderly framework; and the interrelationships among important factors are revealed through the entire process.

9) The methodology functions as a technology transfer framework in concert with its use in planning and development function.

ACKNOWLEDGEMENTS

The work upon which this paper was based was supplied by the Office of Technical Assistance, Economic Development Agency, U.S. Department of Commerce, Grant No. .99-6-09555. The support and assistance of the San Carlos Apache Indian Tribe is gratefully acknowledged. The senior author wishes to express his thanks to Prof. L. Duckstein for his helpful advice and encouragement.

REFERENCES

Anderson, A. T. (1976) Nations Within a Nation. Prepared by A.T. Anderson, Special Assistant to Amer. Indian Policy Review Comm., courtesy of Union Carbide Corp.

Arizona Water Quality Council Hearings. (1976) December 16, Tucson, Arizona.

A. Biswas, Socio-Economic considerations in water resources planning, Water Resources Bulletin, 9. No. 4.

Bogardi, I. (1975) Uncertainty in Water Resources Decision Making, Proc. U.N. Interregional Seminar on River Basin Development, Budapest, Hungary.

Bokhari, S. M. H. (1975) Ex-post Evaluation of River Basin Development in Pakistan, Proc. U.N. Interregional Seminar on River Basin Development, Budapest, Hungary.

K. L. Chaemsaithong, L. Duckstein and C. C. Kisiel, Alternative water resource systems in the Lower Meking, Journal Hydraulics Div., 100, No. HY3, American Soc. Civil Engineering, March, 1974.

Dalkey, N. C., Lewis, R. and D. Snyder. (1970) Measurement and Analysis of the Quality of Life: With Exploratory Illustrations of Applications to Career and Transportation Choices, Rm -6228-DOT, The Rand Corp., Santa Monica, California.

K. David and L. Duckstein, Multi-criterion ranking of alternative long-range value resource systems, Water Resources Bulletin, 12, No. 4, American Water Res. Assoc., August, 1976.

D. R. Davis, C. C. Kisiel and L. Duckstein, Bayesian Decision Theory Applied to Design in Hydrology, Water Resources Bulletin, 8, No. 1 (1972).

Duckstein, L. and C. C. Kisiel. (1971) Collective Utility: A Systems Approach to Water Pricing Utility, Proc. Int'l. Symp. on Mathematical Models in Hydrology, Warsaw, Poland.

Duckstein, L. and C. C. Kisiel. (1972) Cost-Effectiveness Approach: An Example of Water Quality Control in a River Basin. Public Response to the Proposed Principles and Standards for Planning Water and Related Land Resources and Draft Environmental Statement, U.S. Water Resources Council, July, 1972.

Duckstein, L. (1975) The Role of New Technologies for Improved Water Management and Related Effects on Water Law Systems. Presented at International Conf. on Global Water Law Systems, Valencia, Spain, September, 1975.

Hewapathirane, K. and G. White. (1973) Obstacles to Consideration of Resources Management Alternatives, First International Conf. on Transfer of Water Resources Knowledge, E. I. Schulz, V. A. Kolzer and K. Mahmood, eds. , Water Research Publications, Fort Collins, Colorado.

Holub, H. (1972) Some Legal Problems of Urban Runoff, Hydrology and Water Resources in Arizona and the Southwest, Arizona Academy of Sciences, Prescott, Arizona.

Kazanowski, A. D. (1968) A Standardized Approach to Cost-Effectiveness Evaluations, J. M. English, ed., In Cost-Effectiveness, Wiley, New York.

Kazanowski, A. D. (1972) Treatment of Some of the Uncertainties Encountered in the Conduct of Hydrologic Cost-Effectiveness Evaluations. International Symposium on Uncertainties in Hydrologic and Water Resources Systems, U. of Arizona, Tucson, Arizona.

Keeney, R. L. (1973) Concepts of Independence in Multi-attribute Utility Theory, In Multiple Criteria Decision Making, J. L. Cochrane and M. Zaleny, eds., U. of South Carolina Press, Columbia, South Carolina.

Kisiel, C. C. and L. Duckstein. (1972) Economics of Hydrologic Modelling: A Cost-Effectiveness Approach. Proc. International Symposium on Modelling of Water Res. Systems, Ottawa, Canada.

S. C. Ko and L. Duckstein, Cost-Effectiveness Analysis of Waste Water Reuses, J. of Sanitary Eng. Div., 98, No. SA6, Amer. Soc. Civil Engineers. (1972).

Loftin, A. D. and Assoc. Inc. (1977) Study of Existing Electrical Systems. Report to San Carlos Apache Tribe, January, 1977.

Norvelle, M., Percious, D., Brooks, W. and K. Foster. (1977) The Role of the University in Applied Technical Assistance. Paper presented at UNITAR Conf. on Alternative Strategies for Desert Development and Management, Sacramento, California.

M. L. Popovich, L. Duckstein and C. C. Kisiel, A Cost-Effectiveness Analysis of Municipal Refuse Disposal Systems. J. Environmental Engr., 99 (1973). Amer. Soc. of Civil Engineers.

B. Roy, Problems and Methods with Multiple Objective Functions, Math. Progr., 1. No. 2 (1971).

F. Szidarovszky, I. Bogardi, L. Duckstein and D. R. Davis, Economic Uncertainties in Water Resources Project Design, Water Resources Research, 12, No. 4 (1976).

U.S. Army Corps of Engineers. (1973) HEC-2 Water Surface Profiles, Hydrologic Engineering Center, U.S. Army Corps of Engineers, Davis, California.

W. Viessman, Jr. and K. E. Stork, Changing Times for Water Research and Technology Transfer, Water Resources Bulletin, 9, No. 4 (1973).

Water Resources Council. (1973) Principles and Standards for Planning Water and Related Land Resources, Fed. Register, 38, No. 174.

White, G. (1973) Prospering With Uncertainty in Floods and Droughts, First International Conf. on Transfer of Water Resources Knowledge, E. I. Schulz, V.A.Kolzer and K. Mahmood, eds., Water Res. Publications, Fort Collins, Colorado.

Wiener, A. (1973) Water Resource Development Policies and Transfer of Knowledge from Developed to Undeveloped Countries, First International Conf. on Transfer of Water Resources Knowledge, E. I. Schulz, V. A. Kolzer and K. Mamood, eds., Water Res. Publications, Fort Collins, Colorado.

Policy Considerations for Short-Term
Water Management in Arid Regions

F. Harvey Dove

ABSTRACT

In the absence of a long-range water plan, policy directives that take into
account a variety of non-hydrological, as well as hydrological, elements offer
the only viable alternative for short-term management of ground-water resources
in arid regions. In the past, recommendations made by hydrologists concerning
the potential for development of water in deserts have generally emphasized only
the hydrological aspects, whereas a number of economic, social, and political
constraints really must come into play in order to define the essential boundaries of the needs and sources problem. It is these latter constraints that
almost always reshape the purely hydrological recommendations into the real-world
policies that planners adopt for short-term water management. Recognition of
this fact can go far toward improving the ways in which hydrologists carry out
their investigations and present their findings to decision makers. This paper,
with no attempt to be exhaustive, develops a framework of considerations of this
kind, with current examples, for the formulation of water policy in arid regions.

INTRODUCTION

In the absence of a long-range water plan, which rarely is available for
arid regions, short-term policy directives offer the only viable alternative for
the management of water resources when new desert developments are being contemplated. The success of the policies, however, depends upon a prior understanding of the physical, economic, social, and political aspects of the problem.
In fact, the perception of the problem itself is the actual input to the policy-
making process. Unfortunately, there may be a vast difference in how the problem
is perceived. That is, the "incongruity between how those who make policy identify needs and define problems and how those who are affected by policy identify
needs and define problems can be significant" (Jones, 1970).

533

There are many examples of desert water-development schemes that technically seem perfect to hydrologists who recommend them, but which after periods of
implementation and operation turn out to be almost contrary to what the local
people or the government thought ought to be done. These schemes, which may have
been characterized as nomad resettlement projects, modern agricultural enterprises, or municipal water improvements, were often based only upon studies of
the availability of favorable water and soil conditions. Little or no attention
was given to the attitudes of the nomad, the effects that his resettlement might
have on the total society in which he lived, the attitude of the old farmer who
traditionally prefers date palms and alfalfa to cultivating new types of crops
or adapting new agricultural techniques, the attitude of younger men in an
established farming community who conceivably might have adjusted to new agricultural techniques but who have vanished to secure better paying jobs in the
larger cities, or simply the lack of skilled manpower to either operate or maintain sophisticated western equipment. One reason for the blind spot in problem
perception is that the state of the technical art of investigating water availability is considered to be more advanced than the softer arts that relate to
economics, politics, and other social studies.

The hydrologist has a variety of deterministic and stochastic tools at his
command which enable him to talk of water inputs, water outputs, and effects on
hydrologic regimens. However, he is almost always lacking in a background of
knowledge pertaining to the other so-called softer disciplines, although it is
these softer disciplines which in truth implement the solutions that are derived
through the technical approach. Also, scientific procedures for defining water
availability and movement are practically standardized all over the world,
whereas applied economics, politics, and sociology have a tendency to vary with
different cultures and to change with the passage of time.

To establish a point of departure, Fredrik Barth (1962) suggests that we
review our knowledge about the whole problem and ask two questions: "Is this
knowledge sufficient to give a basis for action? And if so, are there special
problems in translating such knowledge into action"? Since the hydrologist is

often required to recommend alternative uses for the water whose availability he
has determined, an appreciation for the other aspects of the problem may better
enable him to effectively communicate vital and relevant information to the
decision maker contemplating water-management policy, especially on a short-term
basis. The efficiency approach to water policy, which is suggested in this
paper, is general in application, but the risks involved with allocating a non-
renewable resource, which is typical of ground water in arid regions, can make
the water management problem more acute. The focus of this paper is toward a
better appreciation for consideration of that kind.

It is still a widespread belief that desert regions of the world are of
necessity short of water, because the layman interprets the lack of precipitation
and the absence of rivers as clear indicators of water scarcity. All too often,
the truth is almost the exact opposite. Ambroggi (1966), for example, points out
that the seven major basins of the Sahara desert in North Africa are estimated to
contain some 15,000,000 million cubic meters of ground water, although the total
annual recharge to this system is perhaps not much greater than 4,000 million

cubic meters a year. At the time Ambroggi made this estimate, the level of water consumption in the seven basins was only about 2,000 million cubic meters a year, which if sustained at that rate, would mean that the ground-water resources would last for more than 7,000 years even if not another drop of water were to fall in the Sahara. Another similar example pertains to the central region of the Saudi Arabian desert. There, Khatib (1974) conservatively estimated that the ground-water storage is in excess of 700,000 million cubic meters, compared with a rate of water withdrawal in 1968-70 of only 1,600 million cubic meters per year. Again, if rainfall were to cease and the rate of withdrawal to remain the same, the volume of water in storage would supply the water demand for a period of more than 350 years.

After a distinction between potential and available water supplies is made, and practical limitations are established on how much water can be withdrawn, it is not at all apparent that in deserts of this kind policies should be written to restrict short-term use of valuable water resources. Almost all policy making places a greater emphasis on near-term solutions rather than on long-term solutions. Thus, it seems inherently illogical to stress long-term conservation of the water for the benefit of as yet unborn generations instead of trying to improve the standard of living of people in the present age. Water policies used to guide developments in Saudi Arabia during the present five-year plan include two statements based upon perceived resource availability (Dove, 1975); they are:

1.   "Immediate urban, industrial, and agricultural demands at inland
     locations will be supplied by ground water."

2.   "Immediate urban and industrial demands at coastal locations will
     receive supplies from accelerated desalination developments."

Of course, not all arid regions are like the Sahara and Saudi Arabian Deserts, and while a more conservative water policy may be in order, the resource should be utilized in the most efficient manner. White (1962) suggests a physical indicator of efficiency based upon a comparison between the volume of water actually used and the amount which would have been used if the most advanced technology were to have been applied. Using this criterion, the method of flood irrigation for agricultural development would have a much lower physical efficiency than sprinkler or trickle irrigation techniques. In a similar manner, a desert city without a sanitary sewer system and no possibility of effluent reuse would approach zero efficiency.

ECONOMICS

The benefit-cost ratio is one of the few readily available tools for determining economic feasibility as an element in water policy decisions. This method of analysis is subject to manipulation, especially in the assessment of benefits, and is often too sophisticated for use by governmental agencies struggling to perform their administrative function, even by those who can afford high level consultants from the outside. Although economic feasibility is being encouraged as a basis for future project decisions, benefit-cost analysis has found little application to water projects in Saudi Arabia. An awareness does exist, however, of the consideration of economics in Saudi Arabian policy as seen in the following directives:

1. "New industries which are large users of water will be restricted to coastal areas unless other locations are economically and socially feasible."

2. "Mining operations will receive water from the most economically available source after due consideration of other needs."

Economic theory is predominantly a product of the western countries, and as such, is a reflection of the methods, values, and preferences common to those countries. In fact, the basic assumptions upon which the theories are developed may not have any real application in other countries. Schumacher (1973) points out that the western economic perception of fossil fuels as capital rather than as income is not concerned with conservation of this non-renewable resource but is concerned with maximizing instead of minimizing the current rate of use. This same approach may be easily transferred from fossil fuels to ground-water reservoirs and applied as an economic basis in water management policy for the allocation of water.

"If the efficiency criterion alone were to determine investments in water facilities, a given supply of water in a basin would be allocated among the various uses so as to obtain the optimum return from the combination of different uses and amounts" (White, 1962). It should be noted that full optimization of water use is a theoretical ideal to be used as a goal and is limited by the difficulties in placing a price or market value on everything and by the inherent motives within people, organizations, and governments. Modern economics as a social science incorporates a high degree of sophistication into economic analysis through the application of statistical mathematics and computer modeling. This sophistication leads to highly trained manpower requirements for governmental agencies and could in itself be the obstacle in its application. Nevertheless, water management policy developed through considerations of economic efficiency would incorporate a useful economic indicator along with the technical measure of physical efficiency.

POLITICAL CONSTRAINTS

Water policy in developing countries is probably much more affected by political considerations than by those of economic feasibility. In a basic sense, it is usually difficult to find any great fault in what might be called sweeping national or regional policies. Political figures tend to be able to write rather simple policies such as "the water resources of this country shall be developed in the best way for the overall benefit of the people in our society." The world's laws and statutes relating to water are commonly couched in language of this kind, sound very clear and understandable to even laymen, and contain in themselves really no contradictions or points of issue. It is when these sweeping generalities begin to be translated into specific plans that the true problems emerge.

When a government has written a law or established a policy in broad generalized language, the next step is to translate this policy into action by referring it down to a ministry or agency at some lower level. Here is where the difficulty actually begins, for the entrustment of the enactment of the policy may be delegated to either a particular agency (as for example the Ministry of Agriculture) or may even be broken up into several components and turned over to a series of ministries for implementation.

Each ministry, usually displaying the normal human frailties of wanting to build an empire and to compete with other ministries for funds and resources, then independently starts to construct what it conceives to be the purpose, plan, and implementation of the policy directive. If the ministry is staffed essentially with agricultural specialists, it is a foregone conclusion that the national policy will begin to have an agricultural flavor to it. If the implementation is entrusted to an environmental protection agency (which still of course is mainly in the province of the more developed countries), then it is almost certain that the policy will place more stress on the environmental effects of the water withdrawals than on the benefits to be achieved by it.

Because these are the routes most often followed in the past by governments at all levels, even by the most developed ones, it is evident that no overall coherent and logical policy can emerge. Instead, the agency with the greatest clout will in all likelihood be the one whose programs will prevail. Furthermore, because societies often place a higher priority on the purely technical or scientific disciplines rather than on the many disciplines belonging to social studies, there is a great tendency for the technical view to prevail, commonly with complete disregard of the valuable inputs that could have been made from the softer sciences.

There is of course an understandable basis for behavior of this kind. One must admire the engineer or scientist who can produce reams of factual data, statistics, graphs, and illustrations to support his viewpoint on water availability or water developments. Contrariwise, there is an understandable reluctance to accept the sometimes vague or conflicting views of the sociologist or the political scientist who is dealing with a much more complex problem and who does not have readily available to him a set of modern analytical tools which can quantify human reactions of these kinds.

The definition of or measure of political efficiency is not as easily attainable as the definition of physical or economic efficiency. Possibly the indicator could be some kind of a quantitative assessment of governmental stability, governmental strength, administrative effectiveness, or simply what works. Ingram (1972) has identified five decision rules as essential ingredients for political viability and the governing of the relationship between policy makers in the United States. The five decision rules are based upon measures of local support, agreement, mutual accommodation, mutual noninterference, and fairness and equity. Undoubtedly, some or all of the same rules can be applied to other cultures, and there may be some new ones besides. However, there would be the early necessity to determine how the specific decision process works in order to postulate some measure of political efficiency. One important observation noted by Ingram is that these decision rules change with time, and to a large degree can reflect the changing perception of society in terms of national goals. For example, "ecology" and "environment" have rivaled "growth" and "development" as political considerations in water development and have introduced a form of discomfort and uncertainty into the once-traditional policy decisions.

SOCIAL CONSIDERATIONS

In a number of cases in desert regions of the Sahara, well intentioned technical people (sometimes from bilateral or international aid programs) have designed high-capacity water wells along nomad trails in the belief that this would improve the raising of livestock and the standard of living of the nomads. All too often, the end result was that increasingly large numbers of nomads and herds of cattle converged on these sources of water, only to turn the regions around the wells into wastelands as the cattle consumed every bit of available forage. Finally, the nomads could no longer move toward these water points simply because the cattle could not find enough food within miles of the wells to be able to survive.

In still other cases, modern water wells installed along cattle routes conflicted directly with the traditional nomadic patterns of camping at convenient water holes. To their chagrin, the hydrologists then learned that instead of taking water from the modern water wells, the nomads preferred to continue using the old water holes or to enjoy a pleasant evening with each other digging new water holes in the traditional fashion. Sometimes, the nomads also took advantage of the situation by breaking off of the well pumps and casings to be fashioned into knives, other tools, implements, and ornaments.

Still another example is given by the intensive desire of some foreign experts, mainly from developed countries, to install the most modern types of water plants in desert communities. Highly sophisticated western equipment such as valves, pumps, and automatic control systems would be imported and installed in these plants, with hardly a hope that anyone in the local community could either operate them or repair them in the event of breakdown. It often would have proved infinitely more sensible and far less expensive to have employed human labor for such tasks as opening and closing valves, adding chemicals to the water, etc. Moreover, it would have been far cheaper, would not have called for the use of foreign exchange, and actually would have improved the standard of living by providing new jobs to people who otherwise would have been unemployed.

Gilbert White notes that "the traditional view of primarily agricultural use of limited water supplies seems destined to strenuous challenge." There are obvious economic overtones to this statement, but there are also social considerations as well. Short-term water policy in Saudi Arabia includes some awareness of the social implications, in the directive:

"Increased water use by agriculture must first demonstrate that the long-term usage is in the public interest."

There has been a trend in the United States to involve the general public in policy decisions. The efforts have met with mixed success, but often the voice that prevailed was not that of the general public but that of the activist. A sampling approach is probably a more reliable means of determining public values and preferences; however, the initial design of the questionnaire is a critical element in the retrieval of valid information.

Unlike in the political context, where a characterization of "what works" could conceivably be the measure of political efficiency, the concept of what constitutes social efficiency is even more elusive. Certainly man, made in the image of God with spirit, soul, and body, cannot be reduced to a set of numbers and fed into a computer. Possibly some characterization of social well-being could provide the answer, but "the major theoretical problem is one of evolving an analytical framework and model of society" (Brown, 1976). In the past, social scientists have too often avoided policy and problem-oriented research and have been satisfied to objectively and impassively note some cultural phenomena. This attitude has left the door open for the physical scientist to try his hand at mathematical modeling and social quantification.

Attempts have been made to develop and test a meaningful set of social indicators specifically for water-resource development (Roefs, 1972). As Brown has noted, "the key is to find sensitive linkages between social well-being parameters and water-resource development." Typical social elements pertaining to a social profile of the culture could include evaluations of education, health, population, ethnic composition, employment, housing, and religion. With the wide variations in cultures across the globe, it is not obvious at this point how the combination or weighting of social elements like those listed above could be used to arrive at a valid definition of social efficiency. One can only agree with Gilbert White: "the social dimension of arid zone research is one offering a great challenge."

CONCLUSIONS

No simplistic or incomplete approach can be taken to the formulation of sound water policy. Instead, the economic, political, and social aspects of the problem, as well as the physical aspects, must be considered to the best possible degree. Water policy based upon an efficiency approach is possible in the physical and economic realms, but much additional definition and development will be needed to apply the same concept in the political and social realms. Probably the most important variable contributing to the success of any water policy is the characterization of social well-being; yet this is precisely the variable that appears to be the least understood.

The tendency to quickly "do something" in response to perceived problems, without a serious consideration of the whole problem, must be thoughtfully resisted by all responsible parties. However, the need for making short-term policy decisions in desert environments is self-evident, and since the largest unknown is usually not water availability but social well-being, the latter consideration should receive the highest priority in the formulation of policy.

REFERENCES

Ambroggi, Robert P. (1966), Water Under the Sahara, Scientific American, 24, 5, 21-29.

Barth, Fredrik (1962), Nomadism in the Mountain and Plateau Areas of South West Asia, The Problems of the Arid Zone, Proceedings of the Paris Symposium, UNESCO, France, 341-355.

Brown, Leonard R. (1976), Social Well-Being and Water Resources Planning, Water Resources Bulletin, 12, 6, 1181-1190.

Dove, F. Harvey (1975), Developing Water Policy: A National Comparison, Eleventh American Water Resources Conference (to be published in Water Resources Bulletin), New Orleans, Louisana.

Ingram, Helen (1972), The Changing Decision Rules in the Politics of Water Development, Water Resources Bulletin, 8, 6, 1177-1188.

Jones, Charles O. (1970), An Introduction to the Study of Public Policy, Duxbury Press, Belmont, California.

Khatib, A. B. (1974), Seven Green Spikes, Kingdom of Saudi Arabia, Ministry of Agriculture and Water.

Roefs, T. G., et. al. (1972), Test of a Planning Inquiry System: Water and Waste Management in Pima County and Suffolk County, Project Number C-3377, University of California, Riverside, California.

Schumacher, E. F. (1973), Small is Beautiful, Economics as if People Mattered, Harper and Row, New York.

White, Gilbert F. (1962), Alternative Uses of Limited Water Supplies, The Problems of the Arid Zone, Proceedings of the Paris Symposium, UNESCO, France, 411-421.

LONG-TERM PLANNING OF THE USE OF WATER RESOURCES

P. Lagache

Almost a century ago, the population of the earth was one thousand million. Already, there are four times this number of people and by the year 2000 there could well be 6 to 7 thousand million. The desire for improved living standards and, in particular, the awareness that vital resources are limited, presents societies with more and more increasingly urgent problems, such as:

- the development of areas either not yet used productively or in need of improvement,

- the improvement of living conditions: economic and sanitary, cultural in the most deprived areas,

- organization of a harmonious natural environment.

Those responsible for the economic and social development of arid areas are particularly concerned by these serious development decisions:

- decisions which are particularly difficult because they involve the use of resources which are not renewable (oil products and rich mineralores, arable land, etc.) in areas where water resources are themselves limited, and the use of considerable technical facilities and financial resources;

- decisions which are particularly difficult because they concern, in most cases, several social agents, and may have effects on many sectors of the national and/or regional economy.

Also, the speed, diversity and scale of changes which affect the social, economic and technical environments of the various countries of the world, whatever their present degree of industrialisation, have an influence on the future of arid areas.

As a framework for coherent decisions, long-term planning of water resources is therefore a particularly useful, even necessary, development aid in order to guide and stimulate the economic and social activities of these areas which are still deprived at present.

We intend, therefore, to consider some of the problems confronting the preliminary studies for a plan for the use of water resources, viz.:

. definition of the field of study,

. the techniques to be used to draw up a development strategy
for the use of water resources.

DEFINITION OF THE FIELD OF STUDY

For planners, the region is an extensive territorial subdivision which is part of
either a national or international geographical area.

For developers, the area of action is defined as the area where biophysical, economic and social phenomena may be perceived, understood, forecast, used and/or corrected.

The geographical area under consideration therefore becomes a framework for analysis. However, the requirements of the study oblige the developer to adapt this
framework to existing territorial divisions. The management of the plan is usually the responsibility of a regional Authority. Also, the available data, particularly statistics, is usually collected with reference to administrative divisions.

The living conditions particular to arid areas make it necessary to consider the
study area as an operational unit which is both sufficiently large to cover all the
exchanges to be introduced or the traditional exchanges between areas inhabited by
nomads and areas occupied by a sedentary population and sufficiently restricted so
that the situation which is to be changed, within the limits imposed by the available water resources, may be controlled.

These resources are very unequal both in quality and in quantity. In the majority
of cases, they are groundwater resources, and occasionally fossil water, in which
case they are not renewable by natural means. The water is currently used or may
potentially be used for economic and social activities whose development depends
on local or non-local factors, i.e. factors which can or cannot be considered on
a regional level.

A preliminary survey of the field of study is desirable in most cases where a regional development plan has to be drawn up. The Algerian authorities, to whom we
give credit for their efforts to improve development planning methods, allowed us
to carry out this preliminary stage, involving the definition of geographical boundaries and of the field of study to be covered in order to determine all the factors
relevant to the development of the semi-arid regions in the South of Guelma, Tia-
ret and Saida.

It is from this experience, in particular, that we draw the following remarks:

Concerning the Primary Sector

In the first analysis, planning of the primary sector concerns the regional sphere.
All the programmes (agriculture, stock-breeding, mines, etc.) are based on physical resources (very localised geographically).

However, the exploitation of these resources involves problems which may be solely
regional (distance, difficulties of transport, etc.) or both regional and non-
regional (marketing, pre-production and post-production operations, labour, etc.).
If the target of this development is restricted to maximisation of production, it
is of little importance whether the decision is national (most frequent case) or
regional: the problem is simply one of location. On the other hand, if impor-

tance is attached to development of the system of production or to the impact of investment on the region, it is indispensable to consider the decision at the regional level.

Energy production often exploits physical resources (deposits of fossil fuels,use of solar energy) but the requirements to be satisfied, especially in arid areas, are rarely limited to the region (except specific cases: creation of a mining industry, for example). Also, the investment cost is always too great to be met at a regional level.

Whatever the regional effects of energy production, it generally has to be planned coherently at a national level.

Concerning the Secondary Sector

Industries may exist which have a regional character, such as crafts, consumer industries, small-scale building materials industries, service industries (garages, hotels, etc.).

However, the remaining industrial activity has a national character and can only be planned centrally:

- its market is rarely limited to that of the area since economies
  of scale lead to plants being over-sized in relation to the area;

- in order to provide supplies for the plants, it may be necessary
  to have agreements between regions, without which the plants
  would be unable to reach the critical size;

- the inter-industrial links that it creates (at both pre-production
  and post-production stages) and its service requirements are not
  compatible with a very limited economic context;

- a large plant raises problems with regard to specialised labour
  supply and financing which cannot easily be solved within the
  context of the arid areas themselves.

Concerning Other Factors of Socio-economic Development

For the same reasons as in the secondary sector, economic services can only be planned at the local level in exceptional cases:

- the activities are rarely limited to the regional sub-unit;

- economies of scale are considerable (chain-stores, consolidated
  transport firms, etc.);

- with respect to transport, certain infrastructures are not feasible at the regional level (railways, main roads, ...). These
  have to be planned on a national scale. However, there may be
  certain regional requirements which have to be taken into account
  (necessity for communications between regions).

Also, the economic services sector, more than any other, leads to population changes; it is therefore one of those that the regional planner considers necessary to use.

As social services, at the most basic level, are linked directly to the require-

ments of the inhabitants of a region, they must necessarily be planned at the regional level.

This is particularly true in arid areas in order to create acceptable living conditions (health and primary education; in part elementary technical education,
although this is less clear as recruitment and the professional outlets for trained personnel are not usually confined to the region). Secondary and secondary
technical education often has a much less marked regional character. Regional analysis certainly provides important information, but maximum advantage can only be
obtained by placing it in the national context.

The planning of social services always involves a significant, even major proportion of regional factors and data.

However, the planning of social services often involves another aspect: the proportion of national resources which is to be allocated to them (proportion of gross
product or budget resources, depending on the case).

The importance of this financial decision means that it must normally be made at
the national level.

This leads as to consider financial resources which, together with physical and
human resources, are one of the major aspects of planning.

Arid areas   usually have very low money incomes and savings levels, so that most
of the savings available for modern investment comes from:

> . large firms in the modern sector of the economy, which are
> almost always of a national nature and which are consequently
> likely to use their savings according to non-regional criteria;

or

> . taxes, the largest proportion of which are national taxes, since lo-
> cal taxes provide very little revenue.

A considerable proportion of the resources used to finance regional development is
usually obtained from outside the region, whether they be private or public funds,
subsidies, or loans.

It is therefore obvious that the availability of these financial resources can only be assessed at the national level.

With respect to public funds, the possibility of obtaining subsidies and the capacity for running up foreign debts are factors which cannot be analysed at the regional level.

Foreign currency resources are purely national factors materially, because they are
generally collected by national organisations, but also because of their economic
justification.

It is the national unit which exports surplus products after national requirements
have been met and taking into account exchanges between regions.

Special mention should be made of foreign trade links particular to the region: it
may be that a region produces considerable surplus product as a result of production of a speciality or a commodity which it is in a particularly good position to
supply (e.g. oil).

In under-developed countries whose regional economies are too similar for there to be much trade between regions, it is desirable to encourage such trade and, consequently, a certain amount of specialisation in production.

Consequently, in order to analyse a region with respect to its foreign trade relations and draw conclusions concerning regional development, it is necessary to be aware of the national context, its requirements and the directions which it imposes on the nation as a whole and on the various regions.

In order to summarize this approach to the economic phenomena which can or cannot be considered on a regional basis, it may be said that:

. the national sphere, in a strict sense, covers activities for
  which an attempt is made to optimise the use of resources;
  even certain activities in the primary sector are of this type
  (oasis crops for export -dates-, mines, oil deposits, ...) and
  their regional character is limited;

. the regional sphere coincides with activities for which optimi-
  sation of resources is only a secondary aim (modification of
  the agricultural and stock-breeding production systems, modification of relations between social groups and of life styles
  and satisfaction of basic needs).

However, it is necessary to study the definition of regional and national spheres in the light of another criterion: that of decisional levels. Planning necessarily involves decision-making power: should it be taken that spheres which can be considered on a regional basis correspond exactly to regional decision-making power?

Very difficult problems are raised here which can often only be solved within the context of a given State.

Drawing up a long-term plan for the use of water resources in arid areas therefore requires particularly thorough research into the field of study, not only to understand the data necessary for the assessment of these resources and their development under the effects of various hypotheses of requirements, but also to estimate the development of these water requirements which often depend on external as well as local factors.

This research is particularly difficult where arid areas are concerned as the development of activities which require water depends more on voluntary action than on natural tendencies, within a network of numerous biophysical and socio-economic limitations.

This brings us to a new category of problems: those which are raised in drawing up a development strategy for the use of water resources.

TECHNIQUES FOR DRAWING UP A DEVELOPMENT STRATEGY

In defining the objectives for meeting demand, taking into account the volume and quality of water resources, there is firstly a problem of coherency between the physical and economic situations and secondly a problem of forecasting the demand for water which can be met.

The methodological approach which we have adopted for studies concerning plans for the use of water resources in arid areas (particularly for souther Tunisia and the Farah Rud Valley in Afghanistan) is based on Systems Analysis and the Scenario

Method.

Insofar as the problem of water is central, the systems to be analysed may be identified by means of a matrix combining Resources and Requirements. The different
systems which make up the situation being studied can be determined on the basis
of an exhaustive inventory of all the data concerning the existing and potential
use of water resources. An analysis may then be carried out of the relationships
between the variables explaining this use of resources; and from these links, the
natural and socio-economic systems implied by the development of the resource can
be identified.

These systems are therefore defined in terms of their structures (combination of
inter-related variables such as discharge of a spring, irrigable area, number of
agricultural workers, available capital, demand for agricultural products, climatic phenomena, etc.), of one or more aims assigned to them (increase of income per
job, etc.) and of a certain capacity to deal with their internal contradictions in
a durable manner (e.g. yield, quality, product prices).

If the observations concerning water resources are made on an adequate statistical
series, the forecasts do not raise any insurmountable problems. Mathematical simulation models have been developed which can be used to study changes in surface
water conditions: floods, sediment transport, low-water periods, pollution and
salinity characteristics, etc.

Models simulating the conditions of aquifers also exist and can be used to forecast
the reactions of the aquifers to pumping (ERESS model drawn up by UNESCO for Tunisia). It is therefore possible to gain a sufficiently accurate knowledge of available water resources at various dates and according to various hypotheses of use.
Forecasting of biological and economic conditions is obviously much more difficult.
The Scenario Method attempts to hypothesise on the various possible future situations resulting from the interaction of different factors (or socio-economic agents
concerned by the development of the area) and the biophysical and socio-economic
conditions influencing their actions (Scenario methods used by SOGREAH for Southern Tunisia and by DATAR for scenarios d'Aménagement du Territoire).

This method is supported by the analysis of the Water Requirements and Resources
system and identifies the variables which depend on the decision-making power of
the authorities responsible for development (decisional variables). It also aims
to reveal the invariable factors and relatively inertial tendencies of the system.

With knowledge of the motive factors (decisional variables) of the strategies of
the agents and the seeds of change revealed by the systems analysis stage, it is
possible to build various possible scenarios for future development. This is carried out in successive stages followed by an examination of the development of the
elements composing the system in relation to each other, in order to detect possible incoherencies in forecasting or risks of conflict.

This combination of the systems analysis and scenario methods has the advantage of
providing an overall view of possible future situations and therefore of assisting
in the choice of the objective or objectives which, amid limitations of various
kinds, will be able to provide guidelines for the efficient and harmonious development of the use of water resources.

We have dealt with only a few of the problems which are encountered in studies for
drawing up a plan for the use of water resources in arid areas, and particularly
at the level of definition of the field of study, data acquisition and the strategic analysis of development. In this respect, we have mentioned certain methods
of analysis which indicate the continuous effort that has been made in recent years

in the improvement of decision-aid techniques and their use in the preparation of land development plans.

In conclusion, 7 guidelines may be given for improving of the studies of the development of arid areas, with the following aims:

(1) - To examine and weigh systematically all the factors determining the demand for water and to situate activities which require water in a wide and long perspective with respect to competition, regional environment, national strategy, etc.

(2) - To include in the analysis the qualitative variables, which are often more important than quantifiable variables.

(3) - To verify the coherency of measures in the organisation and financing of development.

(4) - To reveal the insufficiencies in available information, making a distinction, in particular, between indispensable information and information which is only secondary with respect to the objectives chosen and to identify priority development actions which would not be questioned in the light of additional information.

(5) - To assess the sensitivity of the decisions to changes in one or other of the selection criteria or in its relative importance, so that the hydraulic development programme will be better adapted to the overall economic and social development strategy.

(6) - To avoid, as much as possible, dangerous and irreversible situations for the future of the natural and/or socio-economic balances which exist at present or are to be created in arid areas.

(7) - To facilitate the exchange of the various points of view involved in the development.

Considerations in the Development and Use of Multisector Models for

Long-Term Water Planning in Desert Regions

Everard M. Lofting

ABSTRACT

    Arid or desert areas usually are characterized by abundant sunshine and
warmth, conditions which are essential to the growth and reproduction processes
of plants. If the soil and available water resources are also of high quality,
then the agricultural base of a desert region can usually be superior to that
of the so-called dry-farming areas. Conditions for rapid plant growth and
multiple croppings virtually assure the success of agricultural ventures. As
the world's population increases, pressures for land and food demands intensify.
A growing awareness of the climatic variability of the dry-farming areas
suggests desert reclamation as a possible solution to these problems.
    In earlier periods the reclamation of arid lands was typically carried out
as a single purpose undertaking. Increasingly arid regions are being developed
not only for agricultural purposes but in order to exploit other resources
such as petroleum or other energy sources, or to serve as sites for population
centers which accomodate secondary and tertiary activities related to the
agricultural and extractive industries. When this is so, planners require a more
comprehensive methodology for estimating the needed levels of resource inputs
for the different economic sectors and the value of product outputs which may
be expected. These more comprehensive analytical methods fall under the heading
of regional input-output or multisector modelling techniques.
    It is possible to develop appropriate regional input-output tables by using
secondary data sources. Typical data sets are available from the United States
Bureau of the Census and from the Bureau of Economic Analysis. These can be
augmented with certain local economic data available at the state and county
level. Satisfactory water data for use with the economic model are harder to
come by and must be developed with local conditions of use in mind. Projections
of total water use through some given planning period must be made,in many
instances, judgementally taking into account possible drought conditions and
other contingencies.
    The Colorado Desert (Imperial County,California) input-output model, presented
in outline as an example, suggests that where absolute constraints on water
supply exist a linear programming formulation may provide policy guidance
for allocating resources among sectors or may suggest required improvements
in water use efficiencies to meet future production levels.

Considerations in the Development and Use of Multisector Models
for Long-Term Water Planning in Desert Regions *

Introduction

    The desert and semi-desert regions of the world, under favorable climatic
circumstances, provided the sites in which the major civilizations developed.
The critical factors were water and agriculture.

*Everard M. Lofting, Berkeley, California, May 1977

Approximately 20% of the land area of the Earth has been defined as arid (Meigs, 1953). In contrast, only 0.1% or 1.45 billion hectares of the Earth's land area is cultivated (Cuny, 1961). Arid or desert areas usually are characterized by abundant sunshine and warmth, conditions which are essential to the growth and reproduction processes of plants. If the soil and available water resources are also of high quality, then the agricultural base can usually be superior to that of the so-called dry-farming areas. Conditions for rapid growth and multiple croppings virtually ensure the success of agricultural ventures. As the world's population increases, pressures for land and food demands intensify. A growing awareness of the climatic variability of dry-farming areas suggests desert reclamation as a possible solution to these problems.

In earlier periods the reclamation of arid lands was typically carried out as a single purpose undertaking. Increasingly, arid regions are being developed not only for agricultural purposes but in order to exploit other resources such as petroleum or other energy sources, or to serve as sites for population centers which accommodate secondary and tertiary activities related to the agricultural and extractive industries. Where this is so, planners require a more comprehensive methodology for estimating the needed level of resource inputs for the different economic sectors and the value of product outputs which may be expected. These more comprehensive analytical techniques have been understood and recommended for several decades. The techniques fall under the categories of regional input-output or regional interindustry analysis. More recently, these types of analyses have come to be categorized as multisector modeling techniques. The emphasis of this paper relates to their direct application and method.

At the national level, income and product accounts are prepared annually in considerable detail. Below the national level, population, personal income by major source, employment, and payrolls are the only standard socioeconomic variables that can normally be assembled for analytical purposes.

In most instances, the economic data that are readily available at the regional level are insufficient to permit a thorough analysis to be carried out. It is usually necessary to resort to various estimating techniques to develop measures that permit a more satisfactory analysis to be undertaken.

During the 1930's Wassily Leontief (1936) pioneered in the development of interindustry (input-output) economics. As a consequence, detailed interindustry tables for the United States economy are now prepared routinely for each census year by the Department of Commerce (Survey of Current Business, 1966),(Survey of Current Business, 1969), (Survey of Current Business, 1974).

At a national level, census data are used to develop interindustry transactions and to reconcile these with the details of the annual income and product accounts. The national tables provide insights into the technological structure of production and have brought a greater measure of empirical analysis and inductive reasoning to many economic problems.

Below the national level, for states and regions, the counterpart of the income and product accounts do not exist and census data on industry sales and purchases by geographic region become more limited. Because of this, regional interindustry economic research has developed more slowly and is characterized by several rather distinct approaches. The essential difference in these approaches relates to the emphasis given to survey techniques and the use of primary data as opposed to the reliance on secondary data sources and methods for deriving regional estimates of interindustry flows by indirect means. The professional controversy over "survey" versus "non-survey" techniques continues but there is increasing evidence that secondary data sources will provide satisfactory input-output tables at the regional level for most analytical purposes (Boster & Martin, 1972).

Regardless of the specific approach used, the increasing need for analyses which require an understanding of economic structures of regions or smaller localities in terms of planning alternatives and benefit-cost considerations makes input-output an almost indispensable tool.

S.V. Ciriacy-Watrup (1954) was one of the first to stress that the input-output system might be useful in promoting the type of regional accounting necessary for analyzing the effect of public investment in water resource projects. Later Folz (1957) noted that although no studies were on record of the interindustry method being applied to a river basin, it seemed probable that such an approach could be the most fruitful of any yet devised for determining the general pattern of growth that might be anticipated for a given region. Later, at an international conference on input-output methods, Nemchinov (1963) placed special emphasis on the use of interindustry models for regional resources development and allocation with particular reference to water, power, timber and minerals. Martin and Carter (1962) pointed to similar possibilities. In a UNESCO publication, S.W. Wilson (1964) emphasized further the value of interregional input-output analysis in water planning:

> Through an input-output matrix, the dynamic processes of change
> can thus be isolated, their effect on income and employment
> spotlighted, and their pressure on water supply determined.

The fundamental aspects of the role that input-output analysis can play in water resource economics relates to its accounting scheme. This scheme immediately provides a basic regional framework within which resource needs can be assessed and compared with their planned availability.

Theoretical Background

(i) Input-output Analysis

The basic theory and underlying assumptions of interindustry analysis are covered comprehensively in several standard texts (Chenery & Clark, 1958), (Dorfman, Samuelson, and Solow, 1958), (Hadley, 1962). To provide some minimal background for present purposes the essentials of the Leontief systems are presented.

Let the matrix of intersectoral transactions be represented as:

$$
\begin{bmatrix}
x_{11} & \cdots & x_{1i} & \cdots & x_{1n} \\
\vdots & & & & \vdots \\
x_{i1} & & x_{ij} & & x_{in} \\
\vdots & & & & \\
x_{n1} & & x_{nj} & & x_{nn}
\end{bmatrix}
$$

The direct input coefficients per dollar of output are given by:

$$\frac{x_{ij}}{x_j} = a_{ij} \qquad (i, j = 1, \ldots n)$$

The matrix of "technical coefficients" of production A is of the form:

$$
\begin{bmatrix}
a_{11}, a_{12}, & \cdots & \cdots & \cdots & a_{1n} \\
\cdots & \cdots & \cdots & \cdots & \cdots \\
a_{n1}, a_{n2}, & \cdots & \cdots & \cdots & a_{nn}
\end{bmatrix}
$$

The final demands and total ouptut sectors are given as:

$$
\begin{bmatrix}
x_1 \\
\vdots \\
x_i \\
\vdots \\
x_n
\end{bmatrix} = X
\qquad
\begin{bmatrix}
Y_1 \\
\vdots \\
Y_i \\
\vdots \\
Y_n
\end{bmatrix} = Y
$$

Gross output less intermediate use equals net output, or
final use

$$X - AX = Y$$

Multiplying by the identity matrix I results in

$$IX - AX = (I-A) X = Y$$

The inverse of any matrix A is defined as that matrix which when
multiplied by the original matrix A will yield the identity matrix I.
Thus

$$AA^{-1} = I$$

By obtaining the inverse of the matrix I-A, the general solution of the
Leontief system is provided.

$$(I-A)^{-1} Y = X$$

The $(I-A)^{-1}$ is known as the Leontief inverse and gives the direct plus
indirect (total) requirements of each industry per dollar of deliveries to final
use.

This model can be used for an analysis of present structure, for consistent
forecasting, and for determining the sector by sector impact of proposed increases or decreases in final demand or gross output.

For projecting water or other resource use, or employment, to some future
time frame, the concept of a factor content matrix is introduced. The Leontief
inverse is premultiplied by a diagonal matrix in which each element of the principal diagonal consists of water use (w_i), or employment (e_i), in physical units
per unit of product output in dollars (i.e. coefficients of use). Thus:

$$
\begin{bmatrix}
w_1 & & & 0 \\
& w_2 & & \\
& & w_3 & \\
& & & \ddots \\
0 & & & w_n
\end{bmatrix}
\begin{bmatrix}
b_{11} & b_{12} & \cdots & b_{1n} \\
\cdot & & & \cdot \\
\cdot & & & \cdot \\
\cdot & & & \cdot \\
b_{n1} & & & b_{nn}
\end{bmatrix}
=
\begin{bmatrix}
w_{11} & \cdots & & w_{1n} \\
\cdot & \cdot & & \\
\cdot & & \cdot & \\
\cdot & & & \\
w_{n1} & \cdots & & w_{nn}
\end{bmatrix}
$$

The factor content matrix W can then be postmultiplied by a vector of
final demands Y' which has been estimated for some future time period -- i.e.,
1980 or 1990, or indeed 2020. The multiplication will yield a vector giving
total regional water use or employment by sector for the future time period.
A grand total can be given by summing the components of this vector.

It can be assumed that the industrial technology of the region represented
by the b_{ij} will change rather slowly in an evolutionary manner (Carter, 1970).
In practice, deriving the coefficients (w_i) for each sector for use with the b_{ij}'s
requires a careful analysis of available data and the use of judgment if there is
reason to believe that these coefficients may change due to various circumstances
over the forecasting period.

Typically, economists believe that resource inputs may vary radically with
price changes (price elasticity of demand) so that the w_i's may not be stable.
If the technology or physical conditions of use are relatively fixed, price may
not substantially influence basic resource use, however.

State of the Art of Forecasting Water Use

   (i) General

Estimates of water use are usually made for the major industry divisions and
the household (residential) sector of the economy. Traditionally, however,
water use categories and economic sector categories have not been aligned.
Specifically, these categories can be listed as follows:

Table 1

Economic Sector	Water Supply-Demand Category
1. Agriculture	1. Irrigation
2. Mining	2. Mineral Industry Water Use
3. Manufacturing	3. Industrial
4. Utilities	4. Thermoelectric Power
5. Trade and Services	5. Commercial
6. Households	6. Municipal (Part)

To bring the Economic Sector categories into agreement with Gross National Product and its components for analysis and projection purposes the following additions to the economic sectors are necessary. This grouping provides an exhaustive, highly aggregated, classification scheme for all productive sectors of the national economy or its geographical regions.

Table 2

Economic Sector	Water Supply-Demand Category
1. Agriculture	1. Irrigation
2. Forestry and Fisheries	2. ------
3. Mining	3. Mineral Industry Water Use
4. Construction	4. ------
5. Manufacturing	5. Industrial
6. Utilities	6. Thermoelectric Power
7. Transportation and Communication	7. ------
8. Trade and Services	8. Commercial
9. Households	9. Municipal (Part)

Gross National Product (GNP) is a scalar quantity that is typically projected by federal agencies in constant dollar terms to various target years (Bowman, 1976), (Kutscher, 1976). The projected scalar values can then be decomposed into the sectoral components that reflect relative growth or decline within the overall control total. The resulting estimates can be termed "consistent". That is, the interdependent nature of the sectors of the economy is usually explicitly (or implicitly) recognized. If the individual sectors were projected on the basis of historical trends or other criteria and the values summed for the target years, results may be inconsistent with more reasonable estimates of GNP based on the material requirements of the projected population and resource availability. Moreover, since water is regional in its occurence, the national control totals can be disaggregated spatially to yield a further consistency for the various sectoral components.

Before dealing with the specific sector demand analysis and projections, some further points should be made regarding the alignment of water-supply-demand categories with the economic sector categories.

Consistent estimates in money terms can be made for GNP and its sector components as indicated above. It is desirable that these be matched (aligned) as closely as possible with water demand estimates in physical terms, i.e., gallons/day or acre-feet per year. This is not an easy problem since the engineers and hydrologists generally charged with the responsibility for gathering or estimating the water data do not choose their classifications to fit precisely with the economic sector specifications. Water uses may be measured or gauged by the amount supplied in a given time period. Further, water may be impounded and supplied to certain joint uses and it may not be known except in the most aggregate way which end-use actually withdrew the water. For example, water may be impounded and distributed by a public water supply system. The water may be supplied to households, industry, municipal buildings, and commercial enterprises. In fact, multiple unit dwellings are frequently considered a commercial use of water and are so classified by many water supply agencies. Thus, commercial and multiple unit dwelling household uses may be inseparable in the supplying agency's records and other estimating techniques must be found to separate these for analytical purposes. The same type of end-use identification occurs for commercial

and industrial uses from time to time. The overall alignment scheme suggested
above is not entirely precise but is probably the most satisfactory for establishing aggregate control totals, given the nature of the basic water use data as
presently compiled in most countries.

(ii) Agricultural Water Use

In the United States very large amounts of water are developed and used in
irrigated agriculture in arid and semiarid areas, particularly in the Southwest.
There is a general belief that increasing the supply price of this water would
result in a much more efficient pattern of use. It is argued that developing
water use coefficients for long-term projection purposes on the basis of present
use rates could grossly overstate estimates of agricultural water use in future
time frames. While there may be some validity to this point of view, the critical consideration is whether autonomous price increases will serve to bring more
efficient use or whether allocations based on the minimum amounts required by the
technological and physical conditions of use would be a more effective procedure
for projecting ultimate levels of use. Since the agricultural sector may account
for the bulk of the water used in any long term management plan for an arid region, some detailed consideration of techniques for calculating the use coefficients for this sector will be appropriate.

(a) The Engineering and Hyrological Considerations of Agricultural Lands

The engineering and hydrological approaches to estimating water use in arid
regions emphasize the fundamental importance of the concept of aridity (McGinnis
et. al., 1968). Aridity is essentially a measure for expressing the dryness of a
region. It is a comparison between water supply and the physical rate of evapotranspiration due to the prevailing climatological factors. Physical scientists apply the term "water need" to this latter concept.

The following definitions which relate water and physical conditions of
aridity are given by Veihmeyer (1964). They are presented here to aid in the
understanding of the physical and technological conditions of agricultural water
use in arid regions.

Evapotranspiration: the process by which water is changed from a liquid or solid
state into the gaseous state through the transport of energy.
Transpiration: the evaporation of water absorbed by the crop and transpired
and used directly in the building of plant tissue, in a specified time. This
does not include soil evaporation.
Evapotranspiration: the process by which water is evaporated from wet surfaces
and transpired by plants.
Consumptive use: the quantity of water per annum used by either cropped or
natural vegetation in transpiration or in the building of plant tissues, together with water evaporated from the adjacent soil, snow, or from intercepted
precipitation. It is sometimes termed "evapotranspiration".
Irrigation requirements: the quantity of water exclusive of precipitation that
is required for crop production. It includes surface evaporation and other
economically unavoidable wastes.
Water requirements: the quantity of water, regardless of its source, required
by a crop in a given period of time for its normal growth under field conditions.
It includes surface evaporation and other economically unavoidable wastes.
Potential evapotranspiration: the evapotranspiration that would occur if there
were an adequate soil-moisture supply at all times.
Wilting point: the wilting point represents the soil moisture at the time when
plants cannot extract water from the soil. When the wilt point is reached,
the plant dies.

Given the physical conditions of an arid region, irrigators will presumably
wish to use that amount of water which will minimize their cost and at the
same time will ensure that the soil moisture is kept substantially above the
wilt point at all times to prevent the possibility of crop loss. Several methods
for achieving this goal have been developed. The Blaney-Criddle method (1958)
is one that is frequently used. Consumptive use for a given crop is estimated
by the Blaney-Criddle relation $U = k_s B$

where   U is the consumptive use of water in a given season;

k_s is an empirical seasonal consumptive-use coefficient for a given crop;

B is the sum of the monthly consumptive use factors for the given season, or

$B = (\frac{tp}{100})$, where

t is the average monthly temperature in $^\circ$F. and

p is the monthly daytime hours as a percent of the yearly total.

k,t, and p are available from detailed tables which have been compiled for a variety of crops and localities for all months of the year.

Average monthly consumptive use u in inches can then be calculated. For example, a typical value for a southern California citrus crop would be computed as 21.41 inches of irrigation water under the climatic conditions that have generally prevailed for this region (Ogrosky & Mockus, 1964). It can be expected that this amount will be used consumptively during the growing season. If, of course, climatic changes occur so that evaporation losses are affected, then irrigation needs will alter accordingly.

Consumption of a total required supply of water can be made by considering typical irrigation field efficiencies; leaching water requirements; other field losses; canal seepage losses; and reservoir evaporation and seepage losses. Average irrigation field efficiencies are given below:

Method of irrigation	Range of efficiency, %
Graded borders.........................	60 to 75
Basins and level borders	60 to 80
Contour ditch	50 to 55
Furrows	55 to 70
Corrugations	50 to 70
Subsurface	Up to 80
Sprinklers	65 to 75

Although drip irrigation values are not given, it might be expected that these would conform to the upper ranges of the other methods.

The amounts of leaching water depend on irrigation efficiency, crop type, soil conditions and amount of salt present in the irrigation water. Leaching water requirements vary with local conditions, but can essentially amount to between 10-35 percent of the initial depth of irrigation water applied. Figure 1 shows grain yields as related to the depth of water used for leaching in the Delta Area of Utah.

Figure 1
Grain yields as related to the depth of water used for leaching
in the Delta Area, Utah (Reeve and coworkers, 1948), (Richards, 1954)

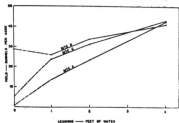

(b) Climatic Variability and its Impact on Forecasting Water Use.

In discussing population pressures, world-wide food demand, and climatic variability, Winstanley (1974) states that:

Probably the most serious problem facing the world concerns our
ability to meet the increasing demand for food...(Figure 2)
...food production must be doubled in about thirty years to meet
the projected demands--and it has taken at least ten thousand years
to attain the present level of production... Evidence is accumulating
which shows that the climates of the earth are changing, and it has
been suggested that they might be changing in a direction which would
have a net adverse effect on world food production...(I.F.I.A.S., 1974:
Rockefeller Foundation, 1974). Probably the main reason for irrigating
and draining the land is to increase food production and one
of the main factors determining the need for irrigation and
drainage is climate.

Figure 2

Projections to 1985 of population, food demand, food production, and food
balance in (a) the developed countries (including Eastern Europe and the U.S.S.R.)
and (b) the developing market economy countries. 1969-71=100. Data source:
U.N. (1974).

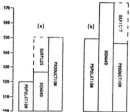

A study conducted at the University of Wisconsin (Bryson & Kutzbach, 1973)
indicates that the Earth's climate may be returning to that of the neo-boreal
era, typically 1600-1850. The study indicates that this was an era marked by
drought and famine. During the last fifty to sixty years the World has experienced the best agricultural climate since the eleventh century but now appears
to be returning to the type of climate which existed over the last four hundred
years. The recent climatic pattern of agricultural optimum is being replaced
by the more normal climate of the neo-boreal era. The implications are abundant-
ly clear that long-term water planning in arid regions should be fundamentally
governed by the climatological factors relating to aridity and possibly increasing aridity. The determination of the w_i's for use in the agricultural sectors
of a multisector model for the base year and for future time frames should be
made in a manner that recognizes the possibility of climatic change. Computation
of the total water supply and projected needs for the region should similarly
be made so as to attempt to minimize drought losses wherever this is possible.

The Colorado Desert Input-Output Study as Example.

(i) Overview

Imperial County, California, consists of 4,284 square miles of Colorado
Desert land situated in the extreme southwest corner of the State. The County
derives its name from the Imperial Valley Land Company, a subsidiary of the
California Development Company which at the turn of the century reclaimed the
southern portion of the desert for agriculture.

The first irrigation project in the Imperial Valley was developed in 1902.
In 1938 the Imperial Dam on the Colorado River was completed to provide additional irrigation water. The Valley is now supplied by the All-American Canal
which was completed in 1940.

Agriculture has been the leading economic sector in the county economy,
although there has been some modest diversification. Recently, attention has
been focused on the potential development of the geothermal energy resources
of the Valley, and there appeared to be a need for assessing the economic
impacts that such a development might have. The multisector analysis was
therefore aimed at determining the county-wide impact of geothermal development
in terms of income, employment, and production levels.

In order to understand in relative terms the impact of certain major
developments in Imperial County, it is important to know what sectors are
present in the economy and how they are interrelated.

The construction of the input-output model for Imperial County(Lofting,1977)was based on secondary data sources. To make the results consistent with national, state, and local data sets, a procedure was devised which would permit the updating and disaggregating of the 1963 Multiregional Input-Output Model (Polenske, 1970) results for the State of California to 1972, and then to allocate, or proportion the 1972 State production totals for the various sectors to the County on the basis of payrolls and production employment.

The United States Bureau of the Census County Business Patterns data suffer from the defect that at the county level many entries are deleted due to the non-disclosure regulations under which the Bureau operates. If there are fewer than three firms in a county, the data must be withheld. A computer program was developed to overcome the disclosure problems by beginning with national summary data and summing and balancing the individual totals provided for each industrial classification. The same procedure was repeated at the state level and thence down to the individual county. Estimates of the withheld data were thus obtained. The data prepared by this process were then compared with county business directory data and where there were differences, the directory data were entered into the County Business Patterns reconstructed base. The resulting county payroll data were taken as a measure of production, and the nation-to-state and state-to-county proportioning of production estimates were made accordingly.

In similar manner the components of final demand were allocated to California counties by means of the Bureau of Economic Analysis County data series which gives estimates of personal income by major source for the industry divisions including Construction and State and Local Government. The County table of interindustry transactions was constructed by means of regionalizing the national coefficients using the Simple Location Quotient (SLQ) technique. This technique has been tested and ranked as "the best of the purely non-survey approached" by Morrison and Smith (1974).

Since the study has as its goal an assessment of the economic impacts of the geothermal sector, it was necessary to add the representative technology of this sector to the county model. The 1972 national table had no geothermal resources sector present so that a basic engineering and financial appraisal of the prospective geothermal industry had to be undertaken and estimates made in dollar terms of its inputs and outputs. This analysis was prepared by Professor Adam Z. Rose, (1976), and the resulting row and column elements were added to the model. The model in its final form showed 85 discrete economic sectors. Gross County Product in 1972 as established by the model was determined to be approximately $450 million.

(iii) Preparing and Reconciling Water and Employment Data.

(a) Employment

County business patterns and local directory data were used to obtain detailed estimates of employment for each of the 85 sectors of the county economy. The California Employment Development Department provides data on covered employment by major industry divisions for each county. The detailed sector estimates obtained from the county business patterns data base were brought into overall agreement with the industry division totals provided by the Employment Development Department.

(b) Water Use

To analyze and project county water use, detailed data by industry were obtained from the California Department of Water Resources (1973). The state-wide estimates for the agricultural sectors were made to agree with the average annual use data for specific crop types for Imperial County given by Po Chuan Sun (1972). The basic data presented by Sun were taken from reports prepared by the Imperial Irrigation District and the County Agricultural Commissioner's Report on Imperial Valley.

Projections of Economic Activity 1975-2020

(i) OBERS Projections

A nationally consistent set of economic projections termed the OBERS projections are prepared jointly by the U.S. Department of Commerce and the

Department of Agriculture for the Federal Water Resources Council. These projections are made for counties and county groups for the entire nation based on the most likely trends of economic activity that can be expected throughout a fifty-year time period ending in 2020. In May of 1975, the OBERS projections for agricultural activity were revised and published as a Series "E" Prime Supplement (1975). The OBERS projections are not intended to be definitive or precise but are designed to provide a consistent frame of reference for public planning agencies, particularly in the field of water resources planning.

For the economic study of Imperial County a more comprehensive and detailed set of projections of economic activity to the year 2020 was prepared. The 1972 eighty-five sector interindustry model developed for Imperial County was used as a basis for these projections.

(ii) Population Projections

The population projections developed by Dr. James Pick (1970) were essentially used to drive certain final demand components of the model. Export demands were projected independently and were then added to the other population oriented components of final demand to obtain an overall bill of goods for each five or ten year time frame from 1975 to 2020. It should be made explicit that the Series I population projections assume no development of geothermal energy - the "base" or "without" case. The Series III projections assume geothermal development and a relatively rapid rate of population growth in the later time frames.

(iii) Final Demands and Gross Output

Specifically the Personal Consumption Expenditures (PCE), Capital Formation and Government Expenditures Final Demand vectors were added together to form a single vector that could be considered to be oriented directly to expenditures serving local needs within the county.

The final demand export vector is dominated by agricultural exports and it is anticipated under the Series III assumptions that geothermal electrical power will also be a major export sector. The agricultural exports were forecast by means of the OBERS Series "E" projections shown in Table 3. The projections of electrical energy exports were based on the "Planning Purpose Projections of Geothermal Energy Installations" proposed by the Regional Industrial Council (1976). These values are shown in Table 4 .

The projected export demands were added to the local demands and a Leontief Inverse (i.e., $I-A^{-1}$) of the eighty-five sector model was postmultiplied by the total final demand vector estimated for each target year to obtain estimates of industrial output by sector in 1972 constant dollars.

Table 3
Agricultural Production
(Thousands of '67 Dollars)

SECTOR	YEAR						
	1972	1975	1980	1985	1990	2000	2020
Livestock	36,504	38,328	41,362	45,179	48,943	56,461	65,822
Index	(1.00)	(1.05)	(1.13)	(1.24)	(1.34)	(1.55)	(1.80)
Crops	133,195	135,000	152,720	164,205	174,986	196,548	231,177
Index	(1.00)	(1.01)	(1.15)	(1.23)	(1.31)	(1.48)	(1.74)

Table 4
Megawatt Capacity

Energy	YEAR					
	1975	1980	1985	1990	2000	2020
Mw Capacity	0	50	500	1,100	2,500	4,500

(iv) Employment Projections

To convert the projections of output into employment terms the vector of base year employment (1972) was used. The eighty-five order vector of employment for 1972 was set up as a diagonal matrix of coefficients and multiplied into the Leontief Inverse. This produced a base year employment interactions matrix. This matrix was then postmultiplied by successive final demand estimates for each target year for each of the population estimates given for the Series I and III projections. Given the eighty-five order employment interactions matrix, a predetermined level of population for Imperial County, and a set of export demands, the level of employment was forecast. As a crude cross check on the validity of the employment projections for the longer term, the resulting employment/population ratio is shown for the years 1970-2020 in Table 5.

Further refinements could be made which would include estimates of productivity increases and the increases in real income over time,but the present results suggest tolerable levels of accuracy.

Table 5
Employment/Population Ratio

SERIES	YEAR							
	1970	1972	1975	1980	1985	1990	2000	2020
I.	.316	.320	.340	.346	.346	.347	.350	.349
III.	"	"	"	.356	.361	.367	.359	.330
OBERS	.320	–	.340	.360	.370	.370	.390	.390

(v) Personal Income Projections

To project county personal income, the employee compensation and property-type income components of value added were scaled to agree with the published personal income figure for 1972. Payments to government employees, which are exogenous to the model, must also be included, so the projections of personal income shown in Table 6 has been increased by 15% to account for this. Table 6 shows total personal income and per capita income for the Series I and III population projections; under Series I, per capita income increases from $4,948 to $7,150. It is clear that the geothermal development alternative will increase the annual total personal income by $700 million above the level of income that may prevail without the development of geothermal resources.

Table 6
Per Capita Income (1972 Dollars)

Year	Personal Income* (1,000,000)		Population (1000)		Per Capita Income $	
	I	III	I	III	I	III
1975	410.	411.	83.1	83.1	4926.	4948
1980	447.	445.	88.8	84.8	5035.	5245.
1985	478.	530.	95.2	90.9	5022.	5831.
1990	509.	633.	101.6	97.0	5009.	6529.
2000	570.	897.	113.4	121.0	5023.	7416.
2020	676.	1377.	136.9	192.6	4934.	7150.

(iv) Formulation of the County Input-Output Programming Model.

To analyze the effects of water constraints the 85 sector programming model was formulated as follows:

Maximize VX   (Objective Function)

Subject to   where $X = \begin{bmatrix} x_1 \\ \vdots \\ x_n \end{bmatrix}$ $Y = \begin{bmatrix} y_1 \\ \vdots \\ y_n \end{bmatrix}$ $h = \begin{bmatrix} h_1 \\ \vdots \\ h_n \end{bmatrix}$

$(I-A)X \leq Y_U$ $WX \leq h$

$(I-A)X \geq Y_L$ $X \geq 0$ $W = (w_{ij} \cdots w_{in})$

and V_i = sectoral value added; X_i = Activity levels; Y_i = Final demands; W_i = Resource use coefficients; h_i = Resource availabilities.

(viii) Results

The 85 sector input-output model in the linear programming format was used with five alternative objective functions. These were: Total Value Added; Employee Compensation; Property Type Income; Indirect Business Taxes; and Employment.

The 1972 OBERS projections for the Colorado Desert subarea show personal income growing from $310 million (1967 dollars) in 1971 to $1.8 billion (1967 dollars) in 2020. This is approximately a four-fold increase. With this project-ed growth as a guide, final demand bounds were set in initial computer runs so that the elements of final demand could be no lower than the prevailing 1972 level, while the upper bounds were permitted to increase by four times the 1972 value. A further constraint, the resource availability constraint (h_i), was provided by the 3 million acre-feet of water allocated annually to the Imperial Irrigation District from the Colorado River.

It was readily determined from the linear programming model that given the 1972 water use coefficients (w_i's) established on the basis of the water data given by Sun (1972), the levels of agricultural output projected for the county by 2020 in the Series E' OBERS projections could not be attained. To achieve the projected level would require an improvement of approximately 35% in the efficiency of water use in the agricultural sectors. It was not determined independently by resort to a technique such as the Blaney-Criddle formula cited earlier in this paper whether the climatological and physical factors prevailing in the Imperial Valley will permit such improvements. Without resorting to a careful analysis of evapotranspiration rates by crop type, the general consensus of those authorities familiar with local conditions is that improvements in the efficiency of water use of the magnitude indicated by the LP model can be sustained.

REFERENCES

BOOKS

Carter, A.P. (1970), Structural Change in the American Economy, Harvard University Press, Cambridge, Mass.

Chenery, H. and P. Clark (1958), Interindustry Economics, John Wiley & Sons, New York.

Cuny, H.,(1961), Les Deserts dans le Monde, Payot, Paris.

Dorfman, R. P.A. Samuelson, and R.M. Solow, (1958), Linear Programming and Economic Analysis, McGraw-Hill, New York.

Hadley, G.(1962), Linear Programming, Addison-Wesley, Reading, Mass.

McGinnis, William S., Bram J. Goldman, Patricia Paylore,eds.,(1968), Deserts of the World, University of Arizona Press, Tucson, Arizona.

Sun,P. (1972), An Economic Analysis on the Effects of Wuantity and Quality of Irrigation Water on Agricultural Production in Imperial County, California, University of California, Davis.

ARTICLES IN BOOKS

Nemchinov, V.S. (1963), Statistical and Mathematical Methods in Soviet Planning, T. Barna, ed., Structural Interdependence and Economic Development, Proceedings of an International Conference on Input-Output Techniques, St. Martin's.

Ogrosky, Harold O. and Victor Mockus (1964), in Ven Te Chow, ed., Handbook of Applied Hydrology.

Richards, L.A. ed., (1954), Improvement and Management of Soils in Arid and Semiarid Regions in Relation to Salinity and Alkali, Diagnosis and Improvement of Saline and Alkali Soils, U.S. Dept. of Agriculture Handbook 60.

Veihmeyer, Frank J. (1964), Evapotranspiration, Ven Te Chow, ed., Handbook of Applied Hydrology, McGraw-Hill, New York.

ARTICLES

Boster, Ronald S. and William E. Martin, (1972), The Value of Primary Versus Secondary Data in Interindustry Analysis: A Study in the Economics of the Economic Models, The Annals of Regional Science, December,pp.35-44.

Bowman, Charles T. and Terry H. Morlar (1976), Revised Projections of the U.S. Economy to 1980 and 1985, Monthly Labor Review, March.

Criddle, W.D. (1958), Methods of Computing Consumptive Use of Water, Proc. Am. Soc. Civil. Engrs., J. Irrigation and Drainage Div., vol.84, no. IR1, pp.1-27, January.

Folz, W., (1951), The Economic Dynamics of River Basin Development, Law and Contemporary Problems, 22: 211, No.2, Spring.

Kutscher, Ronald E. (1976), Revised BLS Projections to 1980 and 1985: An Overview, Monthly Labor Review, March.

Leontief, Wassily, (1936), Quantitative Input-Output Relations in the Economic System of the United States, Review of Economics and Statistics, 18, 105-25, No.3, August.

Morrison, W.I. and P. Smith (1976), Nonsurvey Input-Output Techniques at the Small Area Level: An Evaluation, Journal of Regional Science, vol.14,No.1/

Thompson, R.G. and H.P. Young (1973), Forecasting Water Use for Policy Making: A Review, Water Resources Research, v. 9, No. 4, August, pp.792-799.

Winstanley, D, Brian Emmett, and Gil Winstanley, (1964), Climatic Changes and the World Food Supply, paper presented at the American Society of Civil Engineers' Irrigation and Drainage Specialty Conference: Contribution of Irrigation and Drainage to World Food Supply, Biloxi, Mississippi, August 14-16.

REPORTS

Bryson, Reid, and John Kutzbach (1973), Wisconsin Plan for Climate Research, University of Wisconsin, Madison.

Ciriacy-Wantrup, S.V. (1954), The Role of Benefit-Cost Analysis in Public Resource Development, Water Resources and Economic Development of the West, Report No.3, Benefit-Cost Analysis, Berkeley: Committee on the Economics of Water Resources Development of the Western Agricultural Economics Research Council.

Edmunds, Stahrl, Memorandum to Jeffery Wiegand, Dry Lands Research Institute, July 7, 1976, RE: Planning Purpose Projections of Geothermal Energy Installations.

Howe, Charles W. (1972), Economic Modelling: Analysis of the Interrela tionships between Water and Society, International Symposium on Mathematical Modelling Techniques in Water Resource Systems, Ottawa, Canada.

Lofting,E.M. (1977), A Multisector Analysis of the Impact of Geothermal Development on the Economy of Imperial County, California, Dry Lands Research Institute, University of California, Riverside, California (forthcoming).

Martin, W.E. and H.O. Carter (1962), A California Interindustry Analysis Emphasizing Agriculture, Giannini Foundation of Agricultural Economics, University of California, Berkeley, February.

Meigs, Peveril, (1953), World Distribution of Arid and Semi-Arid Homoclimates, Rev. Res. on Arid Zone Hydrol., UNESCO, Paris, pp. 203-210.

Pick, James B. Population Projections for Imperial County (1970 Base Year), (mimeograph), Dry Lands Research Institute, UC Riverside.

Polenske, Karen R, (1970), A Multiregional Input-Output Model for the United States, Report No. 21, Economic Development Administration United States Department of Commerce, Washington D.C., December.

Rose, Adam (1976), The Impact of the Operation of a 200 MW Flash Steam Power Plant on the Imperial County Economy (A Preliminary Analysis), Dry Lands Research Institute, University of Californis, Riverside.

State of California (1973), The Resources Agency, Department of Water Resources, Division of Resources Development, California Statewide Input-Output Model Base Year (1967), Sacramento, December.

State of California (1974), Department of Finance, Population Projections for California Counties 1975-2020, Report 74 P-2, Sacramento, June.

"1972" OBERS Projections: Regional Economic Activity in the U.S.. Prepared by the U.S. Dept. of Agriculture Economic Research Service, Natural Resources Economics Division for the U.S. Water Resources Council (Washington, D.C.May 1975). Series E' Population Supplement, Vol.1.

The 1958 Interindustry Relations Study, Survey of Current Business, U.S. Department of Commerce, Sept. 1966.

The Input-Output Structure of the U.S. Economy:1963, Survey of Current Business, November, 1969.

The Input-Output Structure of the U.S. Economy: 1967, Survey of Current Business, February, 1974.

Wilson, A.W. (1964), Economic Aspects of Decision-Making on Water Use in Semi-Arid Lands, Arid Zone Research, by the United Nations Educational and Scientific Commission, vol.,26, New York.

MANAGEMENT OF GROUND - WATER RESOURCES IN ARID ZONES

Prof. Dr. Hans Schneider, Bielefeld, Munster

ABSTRACT

Arid zones are characterized by the annual amount of precipitation between 0 and
250 mm p.a. In these zones evaporation is always higher than precipitation.
Evaporation rises up to 4000 mm p.a.

To make human and industrial settlements and agriculture possible, water is indispensable. As there is no surface - water in these regions, except under
special conditions, ground-water everywhere is wanted to be used.

Ground-water is bound to special hydrogeological conditions, as there is the
existence of aquifers-fractured rocks or sands and gravels in a sufficient thickness - and a sufficient annual amount of precipitation. Ground-water reservoirs
only can be managed, if there is an relevant recharge, otherwise they only can be
mined as an oilfield. Therefore it is necessary to investigate the catchment
areas - the outcrops of aquifers - and the possibility of recharge.

In arid zones the knowledge of recharge is very low. Data given by various
authors vary from 0 to 5 % of the annual amount of precipitation. Under these
conditions there is only a small part of ground-water free for management.

In humid zones for example a discharge rate of 1000 m^3/d demands a catchment area
of 1 km^2 in a sandy aquifer by recharge rate of 50 % of 750 mm/a precipitation.
In an arid zone with a precipitation of only 75 mm/a and a recharge of max. 5 %
the same discharge of 1000 m^3/d demands under the same hydrogeological conditions
a catchment area of 100 km^2.

These facts show intensification of using ground-water in arid zones very soon
will pass over the natural recharge and enter in ground-water mining conditions.
When starting mining the ground-water reservoirs, planning engineers must answer
the question of what will happen to all the new founded settlements when the
ground-water resources are mined. When the reserves would be used up, the cultures developed would inevitably have to be abandoned unless surface waters are
collected in very distant areas before being discharged into the sea, and led from

562

there over large distances by canals and pumping facilities to the cultivated
regions in the arid zones, such as has been made possible in the case of decreasing ground-water resources in California.

Considering the large amounts and the considerable energy costs required, any
substitution of the lacking ground-water by desalted sea water is unlikely even in
the remote future.

INTRODUCTION

The arid zones which are also called subtropical arid zones are adjacent to the
tropical summer-rain zones in both the Northern and the Southern hemisphere. They
correspond to the high-pressure zones of the 'horse latitudes' calms. Here the
air masses descend and are heated dynamically. In the course of this process
their temperature rises above the condensation point. Only in rare cases is the
dry weather interrupted by the precipitations overlapping from the neighboring
rain zones. Therefore the annual precipitations remain far below 250 mm. Their
distribution over the oceans is characterized by broad bands, drifting towards the
continents in eastward direction. (Figure I). The arid zones include, for example, the steppes and the deserts in North and South West Africa, in the West of
North and South America and in Australia.

Cool sea drifts enhance the shortage of rain at the Western coasts. Fog and dew
are but a poor substitute. This is most conspicuous at the Chilean and Peruvian
coasts. From here on a zone of low precipitation extends towards the equator in
Northwestern direction, following the Peru current. The arid zone comprises the
zone of steppe climate and the zone of desert climate. The so-called aridity
boundary between steppe and wooded zones runs along the line where the annual
precipitation (mm/a) falls below 250 mm. The arid zones are mainly situated between North latitude 30° and South latitude 30°.

The desert zones cover 18 million km^2 = 12 per cent, the steppe zones 21 million
km^2 = 14 per cent of the earth's surface, i.e. a total of 39 million km^2 = 26 per
cent of the earth's surface consist of arid zones. The high rates of evaporation
are a characteristic feature of these zones. In Mauretania, for example, the
rates of evaporation lie between 2500 and 3000 mm/a, and may even reach up to 4000
mm/a.

GROUND-WATER IN ARID ZONES

The arid zones are not without ground-water resources where the geological conditions permit it to accumulate. So, for example, we find numerous large ground-
water resources reaching even up to a great depth in the Nubian sandstone and in
the Intercalary Continental in the North African steppe regions, forming aquifers
in sandstone and limestone. Population increase and industrial growth in the arid
zones of Northern Africa and in the Middle East are accompanied by an increase in
water demand.

Even in the fifties, the general opinion was still held that a large ground-water
lake lies beneath the Sahara which might be used as a permanent source of irrigation water and for covering the Sahara with vegetation. It was believed that
this would solely require boring an adequate number of wells and to utilize for
the purpose of irrigation the water which partly crops out freely as artesian
water. Even the idea was expressed that these ground-water resources would make

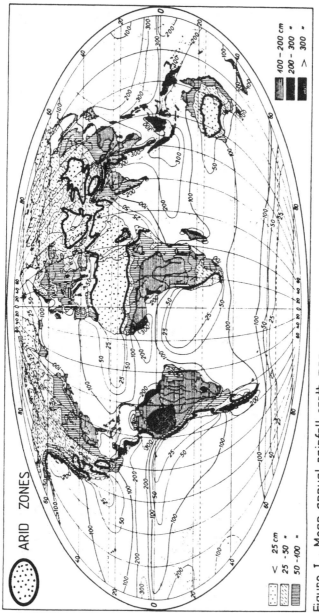

Figure I : Mean annual rainfall earth map

it possible to cover the Sahara with vegetation to the extent that the climate would change considerably. This change would then cause an increase in precipitation that might make irrigation with ground-water largely unnecessary. The assessment of ground-water resources available for use was exclusively based on the results of pumping tests without considering the required ground-water recharge by percolation of part of the rain-water.

With the development of modern ground-water management based on the insight that, at length, the amounts of water withdrawn from an aquifer must not exceed the amounts returned to it by the concomitant process of natural ground-water recharge, doubts arose as to whether the huge amounts of the ground-water reserves of North Africa might actually be utilized to the extent advocated until that time. Since this question is of decisive importance for the development of population, industry and agriculture, a few examples from North Africa will now be cited in order to demonstrate the limits of the utilization of the ground-water resources available there.

THE AVAILABLE GROUND-WATER RESOURCES

In the humid zones, attempts at the assessment of the ground-water resources have for almost 100 years been made solely by means of pumping tests without considering the process of ground-water replenishment. Authors to be cited in this context are Darcy, Dupuit and Thiem.

When it was recognized that these stationary methods did not yield reliable data, the non-stationary methods were developed (Theis, Wiederhold, Jacob, Hantush, etc.). All these methods only give characteristic hydraulic parameters of the aquifer, but do not indicate the amounts of water that may be withdrawn from an aquifer without depleting the ground-water reserves. They only provide basic information on the amounts of the ground-water that may be made available by technical means, but no guidance as to the management of the aquifer in question. Only the ground-water that is actually available may be subject to ground-water management. The quantity of the available ground-water depends on the amounts and seasonal fluctuations of precipitation, soil formation in the catchment area of the aquifer, cultivation and vegetation. Numerous studies have been conducted particularly in Europe in the last 20 years, to investigate the influence of these factors.

Various methods have been applied to determine the portion of precipitation drained by the soil and entering the aquifer. Lysimeters with varying types of soil were used for this purpose. Even in the case of similar soils and similar vegetation, the rates of percolation varied. This revealed that the rates of percolation are considerably influenced by the annual precipitation and its distribution.

Figure II shows the relationship between the rates of percolation and the annual precipitation according to Dyck and Charabellas. The highest rates of percolation are reached in the case of fallow sandy soil. A thicker cover of vegetation will reduce the rate of percolation in this case, too. The finer the soil particles (loess) and the more clay is contained in the soil, the lower are the rates of percolation, with equal precipitation. The relationships indicated apply to unconsolidated rocks only.

These methods may only be used for the aquifers in solid rocks - rugged sandstone and limestone - if their outcrops are covered by strata of unconsolidated rocks. In this case the rate of percolation may be approximated satisfactorily. However,

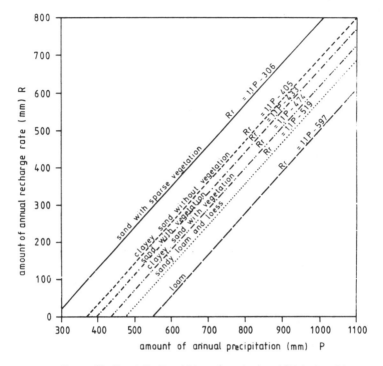

Figure II : Precipitation (P) and recharge (R) in humid
climate in mm/a(according to DYCK and CHARDABELLAS 1963)

if solid rock aquifers are cropping out at the surface without being covered by
unconsolidated rocks, the rate of percolation may only be roughly approximated.

It is relatively easy to assess the rate of percolation in the case of geologically clearly defined aquifers that are fully utilized. In this case, the amounts
of ground-water withdrawn correspond to the mean annual rate of percolation, if
the ground-water level is only subject to seasonal variations and does not show
any continuously decreasing tendency in the case of constant withdrawal. If the
amount of water withdrawn in this case is divided by the outcrop surface of the
aquifer, the mean rate of percolation per km^2 is obtained which may also be expressed as a percentage of mean annual precipitation.

A further aid is the dry-weather flow method developed by Matthess and Thews
(1963). An increase in dry-weather flows in a brook basin indicates areas with
larger ground-water discharges. With the knowledge of the geological structure of
that area, they may be related to certain aquifers. Dry-weather flows, however,
always reflect the minimum recharge, since the higher subsurface flow rates of the
winter and spring season are not taken into account.

Based on the results of measurements carried out by Deutloff (1974), Schneider
(1975) found relationships between precipitation and dry-weather flow similar to
those observed for aquifers in unconsolidated rocks. The question should be asked
now whether one or the other of these methods may also be applied in arid zones.

This question must be answered with no, since in the arid zones regular precipitations are missing. In these regions, occasional strong rainfalls occur, although mostly limited to their edges, the water spreading at the surface and then
immediately being dried up by intensive evaporation.

At the boundaries of these regions, at the foot of the mountains with heavier
precipitation, the Wadis and Shotts may be flooded, the water partly percolating
into the river beds or reaching the edge of the deserts and being drained there in
the detritus. For this reason only logically circumscribed ground-water recharge
is taking place. The figures given in the literature vary between 2 and 5 per
cent of the local precipitation.

This is why we need different criteria to determine the ground-water recharge in
arid zones. The age determinations (C-14-age) carried out in the last 15 years
yielded significant information on ground-waters and oasis waters. For this
reason the results of the age determinations of ground-waters in several regions
of North Africa will be discussed in more detail in the following.

GROUND-WATER IN ARID ZONES OF NORTH AFRICA

Due to the geological conditions, the arid zones of North Africa show a large
number of hydrological basins. Figure III illustrates the various hydrological
basins of North West Africa according to Drouhin (1953). Figure IV shows the
outcrops of the most important aquifers of the Intercalary Continental - mostly
Nubian sandstone according to Drouhin (1953). The map shows that by far the
largest outcrop areas of these aquifers are situated within the zone with less
than 250 mm/a precipitation, i.e. within the arid zone. Only in the Atlas in the
North and in the far South, the outcrops are located in areas with higher precipitation. Only in these areas may any noteworthy recharge of ground-water be
expected. In all other areas no recent recharge has taken place as will be shown
in the following.

Egypt

The main aquifer of Egypt west of the Nile is the series of Nubian sandstones,
partly consisting of solid sandstone and partly of less consolidated rocks of
varying age, ranging from Permocarbonic to lower Cretaceous. The outcrops of
these aquifers are in the South (Figure VI) near Assuan and extending into Sudan.
Towards the North the aquifers are covered by layers of Upper Cretaceous to
layers from Pliocone. From Bahr onwards towards the North an increasing number of
aquicludes occur.

As early as in 1927, Ball demonstrated a uniform ground-water level gradient running from Southwest to Northeast (Figure V). Towards the end of the Pliocene, the
Nile had already deeply cut into the socle of the mountains. The sole of the bed
of the Nile near Assuan was 210 m beneath the present level (about 120 below the
sea level) and at the estuary of the Nile (probably the Riß Pleistocene) 200 m
beneath the present level of the Mediterranean.

Deep down-cuttings of this type appear to have occurred several times in the
course of the glacial periods and to have been restored to the previous condition
within the interglacial periods. For the Wurm Pleistocene, Pfannenstiel assumes
90 m below the sea level. The Nubian sandstone was then more deeply cut North of
Komombo towards Luxor - Qena than nowadays. Where the Nubian sandstone is plunged
into the geologically recent sediments of the Nile valley, Nubian thermal waters

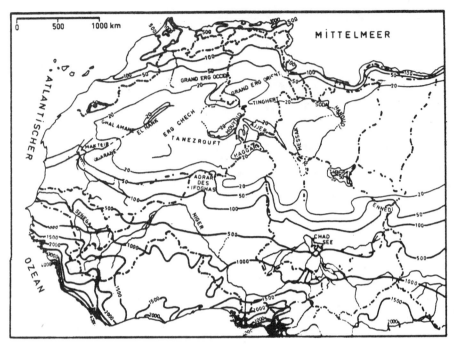

Figure III : Hydrological basins of North Africa, and mean annual
rainfall 1926 – 1950 (according to G. DROUHIN 1953)

are even today entering the Nile. Thus even today there is a hydraulic connection
extending to the Nubian aquifer.

For the periods of deeper down-cuttings of the valley-floor of the Nile, draining
of Nubian waters up to the area of Luxor-Qena may be assumed. In these periods,
at least the upper 200 m of the aquifer may have been emptied. These areas extend
far beneath the aquicludes occurring towards the North on top of the Nubian aquifer.

In the following period, between 25,000 and 35,000 years prior to our age, i.e.
during the high Wurm Pleistocene - there were heavy, though irregular rainfalls.
This is confirmed by the innumerable, now dried up drainage systems spreading over
the whole Sahara and occurring in great quantities around the oasis. The Nile
valley was flooded. This made it possible that Nile water entered into the Nubian
aquifer and from here onwards filled it up far into the West to the present level
again. A further ground-water recharge around the oasis resulted from the peri-
odical flow of waters from the river systems directed towards these areas in the
pluvial period.

Are these assumptions confirmed by the age determined for the ground-waters in
question? Fig. VII shows the C-14-ages of the various regions determined by
Muennich and Vogel (1962). The survey shows that most of the waters withdrawn
from relatively deep wells are between 19,400 and more than 30,900 years old, i.e.
they must have their origin in the pluvial period. Only few waters - Wadi Natrun
and Marsa Matruh - are younger, i.e. from 1,170 to 2,780 years old. They originate in a more recent pluvial periods. Here even recent groundwaters will occas-
sionally occur in aquifers that are very close to the surface.

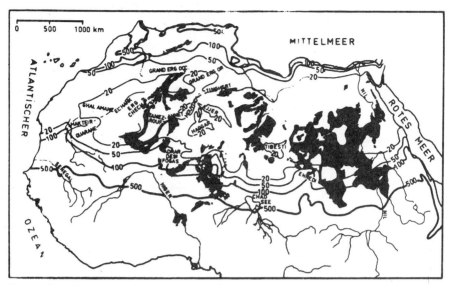

Figure IV: Hydrological basins of North Africa and outcrops of Intercalary Continental (according to G. DROUHIN 1953)

Apart from the few exceptions mentioned, the examined waters do not indicate any more recent ground-water recharge. Before these ground-waters - to be characterized as fossile waters - were used more intensively after 1950, a steady state had developed between ground-water recharge in the South, in Sudan, and the withdrawal or free outcrops of ground-water in the natural oasis. In his studies, Knetsch, 1962, concludes that even the current artificial utilization by means of deep wells is to be regarded as an excessive exploitation of these ground-water resources, considering the speed and quantity of recharge. For this reason he recommends, for artesian wells in particular that withdrawal should be matched with artesian pressure, i.e. should be kept constant by keeping it at a low level. This applies even more to wells equipped with pumps. So, for example, Knetsch and others found a decrease in the yield of many artesian wells by 22 to 86 percent within 1 to 12 years.

Libyan Coastal Area

In the Libyan coastal area, Muennich and Vogel (1962) found relatively old to recent ground-water. The recent ground-water occurs in strata of minor thickness on top of the older ground-water. The relatively old ground-water flows to the coastal area from the Southern higher elevations, flowing over large distances. The C-14-age of these waters ranges from 2,150 to 3,470 years. They originate in the ground-water recharge of the most recent pluvial period about 3,000 years ago. Even the more recent waters of Homs and Suk el Sebt are between 280 and 630 years old (Figure VII).

The North of Lybia and Tunisia is covered by recent transgression sediments of the Miocene and the Paleocene. Here only saline waters and brackish waters are found that cannot be utilized. Only further in the South (Figure VII) is the ground-water of the Nubian sandstone suitable for use.

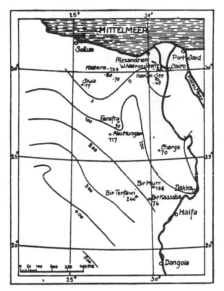

Figure V : Groundwater level in the
          Nubian sandstone East
          Sahara (according to BALL 1927)

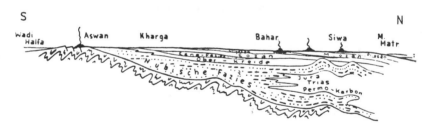

Figure VI : Cross section West Egypte, Nubian sandstone
            (according to KNETSCH et.al. 1962)

Central Sahara

Murzuk Region: The Murzuk basin is one of the largest tectonic units of the Sahara. It covers an area of 900 x 1200 km. In this basin, Cambrian to Jurrassic
or Cretaceous sediments are deposited on Precambrian metamorphic series, gently
slanting towards the centre of the basin (Figure X).

The hydrogeological system of this basin is largely closed. Only in the Northeast
there might be the possibility of subsurface flow to the Syrte basin or towards
the rift valley of the Hon. The sediments are formed by alternative layers of
sandstone and clay-stone. The purely continental series that are of interest in
this context have a maximum thickness of 1,500 to 2,000 m.

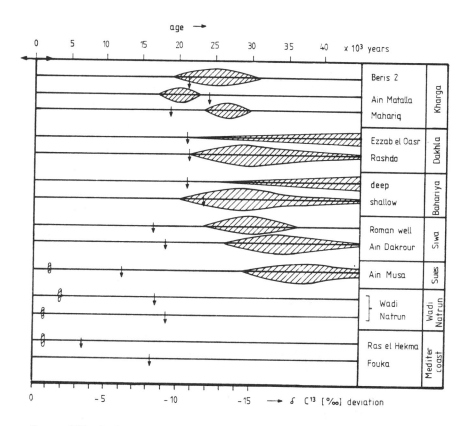

Figure VII : C-14-age of groundwater: Libyan desert
(according to MUENNICH and VOGEL 1962)

Within certain tectonic morphological areas, accumulations of ground-water close to the surface have developed. These zones are linked with oasis that in part form long corridors of vegetation and permit that some regions of the Fezzan are inhabited despite of being situated in the Middle Sahara.

Where aquifers are plunged beneath more compact claystone covers, springs crop out, and when wells are bored, artesian water is found. The Tassili-Plateau is to be regarded as recharge area for the ground-water formed in the Cambrio-Ordovician. The Djebel Acacus may be regarded as catchment area of the water found in the Silurian sandstone (Figure XI).

In Wadi Shati, a ground-water flow directed from North to South is backed up by clay-stones of the Devonian and the Carbonic. Wherever the borings reach the aquifer, the water is under artesian pressure. The considerable higher area of Djebel Gargaf is to be regarded as the recharge area of this ground-water. This is the outcrop area of the sandstones of the Cambro-Ordovician and of the Devonian. Fig. IX shows the age determinations carried out by Klitzsch et al. (1976).

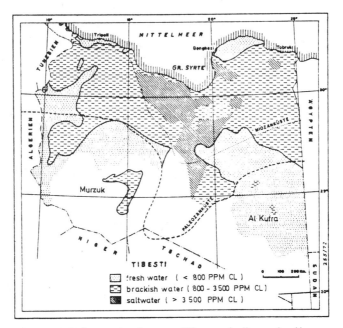

Figure VIII: Groundwater quality and dissemination
of Miocene and Paleocene encroachment,
(according to J.JONES 1964)

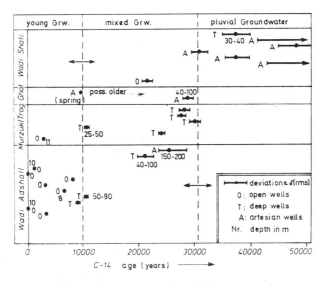

Figure IX : C-14-Age (years) of Groundwater
Central Sahara (according to KLITZSCH 1976)

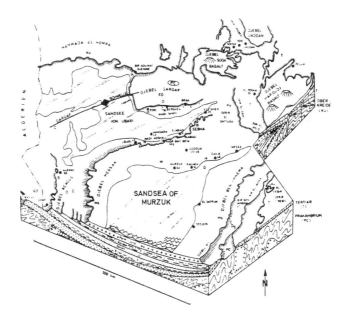

Figure X : Murzuk basin: Geological diagram
(according to KLITZSCH et.al.1976)

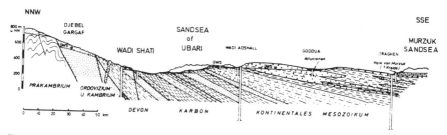

Figure XI : Murzuk basin: Geological cross section
(according to KLITZSCH et.al. 1976)

In the catchment area of Wadi Shati, in Djebel Gargaf, the higher sandstone mountains show a drainage system steeply falling to the centre of the basin in the
South. There are no end weather pits for collecting the surface water. Strong
local rainfalls in this area will run off rapidly and may certainly not lead to
recharge of the aquifer more than 50 m deeper. In this area recharge of ground-
water may only be expected during long and persistent phases of humidity, and
certainly the short periods of humidity following the pluvial periods cannot be
regarded as such. Thus it is not surprising that in Wadi Shati artesian waters
ascending in relatively close vicinity of the catchment area have an age that
shows that they may be classified as originating in the latest pluvial period.
With its age of 21,800 to more than 42,000 years, Wadi Shati indicates the time
when the last ground-water recharge took place in this area.

The waters of Ghat, the Wadi Adshall and the series of oasis of Murzuk that are
also in part artesian waters, originate from several superimposed aquifers that
are separated by aquitards. Here the waters might have mixed in end weather pits
and Quarternary valley sediments. The more recent waters occuring in this area
originate in postpluvial periods of humidity (Figure IX and XIII). These mixed
waters show a slightly lesser C-14-age than the deeper aquifers. They are mostly
not artesian and occur in the centre of the morphologically deep basin areas.
They originate in the first line in postpluvial periods of humidity (Figure IX).
The age determinations of the ground-waters do not provide any evidence of any
considerable portion of recent waters in the ground-water reservoir of the Middle
Sahara.

Klitzsch et al. (1976), after their short comprehensive studies, concluded that
each withdrawal of water in the Murzuk basin - as safe as this may be in other
basins of the Sahara - must gradually deplete the basic ground-water reserves.
This applies in particular to the artesian aquifers, since none of them are recharged at present. Apparently, there was no or hardly any recharge of ground-
water in the post-pluvial periods of humidity, either. Within the Murzuk basin,
Wadi Shati appears to be particularly at risk.

Al Kufra basin: The Al Kufra basin is about 900 km South of Benghazi in the
Lybian Sahara. Today it consists of several no longer interconnected areas of
vegetation. Numerous hills with dead vegetation in their surroundings indicate
the former much larger extension of the oasis area. The main oasis is situated in
a flat depression, the deepest part of which is covered by 2 saline lakes. These
are partly filled with water in winter when the rate of evaporation falls below
the rate of ground-water inflow. The high salt content also found in the ground-
water close to the surface, is due to a gradual enrichment over thousands of years
caused by the continuous inflow of water with simultaneous high rates of evaporation. The utilization of the ground-water reserves by new wells has in the
vicinity of the oasis already produced a lowering of the ground-water level by
several decimeters, so that within few years, an increasing deterioration of the
natural vegetation will have to be expected.

The Kufra basin extends over an estimated area of 150 x 400 km^2. The whole filling of the basin is of continental origin and consists of layers of predominantly
slightly consolidated sandstones, mudstones and claystones, the upper Jurassic-
Cretaceous part of which is known as Nubian series. The main aquifers form the
clastic layers, particularly the sandstones of the Mesozoic. If according to
Dietrich (1976), an average thickness of the basin filling of about 500 m is assumed, this gives in the case of an utilizable pore volume of a total ground-water
reserve of 3 x 10^{12} m^3. The precipitation recently measured in Kufra is so low
that it may be neglected. For the period under review of 36 years, it is lower
than 2 mm/a.

In this case, too, determinations of the C-14-age were carried out in order to
elucidate the origin of the ground-water. The most recent waters occurring close
to the surface showed an age of just under 10,000 years, while the ground-water
resources in greater depth are about 30,000 years old.

These data fit very well into the diagramme of climate changes drawn up by Mc.
Burney, which is also applicable to Kufra (Figure XII). According to the age
determinations there was no ground-water recharge in Kufra in the last period of
humidity about 3,000 years ago. Just under 10,000 years ago, arid climatic conditions in the Sahara followed a period of humidity lasting about 25,000 years, a
period that certainly allowed effective recharge of ground-water. The period of
the present arid climate must have been interrupted by a period of greater humidity about 3,000 years ago. However, it had only a minor influence on the re-

charge, although it was sufficient to create more favourable conditions of life in the Sahara for some time.

In his studies, Dietrich (1976) comes to the conclusion that with an average hydraulic conductivity of 3 m/d, a gradient of 4.5×10^{-4} and a storage coefficient of 0.1, the ground-water has moved in the aquifer by about 70 km in 10,000 years and by about 200 km in 30,000 years. Justly he concludes that the recharge of ground-water 10,000 to 30,000 years ago must have occurred predominantly within the area of the basin itself.

CONSEQUENCES OF INCREASED UTILIZATION OF THE FOSSILE WATER

From the pumping tests, using the test wells established in Kufra, it was calculated by means of an analog model that in the vicinity of the wells the end of the quasi-stable state may be expected in about 10 to 20 years time. First modeling results, however, indicate that the lowering of the ground-water level will take place more rapidly. A lowering by about 80 - 140 m 30 years after putting the pumps into operation and of about 100 to 170 m 50 years after this date is to be expected.

COMPARISON OF THE POSSIBILITIES OF GROUND-WATER RESOURCES MANAGEMENT IN HUMAN AND ARID ZONES

For better understanding of the importance of recharge for the management of ground-water resources, a few essential data will be given in the following. The comparisons are based on a ground-water basin with a surface area of 100 km^2 with a purely sandy aquifer under the various climatic conditions.

In humid climate with an annual precipitation of P = 750 mm, a recharge of R_r = 50 percent P, i.e. R = 375,000 m^3/km^2 x a or, converted to 100 km^2, of R_T = 37.5 x 10^6 m^3/a, will take place. In a climate with P = 250 mm/a (at the edges of arid zones) and an estimated ground-water recharge of R_r = 2 percent of P, a recharge of R = 5,000 m^3/km^2 x a will take place. Calculated for an area of 100 km^2, this gives a total recharge of 500,000 m^3/a (Figure XIII).

In the completely arid area with P = 0 mm/a, the recharge R_T = 0. In this region only fossile water is available for utilization. Assuming, for example, for the Kufra basin with an area of 60,000 km^2 a mean thickness of the utilizable aquifers of 500 m and a utilizable total pore volume of 10 percent, an amount of groundwater of 3×10^{12} m^3 would be stored in the aquifer. This amount, however, may certainly only be made available by 50 percent. This corresponds to an amount of water of 1.5×10^{12} m^3.

According to Kopp (1975), 5,000 m^3 water per ha are required for irrigation purposes for one crop = 500,000 m^3/km^2. If, for example, 30,000 km^2 are to be made available for one crop/year, this would require 1.5×10^{10} m^3/a for irrigation purposes. The ground-water resources to be made available would then last for

$$T = \frac{1.5 \times 10^{12}}{1.5 \times 10^{10}} = 100 \text{ years}$$

If the same area would have to be utilized for 3 crops in succession - vegetables etc. - what would be possible, considering the given climate, if irrigation would

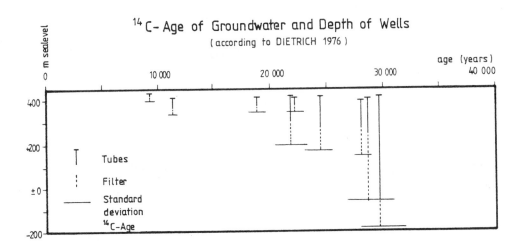

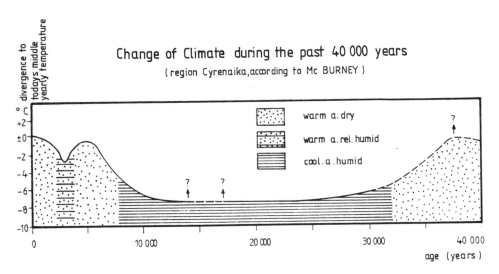

Figure XII : C-14-age of groundwater in the Kufrah Basin and paleoclimatic
conditions

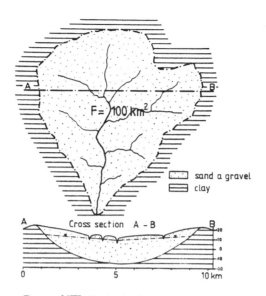

Figure XIII: Hydrological basin (humid climate)

$$P = 750 \text{ mm/a}, \quad R_r = 50\%; \quad R = 375\,000 \text{ m}^3/\text{km}^2\text{a}$$
$$F = 100 \text{ km}^2, \quad R_T = 37\,500\,000 \text{ m}^3/\text{a}$$

Arid Climate

$$P = 250 \text{ mm/a}; \quad R_r = 2\%; \quad R = 5\,000 \text{ m}^3/\text{km}^2\text{a}$$
$$F = 100 \text{ km}^2; \quad R_T = 500\,000 \text{ m}^3/\text{a}$$
$$P = 0 \text{ mm/a} \quad R_T = 0 \text{ m}^3/\text{a}$$

be sufficient, then 2,000 mm = 2,000 m^3/ha = 2,400,000 m^3/km^2 x a would be required in the upper Miserga valley in Tunisia according to Kopp.

For the assumed 30,000 km^2, an annual amount of irrigation water of $7.2 \times 10^{10} \text{ m}^3$ would be needed. The amount of ground-water to be provided, $1.5 \times 10^{12} \text{ m}^3$, would then permit such an intensive utilization for

$$= \frac{1.5 \times 10^{12}}{7.2 \times 10^{10}} = 21 \text{ years}$$

These figures make evident how important it is to use the fossile water judicially. Any utilization will reduce the reserves. Thus the ground-water resources are more or less comparable to mineral oil resources. The more intensively they are exploited, the more rapidly they will be consumed and exhausted.

It is certainly possible to carry out major isolated agricultural projects in the arid zones by making use of the fossile ground-water resources if the withdrawal of ground-water is kept within narrow limits. The withdrawal of major amounts of ground-water has far-reaching area-wide effects. Irrigation projects may thus only be developed to a limited extent. Any covering of the Sahara with vegetation - an idea that even today is taken up in the press as a vision for the future - is bound to remain utopian.

When the reserves, in particular in the locally circumscribed hydrological basins, would be used up, the cultures developed would inevitably have to be abandoned unless surface waters, such as from Kongo and Nile, are collected in very distant areas before being discharged into the sea, and led from there over large distances by canals and pumping facilities to the cultivated regions in the arid zones, such as has been made possible in the case of decreasing ground-water resources in California. Considering the large amounts and the considerable energy costs required, any substitution of the lacking ground-water by desalted sea water is unlikely even in the remote future.

REFERENCES

Ball, J.: The artesian water supplies of the Libian desert. The Geograph. Journ., vol. 70 m 105-118, London 1927.

Degens, E.T.: Geochemische Untersuchungen von Wassern aus der Agyptischen Sahara. Geol. Rundschau, 52/2, 625-639, 3 Tab. Stuttgart 1962.

Deutloff, O.: Die Hydrogeologie des nordwestfalischen Berglandes in der Umgebung von Bad Salzuflen u. Bad Oeynhausen, Fortschr. Geol. Rheinld. u. Westf., 20, 111-194, 12 Fig., 9 Tab., 4 Maps, Krefeld, Marz 1974.

Dietrich, G.: Die Al-Kufrah-Projekte als Beispiel fur die Nutzung fossilen Grundwassers bbr, 27/2, 45-48, Koln 1976.

Drouhin, G.: The problem of water resources in North-West-Africa, Reviews of Research on Arid Zone Hydrology. UNESCO, 9 - 41, Paris, 1953.

Dyck, S. and Chardabellas, P.: Wege zur Ermittlung der nutzbaren Grundwasser- reserven. Ber. geol. Ges. DDR, 8, 245-262, Berlin 1963.

Faure, H.: Le Probleme de l'origine et de l'age de l'eau des oasis sahariennes du Niger. IAH-Memoires, Vol. VII 277-278, Reunion de Hannover, 1967.

Jones, J.R.: Ground-Water Maps of the Kingdom of Libya. USGS, Open-File Report, 11 p, 6 maps, Washington, D.C. 1964.

Klitzsch, E.: Salinitat und Herkunft des Grundwassers im mittleren Nordafrika (Sahara). Geol. Jb., C2, 251-260, 1 Abb., Hannover 1972.

Klitzsch E., Sonntag, Chr., Weisthoffer, Kl. and El Shazily, E.M.: Grundwasser der Sahara: Fossile Vorrate. Geol. Rundschau, 65/1, 264-267, 1 Abb., 1 Tab., Stuttgart 1976.

Knetsch, G.Y.: Geologische Uberlegungen zu der Frage des artesischen Wassers in der westlichen Agyptischen Wuste, Geol. Rundschau, 52/2, 640-650, 2 Abb., Stuttgart 1962.

Kopp,: Produktionspotential des semiariden Tunesischen oberen Misergatales bei Beregnung. Schriftenr. GTZ, Ges. f. techn. Zusammenarbeit, Eschborn 1975.

Matthess, G. and Thews, J.D.: Der Abflußzuwachs als Hilfsmittel bei der Beurteilung von Wassererschließungsmoglichkeiten. Notizbl. hess. L. -Amt f. Bodenf., 91, 231-236, Wiesbaden 1963.

Muennich, K. O. and Vogel, J. C.: Untersuchungen an pluvialen Wassern der Ost-Shara. Geol. Rundschau, 52/2, 611-624, 3 Abb., Stuttgart 1962.

Schneider, H.: Die Wassererschließung (Chapter 2: Spezielle hydrogeologische Probleme in ariden Gebieten (aufgezeigt an Beispielen Nordafrikas), 85-108, 4 Tab., 20 Abb., Vulkan-Verlag, Essen 1973.

Schneider, H.: Beitrag zur Frage der Grundwasserneubildung in Festgesteinen, bbr. 26/12, 437-440, Koln 1975.

GROUND-WATER POTENTIAL IN ARID ZONES THROUGHOUT THE WORLD

Harold E. Thomas

ABSTRACT

The largest arid zones in the world are of subcontinental dimensions: North Africa
and Asia Minor in the Northern Hemisphere and Australia in the Southern. All are
subtropical, and the solar energy received throughout the year is great enough that
the potential evapotranspiration far exceeds the annual precipitation. Thus, a
pattern of perennial water deficiency extends over a vast area, and although there
are variations in degree from place to place, neighboring areas can't contribute
much because they are all in this "deficiency" thing together. Rain is rare, but
does occur and may be locally intense; the usually dry stream channels may then
carry torrential runoff; this water may be absorbed in the dry stream bed and
flood plain or accumulate in depressions as lakes and ponds. At all other times
evaporation is supreme, removing soil moisture and surface water, drying up the
ponds and lakes, leaving residual soluble salts in lake beds or stream beds or in
the soil.

The opportunities for recharge to ground-water reservoirs in these sub-continental
arid zones are as rare as the rains and the surface runoff, unless there is some
aquifer that extends beyond the arid zone and into a region where there is water
for recharge. Otherwise, ground-water replenishment is limited in quantity by the
prevailing water deficiency; also such replenishing water may be of inferior quality because of soluble salts in the recharge areas. Hence, it is likely that any
large reserves of potable water found in these great deserts are "fossil" water not
replenishable under existing climate and geomorphology.

Many valleys and basins, plains and plateaus are arid or semiarid because they are
in the "rain shadow" of mountain ranges, from which, however, they may receive inflow of surface water and/or ground water. The largest of these are in the lee of
the high mountains of Central Asia. In southwestern USA there are more than a hundred arid basins, many of them with no outflow and in the Great Basin of interior
drainage, and extending from the Subtropical High Pressure Belt to 45^0 North Latitude. Utah and Nevada hold that the ground-water potential should be the "safe
yield" that can be sustained by the average annual replenishment. This is similar
to what Iranians have done for centuries with their 40,000 kanats (or ganats) on
the Iranian Plateau.

The arid zones that have proved best for the survival, prosperity, and propagation of Homo sapiens are those that are traversed or reached by rivers and large streams which have been utilized by controlled diversion and distribution of the water as needed. In these hydraulic civilizations, some of which have been operating for thousands of years, ground water has been of minor interest, but it may have been a major contributor to the decline and disappearance of some of these civilizations because of waterlogging, drainage problems, and accumulation of salt residues in the irrigated land. With modern pumps and equipment these problems can be a-voided or overcome, and both the quantity and quality of water can be controlled for most effective use. But this will require comprehensive planning, involving the conjunctive use and storage of ground water and surface water. A prime objective of the California Water Plan (1957) was to store water in times of surplus for use in periods of drought and in places of shortage. It was expected that the State's huge ground water basins would be used both for conservation of local supplies and for seasonal and cyclic regulation of imported waters. But instead, the major federal and State water agencies have specialized in huge dams and huge surface reservoirs which become huge evaporating pans when full. And the ground water potential of the Plan - for cyclic storage - has not yet been achieved.

INTRODUCTION

Around the world in 15 minutes! That's faster than any satellite has yet made it. Fortunately, we do not need to survey cold deserts and thus, can avoid the polar regions where annual precipitation is less than 100 mm. First, we should define aridity not on the basis of precipitation alone, but on the basis of the precipitation deficiency, or the amount by which the potential evapotranspiration (Thornthwaite, 1948, p. 56) exceeds average annual precipitation. Now we can give special attention to the Subtropical High Pressure Belts which encircle the globe in the "Horse Latitudes", 30^0 to 35^0 North and South of the Equator. Arid climate - generally clear skies, light variable winds, warm to hot - is characteristic of many of the lands in these belts, and also of extensive oceanic areas.

In California we are at the moment especially concerned about the "Pacific High" which traditionally spends the summer offshore from California and then moves southward for the winter, along with the sure and the upper middle class. While it is gone we get our annual allowance of rain and snow. But in the past two winters it has not moved southward, but has stayed with us. And that has given us the driest two years in a century.

SUBCONTINENTAL DESERTS

The world's largest arid region is in the Northern Hemisphere: the Sahara, extending across North Africa and embracing about 9,000,000 km^2. A dozen nations share this land, which includes mountain ranges above 3,000 meters, lowlands below sea level, extensive areas of sand, rock waste, plateaus, numerous oases, dry river valleys and gorges, the hottest and driest places on earth, and one through-flowing river - the Nile. To survive, you must depend on ground water, from springs, seeps, wells, oueds and wadis (unless you have the energy for desalination of sea water). Successful wells have been dug or drilled in many places, some shallow and others very deep, some artesian and flowing, some capable of large yields by pumping. Although large areas are still unoccupied, unexplored, and unknown, we are assured the Sahara has vast quantities of potable ground water - a relic of past (Pleistocene) more humid climates, and practically nonreplenishable today. Of course, there is some replenishment, from rare storms that may rain as much as 100 mm on local areas, and cause torrential runoff or be readily absorbed by the sand and other permeable rocks that form the land surface; but the total replenishment in a year would be equivalent only to the small quantities of water being discharged at oases which have sustained village, tribes, and nomads for centuries.

Asia's largest arid region covers about 3,000,000 km^2 of Asia Minor, including most of the Arabian Peninsula and extending northward into Iraq, Syria, Jordan, and Israel. It is separated from the Sahara by the Red Sea, but otherwise it is a continuation of the same arid zone along the same Subtropical High Pressure Belt, and there are similar variations in aridity, down to the ultra-dry and unoccupied Rub al Khali in southeastern Saudi Arabia. Potable ground water occurs in numerous localities, but in others it is brackish or saline, and in extensive areas no water sources have been found. Thus, Kuwait has found two tiny basins containing potable water which it began depleting in 1962, and a more extensive and productive aquifer containing brackish water which is used for washing and cooling and for salt-tolerant vegetation; pure water is obtained by desalination of sea water with energy provided by natural gas. Several other OPEC countries use their gas to desalinize sea water. About 2/3 of the Arabian Peninsula is within the borders of Saudi Arabia (2,200,000 km^2, the largest country on earth without a river), which is making a detailed study of its water resources as a basis for evaluation of the ground water potential.

In the Southern Hemisphere the largest desertic area is in Australia, aggregating 5,000,00 km^2 or more of the western and central parts of the continent. In the eastern part of this arid zone (the one million km^2 drainage basin nominally tributary to usually-dry Lake Eyre) irregular rain and ephemeral runoff resuscitate shallow ground water, temporary lakes, water holes, artificial ponds and reservoirs, and forage for stock. The Great Artesian Basin underlies this area and extends westward into more arid country. It contains mineralized water, generally suitable for sheep and maybe cattle; the water is obtained from wells as great as 2;000 meters deep, and is recharged by rain on more humid lands in eastern Australia.

In these three subcontinental arid zones, ground water should be considered a non-replenishable resource - not that it is unwilling to be replenished, but because there is not the wherewithal for replenishment. As a rule we can evaluate ground water as a "fossil" resource analogous to the fossil fuels: its potential is set by the quantity and quality that we find now in aquifers that can be delimited and delineated. If in some areas future studies or events prove that the ground water is replenish significantly, that will be a welcome benefit and extra dividend for the water users there.

In Asia east of the Arabian Sea, the Thar Desert occupies about 500,000 km^2 of Pakistan and India. The climate is arid (as is to be expected in the Subtropical High Pressure Belt), but this region receives water in rivers from the Himalaya Mountains. Much of the lowland is the alluvial plain and delta of the Indus River, abandoned in historic time as the river shifted to westward. Salinity is characteristic of much of the lowland soil, and salts are residuals from evaporation of shallow ponds and ground water.

The Kalahari Desert, extending over 300,000 km^2 in South Africa, Southwest Africa, and Botswana, is a broad plateau which supports a fair cover of vegetation. Perennial streams from humid lands to the East are absorbed in the extensive dune area in the eastern part of the desert. Probably there is significant ground water, and some replenishment, in this desert.

COOL COASTAL DESERTS

Cool coastal deserts occur along the western coasts of three continents and owe their aridity to the Subtropical High Pressure Belt. But their climate is modified by cold ocean currents which cause high humidity and frequent fog, and these are restricted by barriers of high mountains or plateau along the coast, so that the cool coastal deserts are typically narrow and elongated.

The Atacama Desert in South America extends about 1500 km along the North Coast of Chile, between the Pacific Ocean with its cold Humboldt current and the Andes Mountains. This barren strip includes sandy plains, bluffs, plateau, and saline basins up to altitude exceeding 4,000 meters; and it has economic deposits of nitrates, borax and metallic ores, but a dearth of freshwater or even of fertile soil - which had to be imported for the early miners. The largest town, Antofagasta, has a 350 km pipeline for its water supply.

The Namib Desert of southwest Africa is more than 2,000 km long, bordered by the Pacific Ocean with its cold Benguela current. It has extensive dunes, gravel flats, and eroded rock platforms up to 1,000 meters altitude and 80 to 150 km inland. Potable ground water is found chiefly as the underflow of streams, all of which are ephemeral (Logan, 1970).

"RAIN-SHADOW" DESERTS

Asia has several arid zones in valleys or plateaus which are bordered by high mountains that intercept the moisture in the prevailing winds and wring the air dry before it reaches the lower land to leeward. These lower lands are in a "rain-shadow" which is rainless because of the orographic condition.

The Iranian Plateau (200,000 km^2) has an altitude averaging about 1,000 meters, and extensive areas receive less than 100 mm rain annually, partly because of the Subtropical High Pressure Belt, and partly because of the orographic aridity imposed by the Zagros and Elburz Mountains which rise about 2,000 meters above the west and north sides of the plateau. The lowlands do have salt lakes and sand dunes, but they also have permeable sediments and inflow of water from the surrounding mountains, so that ground water is plentiful and has been used for centuries for irrigation of the lowlands. Kanats are the traditional Persian system of irrigation; and there have been as many as 40,000 on the Iranian plateau. Until 1930 Teheran obtained its water supply from 12 kanats that yielded 800 liters per second for the 200,000 inhabitants.

The Kara-Kum (Turkmen SSR) and Kyzyl-Kum (Uzbek SSR) are arid lowlands east of the Caspian Sea and the Aral Sea, which are east of the Caucasus and Elburz Mountains. This arid zone receives inflow from the towering Hindu Kush to the southeast, via several large rivers, of which the Amu Darya is tapped by the Turkmen Canal (started in 1953) to carry 4 km^3 of water annually 1,000 km to irrigate 700 km^2 of land in the Kara-Kum arid zone. This importation could increase either the ground water potential or the ground water problems in this arid zone.

Farther east and north is the Takla Makin Desert, some 500,000 km^2 of the Sinkiang province in northwest China. It is in the Tarim Darya basin (between the Tien/ Shan Mountains rising to 7,000 meters and the Kunlun Mountains peaking at 8,600 meters). This desert includes extensive areas of bare drift sand and moving dunes driven by northeast winds. These are apparently successful in thwarting the eastward flow of the Tarim and its tributaries, which bring much water from the surrounding ranges, only to lose it in the sand or in lakes between the dune ridges. Although this desert is rainless, I believe the major deterrent to human occupancy is not water deficiency, but inhospitable environment. Winter is the only time when you can safely cross this desert. At any other time the desert may be enveloped in an impenetrable dust haze, which chokes and smoothers every living creature.

The Gobi Desert, still farther east and north, is a broad shallow depression 1,000 to 1,500 meters above sea level, between the Altai Mountains and the plateau of Mongolia and the Nan Shan and Greater Khingan Mountains of China. It has an area of about 1,300,000 km^2, but only a quarter of this area is truly waterless; the southern part, including the Ordos Desert inclosed within the great north bend of the Hwang-Ho, and the Ala Shan Desert farther west, consisting of sand wastes and barren gravel, alternating with vast areas of saline clay, all uninhabited. The rest of the Gobi is also treeless, but water is found in wells and in shallow lakes; the southeastern part (Inner Mongolia) is a steppe suitable for pasture, and groundwater is at rather shallow depth (less than 10 meters).

CRADLES OF HYDRAULIC CIVILIZATIONS

In several of these arid zones I have mentioned the inflow of surface water, because that inflow is a fundamental and obvious factor in the ground water potential of the arid zone. It also has a broader significance in an area where solar energy is great enough to evaporate more than all the water received by precipitation, and therefore, an area of perennial water deficiency: that inflowing water is coming from an area of surplus, and is therefore a treasure to anyone seeking to reclaim any part of the desert for human occupance by overcoming its water deficiency.

Several rivers, flowing into or through arid zones, have sustained peoples throughout recorded history. The Nile, rising in the tropics of central Africa, traverses the Sahara, and has provided water for irrigation in the Nile Valley for more than 5,000 years. And the rivers flowing southward toward the Arabian arid zone have been used for irrigation even longer: along the Euphrates River in Sumeria beginning before 4000 BC, and subsequently in Babylonia; along the Tigris River at Nineveh in Assyria before 700 BC.

Irrigation doubtless had small beginnings in such places as lower Mesopotamia, by digging into the river's natural levees to distribute water to selected places on the flood plain. As the irrigated lands and the people sustained by them increased, the diversions must increase, and also the hazards of disaster by flooding from the mainstream. There was need for coordination of efforts. It became necessary to create large-scale enterprises, to have a universal draft (corvée) that could quickly assemble a large force of the ablebodies. Big productive water diversions (for irrigation whether by basin flooding or ditch and furrow) were frequently accompanied by big protective works (for flood control). The resulting agrarian economy, a hydraulic civilization, required the cooperation of large numbers of people in the maintenance of dikes and levees, and other control structure and ditches, and in the irrigation and intensive cultivation of crops. For this, there had to be planning, record keeping, communication, and supervision - organization in depth, and thus a massive and permanent bureaucracy and a monolithic society.

With a minimum of labor-saving tools and animals and a maximum of human labor, the life of a hydraulic farmer was one of unending drudgery. Nevertheless, the hydraulic civilizations of the Near East, North Africa, India, and China maintained themselves for thousands of years. Prior to the Industrial Revolution it is likely that the majority of all human beings lived within the orbit of hydraulic civilization. And it was rewarding to some, if we may judge by the stories of opulence and high life in such cities as Bagdad, Babylon, Sodom and Gomorrah.

THE "GREAT AMERICAN DESERT"

A century ago most of western North American was viewed - especially by those in the humid eastern half of the continent - as the "Great American Desert". Millions of people now live in the region, and rather than speak of "desert" they would prefer to show how they made it "rejoice and blossom as the rose". And truly the region has remarkable diversity. There is Death Valley, with the highest summer temperature, least annual rainfall (40 mm) and evaporation 90 times as

great, and lowest altitude in the USA; and there are other areas almost as arid.
There are also valleys and plains plateaus and mountains which may be classed as
semiarid, subhumid, and several areas with perennial water surplus. Within the
old "Great American Desert" we can identify arid zones that might fit into the categories that I have mentioned for the rest of the world.

The Subtropical High Pressure Belt extends over Southern California and Arizona
and New Mexico and West Texas, plus the northern parts of Sonora and Chihuahua in
Mexico. This is by no means comparable with the arid zones of subcontinental dimensions on other continents, because the arid areas are dispersed among semiarid
plains and plateaus, and mountains that are snow-capped seasonally or at least occasionally. However, it does include some ground water reservoirs which are large
and productive, but which receive negligible replenishment.

The High Plains of Texas and New Mexico, about 50,000 km^2, has a semiarid climate
(300 to 500 mm of rain, chiefly during the growing season, and sufficient for pasture). Pumping from wells for irrigation of cotton and other valuable crops has
increased the prosperity and population of the area, and about half a million people live comfortably and pump more than 6 km^3 of water annually from wells. The
reservoir originally contained about 500 km^3, practically all "fossil" water, non-
replenished, and the storage has been depleted in populous areas until many wells
have gone dry. Numerous plans and proposals have been made for import of water,
which have brought response from the proposal source area there is no water to
spare. The lower Mississippi River appears to be the only undoubted surplus within reach, and that is quite a long reach: more than 1,000 km horizontally and 1
km vertically.

South Central Arizona has an arid climate, but the Gila River and its tributaries
bring in substantial flows from highlands to the North and East. Sixty years ago
irrigation with water from the first large reservoir (Roosevelt Lake) caused water
logging and drainage problems, but those were soon solved by pumping from wells.
In recent years pumpage from wells has exceeded 5 km^3 annually, far greater than
the replenishment, so that the ground water is being mined. To halt or reduce this
depletion the Central Arizona project has been authorized, but not yet constructed,
to import water from the Colorado River.

A cool coastal desert is formed along the coast of Southern California, and Baja
California in Mexico (Meigs, 1970). The cold California current, cool ocean, and
frequent fogs give welcome relief from the hot desert sun in summer, an exception-
ally attractive climate for more and more people. The small coastal streams and
ground water reservoirs have not been enough to supply the needs of burgeoning
Southern California, which therefore imports water from the Colorado River and
Northern California. Baja California has even less water resources, barely able
to supply the small population before the recent invasion by tourists and those
seeking longer residence for recreation or retirement. Here may be a likely environment to develop and test new ideas and do-it-yourself apparatus for solar energy, windpower, distillation, hydroponics, salt-tolerant vegetation, etc. (Hodges,
1970).

Desert valleys and plains in the rain shadow of mountains are numerous and widely
distributed in the American arid zone, from the Subtropical High Pressure Belt
northward to 45^0 Latitude. Most of these are intermontana structural valleys with
arid or semiarid climate, and many receive surface and subsurface inflow from adjacent mountains. This is the Great Basin, where all the water of precipitation
is returned to the atmosphere, and there is no outflow except possibly by deep
aquifers. There are more than a hundred separate basins in California, Nevada,
Utah, Idaho, and Oregon, some isolated and closed, others integrated into stream
systems which drain into sinks such as Great Salt Lake, Carson Sink, or Badwater

in Death Valley. These are thus truly basins, in that they hold all the sediment (valley fill) and water (as ground water, lakes, or playas). Utah and Nevada limit the development and use of ground water in each individual basin to prevent depletion of the stored resource: they limit withdrawals from wells to a "safe" or perennial yield equivalent to the average annual replenishment.

The American arid zone is traversed by two major rivers, both with headwaters in the Rocky Mountains and both interstate and international streams: the Rio Grande (Rio Bravo in Mexico) flowing to the Gulf of Mexico, and the Colorado River naturally discharging into the Gulf of California. The Colorado is unique among arid-zone rivers in that so much of its course is in canyons, where it must be dammed and then piped and tunneled, and pumped and diverted to places far from the river. Some of these diversions have been mentioned and there are many others in California. In several places near the Pacific Coast pumping from wells lowered the water table below sea level, inducing sea water encroachment. The sea-water advance was halted by injecting Colorado River water into a line of recharge wells along the Coast to create a barrier ridge of fresh water.

If I digress from my theme of ground water potential, and get carried away by instances of exceeding or ignoring or not knowing the ground water potential, it is because that has been a way of life in much of the American arid zone. Ground water development has been almost entirely by private enterprise or private property. The first productive well in a new area - whether flowing or pumped - has often been accompanied by an assurance that it tapped an "inexhaustible" supply; and the boom was on, with more people, more wells, more irrigation. Crisis came later, with loss of artesian pressure, falling water tables, "dry" wells, land subsidence, salt-water encroachment. Then it was time to call in geologists, engineers, physicists, chemists, to find out what went wrong, and then to make laws or bring lawsuits to establish water rights. Some scientists have gone from crisis to crisis, making scientific studies and reports of each, and have become experts at problem-solving.

One of the answers sought has been the ground water potential. For a specific ground water reservoir or aquifer the potential has generally been an "ex post facto" thing, based upon an established pattern of wells (as to location, depth, and yield), historic record as determined by available data, aquifer tests, and the traditions and conventions of the people as to what is right and proper, reasonable and beneficial. Such quantitative determinations have formed the basis for decrees as to water rights, and for administration and control of ground water development and use.

For comprehensive, long-range planning the ground water potential cannot generally be stated quantitatively or specifically, partly because essential data are incomplete or lacking; but chiefly because the ground water potential is dependent upon and varies with other elements of the hydrologic cycle, and the water resources, and especially surface water.

Conjunctive use of ground water and surface water have long been recognized as essential in water planning for California. Studies made during the formulation of the California Water Plan (California Department of Water Resources, 1957) indicated that its objective could not be achieved without full and careful use of the ground water resources. "The answer for the future thus lies in the full development and use of our ground water basins, both for conservation of local supplies and for seasonal and long-term cyclic regulation of imported water supplies" (Berry, 1962, p. 3). This "cyclic" storage would involve the planned use of ground water storage in conjunction with surface storage facilities.

Unfortunately, the idea of "cyclic" storage hardly got off the ground - or into the ground. In the wet year 1970 the reported artificial recharge to the State's ground water reservoirs was about 2 km^3, equivalent to about 10 percent of the water pumped from wells during the year. Although, this is no way to build up reserves for the "future" drought that we are now catching up with, the quantity was large enough to show that artificial recharge is feasible for cyclic storage in underground reservoirs. Conflicts of interest - private interests, public interests, and perhaps the dispersal of public interests into bureaucratic interests - may be partly responsible for this under-achievement. In California private property rights have been asserted and protected, particularly as to ground water and the underground space that may be occupied by it. Federal and State water agencies have acquired rights to most of the surplus (surface) waters in the State, and they are sure they would lose the right to and therefore the control of any water they place underground by artificial recharge. As to the California Department of Water Resources, "there did not exist, and there does not exist today, any means by which the State can involve itself in ground water basin management, so we have to go to surface storage" (Teerink, 1967). And the Federal Bureau of Reclamation lines its canals with concrete where they cross the permeable natural recharge areas of ground water reservoirs, to prevent "loss" by seepage. Eight years after the California Water Plan had been approved by the public and implemented, the California Department of Water Resources reported (1970, p. 72): "Full realization of such integrated surface water - ground water system operations in areas where the ground water resource is available will require legal and legislative action, and social and political acceptance."

Because of these legal and political handicaps, conjunctive storage and use of ground water and surface water depends upon the conjunctive operations of local agencies whose dominant concern is surface water.

Excluded from authority over and responsibility for the ground water potential, Federal and State agencies have specialized in huge dams and huge reservoirs,which may have sufficient holdover storage to ameliorate the effects of one or two years of severe drought. Most of the surplus waters are instreams draining the Sierra Nevada, and most of the reservoirs are in the Central Valley, at altitudes less than 400 meters, where the summer heat and arid climate take a heavy toll in evaporation. Eventually, we may be able to verify Conkling's Law: "No matter how large the /surface/ reservoir capacity, streams of erratic and cyclic flow will yield for useful purposes no more than 50 to 60 percent of the average annual discharge, because the remainder will be lost, over the years, by evaporation from the water surface of the reservoirs necessary to impound the waters of the infrequent years of large discharge" (Conkling, 1946).

SUMMARY

In the vast subcontinental arid zones, rainless and riverless, consider the ground water as a stock resource, minable and not replenishable. If some areas do receive some replenishment and have more water than you predict, some people will be happy. In the cool coastal deserts where ground water is hard to find, make the most of the juxtaposition of cool ocean and mountains, pleasant climate and energy sources and ideas for obtaining fresh water. In arid plains and valleys that receive some inflow of water from more humid regions -whether melting snow, subsurface flow, mountain streams, or large rivers - the ground water potential should not be isolated, for it becomes a part of the larger problem and solution of the overall water-resource potential. In comprehensive water planning ground water has a potential for regulating and extending the varying supply of inflowing water, for warehousing waters in times of surplus which can be withdrawn and moved to places of need. But it can be detrimental in causing problems of drainage and salt accumulation in agricultural lands.

REFERENCES

Bain, J. S., Caves, R. E. and J. Margolis. (1966)   Northern California's Water Industry, Johns Hopkins, Baltimore.

Berry, W. L. (1962)   Ground Water in California's Future, Conf. on Ground Water Recharge, 3d, Berkeley, California, 3.

California Department of Water Resources, The California Water Plan, Bulletin, 3, 272 (1957).

California Department of Water Resources, The California Water Plan -- Outlook in 1970, Bulletin, 160-70, 72 (1970).

H. Conkling,   Utilization of Ground-Water Storage in Stream-System Development, Am. Soc. Civil Engr. Trans., 111, 279-280   (1946).

Hodges, C. N.   (1970) A Desert Seacoast Project and Its Future.   In Arid Lands in Perspective, W. G. McGinnies and B. J. Goldman, eds., U. of Arizona Press, Tucson, Arizona, 119-126.

Logan, R. F. (1970)   Geography of Central Namib Desert.   In Arid Lands in Perspective, W. G. McGinnies and B. J. Goldman, eds., U. of Arizona Press. Tucson, Arizona, 127-144.

Meigs, P.   (1970)   Future Use of Desert Seacoasts.   In Arid Lands in Perspective, W. G. McGinnies and B. J. Goldman, eds., U. of Arizona Press, Tucson, Arizona, 101-118.

Teerink, J.   (1967)   Panel Discussion with Harvey Banks.   Conf. on Ground Water Recharge, 6th, Berkeley, 34-83.

Thomas, H. E.   (1976)   Summary Appraisal of the Nation's Ground-Water Resources - California Region.   U.S. Geol. Survey Professional Paper 813-E, 51.

C. W. Thornthwaite, An Approach Toward a Rational Classification of Climate, Geog. Rev. 38, 55-94 (1948).

PLANNING OF GROUND-WATER DEVELOPING IN THE SAHEL
(WESTERN AFRICA - SOUTH OF SAHARA)

R. Biscaldi, G. Castany, J. Margat, P. Ungemach
Presented by L. Monition
Bureau de recherches geologiques et minieres, Orleans, France

INTRODUCTION

Outlines and Object of the Study

Following many hydrogeological studies performed mainly by French experts, in Western Africa over about 30 years, "le Bureau de recherches geologiques et minieres" undertook a large synthesis of all information relevant to underground resources of the Sahel - Africa South of Sahara (Western Africa). This synthesis has been made at the request of the "Fonds d'aide et de cooperation" of the French Republic and of the "Comite interafricain d'etudes hydrauliques". (C.I.E.H. - Uagadugu, Upper Volta).

The results of this synthesis are shown on 1/1 500,000 scale maps, covering an area of 3,890,000 square km between the isohyetes 100 and 1,000 mm, over 8 states: Mauretania, Senegal, Gambia, Mali, Upper Volta, Niger, Chad and Cameroon (Northern part).

The purpose of these maps is to show ready-to-use data that may serve as a basis for the planning of underground water development and more especially for the realization of regional or national agricultural development projects.

Owning to their small scale, to the value and the density of the data and to the samplified hypotheses, these synthesis maps, drawn to serve as planning tools, are neither to be used for local development nor, all the more for bore-hole siting. They express regional trends and therefore cannot provide accurate numerical information, difficult if not impossible to get, but rather give relative values (the larger the surface involved, the more significative the figures). Therefore, it will be quite necessary, when drawing up preliminary or execution projects, to refer to more detailed studies that can be found in the bibliography, or to gather complementary field data.

Data and Literature Sources

These maps are based on a review of geological, hydrodynamical and hydrogeochemical data drawn from an important bibliography essentially made of unpubllished
reports or of papers of restricted issue.   These data have been complemented by
some verbal information supplied by French hydrogeologists who worked in the
mapped regions.   It is therefore a record of knowledge pertaining to underground
waters of the region considered - up to the year 1975.

Because of the large extent of the studied area and of the diversity of the regions involved, important differences occur in the value and still more in the
density of the data used.   Therefore, in some cases, it became necessary to resort
to dash-lines to represent contours largely interpreted and to special symbols for
areas with insufficient data; some areas are even completely blank - separate
small maps have been commonly used in order to resolve some particular mapping
problems.

A computer (AFSECH programme) has been used for the data handling.

GEOLOGICAL FORMATIONS, HYDROGEOLOGICAL ZONES AND MAIN AQUIFERS

Major Geological Formations of the Mapped Area

The identification of the internal structures and of the boundaries of the major
aquifers, mainly deducted from the lithostratigraphy, is the basis of the underground water mapping.

Geological formations defined in the mapped area range from indurated rocks belonging to the old crystalline and crystalliphyllian basement up to unconsolidated
deposits of the quarternary period.   Continental sedimentation is prevailing except to the West in the Mauritania - Senegal basin.   The major geological formations allowed the delineation of about 30 ground-water regions, the limits of
which are represented by a dark line.   An identification number is assigned to
each region and reported on each map and each sketch as well as in the descriptions given in explanatory notes.

Continous Aquifers and Discontinuous Aquifers

A basic distinction is made between the continuous aquifers and the discontinuous
aquifers.

Continuous aquifers are characterized by sedimentary formations with porosity
(sand, sandstone, etc.) or fissural porosity (karstified carbonate rocks or highly
fissured rocks).   Their characteristics present some continuity in space, although
they may vary owing to the heterogeneity of the water-bearing material (lateral
and vertical variations of facies) and to changes in thickness.   The large extent
of the aquiferous system allows important storage of water.   For practical reasons
every geological formation bearing a continuous underground water body, that can
be tapped in any points, has been mapped as a continuous aquifer.

Discontinuous aquifers as opposed to the previous ones, are characterized by
geological formations made of compact rocks, little or non pervious, but locally
present a secondary permeability created either through physicochemical alteration
or through fissuration or still through fracturation.   Into this classification
fall formations belonging to the basement as well as the consolidated sedimentary

formations such as indurated sandstones, quartzites, massif limestones and dolomites.

Ground-water Regions

Two major ground-water regions can be distinguished after their tectonical structure:

- the crystalline and crystallophyllian region;

- the sedimentary basin region:

    1. Crystalline - crystallophyllian region:
       The basement outcrops in the northern and southern
       part of the mapped area and on the eastern edge of
       the Mauretania - Senegal and Tchad basins
       (Ouaddai).

       In the northern part, from West to East, it con-
       stitutes the massifs of Adrar des Ifoghas, l'Air,
       Maradi, Zinder, while in the Southern part are the
       Libero-Ivoirian and Voltaic shields.

       The basement zones are made of schists, micasch-
       ists, granites, gneiss and quartzites. The
       underground waters occurring in the altered and
       fissured formations, belong to discontinuous
       aquifers. The catchment works only yield small
       water supplies and the "exploitable" underground
       water resources are not very important.

    2. Sedimentary basin region:

       Three large sedimentary basins lie over Western
       Africa. They are filled with unconsolidated
       deposits or with rocks displaying a carbonate
       basis. From West to East the basins are:

       . Mauritania - Senegal basin,
       . Niger basin,
       . Chad basin.

       In the Northern part of Mali, only the Southern
       limit of the large basin of Taoudeni is repre-
       sented.

       Except for the deepest aquifer, which is dis-
       continuous, the main aquifers are continuous; they
       are from top to bottom:

       -   unconfined aquifer in the alluvial formations
           deposited by the Senegal and Niger rivers;

       -   unconfined aquifer in dune sands along the
           sea-shore NE of Dakar and in ergs;

- aquifer in Plio-quaternary clayey sands of the
  Chad basin (Tchad, Niger, North Cameroun).
  They are heterogeneous fluviolacustrine and
  eolian deposits containing deep unconfined
  aquifers;

- confined aquifer in lower Paleocene sands of
  the Chad basin;

- aquifer, generally unconfined, in terminal
  Continental (Post Paleocene) sands and sandstones which display large extent in all the
  basins.

The heterogeneous reservoir is made of clayey
sandstones, of more or less coarse sands and of clay;
lateral and vertical changes in facies are common.

- aquifer in Eocene dolomites and limestones of
  Senegal and sands and sandstones of Mauretania;

- aquifer in Paleocene karstic limestones of the
  Cape Verde peninsula (Senegal);

- confined aquifer in Maestrichtian sands and
  sandstones. This is the major one in Senegal,
  and extends into Mauretania;

- aquifer in lower Cretaceous sandstones of
  Tegema, Telona and Agades (Niger);

- discontinuous aquifer in Infracambrian and
  Primary sandstones and dolomites, along the
  boundaries of the sedimentary basins.

General Principles of Aquifer Mapping

The shallowest aquifer is first mapped. If it cannot be economically developed
for agricultural purposes, the reported data concern the next underlying aquifer
which fulfils the conditions. The lower aquifers, systematically of confined and
continuous type, are either mapped at a smaller scale or represented by isovalue
contour lines coloured in orange.

Maps and Explanatory Notes

Four colour maps, each one composed of three sheets, 113 x 107 cm (West, centre,
East) have been drawn.

- Map 1   : Productivity of aquifers - Initial capacity
             of catchments works;

- Map 2   : Average cost of works and of exploitation of
             underground water;

- Map 3   : Aptitude of water for irrigation (quality of
             water)

- Map 4   : Underground water resources.

Explanatory notes are attached to each map. Maps 1 to 3 will be found in the same booklet and the map 4 in a separate one. References are listed in a comprehensive bibliography.

Object of the Present Note

The main object of this note is to describe the methodology used for the mapping, and also to discuss the qualities and deficiencies of the maps. The major characteristics of the underground water will be presented briefly while special attention will be paid to water resources.

PRODUCTIVITY MAP OF THE AQUIFERS - INITIAL CAPACITY OF THE EXPLOITATION BORE-HOLES

Objective of the Map

This map gives the productivity of aquifers by providing the initial production yields of the exploitation bore-holes (m^3/d).

Definition and Calculation of Productivity

The productivity is an estimation of the yield (in m^3/d) at the beginning of the exploitation of a perfect catchment work tapping a given aquifer (borehole generally). Therefore it does not take into account the possible evolution of the yield with time which is linked to factors such as renewability or exhaustion of the reserves, interference or an ageing of catchment works.

Productivity is a function of the following factors:

- in continuous aquifers: permeability, length of the screen which is itself a function of the thickness of the aquifer. The initial exploitation yields range from 50 to some thousands of m^3/d;

- in discontinuous aquifers: importance of fracturation, type and thickness of surface formations (altered frange or alluvial deposits). The initial exploitation yields are generally low, from some m^3/d up to some 100 m^3/d.

The productivity, P, is calculated after the following formula:

$$P = q_s \times S$$

P = Productivity in m^3/d.

q_s = Specific capacity of a perfect catchment work calculated after 24 hours of pumping $m^3/d.m$.

S = Theoretical drawdown estimated after the thickness of the aquifer and by taking into account the type (confined or unconfined) of the aquifer.

In practice, the application of that formula faces some difficulties (imperfect knowledge of the aquifer or insufficient data about catchment works) or sometimes leads to figures which are technically or economically unrealistic (for example: overestimated lift height or catchment work dimensions inducing technical problems impossible to resolve or excessively high costs). Therefore some type cases have to be distinguished: they correspond to homogeneous aquifers where the hypotheses and the boundary conditions are uniform (they are given, sector by sector, in the sheet description). Three types of aquifers have been defined. Based on the hydrogeological conditions prevailing in a homogeneous area, the criteria have been selected so as to get minimal figures.

Type I   :   continuous aquifer of known thickness, H. The productivity in m^3/d is calculated after the following formula, with the admissible drawn down being estimated at one third of the total thickness of the aquifer:

$$P = q_s \times H/3$$

q_s = specific capacity of the catchment work, in m^3/d

H = thickness of the aquifer, in m.

Type II   :   complex continuous aquifer, confined or unconfined, of unknown thickness:

$$P = q_s \times S_c$$

S_c = admissible drawdown, constant over a given area, in m.

Type III :   discontinuous aquifer

The productivity is the mean of the local yield values.

Mapping Methodology

The local productivity values are extrapolated on the basis of the hydrogeological situation, technico-economical constraints being taken into account. Two ranges of colours have been used so as the continuous aquifer areas (in green and blue) are clearly differentiated from the discontinuous aquifer ones (in yellow and orange). The higher the productivity, the deeper the colour.

In the continuous aquifer zones, the ranges of productivity values overlap partically, which indicates that the boundaries are not precise. In the discontinuous aquifer zones, they represent a mean value which is also representative of areas of similar petrographic nature (schists, gneiss, etc.).

Deep aquifers are represented by orange isoproductivity curves and by an identification number of the same colour placed near the number referring to the upper aquifer mapped. Other information is given by black lines: aquifer in alluvial formations, dry zones and bevelling limits of artesian areas. Hydrogeological cross-sections, on a small scale, are attached to the sheets.

MAP OF THE AVERAGE COST OF EXPLOITATION OF UNDERGROUND-WATER

Objective of the Map

This map gives the average cost of underground water (catchment and exploitation), including investment and operating costs expressed in CFA francs [1] per cubic meter.

Definition and Estimation of the Average Cost of Underground Water

The average cost of water (in CFA francs(1)) pumped from a continuous aquifer, is the production cost of one cubic meter pumped at a level of 10 m above the ground, meaning under 1 kg of pressure; the pumping regimen being supposed with a stabilized dynamic level according to the criteria of average drawdown defined by the productivity map.

Estimation method of that average cost: it is based on the knowledge of two groups of parameters namely the economical and physical parameters of the tapped reservoirs and of the exploitation:

> ° economical parameters: they involve two categories of expenses:
>
> > - fixed investments related to redeemable expenses for structure and equipment: drilling and equipment of the well, pumping station and engine;
> >
> > - variable operating costs including mainly energy costs (motor - fuel).
>
> ° physical parameters of tapped reservoirs and exploitation:
>
> > They are the geometry of continuous aquifers, hydro-dynamical characteristics, piezometric levels, depth of holes, height of lift and productivity. They are materialized by structural maps, piezometric maps, maps of isocurves of manometric heights of lift, and productivity maps.
> >
> > Because of important differences in the density and the quality of the available data, extrapolation became necessary, especially on the boundaries of the major aquifers.
> >
> > These basic data have been discretized after a typology of holes has been defined (depth and yields) and cartographic zones (1/1 000 000 map) have been established on the basis of a uniform grid made of elementary meshes 100 km square. The processing of those data by a computer allowed cost maps to be established. The program used was the AFSECH program.

---

1 U.S. dollar = 250 CFA Francs.

° Numerous calculation tables and abacuses, given in the
  explanatory note, allow the calculation of the different
  cost parameters: yields and depths of holes from different types of catchments, costs of pumping equipment,
  costs of pumping equipment, costs of drilling and of
  equipment, costs of engines, costs of energy. Abacuses
  allow also the investigation of an optimal exploitation
  yield, with an application to the aquifer occurring in
  the Maestrichtian sands of Senegal.

° The average cost C of a cubic meter of underground water
  is calculated by using the expression

$$C = \frac{IF/n + DV}{V} = CuF + CuV$$

and integrating the economical and physical parameters.

C	=	cost of one cubic meter of underground water, in CFA francs,
IF	=	amount of investments for structure and equipments in CFA francs,
n	=	number of years necessary for the redeeming of investments;
DV	=	variable expenses in CFA francs
V	=	annual volume of underground water pumped in m^3,
CuF	=	fixed unit costs in CFA francs,
CuV	=	variable unit costs, in CFA francs.

The average cost of water varies with the productivity of the aquifers, ranging
from 50 CFA francs (low productivity) to 10 CFA francs (high productivity).

Mapping Methodology

The costs are reported on the 1/1 500,000 maps as well as on computer printed
outputs.

Zones of costs have been preferred to curves of equal values, since they allow to
homogenize these variables resulting from the numerous economical and physical
parameters taken into account.

This representation also appeared more convenient than a more classical method
consisting in reporting on each sheet, or even on each reservoir, a specific
sampling in function of the histogram of repartition of local calculated values.

The map concerns continuous aquifers, and only mean values representative of a
mesh about 100 km^2 of surface can be calculated. At such a scale, it is obvious
that the cost values are only very approximate values and not absolute figures.
Therefore they allow comparison of costs between different regions and not local
evaluations.

A range of 5 colours, from blue to yellow, has been used to differentiate the 5 of
average cost values (in CFA francs): below 10, from 11 to 20, from 21 to 30, from
31 to 50 and above 50.

In areas where underground water is not to be used for irrigation purposes, the
lithology is represented by red lines.

The map shows that on large sectors the costs are competitive; in some cases they
do not reach 10 CFA francs, and they generally range between 10 and 20 CFA francs
(aquifer in Maestrichtian sands of Senegal) small maps represent the sectors where
the aquifers overlap.

MAP OF SUITABILITY OF WATERS FOR IRRIGATION PURPOSES

Objective of the Map

This map gives the areas where the water can be used for irrigation purposes; the
quality criteria are based upon hydrogeochemical characteristics. This map provides a comparative picture of the chemical quality of underground water for
agricultural purposes.

Classification of Underground Waters According to Their
Suitability for Irrigation

These categories have been defined by applying American standards (Reverside
Laboratory) based on electrical conductivity, expressing the total mineral content
and the Sodium Absorption Ratio (SAR).

$$SAR = \frac{Na}{\sqrt{(Ca + Mg)/2}}$$

This method leads to the identification of 16 categories, which after regrouping
end up into 5 classes (Table I). The term which qualifies in each class the
degree of suitability of water for irrigation purposes, indicates in a very
schematic way the possible utilization of the water.

For the waters with a total mineral content exceeding 1 g/1 (unsuitable waters) an
extra subdivision has been made: dry residue not reaching 1 g/1 and dry residue
exceeding 1 g/1.

The basic data used are the chemical analyses of the major constituents: Ca, Mg,
Na, K, Cl, So_4, CO_3, HCO_3 and NO_3. Only the analyses (2,350) showing an ionic
balance lower or equal to 10 % have been retained. The density of data is quite
different from one country to another. These results allow the determination of
the S.A.R. and by using the empirical formula of LOGAN the estimation of the
electrical conductivity - direct measurements of conductivity are scarce.

All these data catalogued in a master card file system have undergone computeri-
zation.

Mapping Methodology

Only the continuous aquifer regions have been mapped, the other ones being represented by red lithological lines.

TABLE I - QUALITY CLASSIFICATION OF WATER FOR
IRRIGATION

Category	Quality	Classes	
1	Excellent	C1 - S1	Water suitable for irrigation of most crops on most soils
2	Good	C2 - S1 C2 - S2	Generally, water that can be used for irrigation of crops semitolerant to salt, on soils of good permeability. Major problems for crops too sensitive to sodium and for soils displaying a high ion-exchange capacity (clayey soils)
3	Permissible Nevertheless	C3 - S1 C3 - S2 C2 - S3	Generally, water suitable for irrigation of crops tolerant to salt, on well-drained soils. The evolution of salt content must be controlled. Major problems for crops too sensitive to sodium and for soils of low permeability.
4	Doubtful	C4 - S1 C4 - S2 C3 - S3	Generally, water with high salt content that can suit to come species very tolerant to salt and to very well - drained and leached soils.
5	Unsuitable	C3 - S4 C4 - S3 C4 - S4	Water generally unsuitable for the irrigation but that can be used under some conditions : very pervious soils, very well-leached soils, crops highly tolerant to salt.

The heterogeneous spatial distribution of the available data led to distinguish 3 types of regions, according to the regional density of the data: regions that can be mapped, regions that can be mapped but through important extrapolation or intrapolation, regions without available data or without resources.

A dry residue or conductivity map drawn first, greatly facilitated the elaboration of this present map. By using conjointly this map with an hydrogeological one, extrapolations in zones of insufficient hydrogeochemical data were made possible.

As a rule, the mapping of a typical region was conditioned by the availability of at least two water points, close enough to each other and with similar chemical characteristics.

Zones with aureoles, which essentially result from a purely geometrical construction, do not always apply to the factor water-chemical quality when non linear;
this is why some interpretations are necessary.

A range of colours, 5 for the Eastern sheet and 6 for the Western and central
ones, going from the blue (excellent) to orange (unsuitable) allows the representation of the water quality according to its suitability for irrigation.
Oblique lines indicate a supposed chemical quality.

A serrate line is used for the representation of superposed aquifers.

As a general rule, the ground water is of good quality for irrigation. Owing to
its low mineral content - some local pollution cases being excepted - it is also
suitable for domestic use.

MAP OF UNDERGROUND WATER RESOURCES

Objective of the Map

This map gives, for each zone, the quantities of underground water that can be
used, by differentiating those that are annually renewable by rain water. Thus
the potential ground water resources than can be exploited are pointed out. In
addition, it gives the lithology of the reservoirs, the depth of the piezometric
level and the chemical quality of the water.

Definition and Evaluation of the Potential Ground-Water
Resources That Can Be Used

These potential utilizable resources, major object of the map, expressed for one
square km, supply an order of magnitude of the quantity of water that can be
exploited. A distinction has to be made between the naturally renewable resources
(potential ones) and the "exploitable" resources.

Renewable resources, expressed in m^3/km^2 per year, represent the portion of rain
water that recharges the aquifers.

Its estimation made from average hydrological data is based on the monthly figure
of the mean efficient rainfall (see the following example, table 2).

The estimation of the aquifer recharge which defines the renewable resources is
made by re-calculating in percentage the average efficient rainfall in function of
the run-off factors controlling the catchment basin: morphology, lithology, alteration and fissuration of formations, depth of piezometric level and nature of
vegetal cover. Thus, figures of 5 % and 100 % have been obtained respectively for
granites with kaolin alteration products, and dune sands. This method of infiltration determination having a value of parametric index for an average year and
relatively large areas, is acceptable for the mapping of resources. In addition,
the indexes that have been obtained have underground undergone calibration by data
estimated through other methods: water-balance, analysis of aquifer fluctuation,
etc.

This resources estimation does not take into account any constraints (technical,
economical, ecological), so it cannot provide anything more than a rational hydrogeological basis for further calculation of renewable resources economically
exploitable.

TABLE 2 - EXAMPLE OF A MONTHLY CALCULATION (in mm)

OF EFFICIENT RAINFALL AT OUAGADOUGOU

(UPPER VOLTA)

MONTH	J	F	M	A	M	J	J	A	S	O	N	D
P	0	3	8	17	81	116	193	260	148	37	1	0
E	207	208	231	186	168	154	141	126	140	161	180	184
P - E	0	0	0	0	0	0	52	134	8	0	0	0

P   : average monthly rainfall (mm) over the 1931 - 1960 period,
          provided by the stations of "l'Association pour la Sécurité de
          la navigation aérienne" in Africa and Madagascar (ASECNA)

E   : potential monthly evapo-transpiration (mm) calculated by the
          TURC'S formula from temperature measurements, sunshine and
          relative humidity, recorded at the ASECNA stations.

P-E   : average efficient rainfall, or rainfall excess (mm/year).

      The average efficient rainfall is 194 mm/year at Ouagadougou.

The volume of exploitable resources is obtained by multiplying the average efficient infiltration by the recharge surface of the aquifer figured out by using a
planimeter.

The exploitable reserves expressed in m^3/km^2 represent the volume of water stored
in the aquifer. Supposing the aquifer is not recharged, the utilization of this
reserve would lead to its partial depletion.

The estimation of the exploitable reserves implies conventional hypotheses which
define a utilizable reserve after some technico-economical criteria (drawdown in
wells, maximal pumping depth, etc.). The exploitable volume is - storage coefficient, S, being considered - the volume that is yielded:

      - by the unconfined aquifer for a drawdown of piezometric
        level equal to one third of the saturated thickness;
        however the maximal drawdown cannot exceed 100 m under
        ground-level (economical constraints);

      - by the confined aquifer for a maximal drawdown of piezo-
        metric level reaching 100 m under ground-level (without
        dewatering of reservoir).

TABLEAU 3 - RECAPITULATORY TABLE
OF UTILIZABLE RESOURCES FOR EACH COUNTRY

Country (portion of country included in the mapped area)	Naturally renewable resources (km3/year)	Exploitable reserves (km3)	Mapped surfaces (km2)
Senegal	9,3	89 à 187	195 000
Mauretania	0,3	54 à 116	471 000
Gambia	1	9 à 19	10 000
Mali	13,1	82 à 186	807 000
Niger	4,6	262 à 577	989 000
Upper Volta	6,2	1 à 3	272 000
Tchad	20,6	263 à 514	1 050 000
Cameroun	5,4	11 à 23	96 000
TOTAL	60	770 to 1 625	3 890 000

For both types of aquifers the fixed drawdown is supposed to affect a unit surface of about 1 square km.

The table 3 has been established after the interpretation of the mapped data.

The natural renewable resources are far from negligible. For a surface of 3,890 km2 they are estimated to 60 billions of m3/year, meaning about half of the volume calculated for France, which is seven times smaller in surface. To these resources are to be added the resources that can be drawn from the reserves.

Methodology

The values of the exploitable reserves, expressed in million of m3 per year, have been grouped into 6 classes and mapped as homogeneous zones by using a range of six colours. These colours range from yellow to orange and correspond to value brackets: below 0.05 - from 0.04 to 0.10 - from 0.09 to 0.25 - from 0.20 to 0.50 - from 0.45 to 1.00 - from 0.90 to 5.

The values of the renewable resources, expressed in thousand of m3/km2 for year, are presented in brackets of 0 to 10 - 10 to 25 - 25 to 50 - 50 to 100 - 100 to 150 and over 150, related to the lithology.

The lithology of the reservoirs is represented on the map by conventional symbols.
The depth of the water under ground-level is represented for them confined
aquifers by isobath curves, of orange colour, about 10 m apart; for the areas
where the data are insufficient or inaccurate, these depths are indicated by their
local values. For the confined aquifers, the isobath curves are reported on
separate small maps with allow a comparative study of confined and unconfined
aquifers. Special symbols indicate the artesian areas. Some wells, plotted on
the map, give the local depth of the aquifer roof; the piezometric level and of
the dynamic level corresponding to the aquifer exploitation made under normal
conditions.

The chemical quality of the water is indicated by the total mineral content: two
categories of ground-water quality, one below 1 g/l and the other one over 1 g/l.

CONCLUSIONS

When drawing up preliminary or final projects of regional agricultural development, economists and planners call upon the assistance of various technical and
economical sections. The role of the hydrogeologist is to present clear and
didactic information which would be directly utilizable at that stage of the
enquiry. The set of the four presented maps together with their explanatory notes
provides answers to the questions about groundwaters: productivity of aquifers,
cost of water, quality of water, suitability of water for irrigation and quantity
of water that can be exploited.

The elaboration of these maps and of their annexes, required an original methodology involving the use of a computer. Since this methodology may be applied to
other regions, its publication appeared useful.

HEALTH HAZARDS ASSOCIATED WITH WATER DEVELOPMENT
PROJECTS IN ARID REGIONS

A. W. Nichols and F. L. Lambrecht
Department of Family and Community Medicine
University of Arizona College of Medicine

Abstract

In order to meet the need for food of populations in underdeveloped
countries of the world, large-scale plans are being developed to increase
agricultural productivity of areas often located in marginal or arid
tropical regions.

The implementation of such projects may result in dramatic changes in
the environment, especially when accompanied by large water impoundment
and irrigation systems. Such changes are usually favorable to the introduction, increase or expansion of disease agents, their vectors and animal
reservoirs.

Many insect vectors of important endemic diseases require surface
water to complete their life-cycles. Mosquitoes, for example, are potential
vectors of malaria, filariasis, encephalitides, yellow fever, and dengue,
whereas phlebotomus flies serve as potential carriers of river blindness.
Artificial lakes created by dam building in Africa have influenced the
distribution of both these vectors and simultaneously, been subjected to
invasion by the snail vectors of schistosomiasis.

In addition to the direct impact of desert-development programs upon
human health, the moving and relocation of sometimes thousands of people
and the introduction of equally large numbers of project workers into new
developing areas have created serious problems of both a physical and
psychological nature.

Whereas unmodified arid environments are highly unsuitable to the
transmission of vector-borne diseases in the absence of appropriate vectors
and animal reservoirs, development of a water supply necessary to support
permanent human settlements in these same areas will lead to new and
significant potential health hazards.

Introduction

Man is part of his natural environment and, as such, participates in
all the interactions that occur between him and surrounding organisms,
including parasites. In an unchanging environment a gradual adaptation

603

develops, eventually leading to stable relationships. Although a certain degree of cross-adaptation may be expected in long-established towns and villages, in mobile communities of humans the ultimate equilibrium is seldom established among host, disease vectors and parasites. Any changes in such environments will be followed by changes in relationships and in the consequent alteration of disease patterns. Whereas such changes unavoidably occur under normal circumstances, they are amplified and affect far more people in the event of large-scale developments as are seen in some regions of Africa today.

During the hunter-gatherers' phases of Man, his impact on the African environment was negligible, comparable to many other animal species that roamed the savannahs. The size of prehistoric human populations was kept in check by the amount of food that could be foraged and the size of territorial expanse needed to do so.

The discovery of food production, by agriculture, permitted populations to expand and increase. Once the maximum extension had been reached, however, the practice of subsistence, shifting agriculture precluded further population increase. Population growth was curbed, moreover, by high infant mortality, this due to a combination of parasitic diseases and malnutrition.

The introduction of better medical care and effective control of epidemic diseases once again promoted a substantial population increase. Unfortunately, this increased population growth could not be matched by equivalent food production, when the old system of shifting agriculture remained the major source of available food supplies. This led to the rapid depletion of soils, and resulted in recurrent famines in the marginal land areas.

International awareness of the precarious situation pertaining to population support in Africa was aroused dramatically with the disastrous drought years in the Sahel at the start of this decade. It was clear that if thousands of square miles of African soils were to be saved and millions of people were to be provided with productive farmland, an overall assessment of the affected areas had to be made and means devised to assure a well-planned approach to halt degradation and, if possible, to reverse the process of land depletion.

Water Projects

At present, arid areas are being surveyed. Wide-ranging plans are being developed to utilize soils made infertile by misuse and geography. The absence of water in those areas is being amended by the building of dams and by irrigation systems. People are being displaced, and relocated. New crops are being planted and new lives being started. Up to very recently, however, little attention was given to the implications resulting from changes in the environment and even less to the consequences regarding health.

The justification for building a dam is based upon need for economic development and, very often, upon political motivation as well. The need for urgent development becomes a single, compelling goal and ancillary problems that may be predicted, including those that are health related, may be seen as obstructionist and unwarranted criticism. Even money which may have been allocated to correct health related problems is frequently withdrawn when expenditures related to building of the dam mount to more than expected, as they often do.

Too often ignored is the fact that the health of the population is part of the economy, as much as any other energy source. Failure to attend to the health aspects of large development projects may lead to economic, if not political disaster. Health planning should be part of the overall project from its inception, as much as is the number of bulldozers required. This planning should take effect from the very first day of work and continue long after the completion of the project or until the health services are fully incorporated into the public health structure of the region being developed.

It is realized that health is only one of the many problems that faces development planners. The whole approach to development should be inter-disciplinary. Engineers, economists, anthropologists, political scientists and other professionals should coordinate their planning and activities at all stages. Any planning should also include the opinion of people who will be involved in changes in their areas.

Health Hazards

Diseases, especially communicable tropical diseases, occur in specialized environments. Some of these occur in characteristic ecological niches, due to transmission by intermediate vectors and maintenance in animal reservoirs. With only seasonal fluctuations, but in an otherwise stable human population, such diseases will occur in a recognizable fashion and eventual control may be applied with reasonable success. On the other hand, wherever profound changes are made in the environment, the "even flow" of events is upset, often resulting in harmful consequences.

Agricultural development in arid or semi-arid regions means bringing in new supplies of water, realizing such water sources may develop into a major new health hazard. This is because a great number of parasites or their vectors are dependent upon water as an essential element in their life-cycle. Areas previously unsuitable for the survival of a parasite or its vector because of absence of water, may change into a highly potential focus for parasite transmission when water is made available.

The formation of large lakes following impoundment of streams results in irrigation systems which increase tremendously not only the acreage of surface water but, more importantly, that of shoreline habitats. The lake may spread out into creeks and branches forming marshy areas and seepage swamps. Wave action can produce sandbars and pools, thereby creating diversified types of larval habitats, with a resulting larger spectrum of potential disease carriers.

The creation of artificial lakes and irrigation systems is followed by important environmental chain reactions. These will affect all elements that compose the environment, faunal, floral and mineral. The resulting changes will eventually be felt by the human populations inhabiting those areas, and, ultimately, modify social, economic and disease pattern rela-tionships.

With regard to health, large man-made lakes and irrigation-agriculture schemes often favor the transmission of communicable diseases by: (1) ex-panding the areas and types of water habitats of insect-vectors and water-borne parasites; (2) by developing vegetation patterns suitable to insect-vectors and animal reservoirs of diseases; and (3) by creating a micro-climate enhancing the potential for parasite transmission. Furthermore, the construction of dams and artificial lakes, together with large-scale agricultural development, may: (1) result in population displacement, with the attendant risk of increased communicable disease contacts; (2) increase contact with wild and domestic animals serving as disease reservoirs; and (3) result in the increase or planting of crops that would offer breeding opportunities for certain mosquito vectors. (Fig. 1 & 2)

Diseases related to peridomestic man-made environment

Bamboo poles (for yam-vines)	Bamboo grove	Water storage	Banana trees, pineapple, Dracaena sp., cocoyams

are favorable breeding sites for *Aedes* mosquitoes

(Various virus diseases)

are plants with axil-leaves that are specific breeding sites of *Aedes simpsoni*.

(Yellow Fever)

Grain storage bins attract rodents and their fleas.

(Plague)

Duckpools provide habitat for snail-vector of Schistosomiasis

Fig. 1

In the tropical regions of the world, health hazards following the construction of dams are compounded by the presence of many parasitic organisms and their vectors not encountered in the more temperate zones. The rural environment, with its lack of sanitary control, the relative absence of available intervention technology, and the economic inability to support research and control measures by local government all play a role in the health hazard process.

Specific Diseases

A large number of disease vector species spend part of their entire life cycle in an aquatic environment. These include all the (1) mosquito species (related diseases: malaria, elephantiasis, dengue, yellow fever, encephalitides and a host of other virus diseases); (2) simulium species (related disease: onchocerciasis or river blindness); and (3) snail species (related disease: schistosomiasis).

In addition to the direct relation between water and disease agents, water courses may provide circuitious conditions favorable to the trans-mission of diseases by promoting vegetation cover used as habitats and resting places for a number of disease vectors. A representative example is the riverine habitat of <u>Glossina palpalis</u>, the vector of gambian sleeping sickness. Many insect vectors are more aggressive in moist environments or at the edge between dense vegetation and open spaces (edge effect), such as is typified by riverine gallery vegetation.

Development or increase of the transmission potentials of diseases resulting from water impoundment

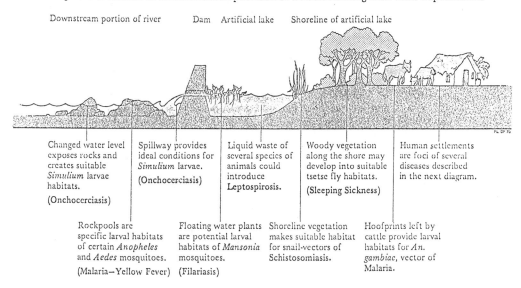

Downstream portion of river   Dam   Artificial lake   Shoreline of artificial lake

| Changed water level exposes rocks and creates suitable *Simulium* larvae habitats.

(Onchocerciasis) | Spillway provides ideal conditions for *Simulium* larvae.

(Onchocerciasis) | Liquid waste of several species of animals could introduce Leptospirosis. | Woody vegetation along the shore may develop into suitable tsetse fly habitats.

(Sleeping Sickness) | Human settlements are foci of several diseases described in the next diagram. |

| Rockpools are specific larval habitats of certain *Anopheles* and *Aedes* mosquitoes.

(Malaria—Yellow Fever) | Floating water plants are potential larval habitats of *Mansonia* mosquitoes.

(Filariasis) | Shoreline vegetation makes suitable habitat for snail-vectors of Schistosomiasis. | Hoofprints left by cattle provide larval habitats for *An. gambiae*, vector of Malaria. |

Fig. 2

The impoundment of a swift-running river, with perhaps only a few potential snail-breeding sites, into a larger lake with stable shoreline and constant water level will tremendously increase snail-breeding areas. This has been witnessed in virtually all dams built thus far in Africa (Gordon Smith, 1975). Snail-breeding is further enhanced in irrigated fields, where snails are now thriving in permanent slow-flowing canals, whereas seasonal drying used to kill them. Schistosomiasis is the most important disease of irrigation projects and is a rapidly growing problem in tropical and subtropical areas of Africa, South America, the Caribbean and Asia. (Fig. 1 & 2)

Simulium flies, vectors of the filarial disease onchocerciasis (river blindness), breed in swift-running and well aerated water required by their larvae. The damming of a river will most likely destroy the flies' breeding sites behind the dam by inundating previous sites, thus making them unsuitable for their larvae because of the decrease in oxygen content. The suitability for Simulium breeding below the dam will depend on water flow and whether the water level is constant or not. If the river level below the dam is lowered and raised sufficiently often, breeding of Simulium will be greatly discouraged. The main risks of breeding sites may be in the spillways of the dam, because the constant running of water will provide ideal conditions of flow and oxygen content. (Fig. 1 & 2)

The other important human filarial diseases, caused by Wuchereria bancrofti and Brugia malayi, are carried by several different species of mosquitoes, each with different larval habitat requirements. Impounded water will normally not be suitable for the main vector of bancroftian filariasis, Culex pipiens fatigans, a species that breeds mainly in water polluted by man and is essentially a peri-domestic mosquito. Malayan filariasis, however, is carried by Mansonia spp., which breeds in floating vegetation. The invasion of man-made and other lakes by such water plants as Eichhornia crassipes, Pistia stratiotes and Salvinia auriculata may vastly increase potential breeding areas of this vector.

In the tropics, several diseases are directly or indirectly related to the presence of surface water near human settlements or where human activity takes place. Infection may occur directly by contact or by ingestion; indirectly by providing a suitable habitat for intermediate vectors of diseases. The following is a brief outline of some diseases classified according to the above categories:·

a. Direct

    i. By ingestion: Contaminated water can be the vehicle of several infectious diseases--typhoid fever, salmonella and shigella dysenteries, amebiasis, cholera, guinea worm, and various helminths.

    ii. By contact: Infections caused by organisms that penetrate the skin and invade the human host during bathing--Schistosoma mansoni, S. haematobium, S. intercalatum and S. japonicum, and many serotypes of Leptospira.

b. Indirect

    i. Many important insect vectors of diseases spend part of their life cycle as immature stages in an aquatic environment: Mosquitoes (vectors of malaria, filariasis, yellow fever, dengue, and several other viruses); Simulium flies (vectors of onchocerciasis); Culicoides (vectors of filariasis).

    ii. Water courses are the permanent habitats of a number of snail vectors of the Schistosoma: mansoni, haematobium, intercalatum, japonicum, and fasciola. Inasmuch as the infective stages of the Schistosomes (cercariae) are water-borne, the aquatic environment thus provides both the habitat of the vector and the means of transmission. Other trematode worm infections are waterborne in other ways: Paragonimiasis (snail and crab); Clonorchiasis (snail and fish); Diphyllobothriasis (conopod and fish); Fasciolopsiasis (snail and waterplant).

    iii. Floating water plants provide suitable habitats for Mansonia mosquitoes, vectors of filariasis, the larvae of which take oxygen from the plants' submerged stems through a modified syphon adapted for piercing the plant cell-wall.

Moreover, the change from a dry area devoid of water surface into wide expanses of impoundment and irrigation may have a distinct effect on microclimates and promote the development of vegetation which may become suitable for disease vectors and reservoirs.

Relocation

Not only will the building of a dam with an irrigation system and later agricultural development have a direct bearing on local health conditions, but it will also directly effect the life and health of thousands of people that must be relocated from areas behind the dam which is filled in by the lake. For instance, in Africa alone the following population resettlements were required for the building of: Kariba Dam (Zambesi River, Rhodesia), 56,000; Volta Dam (Ghana), 70,000; Kainji Dam (Niger River, Nigeria), 41,000; Kossou Dam (Bandama River, Ivory Coast), 80,000; Aswan Dam (Nile River, Egypt and Sudan), 100,000.

For the larger majority of a population in Africa, the compulsory relocation is a traumatic experience. This is enhanced in almost all instances by lack of initial planning and subsequent breakdown of organization. The psychological stress, as described by T. Scudder for the Kariba Dam settlements, is far more important than anyone suspected. This includes anxiety, social disruption, stress of adaptation, and relationship with authorities. The stress was more pronounced among the women, who were emotionally attached to their gardens and shelters. The people were also unhappy about leaving graves and abandoning fields which had been passed down through many generations.

The physical stress of relocation is composed of many facets that include inadequate food supply during the first phases of resettlement, inadequate water supplies, land tenure problems, and changes in the nature and incidence of diseases. Very often relocations are organized in compact settlements to lower cost of transportation and social services. This results in high population density to which many rural villages are not accustomed, and promotes disease transmission. Many times transportation schedules do not work out according to plan and people loaded on lorries and buses spend days on the road instead of hours, often with the loss of chickens and other livestock, and resulting in utter confusion at arrival.

The psychological and physiological trauma of relocation is well illustrated by the death of 55 relocatees in a four months' period, mostly women and children, among people resettled at Lusitu following construction of the Kariba Dam. Gadd et al, described the occurrence as one of sudden onset with vague symptoms leading to death as early as six hours and no more than fifty-two hours, "cases" occurring in groups of the same village or same family. Several cases of poisoning followed in the next months because, although Tonga women had a profound knowledge of edible wild plants in their old areas, this was not the case in the new homes. Driven by hunger, women began to experiment with plants of which they had little knowledge.

Untested water supplies, or hastily built boreholes, may lead to the consumption of contaminated water and foods. For mixed farmers with cattle, the herd's daily water requirement posed an additional problem.

T. Scudder, in his description of the Kariba Dam resettlement scheme also mentions the socio-cultural stress which, he feels, plays a major role in the subsequent success of the program. He feels that the most important strategy is to help people "to get back on their feet" at the earliest possible date, while waiting for some time before preparing the way for more rapid development. This may be done after it is felt that the transition period has been successfully weathered.

Conclusion

In conclusion, we can say that while engineers, economists, politicians and agriculturalists build dams and fields, anthropologists and health workers must look after the people who are affected by this process and who will make the system work. It is imperative that health hazards are recognized during the planning stages of projects and that methods to reduce health risks are considered and applied. Any cost must be incorporated into overall budgetary allocations for the project and provision made for its retention in spite of coast overruns elsewhere in the project.

References

Gadd, K. G., L. C. Nixon, E. Taube & M. H. Webster (1962). Centr. Afr. J. Med., 8:491-508.

Scudder, T. (1975). In "Man-Made Lakes and Human Health," N. F. Stanley and M. P. Alpers, eds., Academic Press, New York.

Smith, C. E. G. (1975). In "Man-Made Lakes and Human Health," N. F. Stanley and M. P. Alpers (eds.), Academic Press, New York.

AN OVERVIEW OF DESALINATION SYSTEMS FOR BRACKISH WATER
AND THEIR APPLICABILITY TO DESERT CONDITIONS

K. C. Channabasappa
U.S. Dept. of the Interior

ABSTRACT

The need to provide food and fiber for the expected population growth has created
an unprecedented demand for expansion of the land areas that are suitable for
human settlement and agriculture. Unfortunately, most of the land areas of our
planet that are climatically suited for human habitat have already been occupied
and are presently overcrowded. Arid regions appear to be the only available land
areas for expansion. One-quarter of our planet is arid. In recent years, these
have become the most vital areas of the world due to the vast mineral resources
contained in them which are urgently needed to improve the socio-economic standards of the world's population. Two basic requirements essential to the development of these mineral resources are: people and water. Both are in short supply
in desert areas. While it may be possible to move people to deserts through
economic or other incentives, it is economically prohibitive to transport water
from other parts of the world.

In the arid areas, there are no dependable surface water supplies available in
desert regions to support human habitation required to manage, mine, transport and
convert natural resources to useful products. It is, therefore, important to make
the most efficient use of limited quantities of brackish and other saline waters
locally available which in their natural state are unfit for human consumption and
industry. Desalination of these sources to fresh water quality is required. Of
the many desalination techniques that are in commercial practice, membrane processes appear to offer the best promise for use in arid regions. The unique
advantages of these methods are: (1) low energy requirement, (2) ease of plant
construction and operation, and (3) operability with any source of energy - oil,
coal, solar, wind, geothermal.

Since vast areas of arid regions do not contain oil, energy sources to provide
water for human support must be based on the two most abundant sources of energy
available in these regions - sun and wind. Recognizing the importance of energy
sources to develop desert areas for human habitation and agriculture, a concentrated effort is being made to develop an economical and practical technology for
harnessing wind and solar energy. Results to-date indicate that solar and wind
energy could be harnessed at reasonable costs.

611

The technical feasibility and economics of coupling solar and wind energy power producing devices with membrane desalination plants has been extensively studied. Results indicate that it is possible to convert the brackish and other saline sources of the desert regions to fresh quality waters at a cost of 70¢ to $1.00/ 1,000 gallons.

INTRODUCTION

Nearly one quarter of our planet is classified as desert. While in the past, very scant attention was given to develop these areas for human habitation, the discovery of vast mineral resources, including oil, phosphate, uranium, etc. over the past 25 years had made it mandatory to develop these areas suitable for human living. The problem is of major importance for the future survival of mankind. Without the mineral resources of these deserts, civilizations as we know today, will gradually recede back to the "dark ages."

Another importance of the desert areas is that they are the only places still available for settlement of the projected population growth. The worlds' population which is now approaching 4 billion mark is estimated to grow at an annual rate of 1.9%. At this rate, world's population will exceed 6 billion within the next 25 years and may be nearer 8 billion. To support this population, a two-fold increase in food and fiber production will be needed. Desert areas offer the best promise for expansion of agriculture and human settlement.

A primary requirement for proper development of the desert areas is adequate supplies of water to support agriculture, industry, and population required to mine, manage and transport the natural resources contained in them. Because of the very limited rainfall, they have little or no dependable water supplies. Long distance transportation of water by sea or inland is economically prohibitive and may be politically unwise. The only alternative is to efficiently utilize the limited surface and ground water resources available in the region. Unfortunately, in many instances, these sources are highly saline and cannot be used directly for human consumption and industry without prior treatment by desalination techniques to convert them to fresh water quality supplies.

Large-scale conversion of brackish or sea waters to potable quality requires considerable amounts of energy. While some desert areas, such as Middle East, do contain large deposits of oil, a great majority contain little or no oil. However, they are blessed with two other energy sources in unlimited quantities - sun and wind. In some deserts, geothermal energy resources are also available. Any desalination technique selected for use in desert areas must be readily adaptable to use of solar, wind, and geothermal energy resources.

Development of desalination techniques has been in progress in many nations over the past 20 years. As a result of these efforts, a number of techniques have been developed for practical application. The most advanced techniques suitable for application in desert areas are membrane processes (reverse osmosis and electro-dialysis) since they can be operated where required even with solar, geothermal, and wind energy. A brief description of these processes, their status including economics, and their applicability to desert conditions is discussed in the following pages.

Membrane Processes

Membrane processes are rapidly becoming the preferred desalination methods for brackish water desalination. Major advantages of membrane processes are:

1. Ambient temperature operation
2. Requires one-third to one-half energy compared to distillation
3. No need for expensive metallic components
4. No thermal pollution
5. Amenable for operation with any source of energy - electricity, solar, wind, geothermal, etc.
6. Flexibility to obtain desired quality water
7. No restriction on plant siting
8. Flexibility in production rate to meet varying demands
9. Spare parts can be manufactured and assembled locally
10. Component manufacturing techniques are simple and capital cost of manufacturing equipment is relatively low
11. Plant capital costs are low compared to distillation
12. Plants are easy to install, maintain, and operate
13. Developing countries may be able to become familiarized in the component manufacturing techniques within a short period of time

These advantages have provided the greatest incentive for membrane processes to become "work horses" for brackish water desalination. For example, at the end of 1971, there were only 94 membrane desalting plants of 25,000 gpd or larger with a combined capacity of 23 million gallons per day (mgd) in operation (1), while during the three-year period between 1972-1974, over 370 new membrane plants with a total capacity of 77 mgd were constructed (2). Since 1974, several large plants with an estimated additional desalting capacity of about 150 mgd have been either placed in operation or under construction; the largest plant under construction is in Riyadh, Saudi Arabia, which will have a desalting capacity of 31 mgd.

The two membrane processes that are in commercial practice are: (1) electrodialysis (ED), and (2) reverse osmosis (RO). The ED process has been in commercial use for over 20 years, while RO has become commercial only during the past 5-7 years. Despite its recent entry into the commercial desalination field, acceptance of RO membrane plants as brackish water desalination devices is rapidly gaining momentum. Both RO and ED processes are based on the use of special membranes to achieve solute-solvent separation in saline waters. In the case of RO, fresh water diffuses through the membrane leaving the salt behind while in ED, demineralization of saline solutions takes place by the passage of salt through the membrane. The driving force employed to cause solute-solvent separation in RO is hydraulic pressure. In ED, electric current acts as the driving force.

The principles of RO and ED processes are illustrated in Figures 1 and 2.

MEMBRANES

Though the principle of utilizing reverse osmosis for desalination of saline solutions had been demonstrated in the laboratory at the University of Florida as early as 1954, its application to commercial desalination was hindered by the

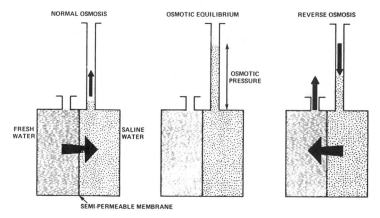

PRINCIPLE OF REVERSE OSMOSIS

FIGURE 1

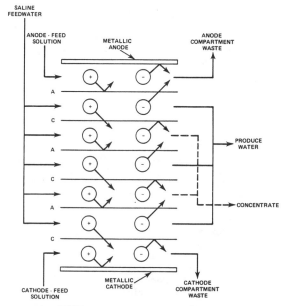

ELECTRODIALYSIS STACK SCHEMATIC

FIGURE 2

non-availability of an RO membrane that had reasonable water fluxes. This difficulty was overcome in 1960 by Dr. Loeb and Dr. Sourirajan through their pioneering research work at the University of California, which culminated in the development of a RO membrane from cellulose diacetate, which had water fluxes of 5 to 8
gallons per square foot per day (gfd) and up to 90% salt rejection when tested

with a 5000 per sodium chloride solution at 800 psi. Since 1960, significant advances have been made in the brackish water RO membrane technology. In addition to improving fluxes of the cellulose acetate flat films by as much as two to three times compared to Loeb's membrane, a family of new membranes have been developed. These include hollow fine fibers prepared from cellulose triacetate and polyamide polymer systems (3).

Recent developments in RO membranes include a new family of membranes designated as "ultra-thin composites" prepared from sulfonated furfuryl alcohol, and poly-ethylene oxide isophthalamide copolymer (4).

The ED membranes are prepared by copolymerization of styrene and divinyl benzene followed by chemical treatment to cause cation and anion exchange properties. Recent developments include ED membranes and associated components for high temperature operation up to 180° F. Since many of the ground waters in the desert regions are generally above 120° F, operation of ED equipment at high temperatures will be required. The membranes will also permit demineralization of geothermal brines. High temperature operation has several benefits on the electrodialysis process. With increasing temperature, membrane resistance and fluid velocity are lowered and the rate of salt diffusion is increased. In process economic terms, this would result in considerable savings of both capital and operating costs.

Membrane Equipment

Several methods of packaging reverse osmosis and electrodialysis membranes have been explored over the past 10-15 years. Of these, only three configurations in RO and two in ED have been advanced to commercial application. These include: spiral wound, tubular, and hollow fine fiber in RO and tortuous and sheet flow designs in ED.

A schematic drawing of the spiral wound reverse osmosis module is shown in Figure 3. The module essentially consists of a number of membrane envelopes, each having two layers of membrane separated by a porous, incompressible backing material. These envelopes together with brine side spacer screens are wound around a water collection tube. The pressurized brine flows axially along the side of spacer screen – pure water flows through the membrane into the porous backing material and then to the central product collection tube.

The tubular design combines two functions in one, in that it uses the surface of the tube as a support for the membrane and the tube wall as a pressure vessel. Normally, the membrane is placed on the inner wall of the tube, and the saline water, under pressure, flows inside the tube. Product water passes through the membrane to the tube wall, where arrangements are made to transfer the product water, now at low pressure, to the outside of the tube. Figure 4 illustrates the operation of a tubular reverse osmosis system.

Modern technology has made possible the preparation of reverse osmosis membranes in the form of fibers. The fibers are hollow and range in diameter from 50 to 200 microns (approximately 0.002 to 0.01 inch). Since they can withstand very high pressures, the fibers function both as desalination barriers and as pressure containers. In an operating hollow fiber reverse osmosis unit, the fibers are placed in a pressure vessel with one end sealed and the other end open to a pro-duct water manifold. The salt water, under pressure, flows on the outside of the fibers, and the product water flows inside the fibers to the open end where it is collected outside the vessel as shown in Figure 5.

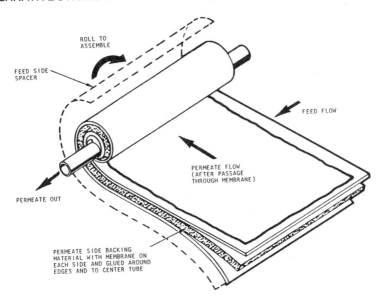

SPIRAL WOUND REVERSE OSMOSIS

FIGURE 3

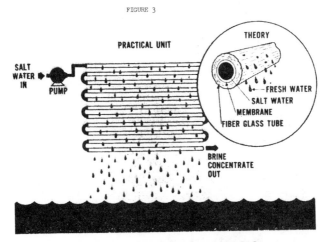

TUBULAR REVERSE OSMOSIS

FIGURE 4

While each of the above described RO membrane configurations have a unique advantage of their own, economic considerations have favored construction of reverse osmosis plants based on spiral wound and hollow fine fiber designs. Except for specialized applications such as industrial wastes treatment, tubular design RO plants are not much favored because of high capital cost.

The major differences in the electrodialysis equipment designs is in the length of fluid flow path and thickness of the plastic spacer that separate the membranes. A sinuous, tortuous flow path is used in one commercial stack design. The total path length of this sinuous design is four to five times greater than a sheet flow

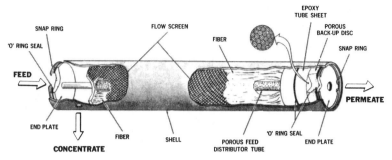

CUT AWAY DRAWING OF PERMASEP˙ PERMEATOR

HOLLOW FINE FIBER REVERSE OSMOSIS

FIGURE 5

stack which is the other hydraulic configuration commonly used. In the tortuous path design, thicker spacers up to 40 mils are used, while in the sheet flow design, spacers are approximately 20 to 30 mils in thickness (5).

The extent of pretreatment required is also related to membrane equipment design. Hollow fine fiber systems require the most, and tubular and ED plants with polarity reversal, the least. In general, the following pretreatment schemes are generally employed:

1. Surface waters
   Chlorination-Coagulation-Sedimentation-Sand filtration

2. High hardness waters
   Lime or lime-soda softening-Sand filtration

3. Low hardness waters
   Sand and maganese-Zeolite filtration

If organics are present in excess concentrations, activated carbon is used as an additional filter. In addition to the above, the feed water is treated with sulfuric acid for pH adjustment, when required, and sodium hexameta-phosphate to minimize $CaSO_4$, iron, and manganese hydroxides from forming scale deposits.

Low-Pressure RO Membranes

The commercial brackish water RO plants operate at 400-500 psi. Desalination of a 3100 ppm TDS brackish water to 240 ppm, at 70% product recovery requires about 8-10 kWh/1,000 gallons of product water. Asssuming an energy cost of 2¢/kWh, this amounts to 20¢ which is about 25-30 percent of total desalting cost. Because of the soaring fossil fuel costs, in the coming years, the energy costs in RO plants may amount to as much as 50% of the total operating cost. Recognizing the need to reduce energy costs in RO desalination plants, a new family of low pressure membranes that have desalination performances (at pressures of 250 psi and below) approximately equivalent or better than those exhibited by present day commercial membranes (400-500 psi) are being developed for commercial application. Successful development of these low pressure membranes for commercial application may reduce the energy cost in RO plants by about 40%.

Dry RO Membranes

To maintain desalination properties of commercial membranes, it is critical to keep them in a wet state at all times, including during shipping, storage, and plant shutdown. If allowed to dry, they tend to lose their solvent or plasticizer component by evaporation and become brittle. Because of this, commercial membranes are shipped in containers with water and are stored in a temperature and humidity controlled room until ready to use.

Process Improvements

Over the past 7 years, significant advances have been made in reducing the capital and operating costs of brackish water desalination in both RO and ED plants. Of major importance are development of: (1) large size, high desalting capacity RO and ED modules; (2) development of membranes that provide significant savings of energy; (3) development of techniques and systems suitable for pretreatment of brackish and sea waters; and (4) optimization of design and operating parameters to ensure trouble-free operation of the plants with minimum labor.

RO Membrane Modules and ED Stacks

Prior to 1970, membrane equipment was generally small and their maximum product water capacity was less than 1000 gpd. Over the past 5 years, U.S. industry, with financial support from the Federal Government, has developed manufacturing and quality control techniques to fabricate large RO membrane modules and ED stacks with product water capacities up to 200,000 gpd. Figures 6 and 7 show the photographs of 2" dia, 4" dia, 8" dia, and 12" dia spiral wound membrane elements and a large electrodialysis stack, respectively.

Pretreatment

An inherent disadvantage of any membrane process used in solute-solvent separation is the tendency of membranes to become fouled with any particulate or collodial matter present in the feed solution. In membrane processes, the particulates form a very thin layer on the membrane surface, thereby preventing direct contact of the saline solution with the desalination barrier. This reduces the diffusion rates of fresh water (in RO) or salt (in ED) through the membranes. In addition to particulate matter, a number of salts ($CaSO_4$, $CaCO_3$, Silica, etc.) present in saline solutions reach supersaturation levels during desalination process and precipitate out as scale deposits on membrane surfaces. To maintain the plant desalting capacity at a fairly constant rate, it is critical to maintain the membrane surfaces fairly free of any deposits. This can be accomplished in two ways. First, the membranes can be chemically cleaned as frequently as required. This results in excessive plant down time, and higher plant operating costs contributed by cleaning chemicals and additional labor required for this purpose. A second approach is to pretreat the saline water prior to its use as a feed to membrane plant for removal of fouling and scale-forming constituents.

Because of the compactness and versatility to treat any type of water, ranging from swamp waters to radiologically contaminated waters, the U.S. Army has

SCALE-UP OF SPIRAL WOUND REVERSE OSMOSIS ELEMENTS

FIGURE 6

selected RO hardware as a major water treatment device for field use by U.S.
troops (6). The specifications call for self-contained, packaged system, designed
for air and land transport, responsive to quick start and stop operation and capable of producing potable water from sources that are close to the troops location. Commercial membranes are unsuitable for this purpose, since the equipment
use is sporadic and membrane modules will have to be stored in wet state for
long periods. To be responsive to U.S. Army requirements, membranes that could be
shipped and stored in dry state at relatively high temperatures prevailing in
desert areas without loss of their functional properties have been developed.

Yuma Membrane Desalting Plant

On August 30, 1973, representatives of the U.S. and Mexico signed a joint agreement which calls for the U.S. to deliver 1,360,000 acre-feet of Colorado River
water to Mexico at the Morelos Dam. The salinity of the water shall not exceed
115 ppm ± 30 ppm over the annual average salinity of Colorado River waters arriving at the Imperial Dam (7). To meet terms of this agreement, the U.S. Congress
has authorized construction of a membrane desalting plant at Yuma, Arizona, to
desalt irrigation drainage flows of 3,100 ppm TDS flowing through the Wellton-
Mohawk irrigation district. A proposed scheme for meeting the treaty requirements
are shown in Figure 8.

The plant will be approximately 100 mgd in size and may include both RO and ED
plants. The plant will be designed for 70%-75% product water recovery. The plans
call for completion of plant construction by the end of 1980 with operation starting in 1981. When built, this will represent the world's largest brackish water
membrane desalting plant. At present, field testing of both RO and ED equipment
on a pilot plant scale is being conducted to obtain plant design data. Typical
performance of RO and ED equipment on a 3100 ppm brackish water is given in Tables
1 and 2.

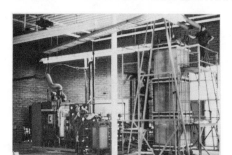

FIGURE 7   LARGE PROTOTYPE ELECTRODIALYSIS STACK
            APPROXIMATELY 200,000 GPD CAPACITY

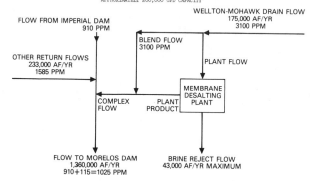

**FLOW DIAGRAM SHOWING PROPOSED SCHEME
FOR MEETING U.S.-MEXICO AGREEMENT**

FIGURE 8

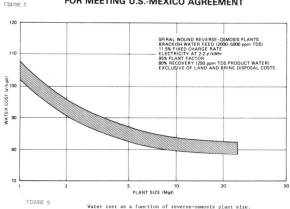

FIGURE 9   Water cost as a function of reverse-osmosis plant size.

Brackish Water Desalting Costs

It is difficult to provide precise brackish water desalting costs. This is primarily due to wide variations in the quality of brackish water available for
desalting. In addition, local factors such as costs of energy, labor, interest
rates, distance between plant location and brackish water source and environmental
regulations on brine disposal have a direct influence on costs. Other factors
that influence desalting costs are plant size and percent product water recovery.
Generalized capital and operating costs curves for RO and ED plants are given in
Figures 9 and 10. The costs are inclusive of membrane modules, piping, instru-

TABLE 1

TYPICAL PERFORMANCE OF RO BRACKISH WATER DESALINATION

YUMA DESALTING TEST FACILITY, ARIZONA

Constituent	Feed mg/l	Overall Product mg/l	Brine mg/l
Calcium	108	0.7	362
Magnesium	90.8	0.5	397
Sodium	951	47.0	2990
Potassium	-	-	-
Iron, total	N.D.	N.D.	N.D.
Sulfate	950	N.D.	3080
Chloride	1244	74.4	4000
Bicarbonate	10.7	6.3	12.7
Total Dissolved Solids	3396	129.0	10843

Physical Properties

pH	5.5	5.2	5.4
Specific Conductance	5605	270.9	16153
Temperature, °C	24.5	28.5	-

Operating Conditions

Product Water Recovery - 69.9%
Feed Pressure - 405 psig
Brine Pressure - 350 psig

Pretreatment - Lime Softening and dual media filtration

TABLE 2

TYPICAL PERFORMANCE OF ED BRACKISH WATER DESALINATION

YUMA DESALTING TEST FACILITY, ARIZONA

Constituent	Feed	Product (Diluate)	Reject
pH	9.00	8.4	7.11
Conductivity,	5192	595	9585
	mg/l	mg/l	mg/l
Calcium	112	12.2	429
Magnesium	78.6	9.9	270
Sodium	921	178.3	2635
Potassium	8.0	1.0	26.4
Iron, total	N.D.	N.D.	N.D.
Sulfate	900	214	2600
Chloride	1128	161	3550
Bicarbonate	26.8	18.5	68.3
Total Dissolved Solids	3182	595	9585

Stage Conditions

Stage 1 current, amps	39	Stage 1 voltage, volts	180
Stage 2 current, amps	18	Stage 2 voltage, volts	125
Stage 3 current, amps	14	Stage 3 voltage, volts	120

Operating Conditions

Reject makeup flow, gpm   8.4
Diluate flow, gpm   22.8
Product Water Recovery, % 63.7
Temperature   20.0° C
Pretreatment — Lime Softening and dual media filtration

mentation, pretreatment, etc., but do not include costs of brackish water pumping and piping to membrane plant, product water distribution and brine disposal (8).

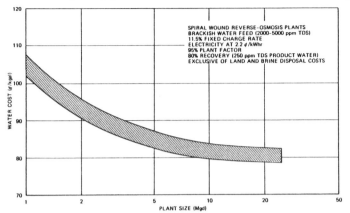

Water cost as a function of reverse-osmosis plant size.

FIGURE 9

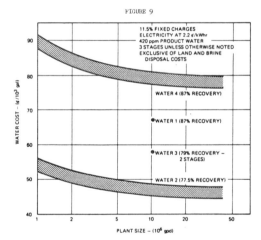

Water cost as a function of feedwater composition and electrolysis plant size.

FIGURE 10

The cost estimates are based on a service life of 3 years for RO membranes and 5 and 7 years for anion and cation electrodialysis membranes, respectively. Pretreatment of feed water was assumed to be sand filtration, injection of 5 to 10 ppm of sodium hexametaphosphate for scale control and adjustment to pH of 6 using sulfuric acid. Since the cost of brackish water desalination by electrodialysis is very sensitive to the composition of feed water, cost estimates for four different quality brackish waters (Table 3) are shown in Figure 10.

Advancement of recent developments to commercial practice such as low pressure membranes, large desalting capacity RO and ED equipment, high flux and high salt rejection membranes, is expected to reduce the brackish water desalting costs by as much as 30% from present levels.

TABLE 3

CHEMICAL COMPOSITIONS OF FOUR TYPICAL BRACKISH
WATERS USED FOR ELECTRODIALYSIS PLANT COST
ESTIMATES

Chemical composition (ppm)	Brackish waters			
	No. 1	No. 2	No. 3	No. 4
Sodium (Na)	886	125	630	900
Calcium (Ca)	118	316	116	250
Magnesium (Mg)	72	69	15	70
Chloride (Cl)	131	67	1,054	1,450
Sulfate (SO_4)	1,943	900	115	590
Bicarbonate (HCO_3)	473	357	78	210
Hardness as $CaCO_3$	590	1,073	354	912
Manganese (Mn)	1	0.10	Nil	0.1
Fluoride (F)			2	
Iron (Fe)	2	1.0	0	0.4
Potassium (K)	16	13	0	5
Nitrate (NO_3)	6.3	19	9	1
Silicate (SiO_3)			17	
Total dissolved solids	3,648	1,800	2,076	3,475
pH	7.6	7.9	8.1	7.3
Temperature, °F	70°	70°	70°	70°
Organics (Chemical oxygen demand)	10	7.9		7

ADAPTATION TO DESERT CONDITIONS

Any desalination technique selected for brackish water desalination in desert
areas must meet certain criteria to be economical and practical. These include:
(1) flexibility for use of any available energy source; (2) short period of plant
construction; (3) ease of operation; (4) high reliability; and (5) feasibility of
manufacturing major components with local labor. As discussed before, membrane
techniques meet all the above criteria and, therefore, are uniquely suited for use
in desert areas.

Use of Solar Energy

Use of solar energy for saline water conversion is not a new idea. In several
parts of the world, solar distillation plants are presently in operation providing
drinking water supplies to the inhabitants of many islands. Typical example
includes islands of the coast of Greece. However, solar energy has never been

used for large-scale conversion of sea or brackish waters. In 1974, U.S. Department of the Interior made a study to determine the technical and economical feasibility of using a solar water pumping system for operation of a 100 mgd brackish
water membrane plant that will be built in Yuma, Arizona to desalt agricultural
wastes. The study considered the technical and economic characteristics of all
applicable solar thermal power systems, solar energy collectors performance as a
function of solar flux and angle, weather conditions, collector orientation ranging from fixed position to various tilt angles and other pertinent component
characteristics. A small prototype solar collector was also fabricated by Arthur
D. Little Inc. of Cambridge, Massachusetts, and field tested at the planned plant
location in Yuma, Arizona (9). A preliminary analysis of several alternate storage methods was also made to determine if storage could be economically used to
allow system operation through the local utility peak periods. A schematic diagram of the solar powered water pumping system studied is shown in Figure 11.
Results of this study are summarized below:

1. Flat plate collectors can utilize diffuse as well as
   direct radiation making such systems suitable for
   providing energy to membrane plants in a wide range
   of geographical areas and climatic conditions.

2. For pumping applications, organic Rankine cycle
   engines can operate with high efficiency at moderate
   boiler temperature levels using a "once through"
   water-cooled condenser.

3. The power system can be assembled for components
   which are either already developed or in the advanced
   stage of development. This would permit near term
   utilization of these systems in desert areas.

4. The flat plate collectors can be mounted in a fixed
   position resulting in relatively simple mounting
   structures and low maintenance requirements.

5. Power system performance tends to optimize at relatively low power levels. This makes it possible to
   assemble systems with a range of power output levels
   from smaller power modules of standard design.

6. The system as shown in Figure 11 can also be used to
   pump water for irrigation and drainage systems.

7. The optimum economic solar thermal power module size
   is about 2 MW. Power costs in such a module range
   from 3.8¢ to 7.7¢/kWh depending upon the use of
   collectors array with or without reflectors, the
   location and prevailing labor cost. Such a system
   could be built at a cost of 1.4 to 1.6 million
   dollars (U.S.) as shown in Table 4.

8. Since the temperature of flat plate collector solar
   thermal module is sufficiently low, hot water thermal
   storage could be used.

The above findings clearly indicate that the solar energy technology is fairly
well established and could be harnessed as a power source over a wide range of
geographical areas and climatic conditions. Compared to climatic conditions

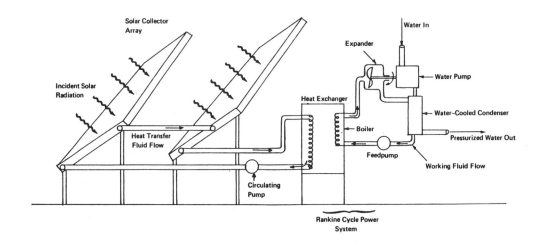

SOLAR-POWERED WATER-PUMPING SYSTEM

prevailing in Yuma, Arizona, the temperatures in desert areas are much higher all year around and there are no constraints with regard to availability of land. More advanced concepts of solar thermal power generation concepts are also being investigated. These include: (1) "power tower" concept where a large mirror field is used to direct solar energy from a large area to a central receiving cavity placed on a tower; (2) solar powered energy systems using parabolic through solar collectors that provide heat at 500° - 600°F to organic turbines. Both these concepts when fully developed will provide an efficient and economical method of harnessing solar energy which could be used for large scale conversion of brackish waters in desert areas.

Wind Energy

The energy crunch resulting from rapidly depleting oil and natural gas reserves in U.S. coupled with the four-fold increase in oil prices by the OPEC countries has promoted intensive studies by the Energy Research and Development Administration of the U.S. Government to develop economical methods of harnessing wind energy. The available wind power in the U.S. is shown in Figure 12. The corresponding annual average potential electric power is shown as the watts per square meter.

Several types of wind energy collectors including horizontal and vertical axis turbines that are capable of producing any where from 100 kW to several megawatts are commercially available or under development (10). The projected power costs as a function of power output, (in kW) for various wind velocities are shown in Figure 13, and the estimated monthly power outputs of some commercial wind gener-

TABLE 4

ESTIMATED POWER COSTS USING A SOLAR ENERGY SYSTEM
GENERATING ELECTRICITY FOR USE IN THE
DESALINATION PLANT

	Collector Array Without Reflectors	Collector Array Using Reflectors
a) Power Module Cost:		
Collector Panels	$ 975,000	$ 700,000
Reflectors	–	88,000
Collector Support	64,500	46,000
Feeder Line Piping	44,700	32,000
Rankine Cycle Engine	390,000	390,000
Generator, Switchgear and Power Lines	127,000	127,000
	$1,601,200	$1,383,000
b) Annual Power Produced (effective)	3,340,000 Kw-hr	3,360,000 Kw-hr
c) Cost of Power:		
Capital Costs	Power Cost (¢/Kw-hr)*	
8%	4.3	3.8
10%	5.3	4.6
15%	7.7	6.7

*Including operational costs of 0.5 ¢/Kw-hr

ators in kW/hr. are given in Table 5. Higher capacity large wind turbines with
power ratings of 500 kW and 1500 kW are presently under development. Their target
performance characteristics and cost estimates are given in Table 6.   The values
presented in the table indicate that wind turbines could be economically used to
produce power at costs as low as 1.7¢ - 2.0¢/kW in a 1500 kW capacity wind energy
system. The manner of coupling wind energy system to ED and RO brackish water
desalination plants is shown in Figures 14 and 15.   The estimated capital and
operating costs for a one mgd electrodialysis plant desalting a 2500 ppm TDS
brackish water feed and operated on wind energy are given in Table 7.

In actual desert areas, where the wind velocities are relatively high, power could
be produced at even lower costs resulting in further reduction of the brackish
water desalination costs.

The foregoing discussions on the status of solar and wind energy systems clearly
indicate that the technology for harnessing wind and solar energy on a reasonable

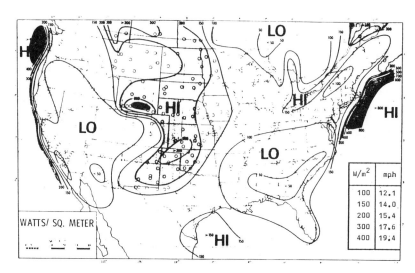

W/m²	mph
100	12.1
150	14.0
200	15.4
300	17.6
400	19.4

WATTS/ SQ. METER

AVAILABLE WIND POWER - ANNUAL AVERAGE

FIGURE 12

PROJECTED POWER COST FROM HAWT

WIND ENERGY SYSTEMS

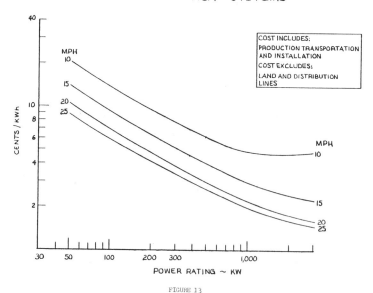

COST INCLUDES:
PRODUCTION TRANSPORTATION
AND INSTALLATION
COST EXCLUDES:
LAND AND DISTRIBUTION
LINES

FIGURE 13

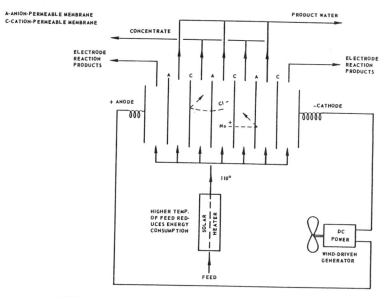

CONVENTIONAL ELECTRODIALYSIS WITH WIND POWER

FIGURE 14

scale is fairly well at hand and these energy sources could be readily used for operation of brackish water desalination plants in areas where conventional energy sources are in short supply.

CONCLUSIONS

At no time in history, the human race has been presented with a challenge of the magnitude and complexity as we are facing today. The inhabitable areas of our planet are becoming overcrowded and known reserves of mineral resources are rapidly dwindling. Since we have no new frontiers to conquer, we must make the most efficient use of all the available land and natural resources to support both the present and projected population growth. The world's population is expected to double to 8 billion in the next 25-30 years and this means we will have to double or triple the world-wide food and fiber production to meet the minimum demands of projected population.

Global climatic changes are another serious threat to human survival. While their effect was less obvious in the past, they have caused serious impact on human lives. The six west African countries south of Sahara, known as Sahel have become the victims of recent climatic change. The failure of the African monsoon beginning in 1968 has driven these countries to the edge of economic and political ruin. Recent drought conditions in western states of the U.S. and in England are other examples of climatic changes. Climatologists predict even more serious climatic changes in the coming decades.

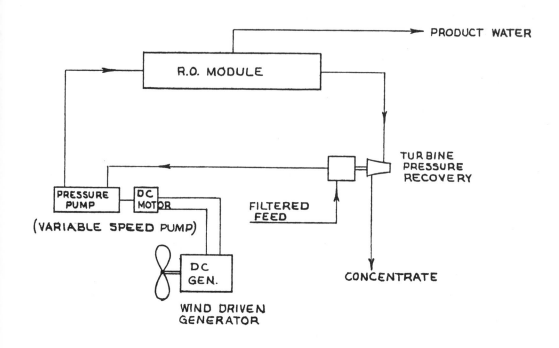

CONVENTIONAL REVERSE OSMOSIS WITH WIND POWER

FIGURE 15

To provide a better world for future generations, we must take steps to bring more areas under cultivation. The only vast areas that are available for expansion are deserts, which cover a quarter of our planet.

While nature has "stored" some of the most valuable mineral resources useful to mankind such as oil, phosphate, uranium, etc. in deserts, they are not blessed with adequate supplies of water, a basic requirement for human settlement and for food and fiber production. Because of the limited rainfall, they have little or no dependable surface water supplies. Long distance transportation by sea or inland is economically prohibitive and is politically unwise. The one and only alternative is to efficiently utilize the limited surface and ground water resources available in the areas. Unfortunately, these sources are highly saline and have to be treated with desalination techniques to prepare them for human consumption and industry.

Desalination methods based on membrane processes appear to offer the best promise for conversion of saline water resources to fresh quality water supplies in deserts. The basic advantage of these techniques is that they are adaptable for use

TABLE 5

Estimated Monthly Power Output

of Selected Wind Generators in Kw-hr

| | Rated Output (Watts) | Average Windspeed (mph) | | |
		8	12	16
Windcharger	200	10-14	19-26	26-36
Sencenbaugh	1000	70-110	130-220	190-300
Dunlite	2000	80-125	160-250	235-370
Electro	6000	200-310	400-620	600-930
Aerowatt	(24	8	11	15
	(130	45	55	80

TABLE 6

Anticipated Performance

Characteristics of

500 and 1500 Kw MOD-1 Wind Turbines

| | 5 0 0 K w | | 1 5 0 0   K w | |
	General Electric	Kaman	General Electric	Kaman
Mean Wind Speed	12	12	18	18
Rated Power, Kw	500	500	1500	1500
Rated Wind Speed, mph	16.3	20.5	22.5	25
Energy Captured, Kw-hr/yr	1.88×10^6	1.3×10^6	6.6×10^6	5.7×10^6
Rotor diameter, feet	183	150	190	180
Rotor Solidity, percent	3	3	3	3
Rotor Speed, rpm	29	32.3	40	34.4
Energy Cost, cents/Kw-hr*	4.2	5.5	1.7	2.0
Capital Cost, dollars/Kw	974	901	449	481
Wind Turbine Cost, dollars	486,000	450,670	674,000	720,800
Plant Factor	0.42	0.29	0.51	0.43

*Assumed 16 percent of capital cost for interest, operations, maintenance, taxes, etc.

of any energy source—oil, natural gas, solar, wind, and geothermal. While some desert areas of Middle East contain vast reserves of oil, all arid regions are blessed with abundance of sun and wind which can serve as energy sources for water resources development. Additional advantages of membrane techniques include: (1) low-energy requirements, (2) short plant construction period, and (3) ease of operation.

The harnessing of solar and wind energy has always been a challenge to mankind. Because of our rapidly depleting fossil fuel energy resources, a concentrated effort is being made by many nations to develop these energy resources for prac-

TABLE 7

Estimated Desalting Costs for 1 mgd

Electrodialysis/Wind Energy Brackish Water Desalination Plant

Ground Rules

Capacity = 1 mgd, Capital Cost $700,000, Feed 2500 ppm TDS
Product Water, 500 ppm TDS, 350 day operation/year
Production, 340,000,000 gallons/year
Average Daily Production, 930,000 gallons

Energy Source - Two 500 Kw G.E. Wind Turbine
Estimated Amount of Energy Produced - 3.76×10^6 Kw/hr/yr.
Cost of Energy - 4.2¢/Kw/hr.
Plant Service Life - 20 years
Interest Rate - 8%

Item	¢/1000 gallons of desalted water
Amortization	12
Energy	40
Operation and Maintenance	10
Contingencies	8
Total:	70

tical application. As a result, some advances have been made in developing commerical equipment capable of generating power from a few kilowatts to as high as
several megawatts.

The technical and economic feasibility of coupling solar and wind energy power to
membrane desalting plants has been adequately studied. Preliminary cost estimates
indicate that with the current state of development, power could be generated from
solar and wind energy at a cost ranging from 1.7¢ to 7¢/Kw/hr. depending upon the
size of the plant, location, wind velocity, prevailing temperature levels, etc.
Technological advances in the next 5-10 years are expected to significantly lower
these power costs.

Based on the current technology, brackish water desalination costs range from 70¢
to $1.00/1,000 gallons depending upon the chemical composition of the saline feed,
plant size, plant location, interest rates, etc. Several technological advances
are currently in progress that are expected to reduce the present day costs by
30% - 40%.

REFERENCES

(1) Desalting Plants Inventory Report No. 5, Office of Water Research and Tech-
     nology, U.S. Department of the Interior, Washington, D.C., March 1975.

(2) Desalting Plants Inventory Report No. 4, Office of Saline Water, U.S. De-
     partment of the Interior, Washington, D.C., March 1973.

(3) K. C. Channabasappa, Status of Reverse Osmosis Desalination Technology,
     Desalination, Vol. 17, pp. 31-67 (1975).

(4) ROGA-UOP, Presentation to Office of Water Research and Technology, U.S.
     Department of the Interior, Washington, D.C., November 1974.

(5) J. R. Wilson, Demineralization by Electrodialysis, Butterwort Scientific
     Publications, London (1960).

(6) R. P. Schmitt, Reverse Osmosis & Future Army Water Supply, A.S.M.E. Pub-
     lication, 74-ENAs-6, The American Society of Mechanical Engineers, United
     Engineering Center, New York, N.Y. August 1974.

(7) Colorado River International Salinity Control Project, Special Report, U.S.
     Department of the Interior (Bureau of Reclamation), Washington, D.C., Sep-
     tember 1973.

(8) S. A. Reed, The Impact of Increased Fuel Costs and Inflation on the Cost of
     Desalting Sea and Brackish Water, Oak Ridge National Laboratory, ORNL-TM-5070
     (1976).

(9) Arthur D. Little, Inc., A Design Study to Determine the Technical and Eco-
     nomic Feasibility of a Solar Water Pumping System for the Welton-Mohawk
     Desalination Plant, R&D Report to Office of Water Research and Technology,
     U.S. Department of the Interior (1974).

(10) Williamson Engineering Associates, The Application of Wind Energy Systems to
     Desalination, R&D Report to the Office of Water Research and Technology, U.S.
     Department of the Interior (1977).

DESALTING FOR INTERNATIONAL
WATER QUALITY IMPROVEMENT*

by

Manuel Lopez, Jr.

Regional Director, Lower Colorado Region
Bureau of Reclamation
U.S. Department of the Interior
Boulder City, Nevada

and

K. M. Trompeter

Desalting Specialist, Lower Colorado Region
Bureau of Reclamation
U.S. Department of the Interior
Boulder City, Nevada

SUMMARY

The background and primary features of Title I of the Colorado River
Basin Salinity Control Project are described. The key feature is the
Yuma Desalting Plant. This membrane plant, with a capacity of over
96 million gallons per day (Mgal/d) (363 400 cubic meters per day
[m^3/d]), will be the largest plant of its type in the world. The
desalting plant process selection, design, and results of major
studies undertaken in support of design and construction are presented.
The Preliminary Engineering Analysis results and the Yuma Desalting
Test Facility programs, including pretreatment and desalting test
results, are presented. The water recovery, plant split, plant and
river operations, and plant and equipment arrangement are described.
Pretreatment descriptions include sludge handling facilities for the
partial lime system. Capital costs plus operation, maintenance,
replacement and power costs are presented. High recovery investigations and energy saving features are also discussed. Integrated
operation and maintenance of the plant are discussed and the current
status of the project is presented.

*Paper presented at:
    United Nations International Conference on Desertification
    of the World; Sacramento, California, May 31-June 10, 1977

DESALTING FOR INTERNATIONAL WATER QUALITY IMPROVEMENT

INTRODUCTION

The Colorado River, over a large portion of its course from the headwaters in the Rocky Mountains of Colorado to its mouth on the coast of the Gulf of California, traverses the vast arid country of the southwestern United States. In this region of very little precipitation, the river and its tributaries are the primary source of water for millions of people in several states. Consequently, the river system has been very highly developed, and the waters are used for a wide variety of purposes, including irrigating crops, producing electric energy, providing recreation, sustaining environmental values, supporting industry, and supplying daily domestic needs.

In some cases, substantial diversions are transported over large distances in order to supply agricultural, municipal, and industrial needs. These uses place varying demands on both the quantity and quality of water over an extensive area which covers portions of Colorado, Utah, Arizona, Nevada, New Mexico, Wyoming, and California. The significance of this resource is reinforced by the fact that the Colorado River is an international stream and has become an important factor in the United States' relations with the Republic of Mexico.

The salinity of the Colorado River, particularly in the Lower Basin, is increasing, and great concern over this rise has been expressed by those who depend upon the river's waters. Even under existing conditions, salinity causes economic damage to users, and the amount of this damage will increase as the salinity levels rise. Some irrigators are already resorting to special practices by growing salt-resistant crops, and municipal and industrial users are faced with the necessity of expensive water treatments. Some areas have drainage problems which could become more acute as waters of higher salinity are diverted.

The average salinity of the Colorado River at its headwaters is currently less than 50 milligrams per liter (mg/l). This progressively increases downstream, reaching a present condition of 825 mg/l at Imperial Dam. Tentative projections indicate that, in the absence of a control program, salinity levels of 1,250 mg/l could occur at Imperial Dam by the year 2000. Such increases would be adverse to agricultural users in the Imperial, Coachella, Gila, and Yuma Valleys in the United States, and in the Mexicali Valley in Mexico, and would cause further economic losses for the very large block of domestic users in California and Nevada. Water users in the Phoenix and Tucson areas would be similarly affected upon completion of the Central Arizona Project.

Recognition of this problem has resulted in the formulation of a basinwide program designed to maintain the quality of the river at or below the present salinity levels.

In accordance with a 1944 Treaty between the United States and the Republic of Mexico, the United States must deliver 1,500,000 acre-feet per year (1850 x 10^6/cubic meters per year) of Colorado River water to Mexico. Approximately 1,360,000 acre-feet per year (1678 x 10^6/cubic meters per year) is delivered upstream from Morelos Dam, Mexico's diversion structure, and the remainder of approximately 140,000 acre-feet per year (173 x 10^6/cubic meters per year) is delivered at the Southerly International Boundary in the vicinity of San Luis.

The salinity of the Colorado River water delivered to Mexico reached a level in 1961 that was unacceptable to the Government of Mexico. This was due to the discharge to the river of highly saline pumped drainage from the Wellton-Mohawk Division of the Gila Project via the Main Outlet Drain Extension (MODE), and a reduction of excess flow in the river. A complaint by Mexico precipitated several legal agreements, all aimed at improving the quality of the water delivered to Mexico at the Northerly International Boundary (NIB). In 1973, Minute No. 242 of the International Boundary and Water Commission (IBWC) became effective. This Minute requires that Colorado River water delivered to Mexico at the NIB must have an average annual salinity no greater than 115 parts per million (p/m) ± 30 p/m (U.S. count) over the average annual salinity of waters arriving at Imperial Dam. In order to fulfill this commitment, all Wellton-Mohawk drainage is currently being bypassed to the river below Morelos Dam (Mexico's diversion structure) and replaced by waters from above Imperial Dam. In order to recover a part of this bypassed (wasted) water, and to assist in meeting the differential set forth in Minute No. 242, a desalting plant with appurtenant works - the Desalting Complex Unit (DCU) - was proposed to desalt the Wellton-Mohawk drainage.

The DCU is specifically intended to fulfill the United States' obligations to Mexico in accordance with Minute No. 242. It has the correlative purpose, in conjunction with the Bureau of Reclamation's Colorado River Water Quality Improvement Program, to improve, enhance, and protect the quality of water available in the Colorado River for use in the United States and Mexico.

The DCU was authorized for construction by the Act of June 24, 1975, Public Law 93-320, Colorado River Basin Salinity Control Act - Title I - Programs Downstream from Imperial Dam.

In its entirety, Title I of the Colorado River Basin Salinity Control Act provides authorization for construction of the works and measures as shown on Figure 1, to control the salinity of Colorado River water delivered to Mexico upstream from Morelos Dam within the limits established by Minute No. 242. Specifically it authorizes construction of a protective and regulatory well field within a five-mile (8.05 kilometer [Km]) zone of the Arizona-Sonora Boundary near San Luis, authorizes replacement of the first 49 miles (78.9 Km) of the unlined portion of the Coachella Canal with a concrete-lined canal to effect a saving of approximately 132,000 acre-feet per year (163 x 10[6]/cubic meters per year) of water to be credited against water lost as bypassed water and reject from the desalting plant, and authorizes construction of a desalting plant to treat approximately 129 Mgal/d (488 000 m³/d) of Wellton-Mohawk drain water using advanced technology commercially available, plus appurtenant works and measures.

The Yuma Desalting Plant, to be located on the MODE, is the key feature of the project, with the product water being delivered to Mexico via the Colorado River, and the reject being diverted to the Santa Clara Slough in Mexico near the Gulf of California.

DESIGN DATA STUDIES

Prior to designing the Yuma Desalting Plant, it was recognized that process and equipment selection must be based on well defined and demonstrated performance characteristics. This was necessary to ensure reliability of production in a complex and water-short hydrologic system, and cost effectiveness of a plant of unprecedented size.

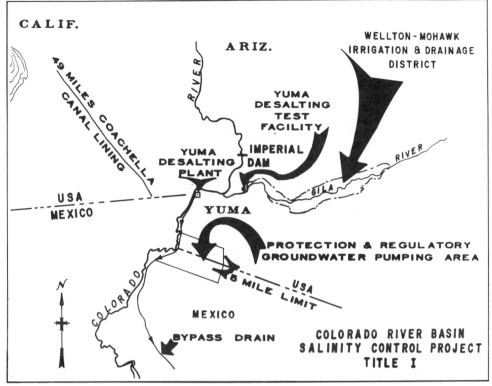

FIGURE I

After identification of design data needs, and development of an
intensive schedule culminating in full plant production in 1981, a
four-fold investigation program was undertaken. The objectives were
to (1) determine industry capability and experience, (2) test processes
and equipment on actual plant feedwater, (3) evaluate applicable experience
of operating desalting plants, and (4) explore higher production technology
to reduce the cost of treated water. This program was initiated in
the fall of 1973. The first investigation, a preliminary engineering
analysis of desalting equipment technology, experience, and industry
capabilities, as applied to the Yuma Desalting Plant, was performed by
Burns and Roe, Inc., under contract to the OWRT (formerly Office of Saline
Water) and resulted in the report "Preliminary Engineering Analysis, Yuma
Desalting Plant, November 1974." This study included a preliminary
plant design. A second effort was construction and operation of the
Yuma Desalting Test Facility near Yuma and adjacent to the MODE to
permit evaluation of pretreatment requirements and equipment as well
as membrane desalting equipment on actual water conditions. This
facility was designed and constructed by the Bureau of Reclamation and
is being operated under contract by Burns and Roe Industrial Service
Corporation. Third, in a separate but supportive effort, the OWRT
initiated a contract for a detailed review of the operational history
of 11 commercial plants. The fourth activity was the issuance and award
of contracts to develop larger membrane equipment. The objective of
this is to develop reverse osmosis (RO) membrane modules and

electrodialysis (ED) stacks that have approximately double the desalting capacity of currently available commercial units. If such large membrane units could be sufficiently proven in time to be incorporated in the Yuma plant, it was projected that the capital costs could be reduced 10 to 20 percent and operating costs by 5 to 10 percent.

The work undertaken on this project is based on prior and ongoing developments which have given favorable results in limited pilot plant operations on brackish water. Under previous OWRT contracts, prototype units of 12-inch diameter by 34-inch wide spiral wound RO membrane elements and 12-inch diameter hollow fine fiber RO bundles had been fabricated and tested successfully at the OWRT Test Facility in Roswell, New Mexico.

Contracts were awarded to Ionics (ED), and DuPont, Dow, and ROGA (RO) for prototype equipment. Testing was completed at Roswell in late 1976, and the equipment is currently operating at Yuma.

Preliminary Engineering Analysis by Burns and Roe, Inc.

The Preliminary Engineering Analysis was undertaken to provide better data on the state of the technology in membrane desalting and develop a preliminary plant design. Also the capacity of manufacturers to respond to procurement of 100 Mgal/d of equipment was unknown. Technological developments were continually being made and it was believed that these should also be identified and evaluated.

The statement of work provided for a preliminary engineering analysis including outline specifications, single-line drawings, flow diagrams, piping and instrumentation diagrams, artist's conception, and a preliminary cost estimate for a membrane desalting plant.

Seven major United States companies were surveyed covering the entire range of ED and RO process applications. Operating experience on desalting brackish waters was essentially equally split between the two processes with (as of July 1974) approximately 82 Mgal/d of total installed capacity. In addition, there were many industrial applications removing dissolved solids or other impurities from process water. Within the United States, the oldest operating ED plant was 12 years old, while the oldest RO plant had been in service five years. The largest plant in the United States in service in 1974, had a capacity of three Mgal/d. Responsive proposals had been received for plants as large as 75 Mgal/d, although no contracts of that size plant were awarded. Construction of a five Mgal/d RO plant in southern California has since been completed (early this year), and production operation is being stabilized.

The desalting capabilities of each manufacturer's equipment, in terms of product water salinity and recovery, were found to be such that each type of equipment could operate successfully in the Yuma Desalting Plant.

The preliminary engineering analysis recommended that the final plant configuration for the Yuma Desalting Plant be determined by competitive bidding which would allow individual manufacturers to optimize their equipment offerings. It was stated to be highly probable that more than one manufacturer would be supplying equipment for the final plant.

Accordingly, a design and cost estimate was developed for a plant
equally split, on the basis of salt removed, between hollow fine fiber
RO, spiral wound RO, and ED. For this base case design capital costs
were estimated at $130,360,000. This includes $83,236,000 for the
membrane plant and structures, $38,534,000 for the pretreatment plant
and structures, and $8,590,000 for other site facilities. Also included
in the cost are Government costs (owner's costs). These are approximately
30 percent of the totals. The total annual costs including capital
recovery and operating costs were estimated at $21,178,000 per year,
which results in a water cost of $0.58 per thousand gallons. This
estimate did not, however, include intake facilities, power supply
costs, or disposal facilities which were excluded from this study.

Yuma Desalting Test Facility

The Yuma Desalting Test Facility (YTDF) was built to test pretreatment
processes and to test membrane desalting equipment potentially to be
supplied for the Yuma Desalting Plant by utilizing actual irrigation
drainage feed water. Approximately 1,100 gallons per minute (gal/min) of
pretreated feed water can be produced for membrane testing.

After completion of the test facility expansion in 1973, desalting
membrane manufacturers were invited to bring test units to the facility.
Initially, seven manufacturers responded. These were (1) Fluid Systems
Division of Universal Oil Products (spiral wound RO), (2) Envirogenics
(spiral wound RO), (3) DuPont (hollow fine fiber RO), (4) Dow (hollow
fine fiber RO), (5) Dow Asahi (sheet flow ED), (6) Ionics (tortuous path
ED), and (7) Westinghouse (tubular RO). All of these units except
Westinghouse have been in continuous operation since they were connected
to the feed water in late 1973. Other manufacturers later expressed
interest and two, Aqua Chem (sheet flow ED) and Hydranautics (spiral
wound RO) were put online in late 1975.

The overriding constraint on the testing is that the testing be
for the benefit of the Yuma Desalting Plant and, thus, units which are
clearly inappropriate for use in the plant are not tested.

Results of early pretreatment testing at the Yuma Desalting Test
Facility indicated that aluminum sulfate (alum) performed poorly on
the Wellton-Mohawk drainwater. Although iron removal was usually complete,
alum failed to consistently remove manganese. Also, the filter cycle
times were normally less than 10 hours before backwashing was required.
The combination of alum with various polymeric coagulation aids did
not improve the results. Combining potassium permanganate ($KMnO_4$) or
lime ($Ca[OH]_2$) with the alum removed manganese as well as iron to below
the detectable level. The use of potassium permanganate as a pretreatment
process was also investigated. All water quality requirements were
normally met including complete removal of iron and manganese compounds.
Even with the use of dual media filters (sand and anthracite), the
filter cycle times varied from one to 30 hours indicating an instability
of the process performance. A specific problem was encountered in that
permanganate would periodically break through the sand filters. The
presence of permanganate in the pretreated water is unacceptable in
view of its chemical action on the desalting membranes. Since potassium
permanganate removes none of the calcium or alkalinity, the chemical
(acid) costs for alkalinity control would be high, and product water
recovery would be limited by calcium sulfate solubility. All of these
factors led to a decision to rule out potassium permanganate as a
pretreatment process.

Partial lime softening was the pretreatment process which most economically met the water quality requirements considered to be necessary for all the membrane desalting processes. Lime treatment produced a 90 percent reduction in alkalinity with a commensurate reduction in the acid requirement and a 45 percent reduction in calcium. It also results in iron and manganese removal to less than detectable levels, organics removal to less than three mg/l (as Total Organic Carbon), and particulate removal to levels below those required by the membrane manufacturers.

An important operating parameter developed during these tests was the use of a filter plugging factor value to assess water quality. This value is obtained by measuring flow volume through a 0.45 micron filter pad for fixed time periods. The ratio of initial time and time after 15 minutes of flow subtracted from 1 is the plugging factor. Present membrane manufacture requirements limit this value to 60 percent or less. No lime softening system has ever been operated using this parameter before.

In all the test operations, the raw drainwater has been treated with approximately three p/m of chlorine gas which generally leaves a residual chlorine reading of about 0.1 p/m in the clear wells. The chlorine is injected with the purpose of preventing adverse biological activity in the plant. This has worked satisfactorily to date.

Preliminary test results from the YDTF indicate that all membrane systems currently being tested will perform satisfactorily on lime-softened Wellton-Mohawk drainwater. Typical data are summarized on Table 1.

The performance of the RO Systems is generally stable. Testing indicates that a flux decline slope (m) of -0.02 to -0.06 can be expected under normal operating conditions.

RO salt rejection was stable in the 90-98 percent range when no mechanical leakage existed.

Scaling tests performed on several RO units indicate that operation with any degree of confidence at the higher product water recoveries (80 percent or above) will require the addition of sodium hexameta-phosphate (SHMP) or some other scale inhibitor. There also appears to be some potential for silica scale when operating at elevated product water recoveries.

The testing has also indicated that the evaluation of performance, or the optimization of operating conditions, must be based on long-term tests since some operating problems have not occurred until around 3,000 to 5,000 hours.

Good performance data and energy consumption data have been obtained from a small low-pressure RO system. However, the reliability and stability of low-pressure modules must be studied further by long-term performance tests such as the Dow LP unit presently online.

The ED testing to date has, to a great extent, been with small packaged units or partial stacks. However, some large-unit data has been acquired. In general, current efficiencies and cell-pair resistances have supported manufacturer predictions. Current efficiencies in the 80-85 percent range have been attained, with resistance in the 80-120 ohm-CM^2/RCP range. Some calcium carbonate scaling has been noted under

Table 1
TYPICAL DESALTING UNIT PERFORMANCE
Yuma Desalting Test Facility

Manufacturer/ Unit Designation	Productivity* (gal/day)	Product Salinity* (p/m)	Recovery Rate (percent)	Operating Pressure (lb/in^{2}g)	Operating Time (hours)
REVERSE OSMOSIS					
Dow 20K	73,500	100	75	375	12,000
Dow 4K	14,000	150	85	350	12,000
Dow LP 1/	76,000	100	70	275	3,000
DuPont B-9	32,000	150	80	420	16,000
DuPont LROM 2/	110,000	100	75	400	4,000
Envirogenics 17K	20,000	350	70	400	12,000
Envirogenics Cluster	53,000	200	50	400	3,000
Fluid System 5K	4,500	250	85	400	8,000
Fluid Systems 175K	176,000	350	85	425	3,000
Fluid Systems 12 inch	175,000	200	70	450	4,000
Hydranautics	20,000	180	72	400	12,000
ELECTRODIALYSIS					
Ionics MkIII	32,000	600	70	N/A	8,000
Dow-Asahi SV1/2	5,000	N/A	N/A	N/A	10,000
Aqua-Chem	4,000	N/A	N/A	N/A	2,000

```
*    Average values
1/   Dow LP - Low Pressure
2/   DuPont LROM - Large Reverse Osmosis Module
```

certain operating conditions, although this was not completely
unexpected. There are presently two large scale, multi-stage electrodialysis units in operation at the test facility.

Commercial Plant Operating Experience

and summarize the design, operation, and maintenance of 11 commercial membrane desalting plants. The plants surveyed ranged in size from 2,500 gallons per day (gal/d) to 2.0 Mgal/d and included four ED and seven RO installations. In the report, emphasis was placed on operating experience and problems including brackish water supply and pretreatment, membrane scaling and fouling, equipment and material failures, and field modifications.

The general conclusions in the report are set forth in two categories: plant design and operations. Considerations of most interest are as follows:

- Raw water pretreatment systems have not been satisfactory as installed in the majority of these plants. Acid injection systems (or the lack thereof) are responsible for most problems.

- Significant improvements in RO membrane performance have been made during the time span covered by this report.

- Corrosion of copper alloys, stainless steels, aluminum, cast iron, and carbon steels has occurred in raw water service at one plant or another. No universally applicable set of materials selections seems possible because of the great variety of raw water constituents and concentrations encountered.

- Scaling of ED membranes by calcium carbonate or calcium sulfate has occurred to some extent at all ED plants. This problem has been effectively dealt with by feed water pretreatment, chemical and mechanical cleaning, and polarity reversal operation.

- Organic fouling of both RO and ED membranes has been a more persistent problem than scaling. Chemical cleaning with caustic brine solution has been moderately effective against organic slime in ED plants. Enzyme detergent flushing has been used at some RO plants.

- Unit treatment costs have generally been much higher than predicted. In addition to the rapid cost escalations which occurred during the operating period surveyed, high unit costs are attributable to low plant load factors caused by the following: (1) low water demand, (2) raw water shortage, and (3) inadequate pretreatment systems.

- The rate of membrane performance deterioration with time is low for both ED and RO, if membranes are kept clean.

The report's general recommendations, based on the membrane desalting plants surveyed, are set forth in two categories: planning and design, and operation.

- Planning and design must include an accurate analysis of the quality and quantity of the raw feed water source before the specification of a treatment plant. This is the single most important technical consideration in the planning of a treatment facility.

- Extreme care should be exercised in the design of the raw water pretreatment systems, especially with respect to the equipment selected to control and monitor feed water chemistry and the safety hazards of some unlimited injection systems.

- To assure adequate service and parts availability, maintenance requirements should be fully assessed before purchasing a treatment system or entering into a maintenance contract.

- A piped-up or quickly connected cleaning system is desirable for both RO and ED plants.

- Operation of plants should be continuous, 24 hours a day, to minimize startups and unit water treatment costs. This is predicated upon adequate demand and storage facilities.

- Plant operation may be partly unattended if adequate supervisory instrumentation is installed.

- Detailed accounts of operating costs should be kept, even for small plants, to follow periodic determination of unit treatment costs for comparison purposes.

- A complete chemical analysis of raw and product waters should be made several times each year.

YUMA DESALTING PLANT ADVANCE PLANNING DESIGN

Based on the recommendation of the Burns and Roe Preliminary Engineering Analysis, the results of work at the Yuma Desalting Test Facility, an evaluation of drainage feed water from the Wellton-Mohawk Division, and flow projection of the lower Colorado River system, an advance planning design for the Yuma Desalting Plant was completed by the Bureau in early 1976. The purpose of this design was to scope the types and size of structures and equipment required for further planning of equipment procurement and contract scheduling. Based on hydrologic and desalting equipment data available at that time, a plant with a product water capacity of 108.5 Mgal/d (410 700 m^3/d) at 386 p/m was selected. The parameters of this design were used in preparation of a Request for Proposals for membrane desalting equipment. Contract negotiations resulting from those proposals are now in the final stages.

During the winter of 1976-1977, hydrologic operation studies were updated to reflect current projections of Colorado River Basin hydrology and effects of drainage reduction programs in the Wellton-Mohawk Division. These studies resulted in a final design plant production capacity selection of 96 Mgal/d (363 400 m^3/d) at 254 p/m. This capacity will be modified slightly later this year to reflect the warranted product quality and quantity of membrane desalting equipment for which supply contracts are signed with manufacturers. The number and types of processes to be included, and the number of suppliers, have not been predetermined and are not restricted in any way within the proposals received.

The following is a description of layout, equipment, processes, and other elements of the 108.5 Mgal/d (410 700 m^3/d) advance planning design. A hydrologic schematic of the Colorado River system below Imperial Dam, including the desalting plant flows, is shown on Figure 2.

Site

The proposed desalting plant site is about four miles west of Yuma, Arizona, immediately adjacent to the MODE, on a 60-acre (24 hectare) rectangular plot.

Figure 3 shows the site plan for the desalting plant and associated pretreatment used for planning purposes. This design is based on 154.2

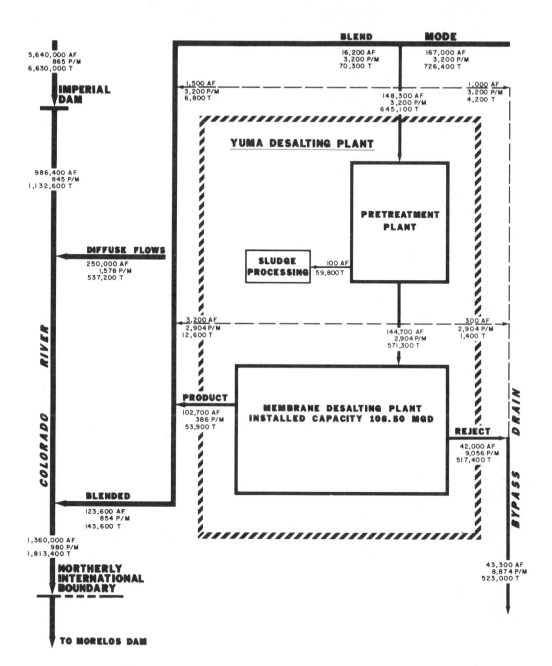

HYDROLOGIC SCHEMATIC
FIGURE 2

Mgal/d (583 600 m³/d) flowing in the MODE and 138.8 Mgal/d (525 400 m³/d) diverted to the pretreatment plant.

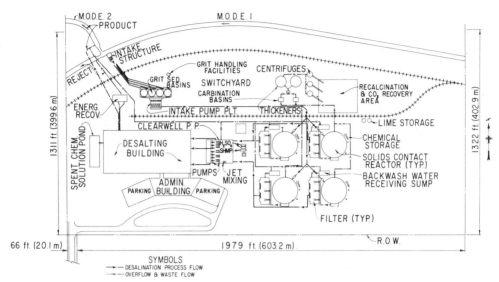

SITE PLAN

YUMA   DESALTING   PLANT

FIGURE 3

Raw feed water will enter the site near the northwest corner and product water and reject stream will leave the site at the same corner. Highway access will be from a new road to be constructed to the south-west corner of the plot. A railroad spur will enter the site near the northeast corner and will lead into the chemical storage area. Service roads will provide access for plant operators to the diversion turnout area and to the product and reject stream discharge area along the drain.

Processes

The total desalting system is divided into two processes: the pretreatment process and the desalting process.

Pretreatment - An intake structure equipped with a trash removal system is located at the diversion point on the MODE to remove the larger solids and floating material. The water then flows to grit sedimentation basins where the heavier suspended solids (greater than 105 microns) are removed.

As the water leaves the grit sedimentation basins, it passes through a traveling screen which removes additional suspended material.

The water is then pumped to four 175-foot (53.34 m) diameter solids contact reactors for partial lime softening. Sufficient lime (250 mg/l) is added to increase the water pH to about 9.5. Lime addition is controlled to remove most of the bicarbonate hardness by precipitation of calcium carbonate but to prevent precipitation of magnesium as magnesium hydroxide. Plans also include provision for coagulant (ferric sulfate) injection at approximately 20 mg/l. Part of the sludge formed is recirculated to assist in forming the sludge blanket upon which the chemical and physical reactions depend. Operating at pH 9.5 results in precipitation of iron, manganese, and possibly strontium from the water. The water from the solids contact reactors passes to the dual-media gravity filters where the remaining suspended solids are removed. Prior to entering the filters, the pH of the water is reduced from 9.5 to 8 to stabilize the water, preventing further precipitation in the filters. The sludge removed from the solids contact reactors is dewatered to 60 percent solids by thickeners and centrifuges and then recalcined to recover the lime. The water removed from the sludge is returned to the solids contact reactors. The filters are backwashed regularly to remove the filtered solids. This backwash water is also returned to the solids contact reactor.

The solids contact reactors produce approximately 650 tons/day (590 metric tons/day) of calcium carbonate sludge. This sludge will be recalcined to produce 125 tons/day (113 metric tons/day) of calcium oxide. Since only this amount of calcium oxide is needed for the reactor operation, excess $CaCO_3$ will be disposed of by conventional land fill/settling methods. Carbon dioxide produced by the recalcining operation will be used to adjust the pH of the water going to the desalting equipment.

The filtered water is collected in a clearwell which acts as a short-term storage tank and a sump for the desalting process pumps. This water has had the turbidity reduced to 0.5 Jackson Turbidity Units (JTU) or less and the plugging factor to 40 percent or less. Most of the bicarbonate hardness has been removed and iron and manganese reduced to undetectable levels. The salinity has been reduced from 3,200 to 2,904 p/m TDS.

Desalting - The desalting process begins at the clearwell, where the low-pressure pumps take suction. These pumps forward the water either directly to the electrodialysis process portion of the plant or to the high-pressure reserve osmosis pumps. After the low-pressure pumps, strainers are installed to protect the membrane equipment. As the water flows to the membrane process equipment the pH is adjusted to 6.0.

In the electrodialysis process, the salt is removed from the dilute stream, which becomes the product as it leaves the stacks. The concentrate stream leaving the stacks is split, with 63 percent recirculated to maintain the required 73 percent recovery. The remaining 37 percent of the concentrate stream becomes the reject from the process.

In the reverse osmosis processes, the high-pressure feed flows into the membrane modules. The product flows through the membrane, leaving the concentrated reject to flow through the system. The valve after the last elements controls the recovery by regulating the reject flow from the modules. The high-pressure reject from all the modules is collected and passes through an energy recovery impulse turbine directly connected to induction generators before leaving the site as reject.

The product from the electrodialysis and reverse osmosis sections of the plant is collected and flows to the MODE-2 diversion channel to

be conveyed to the Colorado River. In this channel the product and
the blend flow of untreated drainage water are combined and this blended
flow, at 768 p/m TDS, then flows to the Colorado River. The reject
from both processes with an average concentration of 9,056 p/m TDS
is collected and flows through the Bypass Drain to the Santa Clara
Slough.

The membrane desalting plant consists of a multiplicity of 1 to
5 Mgal/d (3785 to 18 925 m3/d) sections each separately controlled.
The interior layout of the desalting building is shown on Figure 4.
Three pipe trenches carry the main piping to and from the individual
segments of the plant. The segments are arranged along the trenches,
with each type of equipment along both sides of a single trench. At
the end of the building are located the maintenance areas, refrigerated
membrane storage, a test area, installed cleaning system, and the main
product and reject headers. An overhead crane will be provided to
move modules and equipment to and from these areas. Storage space for
a one year supply of membranes is provided.

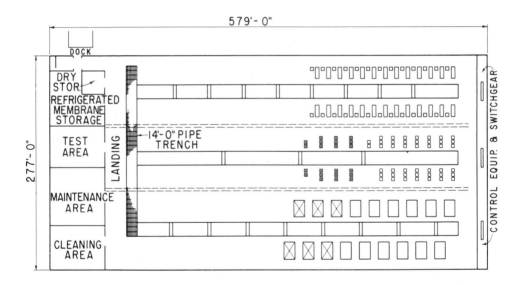

FLOOR PLAN

YUMA DESALTING PLANT

FIGURE 4

The main control center is located in the administration building
adjacent to the desalting building. This building also contains the
administrative offices and visitors center.

The overall control philosophy for the desalting facility is total
facility control from one central location utilizing a digital computer.

The central computer will communicate with smaller remote process control computers in the pretreatment and desalting plant areas. In turn, each process control computer will communicate to each individual process via remote multiplexers located nearby.

Manual control will be provided but will normally be utilized only in emergency situations, or "first-time" operating conditions, due to the size and complexity of this plant. The functions of the central computer will be to monitor and control water flow rates, chemical dosage rates, pH, filter backwash, alkalinity, and plugging factors in the pretreatment area and pressure, voltage, flow rates, pH, salinity and energy recovery in the desalting equipment area. Gate positions along the MODE and at the intake structure will be monitored but not controlled by the facility computer system.

Operation, Maintenance, and Replacement

Although the plant will be computer controlled, continual surveillance will be required to maintain trashracks, conveyors, chemical storage, chemical feed systems, and electronic sensing devices in the pretreatment system. For the desalting equipment, data will be continually analyzed to detect malfunctions in pumps, pressure vessels, rectifiers, stacks, and all associated piping. The replacement of membranes will be accomplished manually using an overhead bridge crane aided by special equipment.

Once the plant is in operation, adjustment of the plant will be required to compensate for changes in temperature, turbidity of the feed, biological activity of the feed, availability of feed water and changes in productivity or power consumption due to aging or fouling of the membranes. Membrane replacement will be initiated when it is determined that elements are not performing to warranted design ratings.

The plant will be manned 24 hours per day with sufficient personnel to handle all but the most serious problems. Staffing will consist of 7 administrative personnel, 8 process engineering, control, and laboratory personnel, 40 operations personnel, and 28 maintenance people.

Costs

Project costs for the Yuma Desalting Plant based on October 1975 prices and 5-5/8 percent interest during construction were estimated from detailed quantity takeoffs for the plant previously described and are summarized in Table 2. Costs shown do not include incoming power costs.

HIGH RECOVERY INVESTIGATION

Public Law 93-320 requires that measures for replacing the Yuma Desalting Plant reject be identified and reported to the Congress. One of the potential alternatives being investigated is increasing the plant recovery to reduce the brine volume.

The best approach to high recovery appears to be modifying the existing pretreatment and desalting plant design. The new design would increase the size of the 70 percent recovery plant to 90 percent recovery by increasing the membrane area approximately 50 percent. A tail end plant, such as a brine concentrator, could be added to the 90 percent recovery design for recoveries above 90 percent. Preliminary studies, however, cast doubts on the economic viability of recoveries above 90-92 percent.

Table 2
CAPITAL COST, ANNUAL COST, AND PRODUCT WATER COST
(October 1975 Prices)
Yuma Desalting Plant

Item	Capital Cost	Annual Costs
Pretreatment	$56,000,000	$3,367,900
Desalting Plant	70,300,000	4,227,800
Control and Operating System	5,300,000	318,800
Appurtenant Works	17,860,000	1,074,000
Total Capital Costs	$149,460,000	$8,988,500
Operation, Maintenance, Replacement and Power		$11,520,000
Total Annual Cost		$20,508,500
Total for plant.product at 386 p/m (102,650 acre-feet/year) (126.6 x 10^6 cubic meters per year)		61.3¢/1,000 gal. ($2.32/m^3)
Total for blended flow to river at 857 p/m (123,600 acre-feet/year) (152.5 x 10^6 cubic meters per year)		50.9¢/1,000 gal. ($1.93/m^3)

Pretreatment modifications would involve removing permanent hardness by using soda ash in the solids contact reactor or by adding an ion-exchange system. It is planned that these processes be tested at the Yuma Desalting Test Facility in the near future.

SCHEDULE AND STATUS

Construction is nearly complete on the Bypass Drain which will carry the plant reject to the Santa Clara Slough in Mexico. The 35-mile (56.3 kilometers) reach in Mexico is completed, and work on the 16-mile (25.7 Km) U.S. section is in the final stages. Prior to the plant coming online, the Bypass Drain will carry all Wellton-Mohawk drainage directly to the Santa Clara Slough.

The Request for Proposals to supply membrane desalting equipment has been issued and proposals were received August 2, 1976. After a period of initial evaluation and negotiation, a notice for best and final proposals was issued to those within the competitive range. These are being evaluated and award of contract is expected in August of 1977. The plant construction contract, including membrane equipment installation, is expected to be issued in May 1978. The plant is scheduled to be online in 1981.

References

1. "Utilization of Waters of the Colorado and Tijuana Rivers and of the Rio Grande" - Treaty between the United States and Mexico, February 3, 1944.

2. Minute No. 242 - International Boundary and Water Commission, United States and Mexico - August 30, 1973.

3. Public Law 93-320 - Colorado River Basin Salinity Control Act, June 24, 1974 (88 Stat. 266).

4. "Yuma Desalting Plant, Preliminary Engineering Analysis" - Colorado River Basin Salinity Control Project, November 1974 - Burns and Roe, Inc., for U.S. Department of the Interior, Office of Water Research and Technology.

5. "Commercial Membrane Desalting Plants, Data and Analysis" - DSS Engineers Inc., September 1975.

6. Request for Proposals - Membrane Desalting Equipment - Yuma Desalting Plant - Colorado River Basin Salinity Control Project, Title I - Arizona - Solicitation No. DS-7186, Negotiated, Bureau of Reclamation, 1976.

7. "Colorado River Basin Salinity Control Project, Plan of Development, February 1976," USDI, Bureau of Reclamation.

8. "Colorado River Basin Salinity Control Project - A Status Report," K.M. Trompeter and F.R. Summers. Fourth National Conference of National Water Supply Development Association, Oklahoma City, Oklahoma.

WATER PROBLEMS IN COALINGA

John W. Masier, Chief
Planning and Investigations Branch
San Joaquin District
Department of Water Resources
The Resources Agency
State of California

ABSTRACT

In Coalinga, California, the only locally available water supply consists of
several deep wells. Since the water produced by these wells has a salt concentration five times higher than the recommended maximum defined under the federal
Drinking Water Standard, drinking water was at one time hauled into town by rail.

Transporting water in tank cars posed the constant threat of bacterial contamination, so in 1959 the city installed an electrodialisis plant to desalinize the
well water and produce its own drinking supply. Yet as Coalinga's population
increased, the demand for potable water began to exceed the plant's output, and
new ways were sought to insure a safe drinking supply.

In 1965, technicians from the University of California installed a reverse osmosis
(R.O.) plant outside Coalinga and began delivering treated ground water to the
community. The R.O. plant proved very reliable, but the output was minimal. The
city's fresh water needs were still not being met, and water hauling was resumed.

Water hauling revived the threat of bacterial contamination, and city officials
also worried about the health risks posed by Coalinga's unconventional water
distribution network. Before 1972, water was piped into homes and businesses
through three taps. The hot and cold taps delivered "hard" ground water, while
the third tap delivered water suitable for drinking and cooking.

"Hard" water posed health problems to the inhabitants of Coalinga in that absentminded or uninformed people would occasionally take a drink from one of the unsafe
taps. In addition, "hard" water was disastrous to pipes and mains. In 1968 the
Fresno County Health Department issued a public statement calling for the replacement of the controversial three-tap system.

In October 1968, the City of Coalinga executed a contract with the U.S. Bureau of
Reclamation calling for an annual entitlement of up to 12 cubic hectometres
(10,000 acre-feet) of water from the Coalinga Canal, a part of the federal Central
Valley Project.

Since the Canal is located 13 kilometres (8 miles) northeast of Coalina, an elaborate conveyance facility had to be constructed to deliver imported water to the
city. The project was completed in 1972. The three-tap system was abolished, and
a guaranteed supply of high quality surface water was insured for the City of
Coalinga.

INTRODUCTION

Many smaller communities in California's dry San Joaquin Valley lack local surface water supplies adequate to meet growing industrial and domestic demands.
Frequently, in such areas, the only large source of locally-available water is
found in deep wells, and this supply may contain dissolved mineral salts that
adversely affect its use for municipal and domestic purposes.

A case in point is the City of Coalinga, a small farming and industrial community
on the west side of the San Joaquin Valley. Coalinga's water outlook was once so
bleak that most businesses in the area seemed destined for economic depression.
The purpose of this paper is to detail the ways in which the citizens of Coalinga
strove to overcome their water problems. By recounting their struggles to provide
themselves with an acceptable water supply - and by outlining the corrective steps
taken to achieve this goal - the paper will demonstrate how technology and per-
severance, applied collectively, can surmount tremendous natural obstacles.

HISTORY

The City of Coalinga has a population of 6,161 and is located near the center of
California, some 110 kilometers (70 miles) southwest of Fresno. In the past, the
surrounding area was noted solely for its oil production. But in recent years,
agricultural activity (mostly cotton) has increased significantly. Prior to 1968,
irrigation was pumped from deep wells, as was the municipal water for Coalinga.
Wells located on the north and east sides of town provided the municipal water,
and these wells ranged in depth from 116 metres (382 feet) to 439 metres (1,439
feet). Each well was pumped from deeper strata than is typical in the central
part of the San Joaquin Valley, and each well delivered water with a dissolved
salt concentration of 2,500 milligrams per litre (about 5 times the recommended
maximum as defined under the U.S. Federal Drinking Water Standard). While this
water was suitable for certain household and other uses, it was not suitable for
drinking.

Until 1958, drinking water for Coalinga was imported in railroad tank cars from
Armona, California, a distance of about 70 kilometers (43 miles). But transporting water by rail constantly posed the threat of bacterial contamination, and in
1959 the city installed an electrodialysis plant to desalinize the natural ground
water and produce its own drinking supply. The plant was capable of desalting 100
cubic metres (28,000 gallons) of water a day, and was the first operational municipal electrodialysis unit in the world. Yet by the early 1960's, as Coalinga's
population increased, the demand for potable water began to exceed the plant's
output, and the city had to look for new ways to provide its citizens with an
adequate drinking supply.

REVERSE OSMOSIS PLANT

In 1965, at the invitation of the City Manager, technicians from the University of California at Los Angeles installed a reverse osmosis plant outside Coalinga and began delivering treated ground water to the community. Reverse osmosis (R.O.) is a process in which desalinized water is produced by pressurizing brine in contact with a semipermeable membrane. The membrane allows water to pass through while rejecting most of the dissolved salts, providing product water suitable for drinking.

The R.O. plant at Coalinga proved to be extremely reliable and trouble-free, primarily because the ground water it refined had a uniform brine concentration and was without microbic contamination. Unfortunately, even when the plants was operating at full capacity (19 cubic metres - 5,000 gallons - per day), and in tandem with the electrodialysis unit, Coalinga's fresh water needs were still not being met, and the city was forced to resume water hauling, this time by truck from Huron, California - some 27 kilometers (17 miles) away. Desite the best efforts of city personnel, water hauling again rasied the threat of bacterial contamination, and fears concerning public health were renewed.

THREE WATER TAPS

Besides the danger of contamination of their drinking water, city officials were also worried about the health threats posed by Coalinga's unconventional water distribution network. Before 1972, water was distributed citywide through dual systems and piped into homes and businesses to three taps. The traditional hot and cold water taps delivered "hard" ground water that was generally suited for sanitation purposes. The third tap delivered imported or, after 1959, "treated" water that was suitable for drinking or cooking. This three-tap distribution system posed health problems to inhabitants of Coalinga in that absentminded or uninformed people would occasionally take a drink from one of the "hard" water taps. Besides this threat, "hard" water was disastrous to plants. Nothing other than salt-tolerant flowers and vegetables could be nourished by ground water, and even hearty crops like cotton had to be germinated with imported water before they could survive ground water irrigation. Moreover, the mains and pipes that delivered "hard" water to homes and businesses experienced rapid deterioration, and the aroma from the sanitation taps was rank.

By 1968, the situation was so bad that the Fresno County Health Department publicly issued the following statement:

> "There is no question, at least in our minds, that the
> City of Coalinga must replace the current dual system
> supply with a single system designed to provide the
> community with an adequate volume of water under a
> reasonable pressure range, meeting the public health
> objectives of safeness, wholesomeness, and potability."

Health investigators were especially alarmed over the possibility of "back-siphonage", a condition in which waste water, or in this case "hard" water, becomes mixed with potable water because of differential pressures between the two systems. Though no cases of "back-siphonage" had ever been reported in Coalinga, officials felt that adequate precautions had not been taken to prevent it from happening. First on the list of precautions was abolishment of the three taps and the corresponding dual water system. Water flow, officials agreed, is easier to regulate when it is confined to one system.

FIRM SURFACE WATER ENTITLEMENT

By 1968, however Coalinga city officials were reluctant to renovate the city's water distribution system, since they were in the process of negotiating a contract with the U.S. Bureau of Reclamation calling for a guaranteed annual supply of imported surface water. Though eager to upgrade their city's water quality and service, Coalinga officials wanted to make certain that the contract with the Bureau was firm before they invested money and effort into a new distribution system.

On October 28, 1968, the City of Coalinga executed a contract with the Bureau calling for an annual entitlement of up to 12 cubic hectometres (10,000 acre-feet) of water from the Coalinga Canal, a federal facility receiving water from the California Aqueduct, the source of which is the Sacramento-San Joaquin Delta. Besides this firm entitlement, the contract also permitted the city to purchase an additional 8.5 cubic hectometres (7,000 acre-feet) of "interim" water each year, as long as supplies were available.

Since the Coalinga Canal is located 13 kilometres (8 miles) northeast of Coalinga, an elaborate conveyance facility had to be constructed to deliver the imported water to the city, and this posed still another problem for the City of Coalinga, this time a financial one. Fortunately, the city was able to secure a low-interest, long-term loan from the State of California under provisions of the Davis-Grunsky Act, and the conveyance facilities became a reality in 1972.

After passing through a water treatment plant near the canal, the water is lifted some 152 metres (500 feet) over a distance of 5 kilometres (3 miles) to a storage reservoir approximately 8 kilometres (5 miles) northeast of the city. From this facility, 305 metres (1,000 feet) above sea level, the water was diverted downhill to another storage reservoir 1.6 kilometres (1 mile) west of Coalinga. Once there, the water was channeled into the city's "hard" water distribution system and piped into homes and businesses. (The old "drinking-water" tap and its separate distribution system were shut down.)

During the period of buildup of the city's demand, large quantities of water are being made available to several oil companies in the area for use in their secondary oil recovery operations involving steam injection. The sale of this water will provide a major portion of the city's revenues during the early years of operation.

END OF AN ERA

The delivery of a guaranteed annual supply of high quality surface water to the City of Coalinga in 1972 marked the end to the community's long-standing water problems. Ground water pumping was reduced sharply, and the three-tap/dual water supply system was quickly abolished. Farmers in Pleasant Valley (the principal farming area near Coalinga) continued to pump ground water to irrigate their established cotton fields, but most inhabitants were soon using imported supplies to tend to their water needs.

Today, in Coalinga, inhabitants are almost totally dependent on imported water. Oil companies and city dwellers alike depend on the supply from the federal project, and even in this, a water-short year, when some farmers are compensating for the lack of precipitation by drilling new wells or deepening old ones, ground water extractions are minimal in comparison to previous years.

Despite the temporary shortages caused by the current drought, Coalinga's current water system should be able to meet the community's estimated. needs through about the year 2010.

RECLAMATION OF MUNICIPAL WASTEWATER BY SOIL FILTRATION

IN ARID REGIONS

Herman Bouwer and R. C. Rice*

ABSTRACT

    Sewage effluent is a valuable water resource, particularly in water-
short areas, where it may be used for irrigation of crops. For public
health reasons effluent used for irrigation of crops consumed by humans must
undergo more treatment than effluent used for irrigation of forage or fiber
crops. Nitrogen, boron, chloride, and other substances in the sewage water
may decrease crop yield or quality. Dissolved salts may restrict irrigation with sewage to salt-tolerant crops and special farming techniques may
be required to minimize yield reductions due to salinity. The value of
sewage water for irrigation and other reuse can be enhanced by additional
treatment to remove bacteria, viruses, suspended solids, and biodegradable
organics, and to reduce the concentration of nitrogen and other constituents. This treatment can be effectively obtained by groundwater recharge
with infiltration basins. Under favorable hydrogeological conditions, 50
to 100 m of secondary sewage effluent may infiltrate per year into the soil.
As this effluent percolates downward to the underlying groundwater, many of
its constituents are removed by physical, chemical, and biological processes
in soil. Thus, the effluent becomes renovated water that can be collected
for reuse with underground drains or by pumping from wells. The renovated
water can be used for unrestricted irrigation (including fruit and vegetable
crops consumed raw by humans), recreational lakes, and certain industrial
applications (cooling or processing water). Reuse for drinking water is
not yet recommended because some of the organic compounds are not decomposed
or adsorbed in the soil, and thus remain in the renovated water. More
research is needed to identify these refractory organics and their effects
on human health when ingested in drinking water.

---

* Director and Agricultural Engineer, respectively, U. S. Water Conservation
Laboratory, ARS, USDA, 4331 East Broadway, Phoenix, Arizona, 85040, U.S.A.

INTRODUCTION

Sewage effluent is an important water resource, particularly in water-short areas. Depending on the treatment it has received, sewage effluent is used for irrigation of crops, parks, and golf courses; industrial applications like cooling; recreational lakes; groundwater recharge; and even drinking water. The concentration of the various chemicals and microorganisms in domestic or residential sewage depends on the quality of the input water and on the water use and dietary habits of the people. Because industrial waste discharges into sewer systems can greatly degrade the quality and reuse potential of sewage effluent, particularly if they contain toxic substances like heavy metals and pesticides, they should be controlled at the source. Average increases in concentrations of chemical constituents in the water due to one cycle of domestic use are shown in Table 1. This table, which also shows the average composition of secondary sewage effluent, applies to mostly domestic or residential sewage in the United States where the sewage production is about 200 to 400 liters per person per day. The data in the third column of Table 1 show the composition of secondary sewage effluent from a primarily residential and light commercial metropolitan area in a desert environment (Phoenix, Arizona) with a sewage production of about 400 liters per person per day. In areas with less per capita water use, concentrations of nitrogen, phosphorus, potassium, microorganisms, and other substances that enter the effluent primarily through human waste, can be expected to be proportionally higher. The nitrogen level in the sewage effluent also depends on the nitrogen uptake of the local population, as affected by protein in diet, nitrate in vegetables consumed, etc. Use of water softeners with sodium-regenerated resins contributes sodium to the water as it goes through one cycle of domestic use.

USE OF CONVENTIONALLY TREATED SEWAGE FOR IRRIGATION

Sewage effluent is often used for irrigation of crops. From a public-health standpoint, the type of crops that can be irrigated this way depends on the degree of sewage effluent treatment (National Academy of Sciences - National Academy of Engineering, 1973). Standards adopted in 1972 by the Arizona State Health Department (Bouwer and Chaney, 1974, and references therein), for example, allow secondary effluent to be used for irrigation of fiber or forage crops not intended for human consumption, and for orchard crops if the effluent does not contact the fruits (overhead sprinklers thus cannot be used). Secondary effluent can also be used as drinking water for farm animals, except for dairy cows.

If the secondary effluent is chlorinated or otherwise disinfected to reduce the total coliform density to less than 5000 per 100 ml and the fecal coliform density to less than 1000 per 100 ml, the effluent can be used for irrigation of crops that are consumed by humans, provided that the food is cooked or otherwise processed to destroy disease-causing organisms. Such disinfected secondary effluent can also be used for irrigation of orchard crops where the effluent does contact the fruit, irrigation of golf courses and cemeteries, drinking water for dairy cows, and providing a substantial portion of water in lakes for aesthetic enjoyment or secondary-contact recreation. If the secondary effluent has received additional treatment (for example, tertiary treatment, lagooning, or groundwater recharge) to reduce the biochemical oxygen demand (BOD) and suspended solids content both

---

*Director and Agricultural Engineer, respectively, U. S. Water Conservation Laboratory, ARS, USDA, 4331 East Broadway, Phoenix, Arizona, 85040, U.S.A.

Table 1. Average composition of secondary sewage effluent in United States and concentration increases due to one cycle of domestic use (from Weinberger, et al., 1966), and composition of secondary sewage effluent from Phoenix, Arizona (coliform and virus data from Gilbert, et al., 1976).

	Average concentration in U.S. mg/1	Average concentration increase mg/1	Phoenix, Arizona mg/1
NH_4-N	16	16	30
NO_3-N	3	2	0.5
NO_2-N	0.3	0.3	1
Organic N			2
PO_4-P	8	8	9
Na	135	70	200
K	15	10	8
Mg	25	7	36
Ca	60	15	82
Cl	130	75	213
SO_4			107
B			0.5
F			2.5
Total salts (TDS)	730	320	1000
Total organic carbon			25
Biochemical oxygen demand (BOD)	25	25	15
Fecal coliform bacteria (per 100 ml)			10^5
Viruses (PFU's per liter)			21

to less than 10 mg/1 and the fecal coliform density to less than 200 per 100 ml, the effluent can be used for unrestricted irrigation (including lettuce, cucumbers, carrots, and other vegetables consumed raw by people); irrigation of parks, lawns, school grounds, private yards, and sport fields; and for providing a substantial portion of water in lakes used for primary-contact recreation.

The above standards were developed only from a standpoint of protecting the public health against outbreaks of diseases that are spread via the fecal and contaminated-water route. In many areas in the world, raw or partially treated sewage is used for irrigation of vegetables and other human food crops. While local populations may develop an immunity to the pathogenic organisms ingested with the harvested crops, tourists and other newcomers to such areas often experience gastro-intestinal upsets. Furthermore, the whole population is vulnerable to outbreaks of typhoid, cholera, hepatitis, and other diseases.

In addition to pathogenic microorganisms, several other substances in sewage effluent affect its suitability as irrigation water. Nitrogen, for example, while being an essential crop nutrient and fertilizer, can have adverse effects when applied in too large amounts or too long over the growing season. In arid regions with less than abundant household use of water, nitrogen contents in secondary sewage effluent can readily exceed 50 mg/l (mostly as ammonium). This means that for every 10-cm irrigation with effluent, the crop receives at least 50 kg N per ha. A crop like cotton may need 120 cm of water for its entire growing season, which would bring 600 kg/ha N into the ground. This is much more than the 100 to 300 kg/ha N normally used for well-fertilized cotton. Irrigating with water that is high in nitrogen delays the maturity of cotton and other crops, decreases the sugar contents of sugarbeets and sugarcane, and lowers the starch content of potatoes. Excessive nitrogen in irrigation water also tends to cause lodging problems in certain crops, and it can adversely affect the quality of fruit and vegetable crops. A recent classification of irrigation water (Ayers, 1975) indicates that no problems need to be expected if the N concentration is less than 5 mg/l. Increasing problems can be expected if the N concentration is between 5 and 30 mg/l, and severe problems if it is above 30 mg/l. Thus, irrigation with sewage effluent may have to be restricted to grasses or other forage crops that can tolerate high nitrogen applications. Other crops may benefit from the nitrogen in sewage when it is used during the early part of the growing season and "regular" irrigation water is used for the rest of the season. If other water supplies are available, the sewage may be blended with regular water to reduce the concentration of N and other constituents.

Nitrogen not used by the crop or not lost from the soil via denitrification, ammonia volatilization, or other process, eventually percolates down to the groundwater. This "deep percolation" water could have nitrogen and salt contents that are considerably higher than those of the effluent, because salts in irrigation water are concentrated in the drainage water. Underground drainage systems may be required to remove this salty water from the aquifer and to dispose of it with minimum environmental impact.

Other irrigation problems may result from excessive salt concentrations in the sewage effluent. Where the drinking water already contains a significant amount of total dissolved solids (TDS), the salt content of the sewage effluent could readily be in the upper part of the 500- to 2000-mg/l range where "increasing problems" can be expected (Ayers, 1975). If the TDS content of irrigation water is higher than 2000 mg/l, severe problems can be expected. This may restrict the crops that can be irrigated with effluent to salt-tolerant types. Drainage to maintain low water tables and regular excessive irrigations to leach salts out of the root zone may be required to prevent salt accumulation in the soil. Also, special farming techniques like planting row crops half-way up the ridge rather than on top of the ridge in furrow-irrigated fields, may be necessary. Deflocculation of clay and resulting reduction in infiltration rate and decline in soil structure due to sodium in irrigation water generally will not be a problem when irrigating with sewage effluent, because the TDS concentration of such water normally will be sufficiently high to prevent deflocculation, even if the sodium adsorption ratio (SAR) is relatively high (Ayers, 1975).

Specific ions in irrigation water can have toxic effects on the crop when certain concentrations are exceeded. Chloride, for example, may produce toxic effects when its concentration exceeds 150 mg/l (100 mg/l if the water is also absorbed by the leaves) and severe problems when higher than 350 mg/l (Ayers, 1975). Sodium may cause increasing problems at SAR-values of 3 to 9, and severe problems at SAR-values above 9 (assuming that the water is absorbed by the roots only). Boron causes increasing problems if the concentration ranges between 0.5 and 2 mg/l, and severe problems when it is above 2 mg/l. Citrus and avocado are among the least boron-tolerant crops; alfalfa, sugarbeets, date palms, and asparagus are among the most tolerant (Ayers, 1975). Concentrations of metals and other trace elements in sewage water normally are below the maximum limits for irrigation water as listed in the report by the National Academy of Sciences and the National Academy of Engineering (1973; see also Bouwer and Chaney, 1974).

RENOVATION OF SEWAGE EFFLUENT BY GROUNDWATER RECHARGE

Before sewage effluent can be used for unrestricted irrigation and primary-contact recreation, it must undergo nitrogen removal, disinfection, and other additional treatment. This advanced treatment is expensive and requires chemicals, energy, technical skill, and complicated treatment plants (Bouwer, 1976). A simpler method for upgrading the quality of conventionally treated sewage is groundwater recharge. For this purpose, secondary effluent, or possibly well-settled primary effluent, is admitted into infiltration basins and allowed to seep to the groundwater. As the sewage water percolates through the soil, bacterial, chemical, and physical processes remove a number of pollutants and impurities so that the sewage has become "renovated" water by the time it reaches the groundwater table. The renovated water can then be collected with drains or pumped wells located some distance from the infiltration basins to allow for some lateral movement (and hence more purification) of the renovated water in the aquifer (Figure 1). Since the optimum design and management of such systems depend very much on the local climate, soil, and groundwater, general design criteria cannot be given. Thus, construction of large systems normally should be preceded by local experimentation with some small basins. An example of such an experimental system in a desert environment is the Flushing Meadows Project west of Phoenix, Arizona, installed in 1967 to determine the feasibility of renovating sewage effluent by rapid infiltration. As a result, wastewater renovation via groundwater recharge is now an integral part of the long-range wastewater management plan for the valley. Average maximum daily temperatures in Phoenix range from 17.8°C in January to 40.4°C in July. Average minimum daily temperatures range from 1.8°C to 23.9°C and average daily temperatures, from 9.8°C to 32.1°C. Rainfall averages about 18 cm per year.

The Flushing Meadows Project consists of six, parallel infiltration basins, 6 by 210 m each (Figure 2). The soil is a loamy sand for the top 1 m, underlain by coarse sand and gravel to a depth of 80 m, where a clay layer begins. The water table is at a depth of about 3 m. Observation wells 6 to 9 m deep are installed between the basins and at various distances from the basins (Figure 2). Secondary sewage effluent is pumped into the basins at one end. The pumping rate is slightly higher than the infiltration rate. The excess water is spilled out of the basins via an overflow structure at the other end, which maintains constant water depth in the basins -- 15 or 30 cm. To determine the effect of basin management

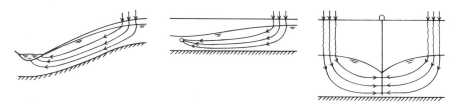

Figure 1.   Removal of renovated wastewater from aquifer by drainage to
surface water (left), interception by drains (center), or
pumping from wells (right).

on infiltration rate and on quality of renovated water, some basins were
vegetated while others were in bare soil (with or without native vegetation)
or covered with a 10-cm layer of fine gravel. Overall, bare-soil basins
with native vegetation were the most desirable.

Infiltration rate decreased during flooding due to clogging of the
soil surface by accumulation of solids and by biological activity. However,
infiltration rate recovered during drying due to shrinking and cracking of
the clogged layer and decomposition of the organic materials. Maximum
hydraulic loading rates (about 100 to 130 m per year at a water depth of
30 cm) were obtained with flooding periods of 2 to 3 weeks alternated with
drying periods of 10 days in summer and 20 days in winter. At this rate,
1 ha of basin area can infiltrate 1 to 1.3 million m^3 effluent per year.
The quality of the renovated water, however, is better when lower hydraulic
loading rates are used.   Thus, there is a trade-off between quality and
quantity of renovated water produced. From a quality standpoint, hydraulic
loading rates of 50 to 60 m per year are preferred. These were achieved
with flooding periods of 9 days alternated with drying periods of 12 days,
and by using a water depth of 15 cm.

Most of the nitrogen in secondary sewage effluent is in the ammonium
form. When this effluent percolates through soil, ammonium is adsorbed by
the clay and organic matter. Before the ammonium adsorption capacity of the
soil is reached, infiltration of sewage should be stopped. During the en-
suing drying period, atmospheric oxygen moves into the upper part of the
soil profile. This produces nitrification of adsorbed ammonium. Some of
the nitrate thus formed is subsequently denitrified in microanaerobic zones
that occur in the otherwise aerobic zone. When flooding is resumed, the
nitrates not denitrified during drying are leached by newly infiltrating
effluent to deeper, anaerobic zones where additional denitrification can
take place. These nitrogen transformations removed about 30% of the nitrogen in the effluent water at hydraulic loading rates of 100 to 130 m/yr, and
60% at loading rates of 50 to 60 m/yr. The nitrogen remaining in the renovated water was mostly in the nitrate form and was concentrated in the
effluent water that infiltrated at the beginning of a flooding period.
When this water reached the intake of an observation well, nitrate peaks
were observed in samples of the renovated water. Rapid-infiltration systems

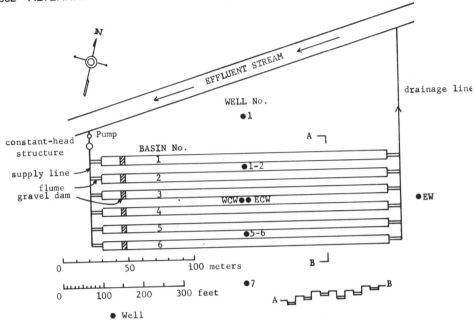

Figure 2. Plan of infiltration basins and observation wells of Flushing Meadows Project.

generally should be managed for maximum removal of nitrogen, because low
nitrogen concentrations in renovated water are important when such water is
used for drinking (recommended maximum concentration of nitrate-nitrogen in
drinking water is 10 mg/l) or for irrigation.

Most of the phosphorus in secondary effluent occurs as PO_4, which is
precipitated as calcium phosphate in the soil and aquifer material of the
Flushing Meadows Project. The PO_4-P content of the sewage effluent in this
project decreased from 15 mg/l in the late 1960's to about 9 mg/l in 1976.
At this low P concentration, the phosphorus removal was 60 to 80% after the
first 9 m of vertical travel. Renovated water sampled from wells outside
the basins (Figure 2) showed more than 90% removal of P.

Fluorine concentrations in the effluent decreased from about 5 mg/l
in the late 1960's to about 2.5 mg/l in the mid-1970's. Fluorides precipitate in the soil as calcium fluoride and with phosphate as fluorapatite.
Irrigation water should not contain more than 1 mg/l fluoride (National
Academy of Sciences - National Academy of Engineering, 1973).

Boron is not removed from the sewage water as it percolates through
the sandy soil of the Flushing Meadows Project. Fortunately, boron concentrations in the secondary effluent decreased from about 0.9 mg/l in the
late 1960's to about 0.5 mg/l in the mid-1970's. Concentrations of most
heavy metals in the sewage effluent were below maximum limits for drinking
or irrigation water (Bouwer and Chaney, 1974, and references therein).
Some metals were immobilized in the soil.

The biodegradable organic matter in sewage effluent, as expressed by the biochemical oxygen demand (BOD), is completely mineralized by aerobic and anaerobic bacteria in the soil. Thus, BOD-values of renovated water were essentially zero. There are, however, also organic constituents in sewage effluent that are not decomposed or adsorbed in the soil. These so-called refractory organics were present at an average concentration of 4 to 5 mg/l (expressed as total organic carbon) in the renovated water from the Flushing Meadows Project. Since carcinogens have been identified in the refractory organics of reclaimed wastewater (Nupen and Hatting, 1975), direct reuse of renovated water for drinking is not yet recommended and more research is needed on the possible health effects of the refractory organics (Ongerth, et al., 1973). The World Health Organization tentatively has set a maximum limit of 5 mg/l for the total organic carbon in reclaimed wastewater used for drinking (Shuval, 1975, and references therein). In-direct use of sewage effluent for drinking, of course, takes place wherever drinking water is obtained from sewage-polluted streams or lakes.

Bacteria in sewage effluent are filtered out at the soil surface or within the soil. Once immobilized, they die relatively rapidly because of antagonistic effects from the indigenous soil microorganisms. Renovated water sampled at a depth of 9 m below the infiltration basins of the Flushing Meadows Project generally contained 0 to 200 fecal-coliform bacteria per 100 ml, compared to about 10^5 fecal coliforms per 100 ml of secondary effluent. Additional lateral travel of about 60 m in the aquifer was necessary, however, to produce renovated water free from fecal-coliform organisms. The secondary sewage effluent contained 1.6 to 75 virus units per liter, but viruses were completely removed after 6 to 9 m vertical movement below the basins (Gilbert, et al. 1976). Virus immobilization in the soil is a function of clay content, pH, and concentration of salt and divalent cations in the effluent. Infiltration of rainfall can remobilize previously adsorbed viruses in the soil and cause them to penetrate to greater depths and possibly reach groundwater (Lance, et al., 1976). Thus, infiltration basins should not be dried when rain is forecast.

Renovated water can be collected from aquifers by natural drainage into streams, by wells if the water table is deep, or by horizontal drains if the water table is shallow (Figure 1). Wells and drains can be located so that all the water that infiltrates as sewage effluent will be collected as reno-vated water. This may be important where it is legally advantageous to intercept all the renovated water before it becomes part of the general groundwater, or where the native groundwater in the aquifer should be pro-tected against encroachment of renovated sewage water. To prevent renovated water from spreading in the aquifer, infiltration and pumping rates should be controlled so that the water level in observation wells at the outer perimeter of the basins (Figure 1, right) remains at the same height as the water table in the aquifer outside the recharge system. For additional in-formation regarding underground flow systems, see Bouwer (1974). Where all the infiltrated sewage water is collected as renovated water, a portion of the aquifer is actually used as a tertiary treatment system for renovating wastewater. Sewage effluent can be renovated with such a system at sub-stantially less cost than by equivalent treatment in a tertiary treatment plant.

SUMMARY

Sewage effluent is a valuable water resource, particularly in water-short areas, where it may be used for irrigation of crops. For public health reasons effluent used for irrigation of crops consumed by humans must undergo more treatment than effluent used for irrigation of forage or fiber crops. Nitrogen, boron, chloride, and other substances in the sewage water may decrease crop yield or quality. Dissolved salts may restrict irrigation with sewage to salt-tolerant crops and special farming techniques may be required to minimize yield reductions due to salinity. The value of sewage water for irrigation and other reuse can be enhanced by additional treatment to remove bacteria, viruses, suspended solids, and biodegradable organics, and to reduce the concentration of nitrogen and other constituents. This treatment can be effectively obtained by groundwater recharge with infiltration basins. Under favorable hydrogeological conditions, 50 to 100 m of secondary sewage effluent may infiltrate per year into the soil. As this effluent percolates downward to the underlying groundwater, many of its constituents are removed by physical, chemical, and biological processes in soil. Thus, the effluent becomes renovated water that can be collected for reuse with underground drains or by pumping from wells. The renovated water can be used for unrestricted irrigation (including fruit and vegetable crops consumed raw by humans), recreational lakes, and certain industrial applications (cooling or processing water). Reuse for drinking water is not yet recommended because some of the organic compounds are not decomposed or adsorbed in the soil, and thus remain in the renovated water. More research is needed to identify these refractory organics and their effects on human health when ingested in drinking water.

REFERENCES

Ayers, R. S. (1975), Quality of Water for Irrigation, in Proceedings, Conference on Irrigation and Drainage in an Age of Competition for Resources, American Society of Civil Engineers, Logan, Utah.

Bouwer, H. (1974), Design and Operation of Land Treatment Systems for Minimum Contamination of Ground Water, Ground Water, 12, 140-147.

Bouwer, H. (1976), Use of the Earth's Crust for Treatment or Storage of Sewage Effluent and Other Waste Fluids, Critical Reviews in Environmental Control, Chemical Rubber Company Press, 6, 111-130.

Bouwer, H. and R. L. Chaney, (1974), Land Treatment of Wastewater, in N. C. Brady, ed., Advances in Agronomy 26, Academic Press, Inc., New York.

Gilbert, R. G., et al., (1976), Virus and Bacteria Removal from Wastewater by Land Treatment, Applied and Environmental Microbiology, 32, 333-338.

Lance, J. C., et al, (1976), Virus Movement in Soil Columns Flooded with Secondary Sewage Effluent, Applied and Environmental Microbiology, 32, 520-526.

National Academy of Sciences - National Academy of Engineering, (1973), Water Quality Criteria 1972, Environmental Protection Agency Report R3-73-033, Washington, D. C.

Nupen, E. M. and W. H. J. Hatting, (1975), Health Aspects of Reusing Wastewater for Potable Purposes, in K. D. Linstedt and E. R. Bennett, eds., Proceedings, Workshop on Research Needs for the Potable Reuse of Municipal Wastewater, Environmental Protection Agency Report 600/9-75-007, Washington, D. C.

Ongerth, H. J., et al., (1973), Public Health Aspects of Organics in Water, Journal of the American Water Works Association, 65, 495-498.

Shuval, H. I. (1975), Evaluation of the Health Aspects of Reusing Wastewater for Potable Purposes in Israel, in K. D. Linstedt and E. R. Bennett, eds., Proceedings, Workshop on Research Needs for the Potable Reuse of Municipal Wastewater, Environmental Protection Agency Report 600/9-75-007, Washington, D. C.

Weinberger, L. W. (1966), Solving Our Water Problems - Water Renovation and Reuse, Annals New York Academy of Science, 136, Art. 5, 131-154.

REUSE OF WASTEWATER
IN DESERT REGIONS

K. James DeCook
Water Resources Research Center
University of Arizona
Tucson, AZ 85721

Paper Presented to Conference on
Alternative Strategies for Desert
Development and Management,
United Nations Institute for
Training and Research and
California Department of Water
Resources, Sacramento, California,
31 May - 10 June, 1977.

ABSTRACT

In a developing desert region, reuse of wastewaters can serve to enhance economic efficiency and improve environmental preservation in water-using activities, through incremental substitution of wastewater for groundwater under limited water-resource conditions. Municipal wastewaters can be used for irrigation of certain commercial crops without advanced treatment, although dilution with other sources of water is necessary to achieve the optimal application of irrigation water and plant nutrients. For most other uses, wastewaters must receive selective advanced treatment to meet water quality criteria. At Tucson and Phoenix, Arizona, municipal wastewaters have been allocated to irrigation of forage, fibre, oilseed, and vegetable crops, and to landscape irrigation and recreational lake replenishment in a regional park; additional allocations of wastewaters are pending for use by the copper mining-milling industry and for power generating plant cooling. Environmental hazards of wastewater disposal are reduced or avoided by planning and management procedures which include control, treatment, and application of wastewaters to beneficial use.

In the formulation of plans for development in desert or arid regions, water resources are implicitly limited and are therefore critical to any planning strategy. Furthermore, virtually any use of water will yield a by-product of wastewater in some form. If not properly included in planning, wastewaters commonly will effect diseconomies or will produce environmental degradation. On the other hand, when wastewaters are efficiently managed they can become a supplementary resource which will serve some water-using activity.

WATER REUSE IN AN ARID REGION

Therefore in a developing country, and more specifically in a developing natural resource community, the recovery, treatment, storage, conveyance and reuse of wastewaters should be considered along with water supply in the earliest stages of planning. In any event, disposal of wastewater without further use should be the last alternative. Such waters commonly cannot be discharged into streams because in a desert region streams are ephemeral and their assimilative capacity is negligible; also, subsurface disposal may be impractical without expensive treatment because of the danger of polluting the ground-water reservoir, which is likely to be the sole source of primary water supply. Accordingly, in such an environment a wastewater discharge should not be regarded as a waste product (with negative value) but rather as a resource with positive value as an input factor for some beneficial reuse.

Wastewater Sources and Potential Uses

Wastewater discharges which emanate from partially consumptive uses and which can be collected at point sources may be classified as municipal (principally domestic) sewage effluents and industrial process or cooling effluents. Industrial discharges are so varied in quantity and quality that generalization is difficult, and each must be evaluated individually. Sewage effluents, however, are common to any area of development where a concentration of resource production, and therefore population, occurs.

Accordingly, for the purpose of this presentation municipal wastewater will be cited as the principal example. This kind of wastewater source commonly is concentrated (1) at a treatment plant served by an integrated (urban or metropolitan) sewerage collection system and employing a conventional primary-secondary treatment, or (2) at a relatively smaller plant serving a local residential or commercial service area, and generally utilizing an oxidation pond or a "package plant" treatment mode.

Appropriate kinds of use to which the treated wastewaters can be applied
are classified herein as follows:

1. Irrigation

    Forage, Fibre, and Oil Producing Crops

    Orchard

2. Industry

    Cooling or Process Water

3. Recreation

    Fishing and Boating

    Landscape (Park and Playground) Irrigation

Matching Source to Use

Conceptually, a specific wastewater can be directed to a particular use
by subjecting it to appropriate treatment. Which treatment processes are
required for each source-to-use combination can be determined by comparing
the quality occurring at the source with the criteria relevant to the use.
The procedure is to identify a quality vector for influent flow to the
intervening treatment process and a quality vector for effluent flow
following treatment. These vectors are simply an ordered listing of all
the significant water quality parameters. The influent and effluent flow
vectors are linked by the parameters which are critical in common to both
the source and the use. These "critical parameters" determine the treatment
process or combination of processes that will be required.

The conventional (primary, secondary) treatment methods applied to
municipal effluents are in many respects standardized and well known,
although there is room for increased efficiency. Primary treatment reduces
coarse, settleable solids, removes some amount of suspended organic materials
by settling, and reduces biological oxygen demand (BOD_5) by as much as 35
percent. Secondary treatment methods (trickling filter, activated sludge,
or oxidation pond) consist of some form of biological oxidation and partial
settling of the residual organic materials, with a resulting cumulative
reduction in BOD_5 of up to 95 percent with the activated sludge process and
somewhat less in the other processes (FWPCA 1968, p. 4-5).

At the advanced treatment level, the general classes of critical parameters and the corresponding treatment processes which effectively act upon
them are as follows, the underlined items being those specifically cited
for examples in Table 1 below: (1) Suspended solids removal--lime or alum
coagulation-sedimentation (=chemical clarification), sand or mixed-media
filtration, and microstraining; (2) inorganic removal--electrodialysis,

Table 1. Matrix of Treatment Methods for Wastewaters Relative to Type of Reuse.

Source \ Use	A_1 Irrigation: Field and Forage Crops	A_2 Irrigation: Fibre and Oilseed Crops	A_3 Irrigation: Orchard	A_4 Irrigation: Produce	R_1 Recreation: Fishing and Boating	R_2 Recreation: Landscape Irrigation
Domestic-Industrial Effluent: City Plant or Separable Source	PRIM SEC	PRIM SEC	PRIM SEC	PRIM SEC CHLR	SEC CHEM FILT CHLR	PRIM SEC CHLR
Domestic-Industrial Effluent: County Plant or Discrete Source	OXP	OXP	OXP SEC	OXP SEC CHLR	OXP CHEM FILT CHLR	OXP SEC CHLR

Source \ Use	I_1 Industry: Cooling	I_2 Industry: Mining and Milling	S_1 Storage (Surface)	S_2 Storage (Subsurface)
Domestic-Industrial Effluent: City Plant or Separable Source	SEC CHEM FILT	SEC CHEM FILT	SEC CHEM FILT	SEC FILT CHLR
Domestic-Industrial Effluent: County Plant or Discrete Source	OXP CHEM FILT	OXP CHEM FILT	OXP CHEM FILT	OXP SEC FILT CHLR

Key to terms: PRIM-Primary; SEC=Secondary, Activated Sludge; OXP=Oxidation Pond; CHEM=Chemical Clarification, generally by alum or lime coagulation and sedimentation; FILT=Sand Filtration; CHLR=Chlorination.

distillation, freezing, ion exchange, and reverse osmosis; (3) organic removal--
activated carbon adsorption, advanced oxidation, foam separation, and
filtration; (4) nutrient removal--precipitation of phosphate, biological
denitrification of nitrate-N, air stripping of ammonia-N, and sand (soil)
filtration; (5) removal of turbidity, color, odor, and toxic substances--
activated carbon adsorption; and (6) removal of pathogenic microorganisms--
chlorination.

Application of these methods to the source-to-use matrix for wastewaters
through the appropriate critical parameters produces a "treatment-method
matrix," as detailed in Table 1.

Unit costs of the appropriate treatment combinations at various plant
scales can be estimated for a specific design and location, but it is
difficult to generalize because of the widely diverse locations of the
world's desert regions. Until recently the unit total costs for primary
and secondary treatment in southwestern United States (Arizona) could be
estimated approximately as follows: At a scale of 2.0 million gallons per

day (mgd), $0.10 U.S. per 1000 gallons or $30 U.S. per acre-foot (ac-ft)
for the oxidation pond, and about the same for primary treatment; at a 10-mgd
scale, these same figures were roughly applicable to the trickling filter
method, but the activated sludge process cost approximately $0.15/1000 gal
or $40/ac-ft.

It is stressed that these figures are given only as an example,
and that actual costs will vary widely with location, size of plant, and
time, because of fluctuations in costs of labor, materials, and especially
energy.

It may be pointed out, incidentally, that treatment plants can be built
in the warm desert regions without the necessity of protection from freezing,
thereby effecting considerable savings in capital expenditure.

The advanced treatment processes are perhaps even more sensitive to the
variable factors mentioned earlier, than are the conventional processes;
no estimate of costs will be attempted here.

AN ECONOMIC EXAMPLE: TUCSON AND PHOENIX

Wherever a concentration of population may develop, a municipal wastewater
supply will accrue concurrently, and its form of treatment and allocation
to reuse will depend in large measure upon the proximity of other water-use
activities. Commonly, irrigated croplands are located nearby, and the
primary-secondary treated wastewater may be utilized in part for irrigation,
with no further treatment needed for many kinds of crops. Also, electrical
generating plants, industrial activities, and parks or recreational areas
may be located in the urban region, and a multiple allocation of wastewater
to these users becomes apparent. The current wastewater production and
allocation conditions in two Arizona population centers--Tucson and Phoenix--
serve as an example.

At Tucson, about 40,000 acre-feet per year of secondary effluent is
produced at the wastewater treatment plant. For more than 20 years, a
substantial portion of the effluent was used for irrigation of forage
crops and cotton. Presently, except for a minor quantity being used on
the City treatment plant farm, it is being discharged to the otherwise
dry channel of the Santa Cruz River, where it flows, at times, as much
as 27 miles downstream. Some of the flow is recharged through the channel
sediments, and is contributing to an increase in nitrate concentration in
the ground-water reservoir beneath and adjacent to the river.

It would seem desirable to salvage this alternative water supply from
non-use and apply it to beneficial use in agriculture, mining, or recreation.

Some or all of the treated wastewater could be conveyed by gravity flow
to the Avra-Marana agricultural area, several miles to the northwest, for
cropland irrigation. The wastewater nutrients which under present conditions
are contributing to pollution of the ground-water basin would be beneficial
to the farmer if he properly used the wastewater in irrigation (Cluff and
DeCook, 1974). This type of use will be assumed in the following illustration.

The farmer can view the prospective wastewater supply as an alternative
to his present pumped ground-water supply, and can make a direct cost comparison,
since these are the only sources of supply in the absence of surface water
in the desert region. This is illustrated in a conceptual way as follows
(DeCook, 1970). To evaluate the effect of imposing a wastewater supply on
the ground-water supply presently available, one may consider the supply
curves for both sources (Figure 1), from which two kinds of effects can
result. First, the wastewater might act to augment the existing supply
and move the water use level Q higher (to the right) under the prevailing
demand curve. However, since agricultural water-use acitivites commonly
are constrained not only by water shortage but by other factors, the added
supply should not be expected to result in a directly proportionate increase
in use by expanded irrigation. Rather, a different kind of effect will
operate, i.e., incremental substitution of the wastewater into the water
supply function in order to minimize costs at an unchanged level of use
Q on the existing acreage.

This substitution in turn can take place in one of two ways, depending
upon the relative shapes of the supply functions; these are illustrated in
the two parts of Figure 1. In the upper diagram, the ground-water supply
curve would be followed up to quantity q_x, and the wastewater supply would
be used for additional quantities up to the level of use Q. The net gain
to the system would be the shaded, roughly triangular area as shown. In
lower diagram the wastewater supply would be utilized up to quantity q_x,
and ground water would supply additional needs to level Q, while additional
wastewater would become a disposal quantity or become available for other
allocation. The net gain under this set of conditions also is indicated
by the shaded area. In addition to the net gain as illustrated for a given
year's use, increasing gains will be realized in subsequent years if the
ground-water supply function shifts upward relative to the wastewater supply
function. This is likely to be the case, because time-related increases in
cost of ground water are primarily a function of increasing pumping lifts
compounded by rapidly increasing energy costs. The controlling factor in
supply cost for wastewater is treatment cost. This may be expected to in-

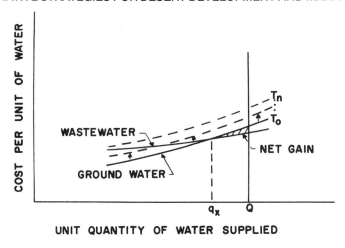

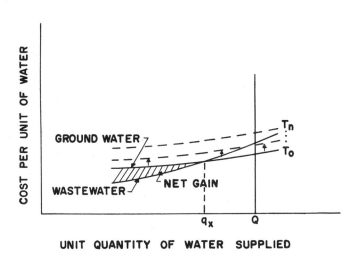

FIGURE 1.   EFFECTS OF SUBSTITUTION OF WASTEWATER
FOR GROUND WATER.

crease with time also, by escalation of material and labor costs for
construction of treatment facilities as well as energy costs; however, such
increases may be tempered by the effects of improved technology leading to
higher efficiency in selective treatment methods, and economies of scale
due to ever-increasing effluent loads.

Two other forms of potential time-related incremental benefit from
the combined supply function can be postulated-- (1) an increasing future
value of the ground water not pumped as a result of the substitution, and
(2) the benefits (or avoided losses) of precluding eventual local ground-
water depletion which might necessitate relocation or extinction of the
agricultural (irrigation) activity.

The relationship illustrated in principle in Figure 1 was plotted
with real data by DeCook (1970) for actual supply conditions near Tucson.
Figure 2 shows the relative costs of supplying ground water and wastewater
to irrigate field and forage crops and cotton. This example assumes that
the ground-water supply is provided from a diminishing stock resource, a
condition described by Jacobs (1968) in his formulation of a ground-water
supply function for Tucson domestic use by diversion from agricultural
use. The "ground water" curve represents the cost of supplying ground
water in 10-mgd increments (numbered "diversion units") from increasingly
distant well fields to a central distribution point, in this case for
agricultural use. The "salvaged water" curve represents treated wastewater, also available for use at the same point. The conveyance and storage
costs are "netted out" in order to compare directly the variable costs for
pumping ground water and for treating wastewater.

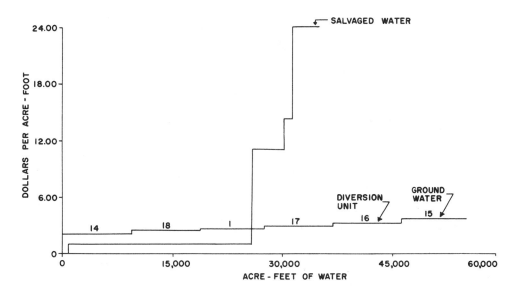

FIGURE 2.   SUPPLY FUNCTIONS FOR GROUND WATER AND SALVAGED WATER TO SELECTED AGRICULTURAL
USES, TUCSON REGION, 1970.

Source: Ground-water data based on Jacobs (1968).

Implicit in this graph is the recognition of treatment of domestic wastes by the water agencies as a service function, since disposal is necessitated in any event. The water users in the example paid only $1.00/ac-ft by contract. Under these conditions the wastewater supply curve is the cost-minimizing function for approximately the first 25,000 acre-feet of supply, at which point the cost veers sharply upward and ground water becomes the least-cost supply.

In the same context as Figure 2, the supply function for wastewater can be extended to all the possible uses considered. In that case the unit costs determining each increment of the supply function would not represent a blanket cost for treatment, but rather the costs of required treatment relative to each use. If higher uses require higher quality, the cost of advanced treatment would be added. Similarly, if alternative sources of water or wastewater become available to the same user, he will find a similar supply function with respect to each source and can determine which source best fits his requirements.

Other potential uses for wastewater in the Tucson region include copper mining-milling processes and irrigation of pecan orchards. The location of these water-using activities is a region 20-30 miles (32-48 kilometers) south of Tucson and about 300 to 900 feet (90-270 meters) higher than Tucson. In that region all uses are presently served by a diminishing ground-water supply, and the wastewater is a possible alternative; however, its conveyance from the Tucson wastewater treatment plant would require capital investment for a pressure pipeline and considerable energy for upgradient pumping. Also, in the case of the mines, it is necessary to determine the type and cost of appropriate advanced treatment processes before the wastewater can be utilized in the copper ore milling process.

An additional actual use of wastewater in the Tucson region is occurring in the recreational field. Randolph Park, a municipal park which includes about 400 acres of landscaped picnic and playground areas, a baseball park, a golf course, and a small lake for scenic enjoyment and fishing, is supplied by the liquid effluent fraction from a satellite "package" treatment plant. This type of plant generally incorporates processes of primary and secondary clarification, an extended aeration treatment, and chlorination. This particular plant is designed for a capacity of about 2.0 mgd, tapping a sewer main and thereby providing a measure of relief to the heavily loaded sewage collection system. Treated effluent from the plant is used in both lake filling and park irrigation, as a partial substitute for high quality ground water.

Turning to the region of Phoenix, Arizona, we see a quite different situation relative to use of wastewater; a multiple allocation of the municipal wastewater there, in a quantity of 75-80,000 ac-ft/yr, has already been implemented. The Buckeye Irrigation Company west of Phoenix utilizes approximately 30,000 ac-ft/yr of the secondary effluent from the large Phoenix treatment plant, through a long-term contract. The wastewater, being less saline than much of the ground water in the district, has been used by the Buckeye irrigators since 1962, in partial substitution for the ground-water supply. This importation of wastewater and concurrent decrease in ground-water pumpage has produced an improved hydrologic balance in the district as well as an improvement in quality of irrigation water. The irrigation company obtains the wastewater at a delivery price not less than $1.50/ac-ft, the exact price being determined monthly according to a stipulated formula. In this instance, as seen earlier, the cost of treatment at the plant is a public cost incurred as a service function and is not borne by the water user. Neither is advanced treatment required by the crops irrigated in this area; therefore wastewater cost to the user is relatively quite low.

An additional benefit is the fertilizer value in the wastewater; it contributes not only to the water requirement but to the nutrient requirement of the crops to which it is applied. The equivalent value of commercial fertilizer was estimated by Cluff et al. (1972) as about $5/ac-ft of wastewater, but in 1977 it is undoubtedly worth much more. In terms of available nitrogen, the secondary effluent actually contains more than that needed by most crops, so that a blend or dilution of wastewater and ground water is most beneficial, the optimal mix depending upon the specific crop-soil-water combination.

Wastewater also is to be used for crop irrigation, under different conditions, in a proposed allocation of effluent from the City of Phoenix to the Roosevelt Irrigation District. The effluent will receive a tertiary treatment by filtration through natural soil using spreading basins and a pump-back well system. Improvement of wastewater quality attained by this method as described by Bouwer (1973) is expected to be adequate to serve unrestricted crop irrigation, recreational facilities, or some industrial uses. As for crop irrigation, the Roosevelt Irrigation District includes acreages of not only small grains, hay and pasture, and cotton, but also a substantial acreage of vegetables such as lettuce and onions, and melons. Presumably any and all of the crops in the District can be irrigated with the product water of the filtration process.

Where municipal wastewater is used for irrigation of field vegetables which are to be consumed by humans without cooking or processing, such as lettuce, appropriate treatment must be applied to avoid the incipient hazards of possible ingestion of viruses, heavy metals, or organic toxins. Where wastewater is used without advanced treatment, as near Phoenix and Tucson, the cropping pattern commonly includes alfalfa, barley, oats, sorghum, and cotton. Additionally, secondary effluent can be used to irrigate certain oil-producing crops for which there is a growing market, as in southwestern United States and northwestern Mexico. These include safflower, linseed, and jojoba, which is a substitute for sperm whale oil and for certain petroleum derivatives, both of which are becoming increasingly scarce on the world market.

A further allocation of wastewater in the Phoenix area has been arranged through a contract option whereby the proposed Palo Verde nuclear power plant west of Phoenix would purchase, treat, and utilize approximately 35,000 ac-ft/yr of wastewater for each of three generating modules, to be completed over a period of ten years.

A fossil fuel-fired generating plant generally has better thermal efficiency, and thus requires less water for cooling, than a nuclear plant (about 14 ac-ft/yr/MWe for fossil-fueled vs. 22 ac-ft/yr/MWe for nuclear). In southwestern United States, an urban population generates approximately three times as much wastewater from its sewage collection system as is required to cool a nuclear generating plant large enough to serve the electrical energy requirement of that population. If the plant is coal-fired, the same urban wastewater production becomes about five times the cooling requirement. In either instance it is evident that in an urban area, local geographical conditions permitting, the total cooling water requirement for power can be supplied and a large fraction of the wastewater will remain for allocation to other uses.

The numerical values in this relationship would be different in other desert regions of the world; the per-capita wastewater production may be lower, but the per-capita energy requirement likely would be lower also. The principle of supplying cooling requirement for electrical energy by use of wastewater would still hold.

ENVIRONMENTAL CONSIDERATIONS

Correlative with the technical feasibility and economic advantages of wastewater reuse, its environmental implications are an important aspect of resource planning, especially in an arid region. It is clear, as stated earlier, that simple disposal in a watercourse without adequate assimilative capacity is environmentally damaging, and in fact will no longer be permitted in the United States under the Federal Water Pollution Control Act of 1972, Public Law 92-500. Neither is underground disposal acceptable without adequate treatment; off-season subsurface storage by spreading or well injection is a useful management alternative, but the injection water commonly requires extensive treatment, especially for injection, to be made compatible with native ground waters.

In the various forms of wastewater reuse, environmental protection or enhancement requires careful wastewater management practices. For example, where treated municipal effluent is applied to field crops for irrigation, analysis of the wastewater, the soil, and the plant material must be made to determine actual nitrogen requirement and utilization by the plant, and the appropriate dilution or proportion of wastewater with ground water must be maintained, so that excessive nitrogen will not be leached through the soil column and into the ground-water reservoir.

Provided such steps are taken and adequate surveillance is made by the wastewater user or wastewater management agency, the wastewater products resulting from resource development become in turn a resource of economic value and environmental acceptability, contributing to the conservation of primary water resources.

ACKNOWLEDGMENTS

Portions of the work cited herein were supported by the State of Arizona and by the Office of Water Research and Technology, U. S. Department of the Interior, pursuant to the Water Resources Research Act of 1964, as amended.

REFERENCES CITED

Bouwer, H., Renovating Secondary Effluent by Groundwater Recharge with Infiltration Basins, in Recycling Treated Municipal Wastewater and Sludge Through Forest and Cropland, W. E. Sopper and L. T. Kardos, eds., The Pennsylvania State University, 1973.

Cluff, C. B., K. J. DeCook, and W. G. Matlock, Technical, Economic and Legal Aspects Involved in the Exchange of Sewage Effluent for Irrigation Water for Municipal Use, University of Arizona Water Resources Research Center, Tucson, December 1972.

Cluff, C. B., and K. J. DeCook, Metropolitan Operated District for Sewage Effluent-Irrigation Water Exchange, Hydrology and Water Resources in Arizona and the Southwest, Vol. 4, University of Arizona, Tucson, 1974.

DeCook, K. J., Economic Feasibility of Selective Adjustments in Use of Salvageable Water in the Tucson Region, Arizona, University of Arizona Ph.D. Dissertation, 1970.

Federal Water Pollution Control Administration, Summary Report, Advanced
    Waste Treatment Research Program, Water Pollution Control Research
    Series Publ. No. WP-20-AWTR-19, 1968.

Jacobs, J. J., An Economic Supply Function for the Diversion of Irrigation
    Water to Tucson, University of Arizona M.S. Thesis, 1968.

Considerations in Planning for
the Disposal of Wastes
in Desert Environments

James J. Geraghty

ABSTRACT

There are two important hydrologic distinctions between arid regions and
humid regions that must be taken into account in planning methods of disposal of
wastewater or other contaminated fluids. First, deserts do not have networks of
streams that can be used to transport wastes far away from their points of
origin, and second, the net loss of water in deserts through evapotranspiration
is generally much higher than in humid areas, so that wastewaters may be lost
entirely to the atmosphere unless they can be stored. The absence of streams
guarantees ultimate degradation of ground-water quality if the wastewaters are
allowed to infiltrate into the earth at high rates through lagoons, pits, septic
tanks, or other similar systems. Exposure of the wastewaters to the atmosphere
for long periods of time will cause precipitation of their dissolved minerals,
which may constitute a benefit in terms of potential recovery of mineral resources,
but which nevertheless results in the complete loss of another vital resource--
water. Assuming that there is no immediate need for reusing the waste water
(say for cooling, washing, or irrigation), the only way to conserve both the
water and its dissolved minerals for possible future use is by injection into
deep wells. Such injection wells can store the wastewater in deep aquifer zones
containing seawater or brines, with excellent prospects for its later recovery
in an essentially unaltered state.

Since the dawn of civilization, man has relied heavily on the presence of flowing water in streams and rivers as a principal method of disposing of the liquid wastes he produces. Most of the world's great cities and industries, for example, are located along rivers that traditionally have been used as natural sewers or drains to transport away waste substances of all kinds. Even where these facilities have been built at inland locations, it is common to construct long sewer mains or drainage channels so that the wastes can continue to be disposed of into surface waters. This happy solution simply is unavailable in most deserts, with the result that a creator of waste in the desert is obliged to turn to other alternative means of disposal.

Basically, there are several choices on what to do with a waste in the desert. Solid wastes can of course simply be stockpiled in dumps or landfills on the ground, which also is the usual practice in humid areas. Liquid wastes can be spread on the ground, discharged into natural topographic depressions, or placed into specially dug pits or lagoons. In some instances, a subsurface sewage-treatment system (like a cesspool, drain field, or septic tank) can satisfactorily serve this purpose. One other alternative, which is discussed in some detail later in this paper, is to dispose of the wastes through deep injection wells.

Inevitably, however, in the absence of some kind of surface-water transport mechanism, any waste discharged in a desert will for all practical purposes remain in the desert, where, if not dealt with properly, it can become a threat to the potability of whatever ground waters may be present. Evaporation rates, of course, are high in deserts, so that waste liquids exposed long enough to the atmosphere will lose their water content and be reduced to dry residuals of salts. If waste liquids are discharged at low rates over large tracts of land, for example, so that evaporation can keep pace with the discharge, little or none of the liquid may have an opportunity of seeping downward to contaminate the ground-water system. Instead, the dissolved substances in the liquids are precipitated out as salts on the land surface and in the soil zone. The valuable water component of the waste is thereby lost forever to the atmosphere, with no possibility of recovery or reuse. Obviously, all efforts should be made to prevent such losses in desert regions where water has a great value.

If, on the other hand, a waste liquid is poured onto the ground continuously on a small piece of desert land, then the liquid will start to move downward through the soil and ultimately will arrive at the water table. Once in the water-table aquifer, the wastes will no longer be subject to evaporation and will begin to travel slowly in the direction of the prevailing ground-water flow. The body of contaminated ground water resulting from this process will not mix or diffuse rapidly with the native ground water, but will stay relatively intact in what hydrogeologists call a "plume" or "streamer" in the subsurface environment. Such plumes may, over long periods of time, extend down-gradient for miles from the point of injection. They are invisible to the naked eye and constitute an ever-present threat to water wells in the region.

Regardless of whether the waste liquid is spread out on the land surface, disposed of into a dry streambed or wadi, or emplaced into pits or lagoons dug specially to handle the wastes, the potential for downward movement of the liquids into the subsurface environment is high. Lagoons in particular, if the rate of inflow is large, constitute a very direct route to the subsurface formations, simply because they provide little opportunity for the fluids to be evaporated before they seep out through the lagoon bottom. Injection of the wastes through a subsurface drain or septic system provides absolutely no opportunity for evaporation to take place, with the result that the full volume of the waste fluid inevitably will drain into the water-table beds.

Disposal of waste fluids into dry stream beds is often thought of as a useful technique in the desert because occasional flash floods tend to carry the wastes miles away from the point of disposal. However, this technique also constitutes a way of contaminating ground waters, because the wastes cannot move very far away before the rate of flow diminishes enough to allow them to seep downward into the soil. To repeat, there is essentially no way of preventing ultimate contamination of local ground water by liquid wastes in a desert environment if there is no stream nearby that could transport the water away over a great distance.

The principal sources of contamination related to waste-disposal practices are:

1. Industrial Waste-Water Impoundments

2. Landfills and Dumps

3. Septic Tanks and Cesspools

4. Collection, Treatment, and Disposal of Municipal Waste Water

5. Land Spreading of Sludges

6. Brine Disposal from Petroleum Exploration and Development

7. Disposal of Mine Wastes

8. Disposal Wells

9. Disposal of Animal Feedlot Wastes

In general, domestic sewage does not contain large amounts of objectionable substances, so that contamination of ground water from this source is of less concern than contamination from industrial sources. The principal constituents of sewage that may create problems are pathogenic organisms (which tend to travel only very short distances in most ground-water environments) and nitrates. By contrast, industries may produce wastes containing large amounts of hazardous or toxic materials which, if they arrive at a water well, may cause serious illnesses or impair the use of the water for many purposes.

Figure 1 depicts schematically how contaminated liquids derived from different kinds of human activities at the land surface can eventually find their way downward into a ground-water system. Many of the pathways shown on the diagram are quite direct, as for example, the movement of liquid sewage from a house through a septic tank downward into the ground water and seepage from the brine residues from oil production that are placed into pits or lagoons. Other routes are more indirect. As shown on the diagram, for example, a waste liquid from a manufacturing plant might first go through a treatment process that produces a sludge residue, which then may be deposited in a dump exposed to the atmosphere. If a heavy rain falls on the dump, a contaminated leachate may be produced that can ultimately seep downward to the water table.

Figure 2 is a cross-section of a hypothetical aquifer system depicting how contaminants derived from various waste sources may move through aquifers. Note that the deep artesian fresh-water aquifer is not insulated against eventual degradation, even though it is overlain by a confining bed. The reason why the threat is real is that most confining zones have enough permeability to allow slow seepage across them, and even if the process takes long periods of time, the deeper water ultimately may become contaminated.

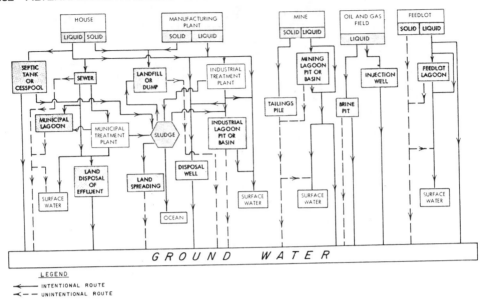

Figure 1 - Routes By Which Ground Water May Become Contaminated as a Result of Waste-Disposal Practices.

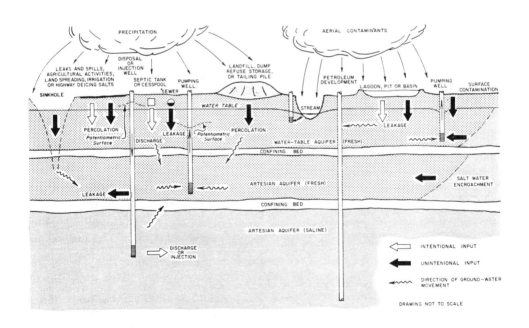

Figure 2 - Cross-Section Showing Mechanisms of Ground-Water Contamination.

Figure 3 illustrates how rainfall may leach contaminants from materials stockpiled on the land surface and transport them into an aquifer in such a way that they are drawn into a nearby pumping well. The principle involved, regardless of whether the waste liquid originates in a stockpile, a lagoon, a septic tank, or any other facility at the land surface, is essentially the same. The time it takes for the well to become contaminated depends on a number of hydrologic factors, such as distance between the waste source and the well, the permeability of the aquifer materials, the pumping rate, the depth of the well, and the presence or absence of confining zones.

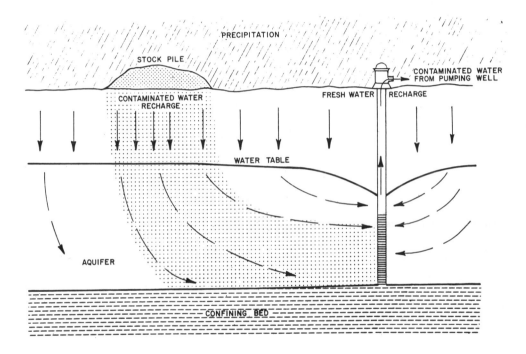

Figure 3 - Contamination of a Water Well as a Result of Leaching of Materials Stockpiled on the Land Surface.

The passage of contaminated fluids through the earth is generally extremely slow. In humid environments, for example, rates of ground-water movement may range from a couple of centimeters to as much as a meter per day, depending on the properties of the aquifer system and on the hydraulic gradient causing water to move through that system. In deserts, on the other hand, where the amount of natural replenishment from precipitation is small or even nonexistent, rates of ground-water movement may be very low indeed -- perhaps on the order of only a few millimeters per day, again depending on the hydrologic situation.

Such low rates of flow mean that a waste injected underground at a particular place may not arrive at nearby water wells for years or even decades. During all of that elapsed time, the fluid will have been moving in the direction of the wells, although water samples collected from the wells will not have detected its presence. The delay can of course instill a great feeling of confidence in the mind of the facility operator, who knows he is disposing of wastes at one point, but who can see no harmful effects on his drinking-water supply. All too often, the contaminated fluid begins to arrive at a well without the operator being aware of it, unless he is meticulously sampling the quality of the water on a continuous basis.

Specific statements cannot be made about the distances that contamination will travel, owing to the wide variability of aquifer conditions and types of contaminants. Also, each constituent from a source of contamination may have a different attenuation rate, so that the distance to which contamination is present will vary with each quality component. Nevertheless, certain generalizations which are widely applicable can be stated. For fine-grained alluvial aquifers, contaminants such as bacteria, viruses, organic materials, pesticides, and most radioactive materials are usually removed by adsorption within distances of less than 100 meters. However, most of the common ions in solution move unimpeded through aquifers, subject only to the slow processes of attenuation.

The shape and size of a plume depend upon the local geology, the ground-water flow, the type and concentration of contaminants, the continuity of waste disposal, and any modifications of the ground-water system by man, such as pumping from wells. Where ground water is moving relatively rapidly, a plume from a point source will tend to be long and thin; but where the flow rate is low, the contaminant will tend to spread more laterally to form a somewhat wider plume. Irregular plumes can be created by local influences such as pumping wells and variations in permeability.

Plumes ordinarily tend to become stable in areas where there is a constant input of waste into the ground. This occurs for one of two reasons: (a) the tendency for enlargement as contaminants continue to be added at a point source is counterbalanced by the combined attenuation mechanisms, or (b) the contaminant reaches a location of ground-water discharge, such as a stream or well, and emerges from the underground. When a waste is first released into ground water, the plume expands until a quasi-equilibrium stage is reached. If sorption is important, a steady inflow of contamination will cause a slow expansion of the plume as the earth materials within it reach a sorption capability limit.

When contamination of a well occurs under the kinds of situations discussed previously, the usual remedy is to stop the method of waste disposal by, for example, halting discharge into a lagoon. Cutting off the input of waste at the source, however, in no way solves the problem, simply because there already is a very long streamer or plume of contaminated fluid in the subsurface environment, which will continue to move toward the well as long as the well remains in operation. If the disposal process has been going on for years, it generally will take many years of continuous pumping before the body of contaminated fluid can even be partly evacuated from the subsurface environment. Thus, a hidden plume of contaminated ground water is, in a sense, like a timebomb ticking slowly away, so that when contamination of a water well begins to occur, it may be too late to do anything about it at all, other than abandon the well and try to find uncontaminated ground water a greater distance away.

In evaluating potential ground-water contamination threats in deserts, at sites where new communities or industries are to be built, it is essential to first determine the natural three-dimensional flow patterns in the ground-water system. For instance, practically all liquid wastes entering such a system first arrive in the shallow water-table aquifer, which may or may not be in use as a source of water for water wells. Or, as is the case in many desert environments, the principal source of drinking water may be a deep artesian aquifer that is at least partly insulated from the shallower beds. Under such conditions, it may be safe to continue pumping from the deep wells without running any risk of immediate contamination by wastes.

Nevertheless, even though the deep artesian beds may be separated from the shallow beds by layers of clay or other materials of low permeability, the potential still may exist for downward migration of contaminants through the confining layers, as mentioned previously. If the potentiometric surface in the deeper aquifer is lowered as a result of pumping, for example, a mechanism may be created for downward movement of the contaminated fluids, as illustrated on Figure 2. Again, depending on the permeability of the intervening layers and on the hydraulic gradient, such downward movement may take place rapidly or may occur over periods of time measured in decades or even centuries. Obviously, operators of industrial facilities in deserts should have a clear understanding of how these processes work and of the particular hydrogeologic factors that govern movement of waste fluids on and below the land surface.

One technique for disposing of waste liquids in the desert, which has the unique advantage of conserving for possible future recovery both the water and its dissolved minerals, is to inject the liquids through cased wells into very deep geologic layers that contain sea water or highly concentrated brines. Deep-well injection of this kind is not yet practiced on a wide scale throughout the world, although there are hundreds of such wells now in operation in the United States for disposal of sewage effluent and industrial wastes. In addition, there are many tens of thousands of such wells used for disposal of brines brought up to the land surface during the production of oil and gas.

In many hydrogeologic environments, the very deepest formations are largely insulated from the shallower ones by thick beds of extremely low permeability. Where such deep beds already contain non-potable waters, it is possible to inject contaminated fluids into them through cased wells, at rates depending on the water-bearing properties of the deeper beds. The rate of injection is governed essentially by the transmissivity of the receiving formation. As an example of an extremely favorable situation in this regard, it is possible in the State of Florida to inject as much as 40,000 cubic meters a day of waste fluid through a single well into deep limestone caverns that naturally contain only salty water. The pumping pressures required for the injection in Florida are extremely low, generally on the order of less than one Kg/cm^2 (14 pounds per square inch). In less favorable geologic environments, higher injection pressures may be needed, with proportionally higher energy costs.

The cost of installing a deep injection well to handle very large amounts of fluid can be rather high, depending on the depth that must be drilled to the receiving zone. The large-capacity injection wells in Florida referred to above, which are about 3,500 feet deep, cost on the order of $1,000,000 apiece, for example. Costs of such wells would of course vary widely in different deserts of the world, depending on depths, rates of injection, and other factors. The wells must be specially designed with multiple casings and cement liners in order to insure that none of the contaminated fluid can escape into shallow potable water zones. Usually, some type of ground-water monitoring is provided to detect any such possible escape.

An interesting feature of deep-well disposal is that the injected fluid does not mix to any significant degree with the native ground water. Instead, it moves radially outward into the injection zone as a discrete body, much in the way a balloon is blown up. The longer the injection continues, the larger becomes the body of injected fluid. In experiments conducted at several places in the United States, it has proven feasible to extract the fluid back out of the same well with a recovery rate of greater than 90 percent.

Injection into the subsurface environment, wherever the hydrogeologic conditions are favorable, is usually the least expensive way of storing large volumes of waste fluids for later recovery. This concept is being increasingly discussed by public officials in the United States who are trying to grapple with growing problems of water shortages, and who are attracted by the idea that sewage effluents or other wastewaters could be stored in the earth and later recovered for cooling purposes, irrigation, or other uses. The methodology is of special interest in the case of industrial wastes, which may contain chemical substances that are not now worth recovering by standard treatment methods but that may acquire a much higher economic value in the future.

In summary, the following concepts or guidelines should be kept in mind in planning for disposal of wastes in a desert environment:

1.   Except in the rare case where a stream is present close to a municipal or industrial facility in the desert, waste fluids cannot naturally be carried very far away from points of disposal, and therefore present a potential threat to the quality of the natural ground water unless they are completely evaporated.

2.   It is relatively easy in most desert situations to allow the sun to evaporate most, if not all, of the water in a waste liquid, leaving the contained minerals behind as dry deposits, but this represents a total loss of the valuable water.

3.   If large amounts of liquid wastes are disposed of continuously on a small tract of land (whether by spreading, discharge into lagoons or pits, or otherwise) the likelihood is high that some or most of the fluids will move directly downward to contaminate the shallow ground water.

4.   Contaminated fluids which have entered the water table will drift slowly along in the subsurface environment as a more or less intact body, and will not mix to any great degree with the native ground water (except after extremely long periods of time).

5.   The predominant direction of flow of a contaminated ground-water body in a desert is toward the nearest water wells. Pumping from these wells creates a lowering of ground-water levels that extends outward from the wells in all directions, and if the waste is somewhere within the zone of influence of the wells, it inevitably will arrive to contaminate the wells.

6.   Injection of waste liquids through deep wells, although it may be more expensive than other alternative waste-disposal methods, can eliminate the threat of contamination of potable ground waters and at the same time preserve in storage water and dissolved minerals that may have an important future economic value.

Treatment of Agricultural Waste Water

Donat B. Brice

Desalting studies were conducted from 1971 to 1975. Several designs of reverse osmosis pilot plants, possessing different types of semi-permeable membrane, were tested for their capability to desalt agricultural waste water. These tests established the technical feasibility of desalting agricultural waste water by reverse osmosis and served as a basis for selecting a larger pilot plant for additional study.

Study began in 1975 with the installation and startup of a 95-cubic-metre-per-day (25,000-gallon-per-day) desalting plant. The economic feasibility of reverse osmosis desalting is being investigated in this study. The tube-type reverse osmosis plant developed by University of California, Los Angeles, was selected for this purpose. The plant has mechanical components and a configuration similar to the small reverse osmosis pilot plant but uses 500 tubes because of a larger capacity requirement.

As a part of the desalting study, a membrane fabrication laboratory was set up at the site to supply desalting membranes to the reverse osmosis plant. The procedure used was patterned after the methods and equipment developed by the University of California, Los Angeles, for making the tubular cellulose acetate membrane. Pilot plant operating personnel manufactured the original membranes needed for the 500-tube reverse osmosis pilot plant.

Irrigation in arid regions will result in a salt buildup in the soil unless measures are taken to remove the salt. In California sub-surface drainage systems are used to carry away excess salts in the drainage water in some of the arid areas. When the salinity of this drainage water becomes too high for reuse by agriculture, it is called agricultural waste water. Such waste water can be treated by desalting methods to produce a good quality water then can be reused for agricultural or industrial purposes. One contemplated use in California for these saline agricultural drainage waters is for cooling thermal power plants. This use serves the purpose of substantially reducing the volume of waste water that must be disposed of and at the same time provides a source of cooling water for thermal power plants in arid regions where other water supplies are not likely to be available for cooling purposes.

The California Department of Water Resources has been operating pilot plants[1] for several years to develop desalting technology suitable for treating agricultural waste water. Since 1971 at a field test site, the Department has been studying the desalting of agricultural waste water by the reverse osmosis (RO) process. The study is being done in two phases (1) to determine the technical feasibility of using the RO process and (2) to investigate the economic aspects of using the RO process. The field test site is known as the Waste Water Treatment Evaluation Facility (WWTEF). It is located in the San Joaquin Valley in western Fresno County near Firebaugh, California.

In 1971, a 24-tube RO unit pilot installation was operated at this site to provide an opportunity to observe the effects of agricultural waste water on the properties of the cellulose acetate semipermeable membranes. Next a 180-tube RO unit was installed with plans to operate the unit for one year to study RO plant and membrane performance and establish optimum levels of desalted water production. However, after four months of operation, the operation was suspended due to severe membrane deterioration and calcium sulfate precipitation problems.

To investigate these problems, a 60-tube unit was installed with provision to add chemical additives to investigate the effect of the additives on the membranes and backing material and to prevent precipitation of calcium sulfate. Successful feed water pretreatment procedures were developed through operation of the 60-tube unit.

In 1972 the federal Office of Water Research and Technology and the Department jointly funded an evaluation study of three proto-type RO membrane units. The three types of membranes were hollow fiber, spiral wound and tubular design. In this evaluation study the three units were operated to determine (1) the life and performance of the semipermeable membrane, (2) the effect of agricultural waste water on the RO process, (3) feed water pretreatment procedures and (4) the product recovery obtainable under various conditions of feed water salinity and treatment.

In 1973 the Department contracted with the University of California at Berkeley (Cooper 1975) to study the bacteriological aspects of membrane decomposition and surface fouling in the RO units. Microorganisms were identified that affected the life and performance of the RO membrane.

---

[1] The primary pilot plants were designed and built by the University of California, Los Angeles, under the direction of J. W. McCutchan.

Various feed water pretreatment procedures were investigated to control the bacteriological effects. It was found that pretreatment by acidification, oxygen removal, or chlorination must be performed to protect the membrane from biodeterioration. It was also found that the membrane composition can affect the susceptibility of the membrane to biological attack.

To establish the maximum fresh water recovery levels that can be attained in a tublar RO unit, when limited by scaling tendency due to calcium sulfate in the feed water, a series of investigations were conducted. With softened feed water[1]/, a 180-tube unit was operated successfully at 90 percent recovery when the total dissolved solids (TDS) were higher than 6,000 mg/l and at a 95 percent recovery when the TDS were 3,000 mg/l. The unit was operated at a 90 percent recovery level from May to October 1974 with an average flux of 815 litres/(square metre-day) and product salinity of 400 mg/l (Department of Water Resources 1976).

Following completion of the Phase I technical feasibility studies in June 1975, preparations were begun for the fabrication and installation of a 500-tube, tubular RO desalter with a 95,000 to 114,000 litres per day capacity. In April 1976 operation of this plant began at the Firebaugh site for the purpose of establishing the economic feasibility of desalting agricultural waste water by the RO process. The Phase 2 studies are intended to provide design information from which desalting costs can be estimated for comparison of reverse osmosis desalting with alternative methods of agricultural waste water disposal. It will also provide information on the feasibility of integrating the RO process into power plant and other industrial cooling systems.

Since the cost of replacing the membranes in an RO system is a significant part of the RO operating cost, emphasis has been placed on trying to reduce this element of the cost. It was believed that a significant reduction in membrane cost could be realized if the membranes were fabricated at the plant site by operating personnel. Therefore, as an adjunct to the tube-type RO plant operation, a membrane fabrication laboratory was set up, with the assistance of the University of California, Los Angeles, to make the initial 500-membrane tubes and to provide replacement.

It was decided to establish the laboratory because the manufacture of the tubular cellulose acetate (CA) membrane was determined to be a relatively simple process, requiring the use of only semiskilled labor and off-shelf equipment. This manufacturing capability made the RO operation at the site self-sufficient in membrane production and demonstrated that such manufacturing could be done on-site.

The CA membrane used in the tube-type RO plant is made by a process developed and patented by the University of California, Los Angeles (LOEB 1966). The membrane is prepared from a solution consisting of cellulose acetate, formamide, and acetone, mixed in a typical ratio of 23:27:50 percent by weight. A viscous liquid resulting from this mixture is cast into a tubular-shaped film. The cast membrane is then fabricated into a working assembly (Figure 1) and posttreated to develop its salt-rejecting property.

---

[1]/ T. Vermeulen and G. Klein, University of California, Berkeley, assisted in the design specifications for the softening unit.

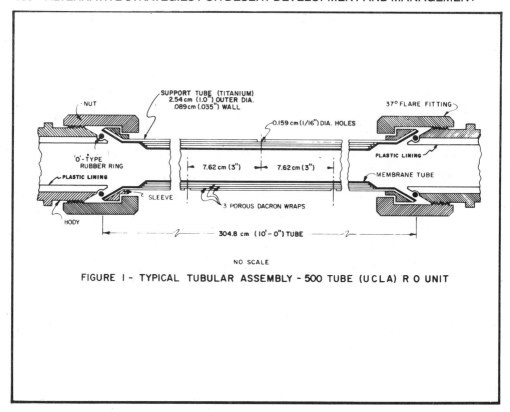

FIGURE I - TYPICAL TUBULAR ASSEMBLY - 500 TUBE (UCLA) R O UNIT

The membrane is batch-cast using a specially designed apparatus
(Figure 2). A casting tube is used to form the membrane into a tubular
shape. The bottom of the tube is charged with casting solution, and a
casting bob is inserted to hold the solution in place. A winch-driven
chain is used to pull the bob upward through the tube at a rate of about
150 millimetres per second. The bob pushes the casting solution ahead
of it leaving a thin wiped film of solution of the inner wall of the
tube.

The casting tube is immediately dropped into a chilled water well
located below the casting apparatus. The well water is held at a
temperature of 1°C to gel the solution. The casting tube is then transferred to a shrink tank containing hot water at 80°C. The shrinking
process allows the membrane to be removed from the casting tube.

After removal from the casting tube, the membrane is wrapped in
three layers of dacron cloth and inserted in a titanium support tube.
The ends of the membrane are trimmed, plasticized, and flared to conform to a flared tube connector.

The completed assembly (see Figure 1) is installed in a curing
loop through which hot water at a pH of 4.5 is circulated for 15 minutes.
Citric acid is added to the water to maintain a pH of 4.5. There is a
slight variation in the cure temperature because of heat loss in the
curing loop. This curing process develops the membrane's salt-rejecting

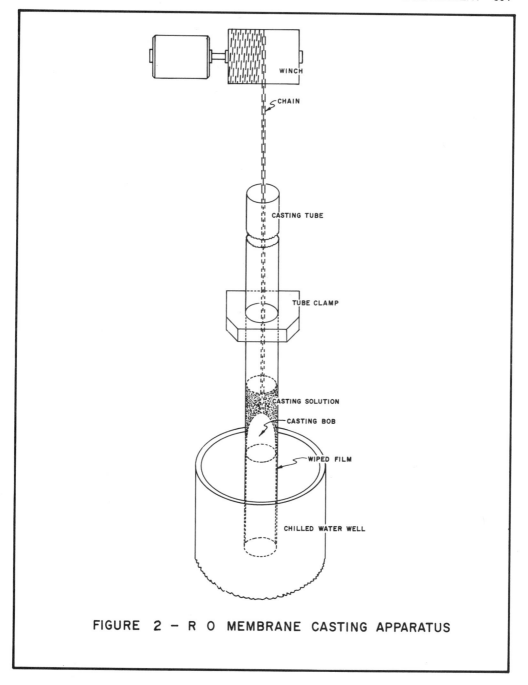

FIGURE 2 – R O MEMBRANE CASTING APPARATUS

property, and a water temperature of 90°C gives the membrane an inter-mediate permeability to both water flux and salt. The loop is then flushed with cold water at a pH of about 7.1.

As a final step, the tube assemblies are installed on a test rack where they are proof-tested for defects and desalting performance. Feed water containing sodium chloride (NaCl) solution at a concentration of 5,000 milligrams per litre is passed through the test rack at a flow rate of 0.32 litre per second and 2,800-kPa pressure. Table 1 shows the results of a typical two-day test run.

The cast membrane is composed mainly of cellulose diacetate and has a dense surface layer formed during the casting process and a rela-tively porous sublayer. The total film thickness is about 100 micrometres, and the dense layer has a thickness of about 0.2 micrometre. The thin, dense layer is formed on the side exposed to the air during casting and is the primary barrier to salt passage. This surface must be in contact with the brine to gain full membrane performance.

Another pilot plant program was begun in January 1977 at the Firebaugh site involving an improved distillation system (Sephton 1975) combined with a novel ion exchange system[1]. The energy input to drive the system is obtained primarily from the operation of a cooling tower in which the concentrated salt solution removed from the cooling tower is used to regenerate the ion exchange resin. The pilot plant equipment was designed and built by the University of California at its Sea Water Conversion Laboratory, Richmond Field Station. The pilot plant will be operated during 1977 to obtain data for design and cost estimating purposes of this treatment system, which is especially suitable for treating agri-cultural waste water for power plant cooling.

Table 1

Ro Tube Test Results

Date	Feedwater NaCl (mg/l)	Test Slot No.	Product Water		NaCl (mg/l)	DR[3]
			Flux			
			(cm^3/min)[1]	(gfd)[2]		
3/12/76	4800	1	84	14.3	310	15.48
		2	85	14.4	270	17.78
		3	84	14.3	380	16.84
		4	92	15.6	345	12.63
		5	93	16.0	380	13.91
3/16/76	5200	1	117	19.9	470	11.06
		2	115	19.6	410	12.68
		3	115	19.6	460	11.30
		4	122	20.7	490	10.61
		5	122	20.7	410	12.68

[1] Cubic centimetres per minute (per tube).

[2] Gallons per square foot per day (of membrane area).

[3] Desalination ratio = $\dfrac{\text{salt concentration of feedwater}}{\text{salt concentration of product water}}$

[1] The ion exchange system was developed by T. Vermeulen and G. Klein, University of California, Berkeley.

References

Cooper, R. C. and Richard, M. G. (1975), "Prevention of Biodeterioration
   and Slime Formation in Reverse Osmosis Units Operated at Firebaugh,
   California", University of California, Berkeley, June 1975.

Department of Water Resources (1976), "Agricultural Waste Water Desalination
   by Reverse Osmosis - Technical Aspects", The Resources Agency, State
   of California, Bulletin 196-76.

Loeb, S. (1966), "A Composite Tubular Assembly for Reverse Osmosis
   Desalination", Desalination 1, 1-100.

Sephton, H. H., (Principal Investigator), "The Use of Interface - Enchanced
   Vertical Tube Evaporation, Foam Fractionation, and Ion Exchange
   to Improve Power Plant Cooling with Agricultural Waste Water",
   UCB-Eng-3841, University of California, Berkeley, February 14, 1975.

A CONTROLLED, ENVIRONMENT, ECOLOGICAL WASTEWATER RECLAMATION,

BIO-FUELS, AND AQUACULTURE FACILITY

DESIGNED FOR THE CHEMEHUEVI INDIAN TRIBE

By

Steven A. Serfling & Dominick Mendola
Solar AquaSystems, Inc.
Encinitas, California

ABSTRACT

A total systems engineering approach to wastewater reclamation, bio-fuels production, aquatic foods culture, natural energy utilization, and total energy conservation has yielded an integrated design for a controlled environment aquafarm for California's Mojave Desert Chemehuevi Indian Tribe. The Tribe is presently carrying out a major effort to build an environmentally, economically and esthetically sound community, in order to reestablish their tribe on their recently regained reservation land on the western shores of Lake Havasu. Of primary importance in achieving this goal is construction of a wastewater treatment facility to eliminate the present lake and drinking water pollution problems, and development of aquaculture and agriculture production facilities for creating tribal employment and a long-term economic base.

The treatment, reclamation, and aquaculture systems consist basically of four high rate, aerated lagoons (0.4 ha. each), covered with greenhouses to maintain a controlled, high temperature environment for maximizing productivity and reliability. This design minimizes mechanical equipment and allows for low operating and maintenance expenses. Solar energy, wind energy, and energy efficient systems will be utilized to minimize operating expenses.

The proposed facility will incorporate special system designs to overcome particularly difficult design criteria. Flow rates are expected to fluctuate widely from less than 3.8×10^5 liters per day during midweek, winter periods, up to peaks of 5.7×10^6 liters per day during spring, summer and fall weekends with tourist influxes of 10-15,000 people per day. Weather fluctuations are severe, with 51° C summer heat, 2°C winter nights, and up to 144 Km/hr. windstorms.

Revenues derived from the production and sale of reclaimed water, dissolved nutrients for fertilization, and cultured freshwater shrimp, catfish, and Tilapia will provide on-site, long-term employment for members of the Tribe. Construction funds for the project are pending from the U.S. Economic Development Agency, and overall project coordination is being provided by the State Economic Opportunity Office.

694

PROJECT DESCRIPTION

The Chemehuevi Indian Tribe is presently carrying out a major effort to build an environmentally, economically, and aesthetically sound community in order to reestablish their Tribe on their recently regained Reservation land on the western shores of Lake Havasu. Of primary importance to achieving this goal is the construction of a wastewater treatment facility to eliminate the present lake and drinking water pollution problems, and development of aquaculture and agriculture production facilities for creating Tribal employment and a long-term economic base.

To fill these as well as additional needs, a unique project has been designed to completely integrate solutions for the Tribe's needs for improved wastewater treatment, aquaculture of fish and freshwater shrimp, reclamation of water for agriculture irrigation, recycling of waste nutrients for food production, minimization of energy requirements and operating expenses, production of bio-fuels (methane) for electricity and space heating, and job creation (see Figure 1).

The unique features and advantages of the proposed project are:

The treatment, reclamation, and aquaculture system consists of four(4) high rate, aerated lagoons, covered with greenhouses to maintain a controlled high temperature environment for maximizing productivity and reliability. This design minimizes mechanical equipment and allows for low operating and maintenance expenses.

PROPOSED CHEMEHUEVI INDIAN WASTEWATER RECLAMATION, AQUACULTURE,& BIO-FUELS PRODUCTION FACILITY
·LAKE HAVASU, CALIFORNIA·
FIGURE 1

(1) Raw Sewage Influent
(2) Solar AquaCell Cell #1 (Primary and Secondary Sewage Treatment)
(3) Ozone Generator and Contact Chamber
(4) Solar AquaCell #2 (Tertiary Treatment and Culture of Fish and Shrimp)
(5) Solar AquaCell #3 (Fish and Shrimp)
(6) Solar AquaCell #4 (Fish and Shrimp)
(7) Operations Building
(8) Effluent Holding Reservoir
(9) Windmill Effluent Pumping Station to Upper Reservoir
(10) Bio-fuels Digestor and Methane Generator
(11) Sludge and Compost Drying
(12) Hatchery and Processing Lab
(13) Upper Reservoir and Fish-out Lake
(14) Windmill Effluent Pumping Station (to Agriculture)
(15) Administration Building
(16) Mobile Home, Camping Parks and Lake Havasu Community
(17) Lake Havasu

To increase treatment efficiencies and nutrient recycling ability, water hyacinths will cover the surface of all lagoons and be regularly harvested for use as bio-fuels, organic compost, or livestock feed.

The wastewater treatment facility has been designed to provide secondary and tertiary quality treatment over a wide range of loading rates, projected to fluctuate weekly from 3.8×10^5 to 5.7×10^6 liters per day, and reclaim water for crop irrigation in a water-short desert area.

An ozone purification system will eliminate pathogens and allow reuse of the nutrient rich water for direct culture of fish, freshwater shrimp, and high protein aquatic plants.

Solar energy, wind energy, and energy efficient systems will be utilized to minimize operating expenses.

Methane, produced from anerobically digested sludge and water hyacinths, will be used to generate electricity to supply all of the project's electrical and fuel requirements.

Revenue derived from culture of fish (catfish, Chinese carp, Tilapia) and freshwater shrimp will offset operating expenses for waste treatment, provide long-term employment, and generate profits to the Tribe.

The controlled environment AquaCell greenhouse systems and temperature regulating ability of the lagoon and sprayer systems will maintain relatively uniform air and water temperatures and, thus, allow continuous year-round high performance even during the 51° Centigrade summer heat, 2° Centigrade winter nights, or 128 Km/hr. winds, common to the area.

The project has been designed to utilize semi-skilled labor and natural processes in place of conventional mechanical, chemical, and electrical methods in order to maximize employment opportunities to the Tribe.

Site Location and Physiography

The encompassing area is approximately 12 acres situated on high ground (800-750' contours) between two line washes in the southeast quarter of Section 25 of the northernmost half of the Chemehuevi Indian Reservation, Lake Havasu, California (see USGS Map - Castle Rock & Havasu Lake, N3430 - W 11422.5/7.5 and N3422.5 - W 11422.5/7.5). The proposed site is approximately 1.2 miles northwest of Havasu Landing on the west shore of Lake Havasu, California and approximately 4,000' due west from the nearest lake shore (see Figure 2).

Physiographically, the property can be described as high desert alluvium, broken intermittently with major and minor seasonal washes, and strewn with allochthonous stones averaging less than 2 inches in diameter. The vegetation is desert scrub with some smaller cacti interspersed. This area can be described as seasonally variable due to major and minor scouring and deposition of substrates carried by runoff flood waters.

AQUACELLS VS. CONVENTIONAL TREATMENT SYSTEMS

The Solar AquaCell System consists basically of an aerated lagoon containing rapidly growing water hyacinths and biologically active substrates, covered by solar heated greenhouses (see Figure 3). It has been designed to combine the

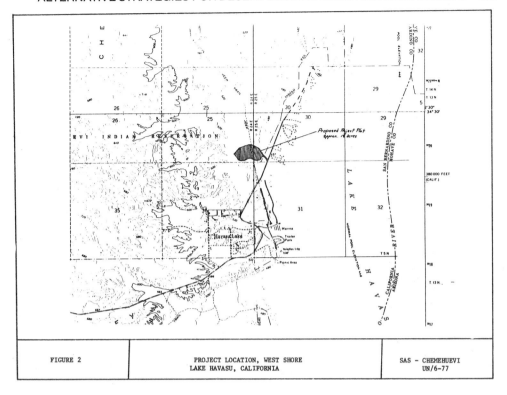

| FIGURE 2 | PROJECT LOCATION, WEST SHORE LAKE HAVASU, CALIFORNIA | SAS - CHEMEHUEVI UN/6-77 |

best features of low construction and operation costs of aerated lagoons, with the control, reliability, advanced treatment capability, and reduced land requirements of conventional, high-technology treatment plants. By trading off expensive concrete, steel, chemicals and electricity, for natural ecological processes utilizing earthen ponds, greenhouses, hardy pollution consuming plants and invertebrates, and solar energy; this process has demonstrated the ability to convert raw wastewater into high quality, reclaimed water for one half to one quarter the treatment expense of conventional methods.

Conventional sewage treatment processes, which have proven adequate with no major design changes for over fifty years, are now recognized as unable to meet the present Federal Water Pollution Control Act Amendments without extensive modification, additions and extreme construction expense. Conventional water treatment systems are also costly to operate, have high electrical demands and consume precious natural resources including fossil fuels, chemicals and water. Conventional treatment processes are incapable of removing or detoxifying the majority of the most harmful components of modern day wastewater, e.g. pesticides, phenols, heavy metals and a host of complex domestic and industrial chemicals now recognized as potentially carcinogenic. (In contrast, biological lagoon systems containing plant components have proven capable of doing this.)[1] Furthermore, they were never intended or designed to fulfill the need for reclaiming the water or for converting waste nutrients into valuable by-products to help reduce operating costs.

FIGURE 3

EVOLUTION OF WASTEWATER LAGOONS

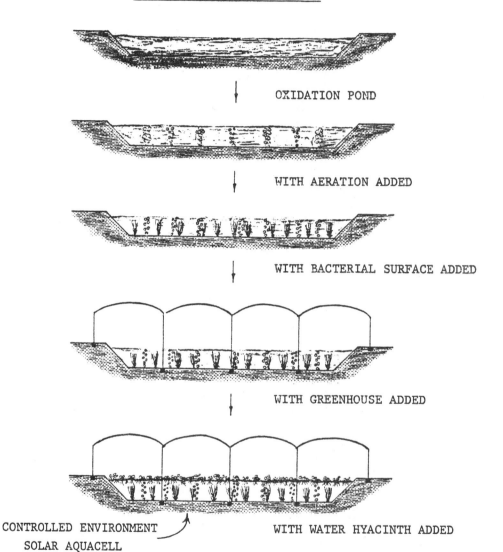

OXIDATION POND

WITH AERATION ADDED

WITH BACTERIAL SURFACE ADDED

WITH GREENHOUSE ADDED

CONTROLLED ENVIRONMENT
SOLAR AQUACELL

WITH WATER HYACINTH ADDED

POTENTIAL ADVANTAGES OF ECOLOGICAL TREATMENT SYSTEMS

Most wastewater treatment systems are essentially "biological", since even conventional, high technology facilities such as trickling filters or activated sludge, are actually dependent on maximizing the survival, productivity and "harvesting" of bacteria. However, ecological theory and practice have clearly demonstrated that monoculture systems, e.g. bacteria only, are inherently less stable and efficient than multi-species, polyculture systems containing a variety of invertebrates, bacteria and sludge grazers, algae and higher plants. In the sanitary engineering profession, Drs. William Oswald and Ross McKinney are among those who first recognized this principal, stating that treatment lagoons, if properly designed and managed, "are capable of meeting any desired effluent criteria currently in effect or that might be adopted in the future", and that "complete waste treatment (tertiary) is attainable in ponds...with economy, dependability, and simplicity in operation and maintenance...to greatly modify the planning of many of the world's waste management schemes in the warm and arid regions of the world", (Oswald, 1973)[2][3].

In recent years, ecological lagoon wastewater treatment systems using aquatic macrophytes (e.g. water hyacinths) and higher aquatic organisms (e.g. invertebrate detritivores and fish) have proven successful at capturing nutrients, concentrating chemicals metabolizing organic compounds, detoxifying dangerous synthetic chemicals, and eliminating pathogenic bacteria and viruses. For example, research by the National Aeronautics and Space Administration in Bay St. Louis, Mississippi, has recently demonstrated that one acre of water hyacinths can remove over 1590 Kg. of nitrogen and 363 Kg. of phosphorus per year, as well as remove and metabolize over 8170 Kilograms of phenol, a toxic organic pollutant and absorb 44 Kg. of trace heavy metals per acre per year. [4]

In spite of the increased treatment efficiency of lagoons using improved techniques, ecological lagoon systems still suffer from the lack of reliability due to seasonal fluctuations and excessive land requirements. To overcome these problems, Solar AquaSystems has developed a covered controlled environment, high temperature, ecologically managed system which is dependable year-round and is capable of treating the wastes of 5,000 - 10,000 people per acre. It is important to recognize that this level of treatment efficiency is presently achieved by properly managed, warm, aerated lagoons, as shown in Table I. The important new achievement of the controlled environment, Solar AquaCell lagoon system is continous performance and year-round dependability.

TABLE I

Comparison of Waste Loading Rate Capabilities
(Refs. 5-10)

Type of Treatment	# of People's Wastes Per Acre (Secondary Treatment)	BOD Load (Pounds Per 1,000 Ft.^{2})
Oxidation Lagoon	50-300	.04
High-Rate Oxidation Pond	1,000 - 4,000	3.4 - 7.0
Hyacinth Lagoon	1,000	1.0
Aerated Lagoon (summer)	4,000 - 8,000	3.0 - 6.5
Solar AquaCell (year-round)	5,000 - 10,000	3.5 - 7.0
Activated Sludge or Trickling Filter	20,000 - 50,000	17 - 35

A Description of the AquaCell System and Process

The Solar AquaCell System is a composite of the best design aspects of four well proven technologies: (1) aerated lagoons for wastewater treatment, (2) culture of floating aquatic plants, e.g. water hyacinths, for wastewater nutrient removal, (3) polyculture of micro-invertebrates, fish and shellfish for maximum removal and bio-concentration of nutrients and organics from wastewater[11] and (4) solar-heated, air-insulated greenhouse for environmental control. A comparison of this process with conventional methods is shown in Figure 4.

The Solar AquaCell system and process consists of: (1) rectangular one to two acre ponds covered with greenhouse-type structures for heat retention,

FIGURE 4

COMPARISON OF THE MAIN TREATMENT PROCESSES OF
CONVENTIONAL SYSTEMS WITH THE SOLAR AQUACELL SYSTEM

1. Conventional Treatment Plant (Activated Sludge, Trickling Filter, Etc.)

SEWAGE ⟶ BACTERIA ⟶ SLUDGE AND SECONDARY EFFLUENT

2. Conventional Treatment Lagoons (Aerobic, Anaerobic, Facultative, Aerated)

SEWAGE ⟶ BACTERIA / PHYTOPLANKTON or MACROPHYTES ⟶ POND SLUDGE AND COMPARABLE SECONDARY EFFLUENT

3. The Solar AquaCell System

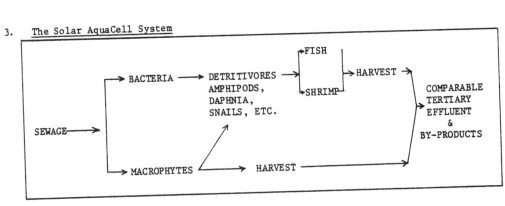

SEWAGE ⟶ BACTERIA ⟶ DETRITIVORES AMPHIPODS, DAPHNIA, SNAILS, ETC. ⟶ FISH / SHRIMP ⟶ HARVEST ⟶ COMPARABLE TERTIARY EFFLUENT & BY-PRODUCTS

MACROPHYTES ⟶ HARVEST

(2) solar heat exchange troughs, panels, and sprayers for increasing influent water temperatures, (3) artificial habitats and substrates ("bio-grass") for increasing biologically active surface area in ponds, (4) channels for control of distribution and movement of influent and effluent water, (5) channels for the controlled polyculture of aquatic plants, invertebrates, fish and shrimp at maximum processing efficiency levels, (6) aeration systems to maintain desired oxygen levels and meet BOD requirements and (7) a subterranean drainage and sand percolation system for final polish of pure, reclaimed water,

Ozone is utilized after the first stage AquaCells to completely purify the water and, thereby, allow use of the dissolved nutrients for production of food by-products (AquaCells 2,3, & 4). All biomass harvested from the first pre-ozone AquaCells is placed in an anaerobic digestor for production of methane and electricity (see Figure 5).

Figure 6 shows the many possible products produced from wastewater through the different pathways, i.e. bacterial, aquatic plants, etc. The final effluent from the Solar AquaSystems treatment will have water quality high enough for many reclamation uses, depending on retention (treatment) time. Chemical analysis of effluent quality from a pilot facility operated by Solar AquaSystems in Solana Beach since September, 1977, has demonstrated that secondary quality water suitable for irrigation can be produced from raw sewage after 2 days retention time (1 acre/1 million gallons/day) or potable quality water produced after 4-6 day retention time (2-3 acres/1 MGD).

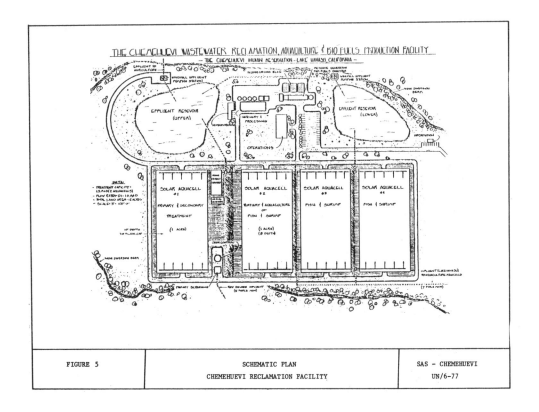

FIGURE 5	SCHEMATIC PLAN	SAS - CHEMEHUEVI
	CHEMEHUEVI RECLAMATION FACILITY	UN/6-77

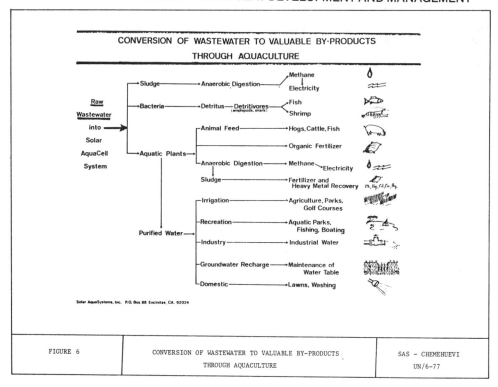

| FIGURE 6 | CONVERSION OF WASTEWATER TO VALUABLE BY-PRODUCTS THROUGH AQUACULTURE | SAS - CHEMEHUEVI UN/6-77 |

Energy Savings Potential

The system uses low energy processes, including solar radiation for heating the culture water, inexpensive, air-inflated plastic films for insulation and control of the environment, efficient aeration systems and water distribution methods using gravity to reduce electrical pumping requirements. Abundant wind resources will be used to turn wind powered water pumps, which will deliver effluent to agriculture sites 2 km. to the north. Methane is also produced by anaerobically digesting the sludge as well as aquatic plants raised in the system, to provide up to 100% of the electrical energy requirements for the water treatment facility. The need for expensive chemicals is also eliminated.

Construction and Operating Expenses of Wastewater Reclamation

Costs of construction and operation will vary, depending on size of facility, location, and degree of treatment required. In general, both capital construction and operating costs will average 50% less for secondary, and 25% less for advanced tertiary quality treatment in comparison to conventional technology. Total treatment costs for one MGD volume or greater will range from 15-25¢ per 1,000 gallons for secondary quality water suitable for irrigation, and 35-50¢ per 1,000 gallons for advanced tertiary or potable quality water. Furthermore, revenues obtained from the sale of by-products, particularly reclaimed water in water short areas, can even reduce net operating expenses to zero.

Participating Agencies

The Chemehuevi project will be constructed in 18 months with funds granted by the
U.S. Department of Commerce, Economic Development Agency. Assistance with
organization, funding, training, and operation of the project is being provided
by Kit Cullen, Dave Pollard, and Nick Bishop of the State of California Office
of Economic Opportunity.

The project will be the first of its kind in the world and will be a model for
ecological wastewater reclamation and reuse, not only for arid regions, but temperate regions as well. It will demonstrate that wastewater in the desert is a
highly valuable commodity; one, that if properly handled can yield useful by-
products including the most precious of all resources, life sustaining water.

REFERENCES

1. Tourbier, J. and Robert W. Pierson, Jr., Editors, BIOLOGICAL CONTROL OF WATER
   POLLUTION, Univ. of Pennsylvania Press, 1976

2. McKinney, R. et. al., TREATMENT LAGOONS - State-of-the-Art, Environmental Protection Agency Report, 1975.

3. Oswald, W. J., Complete Waste Treatment in Ponds, In PROGRESS IN WATER TECHNOLOGY,
   Vol. 3, 1973, pp. 153-163.

4. Wolverton, B.C., Barlow, R.M. and McDonald, R.C., Application of Vascular
   Aquatic Plants for Pollution Removal, Energy and Food Production in a
   Biological System, in BIOLOGICAL CONTROL OF WATER POLLUTION, pp. 141-
   149, 1976.

5. Bagnall, L. et.al, Feed and Fiber from Effluent-Grown Water Hyacinth, in WASTE-
   WATER USES IN THE PRODUCTION OF FOOD AND FIBER - PROCEEDINGS, Environmental
   Protection Agency Report No. EPA-660/274-041, 1974.

6. Boyd, C.D., Freshwater Plants: A Potential Source of Protein, ECON. BOT., Vol 22,
   pp. 359-368, 1968.

7. Cornwell, David A., et.al.; 1977. Nutrient Removal by Water Hyacinths, in
   J.W.P.C.F., January, 1977, pp 57-64.

8. Del Fosse, Ernest S., 1977. Waterhyacinth Biomass Yield Potentials. Paper
   presented at the Institute of Gas Technology, Orlando, Florida, Jan. 1977,
   29 pp.

9. Dinges, Ray, A Proposed Integrated Biological Wastewater Treatment System, in
   BIOLOGICAL CONTROL OF WATER POLLUTION, pp 225-230, 1976.

10. Robinson, A.C., et.al., 1976. An Analysis of the Market Potential of Water
    Hyacinth-Based Systems for Municipal Wastewater Treatment, Batelle Columbus
    Laboratories Report. NTIS #N76-28679, 235 pp.

11. Serfling, Steven, A., Recycling of Nitrogenous Wastes and Role of Detritus in a
    Closed-Cycle Polyculture Ecosystem, in WATER SCIENCE & ENGINEERING PAPERS,
    4501, March, 1976, Department of Water Science & Engineering, University
    of California, Davis.

NEW WATER RESOURCES FOR DESERT DEVELOPMENT FROM ICEBERGS

HRH Prince Mohamed Al-Faisal
Governor: Saline Water Conversion Corporation
(Presented by) Dr. Shawkat Ismail

INTRODUCTION

In the past when the world population was relatively small, people used to look at water resources from rainfalls and rivers with no fear that a day might come when they have to look for other water resources. At that time the inhabitants of the Arabian Peninsula were used to move from one place to another searching for water patiently at the time when most of the deserts in the world were uninhabited.

It was quite natural for people to build their villages and towns near to rivers and in places where underground water supply were available. There was at that time some sort of content among people until man started to look for minerals and petrol in remote places. It happened that most of the places which are rich of such resources are found in regions lacking natural soft water.

The first prospectors who went in search of the valuable raw materials in arid areas were principally concerned with the immediate financial gain and the matter of water supply was secondary in importance at first as long as water could be transported regardless of the cost. Consequently labour cost were extremely high to encourage them putting up with the severe conditions of life in these regions as they had to use as little as possible from the water provided. Since then water was felt to have a price.

In the past desert regions were usually cut off from the rest of the world by the difficulties of transport but nowadays means of transportation from one place to another are becoming easier and faster as the factor of time has an important role in almost all projects.

The sudden world interest in the exploitation of desert lands is caused by the fact that each nation wishes to raise the standard of living of its own people and to provide for the increase of population. It is a' race that all nations are running in for the sake of having a better life or in fear of being left behind. Desalination of sea water and brackish water was accordingly introduced by scientists to help in carrying out many valuable projects in arid areas. Desalinization has become nowadays an established science and new techniques are continuously been discovered but still the cost of the produced water is considered rela- tively high. We must not only relate the great interest for the call of more soft water resources which is raised everywhere upon the sudden drought which struck several European countries during last year but also for the continuously in- creasing demand for soft water from modern industries. Today with the energy crisis becoming more serious one should look over other means which consumes as little as possible from fuel for providing arid areas with water.

WATER SUPPLY THROUGH TRANSPORTED ICEBERGS

It is my belief that all the world resources are made to be exploited by mankind along the life time of our universe. Some of the earth resources have already been discovered whilst other resources whether known or unknown to us will be utilized by the present and following generations.

We usually search for water in its liquid form: being easier to store, to trans- port from one place to another and above all being in the form required by all the creatures on earth and for cultivating our lands.

The amount of soft water on earth represents about 2% of the total existing water of the earth. Only 1% of this soft water is available in the liquid form whilst the rest is found in an icy form. The 99% of the soft water which is in the form of ice is distributed as follows:

> 90% in the Antarctic Region,
>
> 8% in the Arctic Region,
>
> 1% for all the snow covering mountains and lands excluding north and south
> pole zones.

From above, one can see that soft water in its liquid form represents a very small amount relative to what the Antarctic is having in the form of ice. One of the procedures to be taken before exploiting such new water resources is to make a preliminary study and seek means of transportation in order to select the most economical size of icebergs to be towed. A French Engineering Organization had already made several feasibility studies on the subject based upon reliable infor- mation about icebergs of the Arctic and Antarctic. They came up with the con- clusion that it is a promising and successful project and within the capacity of mankind by applying the advanced technology already in hand. Different volumes of tabular icebergs were found in a number of places in both the Antarctic and Arctic off their coasts.

The icebergs of the Artic were found not easy to reach nor to transport outside the regions of their formations. On the other hand icebergs of the Antarctic are found in a number of ice shelves in an enormous amount, easy to reach and to tow away. The following ice shelves were found to be the most suitable ones for the hunting of different sizes of icebergs for the purpose of supplying soft water to different arid zones in the World.

1) The Amery Ice Shelf: It looks over the Indian Ocean.
                                 It can supply Australia with icebergs.

2) Ross Ice Shelf: It looks over the Pacific Ocean.
                                 The most suitable one for providing the arid areas along the coast of South America with icebergs.

3) Filchner Ice Shelf: Looking over that part of the Atlantic Ocean South to
                               the American Continent.

This ice shelf can provide the Namib desert along the South Western Coast of the
African Continent with Icebergs.

The studies being made on regions outside the eastern coast of the Antarctic near
to the Amery Ice Shelf, had shown that the average length of the tabular icebergs
floating there was around 1100 metre. It was quite normal to find also icebergs
of 21 kilo-metres long. The largest tabular iceberg which was seen up till now
was 350 km long, 50 km and had a depth of 250 metres. Most of the icebergs which
were found had a depth between 170-280 metres.

It was noticed that the distance between a floating iceberg and the nearby ones
gradually deceases as one gets nearer to the inner edge of the ice shelf coming
from the ocean. Accordingly it will be easier to select the floating iceberg from
the outer edge of any ice shelf.

In fact by means of the advanced photo-taking technique from space satellites with
their orbits set around the Antarctic region will help in selecting the most suitable icebergs to be towed. For example one of the photos which was taken by
satellites set by the "Earth Resources Technology Satellite" scheme was analysed
for one of the zones where floating icebergs were gathering, the photo came out so
clear and sharp to the extent that floating icebergs of lengths as small as 100
metres were easily identified.

Accordingly with the help of such available information it will be easy to direct
the towing ships straight to places where the largest number of suitable floating
icebergs are located.

It was found that the annual production from such floating icebergs at the Antarctic exceeded 10,000 which melt in the nearby oceans without any use to mankind,
not even a small portion of it.

If optimum economy is required one has to consider very carefully the effect of
the right size, shape and location of the iceberg to be towed on the economical
condition of the Project as a whole.

                              THE HUNTING OF ICEBERGS

A detailed analysis of the satellite pictures taken of the Antarctic zone will be
carried out by specialists in order to locate the first iceberg to be towed. The
size of that iceberg should not be less than 100 million cubic metres and of a
length which ranges between 1200-1400m, width around 300m and thickness of about
250m. A number of towing boats with a command ship having landing facilities to
accommodate two helicopters and a tanker to supply the fleet with fuel will be
prepared for the hunting operation.

Once the required iceberg is spotted, helicopters will start on their jobs by
using echo-sounder recorders to record data upon the thickness and detect any
flaws, caves or irregularities during their survey of the whole iceberg.

The recorded data are then carefully studied as extreme safety and shape regularity are the main concern of the expedition commander before possession of the
iceberg is undertaken. Helicopters are then used to airship the different equip-
ments and food supplies required by the first team of technicians who will land on

the surface of the iceberg and have definite tasks to do before the towing operation is done.

The working team will then start to anchor the first towing bollard on the top
surface of the iceberg - at a good distance from the front end. An iceberg of the
size chosen will require installing 3 bollards, the strength of each is made to
resist a pull of 250 tons. Each bollard is composed of several parts to facilitate and secure the anchorage operation. A ring oven with oil burners supplies
the anchorage tubes of the bollards with the heat required to melt the ice during
the sinking operation of the tubes to a depth of more than 6 metres in the ice.
The end loop of the tow-line is then dropped by a helicopter to secure it on the
bollard while the other end of the line is tied to the tug-winch and is released
to stretch over a mile before putting on the brakes of the winch whilst the tow-
ing-tug is moving in order to start the towing operation. As only one tug-boat is
used it will take over two days to reach a speed of 0.25 of a knot. During that
period the other two bollards are then installed in their places to allow for the
other tugs to start towing after which a speed of 0.8 knot may be reached in a
week time. It is just the start of a long journey ahead.

PROTECTION OF ICEBERGS AGAINST MELTING

Icebergs either drifting by themselves or towed will melt completely before reaching the equator unless means of protection against heat transfer are provided.
First of all the bollards are covered by a foam of polyurethane to isolate them
from the sun heat. Shaping of the front end of the iceberg are then followed to
reduce its drag and this is achieved by slicing the forward corners all the way to
the bottom of the iceberg. This will require drilling several wells along the
slicing lines by means of a specially designed torpedo which will melt its way
down. Cylindrical weights are then set to hang into the wells from a wire which
lies along the slicing line. The wire is then heated electrically to a degree
capable to melt the ice underneath it at the rate of 1 metre deep per hour, whilst
the weights help in speeding up the slicing process by pulling the wire down.

The bottom and sides of the iceberg will be protected by plastic film covers.
Those covers will be held in place by means of cables attached to them and fast-
ened to anchorages along the circumference of the top surface of the iceberg at
about 10m away from the edges. The anchorages are just made of pipes 3m. long,
heated and then forced down the ice. Insulation of the anchorages are carried
out by polyurethane foamed in situ.

The top surface of the iceberg is protected from the heat of the sun by forming a
water pond on top of it. This is done by covering a strip of 5m wide with polyurethane foam all round the edges on the top surface of the iceberg. Water will
accumulate in the pond from the melting of the ice of the pond to a certain depth
and from rain. The icy bottom of the water pond will be protected to a great
extent against the heat of the sun by the evaporation taking place in the water
accumulating which consumes around 80% of the sun heat. Accordingly the melting
rate of the bottom of the water pond will be greatly reduced.

As towing of the iceberg proceeds, the border between the Antarctic Ocean and the
other nearby oceans will be crossed where protection against melting of the bottom
and the sides are needed as the sea water temperature will be around 10°C at that
zone.

At the mooring harbour where the iceberg will be delivered a plastic plant is
erected for the production and assembly on special types of drums of the protec-

tive covers for the icebergs which are known here by the bottom blanket and the side skirts.

The bottom blanket and the side skirts are folded and wrapped around the drums in a way which makes it so easy to unfold and stretch along the parts of the iceberg to be covered with. A tug will then tow a train of such drums, which floats on water, from the mooring site to the place where the iceberg protection operation will be carried out.

The covering of the bottom of the iceberg by its blanket is made by the following method. The full length of the folded blanket is unfurled horizontally on the sea surface at the rear end of the iceberg between two tugs, one on each side of the iceberg. The blanket is then let loose between the tugs in order to sink in water whilst the tugs are getting nearer to each other and still holding the blanket from both ends until the blanket reaches a depth below that of the draught of the iceberg.

The straps attached to the top edge of the folded blanket are transferred to their anchorage points at the top surface of the iceberg at its rear edge. The two tugs will then move forward along both sides of the iceberg in order to unfold the blanket underneath the iceberg until the front of the iceberg is reached. During the unfolding process new sets of straps appear which are transferred to corresponding anchorage points along the side and front edges of the top surface of the iceberg.

The skirt drums are of special design in a way that they can change the position of their axes from being horizontal during towing to being vertical during the unwrapping process along the side of the iceberg.

The skirt is made of several strips. Their total lengths are in excess of the peripheral length of the iceberg to allow for overlapping, especially over the front part of the iceberg where the temperature difference between the iceberg and the passing-by water of the sea is high.

Each strip is made of three parts, different in thickness. The top one is so thick to withstand the action of the sea waves. The middle part is rather thin as it stretches down a good length of the immersed part of the iceberg. The bottom part of the skirt is thicker than the middle one in order to exert a pulling force downward by its weight upon the middle part of the skirt and this part extends below the bottom of the iceberg.

The straps fixed along the top edge of the thick part of the skirt are tied to anchorage points on top of the iceberg and by means of winches the top edge of the skirt is lifted up to about 15m above the sea water.

The sides of that part of the iceberg above water will also be covered by plastic strips hanging down the sides and of a capability to reflect the sunrays.

SLICING OF ICEBERGS

The draught of most of icebergs to be towed is just over 200m which raises no problem in open seas. There are some places where the iceberg has to cross in order to reach the mooring site, such places may be of shallow water just as in the case of Bab el-Mandeb straits at the Red Sea entrance. The water there is only about 60m deep which hinders the passage of such size of icebergs. In a case like this slicing of the iceberg is a necessity although it is an expensive oper-

ation. Slicing will be done in the same way which was used when shaping the bow
of the iceberg and every time a slice is completed it loses balance and topples
over the water. The thickness of each slice should be less than the depth of the
sea bed it is going to cross.

During drilling of the holes for the slicing operation the melted ice accumulating
inside should be continuously pumped out otherwise refreezing may occur.

The bottom and sides of each slice will have to be protected against heat transfer
in the same way as with the original iceberg before towing them to their destination. A certain length of the iceberg at the end of the slicing operation and of
a value over that of the thickness of the iceberg will be unstable to slice,
therefore it will be delivered to countries near to the slicing site and in need
of soft water.

Slicing will be useful if there is more than one mooring site to deliver water to.
It helps also in speeding up the melting of the ice at the unloading terminal and
mooring them very near to the shores.

ICEBERG UTILIZATION AT MOORING SITE

The main purpose of transporting icebergs for thousands of miles is to provide
water to places lacking a steady or sufficient supply of water. This applies in
desert areas where programmes for the future are set for their development. One
day might come when it will be realized that icebergs are considered as one of the
most reliable sources for the supply of soft water for arid areas. On the arrival
of the iceberg to its mooring site a floating aqueduct is connected to shore by
means of which soft water is transported from the iceberg to the water network of
the city at that site or to some intermediate water storage facilities.

It remains to find the means of transferring ice into water. The heat received
from the sun is capable to melt over 8cm thick per day of the surfaces exposed to
the sun especially at the Red Sea region. Water accumulating in the pond at the
top of the iceberg will be drained into the aqueduct by using simple means. The
insulating foam at the top edges are left as it is until the depth of the pond is
over 2m after which it is removed. The bottom blanket can be removed as well. To
speed up the melting of the ice; water from the top pond will be sprayed above the
surface of the pond in the air to gain as much heat as possible from the passing
by warm winds, accordingly the melting rate of the icy bottom of the water pond
will get higher. Other means may also be used such as supplying the water of the
top pond with heat from the warm sea water by passing them into heat exchangers.

Breaking ice into small pieces from the top part of the iceberg by mechanical
means and transporting them through the floating aqueduct in a medium of water
obtained from the top pond has to be considered. In the cases where external
energy sources than solar energy have to be applied in the process of speeding up
the melting of the ice a thorough feasability study should be made for each case
from the point of view of comparing the extra expenses of the energy consumed with
the gain achieved from speeding up the melting operation.

CLIMATIC EFFECTS

In some mooring sites where the weather is relatively warm during hot seasons the
presence of a long train of such icebergs will help in bringing down the temperature of the surrounding atmosphere when wind direction is towards land.

It is interesting to know that in places where the humidity of the air blowing by is rather high, the air will lose a great part of its water content on the iceberg resulting in more water yield than the iceberg would give besides lowering from the degree of humidity of the air.

In favourable climatic conditions people usually work more efficiently; this would attract investigators in establishing more new protects in that place resulting in a raise in the standard of living of people living there.

In the field of agriculture activity if water is provided at a reasonable cost to arid areas, farmers would go for the idea of increasing the cultivated piece of land they already have, especially if it proves to be economically sound by using modern techniques in watering their lands.

The preliminary studies concerning the cost of water from transported icebergs show that it is going to be far below that being produced from desalinization plants of either sea water or brackish water.

When the area of the cultivated land is considerably increased it adds to the improvement of the weather condition. This may result in the formation of low clouds raising the possibilities of having higher rainfall rate. In circumstances like this one would expect more land to be cultivated and a chain reaction will follow in time resulting in a complete change of the place from being a dry land to start with to becoming a green liveable land to end with.

FROZEN AND COLD STORAGE OF FOOD THROUGH ICEBERGS

In the cases where there is surplus in the food products in the country utilizing the icebergs such as vegetables, fruit, meat, fish or any other products which have to be preserved in a cold or frozen state for a long time without deterioration, icebergs may help in this concern.

Due to the length of time an iceberg has to stay until melting into usable water is fulfilled icebergs with their huge negative heat capacity can receive containers of food products during that length of time which is usually more than one year after which they can be transported to another iceberg recently arriving at the site and so on for any length of time. Special containers have to be designed in accordance with the method to be used in carrying out the storage process through the iceberg.

Other countries nearby which do not have to import soft water but require facilities for the storage of their own food products in an inexpensive way can enter into agreements with the country having the icebergs to provide them with such facilities.

CONCLUSION

It is more economical to transport bigger sizes of icebergs than small ones. The first iceberg to be towed has to be around 100 million Cu.metre. After gaining the experience required from transporting that size of icebergs it will be easy to handle bigger sizes such as 1000 million Cu. metre, the dimensions of which are 4 Km long, 1 Km wide and 250 m thick. For example the estimates for the costs of water delivered to Jeddah based on the 1976 prices were:

20.6 Cts/m^3 for the large icebergs
52.7 Cts/m^3 for the small experimental iceberg.

Slicing actually is very costly as it represents nearly 70% of the total cost. So one can see how cheap water would be from icebergs if slicing is not required and especially when large sizes are transported.

Towing of huge icebergs will require special seagoing tugs. As it is more economical to reduce the number of tug-crews by using the minimum number of tugs, therefore more powerful tugs have to be built as the available classical tugs will not be suitable except for the first experimental small iceberg.

Water current routes which run from the Antarctic may be used for the transportation of the icebergs. As part of the route is cold, it may lessen the melting losses and at the same time lowering the towing costs as a less number of tugs will be used. A complete study will be made on this point as the current routes are longer in some places than the direct one.

At present scientists are doing their best in finding out simpler means for the transportation, protection against melting during the crossing journey, slicing and speeding up melting at the mooring site of the iceberg in order to cut down the cost of water delivered to a minimum. The studies so far made are encouraging. We have great hopes that their efforts will blossom in good results very soon, adding, to what they have already offered, enough reasons that the time has come to exploit such huge water resources for the benefit of all nations and for the generations to come.

Nature is offering us a very good chance in solving our water problem as we do not have to cut parts from icy places there and push them into the sea. We have thousands of tabular icebergs floating in the sea water there and are already separated from each other. It is not going to disturb the balance of the earth at the south pole as the amount to be transported is considered so minute relative to what the Antarctic is having. Above all, what we are hunting are actually icebergs which were eventually going to melt by themselves if left to drift in the nearby Oceans among thousands of others that melt every year there.

If we go back into history between the year 1890 and 1900 we find out that small icebergs were towed by ship and sailed from Laguna San Rafael, Chile to Valparaiso and even to Callao, Peru, a distance of 3900 Km.

We have better facilities now than those pioneering groups of people who took that first step. Towing and delivering of icebergs of bigger sizes and for longer distances must not be put off any longer as we have all the facilities we need to realise such valuable projects for the welfare of all people.

Water through icebergs for the development of deserts may be just the start for such vital projects to see light.

WATER SUPPLY FROM IMPORTED ANTARCTIC ICEBERGS

John L. Hult
Application Concepts and Technology Association (ACTA)
P.O. Box 1731, Santa Monica, CA   90406

SUMMARY

There is a large and growing unmet world demand for fresh water. An annual yield
of a billion acre feet (1234 billion cubic meters) of Antarctic icebergs offers
high-quality fresh water for delivery to any deep water terminal. Studies of the
feasibility of using Antarctic icebergs reveal the following:

- There is an abundant supply of icebergs of appropriate
  size and characteristics that is accessible for
  acquisition during the daylight thawing season in the
  Antarctic (January through March of each year).

- The icebergs can be harnessed into trains and be
  brought slowly to deep water near areas of use with
  less energy required per unit of delivered water than
  for desalting, for long distance transfer by aqueduct,
  or for high lift pumping from underground water. The
  ultimate energy requirement is the equivalent of about
  a 10-m pumping lift; however the energy required
  immediately for a small scale delivery using existing
  equipment is equivalent to about a 500-m pumping lift.

- The icebergs must be protected from the melting loss
  to moving seawater for efficient use at any location.
  This can be done by wrapping the iceberg in a plastic
  "wet suit" of adequate strength and controlled insulation spacing and for total delivered costs well
  below those for desalting or long distance transfers
  by aqueduct.

- While the icebergs are moored at terminal locations
  they serve as unique reservoirs with significant net
  accrual of condensation water over evaporation loss.
  The icebergs can be converted for their fresh water
  and cooling benefits at very small costs and with
  benign impacts on most environments in which they
  would be used.

- The principal obstacles to the early exploitation of
  icebergs involve tradition in practices, institutions,
  water rights, vested interests, subsidies, environmental misapprehensions, and the use of water resources to control growth. The least costly way of
  resolving the remaining technological uncertainties
  would be to initiate a pilot program immediately that
  can test and demonstrate alternatives in an operational environment so that efficient designs can be
  selected for operational applications.

A pilot program is recommended to demonstrate the feasibility of bringing Antarctic icebergs to the Northern hemisphere and to comprehensively determine the
environmental impacts to the satisfaction of everyone concerned. Such a pilot
program could most easily and usefully be done to satisfy world wide concerns for
applications by bringing the pilot iceberg to a mooring terminal off the California coast. This would also allow the greatest revenue recovery from first day
covers, souvenirs, and publicity of all types. The earliest pilot program could
use existing shipping to acquire an iceberg in early 1979 for delivery to California by early 1980. Such a program schedule would require authorization in 1977
to begin preparations.

INTRODUCTION

Demand for Fresh Water

As world population soars and development and standards of living increase, the
fresh-water deficiencies in many desert areas of the world are rapidly becoming
acute. Areas of particular concern include the Pacific Southwest United States,
and parts of Mexico, Chile, Australia, the Middle East, and North Africa. In
addition, some areas that are not considered desert may seek additional fresh
water because of the water quality or reliability of some existing water resources, e.g. in Japan and South Africa.

Potential Water Supply

Seawater is by far the largest potential source of fresh water, however the costs
and energy requirements for desalting (by other means than natural evaporation
with solar radiation) tend to make desalting of seawater a last resort. Ground
water may be an attractive potential source of fresh water. However, if ground
water is inadequate or mined excessively (i.e. withdrawn at much greater rates
than it is replenished) or if the pumping energy costs or water quality are not
satisfactory, then other sources of surface water will be in demand. Where more
conventional surface water resources are inadequate, too remote, too expensive or
otherwise unsatisfactory, water from Antarctic icebergs will become attractive.

The total annual yield of Antarctic icebergs that is shed naturally to melt away in the Southern Oceans exceeds a million billion kg per year or 1000 billion cubic meters of melt water per year (1). This is an enormous amount of high quality water that might satisfy many worldwide demands if it could be delivered at attractive costs. It might be used to bring about 100 million hectares of new desert land into full agricultural production and add a significant fraction to the total existing world agricultural yield (2). This water supply should be more reliable than conventional surface supplies that are vulnerable to drought and floods. It would be used to augment and stabilize other water supplies with the Antarctic serving as a giant reservoir for greater assured worldwide supply of fresh water. The feasibility of using Antarctic icebergs is, therefore, an important issue.

FEASIBILITY OF USING ANTARCTIC ICEBERGS

Acquisition

Studies of Antarctic imagery obtained with Earth Resources Technology Satellites (formerly ERTS, now LANDSAT) confirm the abundance of icebergs of appropriate natural characteristics, and their accessibility during the first quarter of a calendar year (3). These studies also show that satellites would be a valuable tool to assist in locating, selecting, and acquiring icebergs to export for fresh water and cooling.

Iceberg Transfer

Harnessing: The harnessing of icebergs presents a variety of new problems that must be considered in arriving at efficient operational solutions. The harness should be able to apply thrust to the iceberg throughout its life for acquisition in the Antarctic, for transfer to the using destination, and for mooring while being converted for fresh water and cooling. The harness must be able to handle the varying thrusts in both magnitude and direction and accommodate the shrinking iceberg as it is melted and finally completely converted.

One method that has been considered is to sink expanding ice anchors deep into the iceberg from the top side so that the refrozen seating of the ice anchors will not be weakened by frozen seawater. This method should require somewhat less harness than other techniques, but the resultant forces on the iceberg are far removed from in-line with the centers of mass and resistance in the vertical plane and are likely to carve havoc in the softer top side of the iceberg. Also this method would interfere with top side operations and have difficulty accommodating the shrinking iceberg while it is moored for conversion. This method would not offer containment during the final breakup of the thinned iceberg.

A more attractive method appears to be one in which the iceberg is surrounded in the horizontal plane by a sling cable network underlaid with tough plastic fabric to distribute the pressures over large areas. The sling would be adjustably tied vertically to keep it in a desirable location. This method is more analogous to that of harnessing a horse, and it can be adjusted to accommodate a shrinking iceberg and to contain any final breakup. The sling can be prefabricated in easily connectable segments to accommodate any iceberg configuration that is encountered. There is a rich spectrum of harness design factors from which to choose cost effective systems including: rope materials (wire, synthetic fibers, etc.), individual rope sizes, methods for adjustment and distribution of stresses among parallel ropes, and methods of coupling to towing engines to aid in controlling towing vehicles.

Energy Required for Transfer: In today's world of limitations, the comparative
energy requirements to obtain fresh water from alternative supplies becomes an
important consideration. This can be determined with sufficient accuracy for a
preliminary assessment from previous studies (1). For the ultimate large transfer
in long iceberg trains from the Antarctic to Southern California in about a year-
long trip, the work done is equivalent to lifting the ice about 10 meters. Most
of this energy is used to overcome skin friction. The Coriolis forces offer great
resistance at the extreme southern latitudes, but their average effect over the
total transfer is about plus or minus 20 percent for the range of possible net
Coriolis fractions from 1 to 0. If the thrust from large conventional propellors
is used to provide the transfer energy at these slow speeds, the efficiency is
only about 5 percent so that the energy consumption would be equivalent to that of
a water pump lift of about 200 meters with pumping efficiencies exceeding 75
percent.

For a small scale program using existing ship propellors for thrust, the form drag
of the icebergs would offer the principal resistance and an energy consumption
equivalent to a water pump lift of about 500 meters would be required. This would
well satisfy a pilot program and initial operations that would do better than
desalting which requires energy consumption equivalent to a water pump lift of
more than 1000 meters.

There are promising techniques for delivering the thrust for transfer operations
with 10 times the efficiency in energy consumption compared with conventional
propellors that could be developed for future large scale transfer operations.
However, existing inefficient propellor systems are available now, less costly,
and adequate for a pilot program and early small scale opeations. They could
offer less costly water than desalting at low altitudes, long range aqueduct
transfers near sea level, or transfers from any source that requires a net pumping
energy more than the equivalent of a water pump lift of about 500 meters in excess
of the altitude of water use.

                              Melting Control

Studies have shown that icebergs will need protection to prevent excessive melting
loss to the seawater if they are exported from the Southern Oceans (1). If they
are unprotected they are likely to dissipate before reaching a Northern-Hemisphere
destination. Even for use in the Southern Hemisphere the losses that would result
if unprotected during transit and while being converted at the destination would
be excessive. The use of icebergs anywhere will justify the costs of protection
to limit melting loss to the sea to less than about 5 percent per year.

The rapid ablation of the ice that is in contact with the seawater is accounted
for by the great heat-transfer rate from the convection of the moving seawater.
If a thin layer of poorly conducting material could be kept next to the ice and
convection in the layer is sufficiently inhibited so that conduction is the principal mode of heat transfer through the layer, the heat flow can be limited to the
desired protective rate. The most attractive way so far conceived for protecting
the icebergs is to trap a quilting of water between the ice and the flowing seawater by means of thin plastic fabric (1). By inhibiting convection in the trapped water, the heat flow is primarily limited to conduction through the quilt of
water. Quilt thicknesses of 3 to 10 cm of trapped water with effective barriers
to convection appear to be relatively easily achievable. If such quilts can be
applied and maintained in the operational environment, they should meet the requirements for protection.

The technical problems of hugging any irregular iceberg surfaces with plastic fabrics of adequate strength to handle the flow resistance, anchoring the plastic fabric to the ice, control of the quilt thickness to shape and smoothen the ice, and the provision of appropriate barriers to inhibit convection within the quilt appear to be solvable. A pilot program could test various protection designs to determine the least costly ones that will satisfy operational requirements. It is estimated that a conservative over design of a protective system for operational use could be provided for less than 2 cents per cubic meter of ice in 1977 dollars for costs of protective materials.

Conversion and Use

Unpublished results of studies of iceberg conversion for fresh water and cooling reveal a number of valuable points (4). The amount of heat required to melt the ice is an impressive feature of the conversion process. In most situations it is likely to require more heat energy than is generated in the environment to satisfy man's other demands. Thus much of this heat energy must be extracted from the environment (radiation, atmosphere, sea, or earth), and a variety of alternative methods of accomplishing this were explored.

Some of the potentially more attractive alternatives involve unconventional techniques for breaking, mining, and transporting ice. However, a more simple alternative that would be very inexpensive and require little development for early application would be to augment the natural heat exchange with the atmosphere by means of sprays. This would expose large surface areas of ice-cold spray to augment the condensation of water vapor and the extraction of heat from the atmosphere for melting the ice. The process could begin during transfer operations and continue for two years at the using destination with melt water storage in pools on the iceberg.

The melt water from the icebergs would be of exceptional quality and could be mixed with other water to dilute the total dissolved solids and improve the quality of all the fresh water. The cooling value of the melt water could be used to increase the efficiency of electrical power generation by about 3 percent. However, if a generating plant is to benefit from a colder coolant it must be designed to take advantage of the colder temperature, and this can be justified only if an assured uninterrupted supply of the colder water will be available for the life of the operations.

Icebergs provide unique reservoirs with significant net accrual of condensation water over evaporation loss. They are relatively invulnerable to earthquakes and other disasters and do not threaten their surroundings with catastrophy. They would have a benign impact on the marine environment, especially when well insulated, even though their presence would be very impressive. They would dry and cool the air passing over them, but any measurable influence should be lost through diffusion within a few times the iceberg dimensions down wind.

For most uses of fresh water, the costs of distribution from the sources dominate the total water costs, and this is still likely to be true for water from icebergs.

Sociological and Political Issues

Providing attractive water supplies from imported Antarctic icebergs is an exciting challenge to the technical community. However, the principal obstacles to the early exploitation of these icebergs are sociological and political in nature.

On the international scene the rules for claiming and exploitng Antarctic icebergs
and the assessment of any liabilities in the process have not been established.
On the national and local levels, expressions of demand for iceberg water must be
exercised. Local water organizations are not inclined to abandon or jeopardize
their contracts, rights, and vested interests for highly subsidized water in order
to seek an unknown water source in the future.

Our laws for water rights and assessments need change to enable a more graceful
transition into the emerging world of limitations (5). The doctrine of prior
appropriation for water is outmoded internationally, in the Antarctic, nationally,
and locally. Much better would be the unrestricted, nondiscriminatory rights of
access to water anywhere subject only to appropriate notice for accommodation and
the obligation for the prorated assessments to assure adequate supply. The open
market could determine the water use subject only to antitrust actions to prevent
the manipulation of supply or demand at public expense by special interests.

If subsidies are desired for some segments within a governmental jurisdiction,
these could be provided directly so that they are not obscured by the complexities
of the pervasive influence of water in our society. Water supply need not be used
as a vehicle to control growth nor as a decisive factor in environmental contro-
versies. The environmental impact of imported icebergs appears benign, but we
should not forgo the opportunity to test and demonstrate the impact for all concerned interests before substantial commitments are made for operational application. A pilot program initiated without delay could test and demonstrate the
environmental impacts as well as a variety of system alternatives so that efficient designs could be selected for possible operational applications. Further
delays encourage research expenditures of marginal value about issues that are
most fruitfully determined in an operational environment.

A PROPOSED PILOT PROGRAM

Objectives

It is recommended that a pilot program be initiated without delay with the following objectives:

- The demonstration of the feasibility of bringing
  Antarctic icebergs to the Northern Hemisphere.

- The comprehensive determination of the environmental
  impacts to the satisfaction of everyone concerned.

It is not proposed to build aqueducts or any land facilities to feed melt water
into operational water systems for the pilot program.

Costs and Financing

The costs of the independent expedition to acquire and bring an Antarctic iceberg
of about 100 million cubic meters (100 billion kg) to the California coast are
estimated to be about $20 million for a minimum pilot program.

The costs of the Coast Guard to monitor the program for about one year during
acquisition in the Antarctic and until delivery to destination should be added.
This would establish operational requirements to insure safety and benign environmental impacts in international waters.

Also to be added would be the destination preparation and monitoring costs. These would include the selection of the terminal site, e.g. Mugu Canyon, the installation of an appropriate set of at least four mooring points, and the establishment of a comprehensive monitoring system of the marine and atmospheric environments to the extent that they can conceivably be impacted by the program. The environments should be monitored preferably for a full year before the iceberg arrival and for at least two years afterwards.

The monitoring of the pilot program could probably best be accomplished to satisfy potential world wide applications and concerns if the California coast is the destination. This would facilitate the use of the best and most comprehensive instrumentation systems for the monitoring, and would encourage the participation of the broadest variety and most active groups with environmental concerns.

The California coast as a destination would also excite the greatest revenue recovery from first day covers, souvenirs, film and literary returns, and publicity of all types. It is estimated that the demand for these products from the first Antarctic iceberg imported to the Northern Hemisphere would yield revenues exceeding 50 to 100 million dollars if brought to California. No other destination for the pilot program would yield comparable revenues.

The Plan

The earliest pilot program could use existing shipping to acquire an iceberg in early 1979 for delivery to California by early 1980. Such a program schedule would require authorization in 1977 to begin preparations for all components of the program.

Existing shipping could be adapted to provide the required towing pull of 400,000 kg with about 45,000 horse power and a fuel consumption for the mission of about 80,000 tons.

The importing operation would involve extensive instrumentation to determine the detailed characteristics of the iceberg at acquisition, to compare the performances of alternative techniques to be tested in the program, and to monitor in detail all pertinent environmental factors along the transit course.

The independent expedition to acquire and bring the pilot iceberg to California could be contracted for on an incentive basis in which the net share in publicity revenues would be determined by the performance in the importing operation.

Any interested government could be invited to participate in the pilot program and gain experience and test information for its peculiar individual applications.

REFERENCES

(1) Hult, J. L., and N. C. Ostrander, Antarctic Icebergs as a Global Fresh Water Resource, The Rand Corporation, R-1255-NSF, October 1973.
(2) Ambroggi, Robert P., Underground Reservoirs to Control the Water Cycle, Scientific American, 236, 5, 21-27, May 1977.
(3) Hult, John L., and Neill C. Ostrander, Applicability of ERTS for Surveying Antarctic Iceberg Resources, The Rand Corporation, R-1354-NASA/NSF, November 1973.
(4) Hult, John L., and Neill C. Ostrander, Antarctic Iceberg Conversion for Fresh Water and Cooling, Application Concepts and Technology Association, unpublished.

(5) Hult, John L., <u>Water Rights and Assessments (Proposals involving Antarctic Icebergs for the Colorado River Basin, California, Mexico and other arid lands)</u>, The Rand Corporation, P-5271, July 1974.

CHARACTERISTICS AND USES OF WATERS PRODUCED WITH
PETROLEUM IN ARID AREAS IN THE UNITED STATES

A. Gene Collins[1]

ABSTRACT

The composition and types of waters produced with petroleum in arid areas of the
United States were surveyed. Arbitrarily, selected areas were those with a mean
annual precipitation of rain of less than 16 inches or 40.6 centimeters. States
with petroleum production in these areas include: Arizona, California, Colorado,
Kansas, Montana, Nebraska, New Mexico, Nevada, North Dakota, Oklahoma, South
Dakota, Texas, Utah, and Wyoming. The study indicated that most of the waters
produced with petroleum can be used in enhanced-oil-recovery operations, and that
some of the waters may be useful after adequate treatment, which may include
desalination, for irrigation, livestock consumption and human consumption.

INTRODUCTION

In this report, subsurface waters associated with oil and gas produced in states
where the mean annual precipitation of rain is less than 16 inches or 40.6 centimeters are considered. Figure 1 illustrates the average mean annual precipitation
of rain (1).[2]

Most oil is produced from geologic basins composed of various types of sedimentary
rocks. Figure 2 illustrates some of the sedimentary basins in the arid areas of
the United States (2). Subsurface oilfield waters originate from meteoric water,
seawater, and juvenile water (water derived from primary magma). The five spheres
of the earth are the lithosphere (rocks), the pedosphere (soils, till and other
surficial materials), the hydrosphere (natural waters), atmosphere (gases) and the
biosphere (living organisms). Oilfield waters are a part of the hydrosphere, and
petroleum is a product of the biosphere.

---

[1]Project Leader, Bartlesville Energy Research Center - ERDA, Bartlesville, OK.
[2]Underlined numbers in parentheses refer to items in the list of references at the
end of this report.

FIGURE 1. - Contour lines of mean average rainfall, inches per year.

FIGURE 2. - Location of the major sedimentary basins in arid areas of the United States

LEGEND

Sedimentary Basin

Deepest part of basin

The total volume of water in the hydrosphere is about $1,338 \times 10^{18}$ liters, and about 8.4×10^{18} liters in groundwater (3). Most of the water, 1300×10^{18} liters, is in the oceans. Less than 50 percent of the total groundwater is in strata below 1 km. The total amount of water in sedimentary rocks and associated with liquid and gaseous hydrocarbons is less than 1.3×10^{18} liters.

All of the petroleum recovered to date has been taken from oil wells drilled into the upper 8 km of the earth's crust. The average thickness of the earth's crust is about 17 km, varying from 5 km under the oceans to about 35 km under the continents (4).

Petroleum hydrocarbons are believed to have originated from organic material in sedimentary material that was produced by weathering and erosion of earth's surface. This eroded material is carried away by water, ice or wind and redeposited, ultimately forming sedimentary rocks. The major sedimentary minerals are clays, quartz, calcite, gypsum, anhydrite, dolomite and halite. Most of the large bodies of sedimentary rocks were formed in marine environments, smaller sedimentary deposits formed in lakebeds, and river flood plains.

CHARACTERISTICS OF OILFIELD WATERS

The amounts and ratios of the constituents dissolved in oilfield waters are dependent upon the origin of the water and what has occurred to the water since entering the subsurface environment. For example, some subsurface waters found in deep sediments were trapped during sedimentation, but other subsurface waters have infiltrated from the surface through outcrops. Some waters are mixtures of the infiltration water and trapped ancient seawater. Also, the rocks containing the waters often contain soluble constituents that dissolve in the waters or contain chemicals that will exchange with chemicals dissolved in the waters, causing alterations of the dissolved constituents (5).

The amounts of dissolved constituents found in oilfield waters range from less than 10,000 to more than 350,000 mg/l. This salinity distribution is dependent upon several factors, including hydraulic gradients, depth of occurrence, distance from outcrops, mobility of the dissolved chemical elements, soluble material in the associated rocks, ion exchange reactions, and clay membrane filtration.

Concentration of ancient seawater could have occurred by surface evaporation, and there are at least three independent processes that can cause major changes in buried, isolated seawater:

    1. Dilution with meteoric or fresher waters that have entered outcrops.

    2. Reactions with minerals in the sediments and sedimentary rocks.

    3. Membrane filtration through clays and shales as a result of pressure and osmosis.

Origin and Composition of Oilfield Water

The origin and composition of some oilfield waters taken from subsurface formations of Tertiary to Cambrian geologic ages has been discussed by Collins (5). Composition of oilfield waters from rocks of various ages varies with depth and diagenetic reactions.

Natural oilfield waters contain sodium, calcium, magnesium, strontium, barium, iron, manganese, sulfate, bicarbonate, chloride, bromide, iodide and dissolved gaes. In addition, they usually contain numerous other minor and trace constituents.

Rittenhouse, et al., (6) concluded that elements in oilfield waters commonly are present in the following concentrations:

```
Percent . . . . . . . . . . . . . . . .Na, Cl
Percent or ppm . . . . . . . . . . . . .Ca, Mg, SO₄
100 + ppm . . . . . . . . . . . . . . . K, Sr
1-100 ppm . . . . . . . . . . . . . . . Al, B, Ba, Fe, Li
ppb (most oilfield waters) . . . . . . .Cr, Cu, Mn, Ni, Sn, Ti, Zr
ppb (some oilfield waters) . . . . . . Be, Co, Ga, Ge, Pb, V, W, Zn
```

They found no relationship between the constituents in the brine and the minerals in the aquifer rocks except for potassium. They postulated that exchange reactions occurred between the clays in the rocks and potassium in the water to control the dissolved potassium.

Compared to seawater, the oilfield waters were enriched in manganese, lithium, chromium, and strontium, and depleted in tin, nickel, magnesium, and potassium. Generally the silicon content varied inversely with the dissolved solids content.

Classification of Oilfield Waters

There are several methods of classifying oilfield waters. However, classification methods developed by Sulin and Schoeller (5) are useful in determining the origin and diagenesis of oilfield waters. Two of the useful characteristics of these classifications are shown in Table 1 and mentioned below.

The Sulin type of water can be SO_4-Na, which is indicative of a continental water; HCO_3-Na which also is indicative of another type of continental water; Cl-Mg, which is indicative of a marine water; and Cl-Ca, which is indicative of a deep, stagnant, subsurface water. Many oilfield waters are the Cl-Ca type, but not all.

The Schoeller index of base exchange (IBE) is useful in determining the predominant reaction of constituents dissolved in the water with constituents in subsurface rocks. For example, if the IBE is a positive number then ion exchanges between the water and rock probably were alkali metals in the water for alkaline earth metals in the rock. If the IBE is negative, the reverse type of exchange probably predominated.

Composition of Some Oilfield Waters from Arid Regions

Table 1 illustrates the composition of some waters produced with petroleum from subsurface formations in Arizona, California, Colorado, Kansas, Montana, Nebraska, New Mexico, North Dakota, Oklahoma, South Dakota, Texas, Utah and Wyoming.

Recently 20° API gravity oil was discovered in the Trap Springs Field, Nye County, Nevada. The productive zone is composed of volcanic rock and is about 4,200 feet deep. According to the operating company geologist, abundant Artesian water of potable quality is found near the surface down to the oil productive zone. A drill stem sample of the water contained 30 mg/l chloride.

The data in Table 1 show the sample number, state, (oilfield, county or legal description in latitude and longitude), geologic formation, depth in feet, specific gravity, pH, milligrams per liter (mg/l) of sodium (Na), magnesium (Mg), calcium (Ca), bicarbonate (HCO_3), chloride (Cl), bromide (Br), iodide (I), sulfate (SO_4) and dissolved solids. Also shown are Sulin brine type and Schoeller IBE for the samples.

Some of the samples contain less than 5,000 mg/l of dissolved solids, and a few contain less than 1,000 mg/l. Most of these samples with relatively low concentrations of dissolved solids are not Cl-Ca type waters. This indicates that they are not produced from a deep, stagnant environment. It also could indicate that they are mixed with a fresher water. The source of the fresher water could be meteoric water entering an outcrop, water from a fresh water subsurface aquifer that has leaked into the saline aquifer because of a poorly completed well or leaking casing, or dilution at the surface after production.

USES OF OILFIELD WATERS

In the early days of the petroleum industry, the waters produced with oil and gas were believed to be entirely useless, and the easiest methods of disposing of them were used. These waters often were allowed to flow by natural surface drainage into streams, lakes, etc. It was not until 1938 that interstitial water was recognized as an intergral part of a petroleum reservoir, and its relation to the origin and migration of petroleum was not even imagined (5).

TABLE 1. — Composition of some oilfield waters from some arid areas of the United States

No.	State	Field, County, or Legal Description	Formation	Depth feet	Specific gravity 60°/60° F	pH	Na	K	Mg	Ca	HCO$_3$	Cl	Br	SO$_4$	Dissolved solids	Other	Brine type	IBE
1	AZ	Apache-Co.	L. Hermosa	--	1.012	--	7,760	20	50	1,520	170	11,600	20	--	21,200	1/	HCO$_3$-Na	-0.62
2	do.	E. Boundry Butte	Hermosa	4,740	1.011	7.8	4,690	--	159	469	4,000	5,120	--	1,470	15,900	--	Cl-Mg	0.04
3	do.	do.	do.	4,842	1.029	6.5	12,600	--	432	987	2,340	20,300	--	1,130	37,800	--	Cl-Ca	0.15
4	do.	do.	do.	5,062	1.042	7.0	18,100	--	624	2,850	1,890	33,000	--	902	57,400	--	Cl-Ca	0.50
5	do.	do.	Pennsylvanian	4,884	1.051	5.8	13,700	--	2,650	8,190	75	42,000	--	1,760	68,400	--	Cl-Ca	--
6	CA	Kern-Co.	Kern	--	1.000	--	210	15	5	70	130	330	2	40	803	--	Cl-Ca	0.23
7	do.	Mt. View	Fruitvale	10,000	1.013	--	5,100	118	82	1,320	600	10,400	51	2,050	17,700	1/,2/,3/	HCO$_3$-Na	-0.76
8	do.	E. Coalinga	Temblor	2,133	1.005	7.1	2,090	--	243	192	1,680	1,340	--	76	7,600	--	Cl-Ca	0.53
9	do.	N. Tejon	Vedder	11,850	1.029	--	7,050	170	800	7,350	--	23,600	97	10	39,400	1/,2/,3/	Cl-Ca	0.08
10	do.	Tejon	Miocene Middle	5,370	1.011	--	5,500	59	89	250	500	9,240	49	10	15,700	2/,3/	HCO$_3$-Na	-1.28
11	CO	Battleship	Dakota	4,822	1.001	8.2	660	--	--	4	990	270	--	10	2,230	--	HCO$_3$-Na	-0.94
12	do.	do.	Lakota	4,766	1.000	8.0	190	--	2	4	350	60	--	60	664	--	Cl-Ca	0.06
13	do.	Powder Wash	Wasatch	4,221	1.009	--	4,520	--	30	280	190	7,410	--	50	12,500	--	Cl-Ca	0.03
14	do.	do.	do.	2,248	1.008	--	4,470	--	30	180	28	7,120	--	10	11,900	--	Cl-Ca	0.03
15	do.	Babcot	Dakota	5,150	1.004	8.2	2,520	--	3	8	1,530	3,040	--	--	7,101	--	SO$_4$-Na	-0.96
16	KS	39.58509°N 101.09299°W	Kansas City Group	--	1.065	7.5	31,900	--	550	1,890	60	51,800	--	3,250	89,500	--	Cl-Ca	0.05
17	do.	39.33389°N 101.26078°W	Cherokee	--	1.035	7.5	17,600	--	245	1,430	--	27,700	--	3,700	50,700	--	Cl-Mg	0.02
18	do.	39.33389°N 101.26078°W	Marmaton Group	--	1.017	8.0	5,700	--	--	897	--	7,060	--	4,500	18,200	--	SO$_4$-Na	-0.52
19	do.	39.33467°N 101.26078°W	do.	--	1.013	8.0	8,610	--	108	405	260	8,330	--	7,900	25,600	--	SO$_4$-Na	-0.83
20	do.	39.33389°N 101.26078°W	Kansas City Group	--	1.024	7.5	11,400	--	--	816	--	15,600	--	4,700	32,500	--	SO$_4$-Na	-0.58
21	MT	Kevin-Sunburst	Ellis	1,460	1.003	--	1,740	--	--	90	2,840	1,130	--	80	5,880	--	HCO$_3$-Na	-0.91
22	do.	Cat Creek	Cat Creek	1,690	1.000	--	354	--	--	--	486	35	--	289	1,160	--	HCO$_3$-Na	-1.03
23	do.	Lothair	Swift	2,005	1.002	7.2	1,300	--	8	30	2,640	510	--	50	4,540	--	HCO$_3$-Na	-0.95
24	do.	Melstone	Amsden	4,250	1.012	7.8	5,540	--	50	330	800	2,460	--	8,550	17,700	--	SO$_4$-Na	-0.90
25	do.	Benrud N.E.	Nisku	7,675	1.033	7.4	16,500	555	132	979	537	26,600	--	1,650	47,000	--	Cl-Ca	0.02
26	NB	Huntsman	Dakota	4,680	1.014	8.5	7,190	--	29	118	710	7,950	--	3,840	19,800	--	SO$_4$-Na	-0.97
27	do.	Gurley	Dakota	4,360	1.044	8.2	22,700	--	121	423	1,220	30,400	--	6,560	61,400	--	HCO$_3$-Na	-0.86
28	do.	Sidney West	Cloverly	4,832	1.006	7.9	3,100	--	15	15	1,960	3,600	--	314	8,990	--	Cl-Na	0.01
29	do.	Lane	Dakota	4,618	1.091	8.0	46,900	200	355	1,420	537	73,000	--	3,800	126,000	--	HCO$_3$-Na	-0.88
30	do.	Sidney West	Cloverly	4,802	1.004	8.2	2,410	--	2	13	1,290	3,020	--	52	6,790	--	HCO$_3$-Na	-1.00
31	NM	Maljamar	San Andres	4,138	1.153	--	77,100	--	952	2,530	240	123,000	--	4,340	208,000	--	HCO$_3$-Na	-1.28
32	do.	Stoney Butte	Mesa Verde	745	1.001	--	659	--	--	251	735	1,080	--	3	24,500	--	HCO$_3$-Na	42.30
33	do.	Kutz Canyon W.	Pictured Cliffs	1,937	1.038	--	19,300	--	132	355	495	1,380	--	435	3,790	--	HCO$_3$-Na	-0.49
34	do.	Fulcher Basin	Fruitland	2,786	1.006	8.2	3,720	--	5	26	2,270	4,200	--	19	10,500	--	HCO$_3$-Na	-0.97
35	do.	Bravehall	Point Lookout	10,070	1.223	6.8	68,000	7,400	4,500	36,900	769	191,000	--	280	309,000	--	Cl-Ca	0.42
36	ND	Wildcat	Duperow	4,408	1.198	6.2	95,900	3,660	6,180	180	500	166,000	--	890	275,000	--	Cl-Ca	0.09
37	do.	do.	Mission Canyon	5,215	1.209	7.2	100,000	--	1,980	9,640	281	182,000	--	440	294,000	--	Cl-Ca	0.15
38	do.	Talley	do.	4,505	1.181	6.2	97,000	--	2,040	7,160	165	169,000	--	634	276,000	--	Cl-Ca	0.11
39	do.	Busil	Charles	2,980	1.172	6.4	54,700	--	1,240	3,320	258	91,500	--	2,950	154,000	--	Cl-Ca	0.08
40	do.	Souris North	Mission Canyon	--	1.129	--	55,900	--	2,630	12,700	115	116,000	--	468	188,000	--	Cl-Ca	0.26
41	OK	Jackson-Co.	Arbuckle	--	1.148	--	60,600	--	2,430	17,400	48	131,000	--	498	212,000	--	Cl-Ca	0.29
42	do.	do.	Wolfcamp	--	1.139	--	56,300	--	3,260	17,500	--	127,000	--	242	204,000	--	Cl-Ca	0.32
43	do.	34.35000°N 98.96666°W	Cisco	--													Cl-Ca	
44	do.	34.31666°N 99.20000°W	Strawn	--	1.055	--	30,800	--	887	3,780	6	56,700	--	78	92,300	--	Cl-Ca	0.16
45	do.	Tangier, W.	Morrow	8,298	1.020	6.3	7,210	37	5	649	334	16,600	438	938	27,300	1/,2/,4/	Cl-Mg	0.94
46	SD	Oak Creek	Minnelusa	1,140	1.001	--	--	--	83	249	190	6	--	747	1,280	--	Cl-Mg	0.98
47	do.	Newell Exp. Farm	do.	4,348	1.003	--	88	--	127	616	137	8,300	--	1,940	11,200	--	SO$_4$-Na	0.0
48	do.	do.	do.	4,348	1.003	--	55	--	117	615	126	79	--	1,850	2,840	--	SO$_4$-Na	-0.62
49	do.	Camp Crook	Deadwood	7,966	1.004	--	874	160	83	242	280	595	--	1,420	8,680	--	SO$_4$-Na	-0.68
50	do.	Buffalo	Red River	8,331	1.003	7.8	1,430	--	34	191	390	1,440	--	1,540	5,190	--	Cl-Ca	0.13
51	TX	Doss	Pennsylvanian	8,900	1.043	7.6	16,600	--	772	2,550	503	29,500	--	3,640	53,600	--	Cl-Ca	0.30
52	do.	Toler	Bend	3,820	1.124	7.5	41,300	--	1,820	12,700	147	91,200	--	405	148,000	--	Cl-Ca	0.22
53	do.	Corsica	Bend	6,000	1.098	4.6	37,800	--	15	9,770	12	75,200	--	577	123,000	--	Cl-Ca	0.21
54	do.	30.90000°N 101.95000°W	--	1,275	1.005	--	2,400	--	95	24	1,000	3,330	--	86	7,000	--	HCO$_3$-Na	-0.54
55	do.	Donnelly	Cisco	8,684	1.117	6.6	59,800	669	972	9,290	120	112,000	262	946	184,000	1/,2/,4/	Cl-Ca	0.17
56	UT	Aneth	Hermosa	5,578	1.154	4.9	43,800	--	4,900	29,800	75	134,000	--	571	213,000	--	Cl-Ca	0.50
57	do.	Chalk Creek	Kelvin	1,945	1.003	6.4	703	--	34	480	183	1,890	--	48	3,340	--	Cl-Ca	0.43
58	do.	Red Wash	Green River	5,415	1.002	8.5	1,280	--	3	6	2,490	86	--	5	4,250	--	HCO$_3$-Na	-0.99
59	do.	Island Unit	do.	4,045	1.049	7.8	23,000	--	269	2,060	195	38,000	--	3,580	67,700	--	Cl-Ca	0.07
60	do.	Wildcat	Paradox	4,950	1.018	8.7	7,180	--	444	876	195	9,400	--	5,880	24,000	--	SO$_4$-Na	-0.37
61	WY	Salt Creek	Wall Creek First	2,742	1.002	--	1,033	--	--	--	1,950	350	--	--	3,330	5,170/170	SO$_4$-Na	-1.10
62	do.	Baxter Basin North	Sundance	4,005	1.011	--	5,710	--	80	309	2,450	4,350	--	24	18,100	--	HCO$_3$-Na	-1.37
63	do.	Glenrock South	Dakota Sandstone	6,186	1.003	9.3	1,740	--	4	6	1,340	1,450	--	170	4,710	--	HCO$_3$-Na	-0.24
64	do.	Arch	Almond	4,793	1.016	7.6	8,790	--	175	206	5,840	11,100	--	24	26,100	--	HCO$_3$-Na	-0.85
65	do.	Gras Diamond	Curtis	3,298	1.017	8.0	7,880	--	171	382	194	9,670	--	4,810	23,100	--	SO$_4$-Na	-0.68

1/ I: No. 1—12, No. 7— 5, No. 9— 19, No. 45— 696, No. 55— 19
2/ Sr: No. 7— 39, No. 9— 218, No. 10—18, No. 45— 27, No. 55— 332
3/ Ba: No. 7— 24, No. 9— 18, No. 10— 3
4/ B: No. 45— 31, No. 55— 36

Use of Oilfield Water To Recover Petroleum

Enhanced oil recovery is defined as the additional production of oil resulting from the introduction of artificial energy into the reservoir (7). Primary oil recovery is defined as the oil and gas produced by natural reservoir energy or forces. Therefore, by this definition enhanced recovery includes waterflooding, gas injection, and other operations involving fluid or gas injection whether for secondary or tertiary oil recovery. Tertiary recovery is any enhanced-recovery operation applied after secondary recovery (8).

Enhanced recovery is applied to an oil-containing subsurface reservoir for the purpose of dislodging oil from the reservoir rock pores and moving it to a production well. Many subsurface petroleum reservoirs contain sufficient energy because of internal pressure to push oil and gas to the surface when first penetrated by a drill bit. However, as the internal pressures become low because of the removal or production of oil, gas and water at the well head, the residual oil in the reservoir cannot be recovered without the application of external or artificial force.

Secondary recovery is any enhanced-recovery operation first applied to a reservoir. Often it follows primary recovery, but it can be conducted simultaneously with primary recovery. The most common secondary-recovery process is water-flooding.

In addition to waterflooding, in which only water is used as the injection fluid, enhanced recovery includes: (1) injection of miscible liquefied petroleum gas (LPG) or the miscible slug process; (2) the enriched gas miscible process, sometimes called the condensing gas drive; (3) the high-pressure, lean-gas miscible process; (4) the carbon dioxide miscible process; (5) caustic solution flooding; (6) micellar solution flooding; (7) polymer flooding; (8) thermal recovery by hot fluid injection; and (9) thermal recovery by in situ combustion (8). In addition, there are variations and/or combinations of these processes, and the name of the process may vary with the author.

Injection water is used in most of these processes but may not be used in every reservoir in which processes 3 and 9 are applied. Because water is used in most enhanced-recovery operations, the quantity and quality of the injection water are very important to the production of this fossil fuel.

Therefore, in most modern petroleum production operations oilfield waters or waters produced with petroleum are utilized to recover additional petroleum. They are re-injected into the formation to maintain reservoir pressure and/or are used in enhanced-recovery operations.

Because the quantity of produced water often is not sufficient to meet the requirements of enhanced recovery, additional water or makeup water is used. This makeup water may originate from another subsurface formation, a river, lake, sewer effluent or ocean. The composition of the makeup water may differ drastically from the natural oilfield water. Mixing the makeup water with the natural oilfield water may result in the precipitation of various solids (7).

Water from surface bodies such as rivers and lakes and shallow subsurface aquifers usually contain only a few grams per liter of dissolved salts. A particular disadvantage of this type of makeup water is that it contains considerable amounts of dissolved oxygen and suspended solids such as sand, clay and plant and animal debris. Salt water from other subsurface formations may contain large quantities of dissolved solids, corrosive gases such as hydrogen sulfide or carbon dioxide and sulfate-reducing bacteria.

Therefore, the problem of disposal of oilfield water in some areas has diminished and probably will diminish even more in the future as the demand for water in enhanced-oil-recovery operations increases. Disposal of some oilfield waters and waste products associated with petroleum production will remain a problem probably as long as petroleum is produced from subsurface strata. The production of excess water from some petroleum reservoirs is more excessive than from others. As a general rule, the water production increases and the oil production decreases with the length of time of production.

Oilfield Waters as a Source of Valuable Constituents

Some oilfield waters contain relatively high concentrations of constituents such as bromide, iodide, lithium (9), and magnesium (5). For example, this year a factory began production to extract 2 million pounds of iodine per year from oilfield waters near Woodward, Oklahoma; sample 45, Table 1 is from Woodward, Oklahoma. High concentrations of iodide ions were discovered in these waters in the 1960's, and at least three papers have described them (10-12).

Bromine has been commercially extracted from oilfield waters produced from the Smack-over Formation in Arkansas for many years. At one time, iodine was extracted from oilfield waters in California. A chapter has been published on valuable minerals in oilfield waters (5).

Other Uses of Oilfield Waters

In some regions brines are used on roads to stablize the clays and gravels, to control dust, and to remove ice. Oilfield waters also may prove useful in Wyoming or Montana in coal slurry pipelines.

The data in Table 1 indicate that some of these waters possibly could be used for irrigation, animal consumption or even human consumption after adequate treatment. It should be remembered that water probably will be available from the subsurface formations that now contain petroleum long after the petroleum has been produced. It will, however, require considerable energy to pump this water to the surface from the deep aquifers (5).

Table 2. - Upper Limits of Dissolved Solids in
Drinking Water for Livestock (14)

Stock	Concentration mg/1
Poultry.	2,860
Swine.	4,290
Horses	6,430
Cattle (dairy)	7,150
Cattle (beef).	10,100
Sheep (adult).	12,900

Irrigation water containing up to 5,000 mg/1 of dissolved solids can be used for some salt-tolerant crops if some leaching rain occurs (13). Some ions in the water are more damaging to crops than others; for example, boron is more detrimental than is sodium. Therefore, a precise analysis of the water followed by an adequate interpretation must be made before any oilfield water is used for irrigation.

Livestock can consume water that contains considerably higher concentrations of dissolved solids than that considered satisfactory for humans. In the western United States, range cattle sometimes drink water which contains about 10,000 mg/1 of dissolved solids. However, these dissolved solids consist primarily of sodium chloride. Waters containing sulfate in high concentrations are undesirable, and waters containing more than 5,000 mg/1 dissolved solids probably will inhibit good growth in cattle. Table 2 shows some upper limits of dissolved solids in water used by livestock (14).

It is difficult to establish exact limits for drinking water for human consumption for any of the constituents commonly found in natural waters, because of differences in tolerance levels of individuals. Limits usually quoted for United States drinking water first were established in 1914 by the U.S. Public Health Service to control the quality of water supplied to passengers by interstate common carriers.

The standards were revised several times, and those in current use date from U.S. Public Health Service, 1962 (15). The concentration of dissolved solids recommended shows an upper limit of 500 mg/l. Here again, the concentrations of the various ions can be critical.

The only way that subsurface waters produced with petroleum could be used for consumption by humans would be after appropriate treatment, and the treatment probably would include some type of desalination. The Office of Saline Water has conducted considerable research on methods to convert saline waters to potable waters (16). Therefore, methodology to desalinate some oilfield waters is available.

CONCLUSIONS

Oilfield waters produced from petroleum wells in arid areas of the United States usually contain relatively high concentrations of dissolved solids. Oilfield waters produced in more humid areas of the United States often contain very similar types and concentrations of dissolved solids. Some of the oilfield waters in arid areas contain relatively low concentrations of dissolved solids. These waters are used in enhanced-oil-recovery operations and may be useful in agricultural irrigation or for consumption by livestock after adequate treatment. Desalination and appropriate treatment of these waters could make them useful for human consumption.

REFERENCES

(1) The National Atlas of the United States of America, U.S. Geol. Sur, Washington, D.C., p. 97.
(2) Supplement to the Oil and Gas Journal, 60, 46, (1952).
(3) Skinner, B. J., Earth Resources, Prentice-Hall, New Jersey, 1969.
(4) Clark, S. P. and A. E. Ringwood, Density Disturbance and Constitution of the Mantle. Rev. Geophys., 2, (1964), 35-88.
(5) Collins, A. G., Geochemistry of Oilfield Waters, Elsevier Scientific Publishing Co., New York, 1975.
(6) Rittenhouse, G., et al, Minor Elements in Oilfield Waters, Chem. Geol., 4, (1969), 189-209.
(7) Collins, A. G., Enhanced-Oil-Recovery Injection Waters, SPE Paper 6603, Society of Petroleum Engineers, International Symposium on Oilfield and Geothermal Chemistry, San Diego, Calif., June 27-29, 1977.
(8) Herbeck, E. F., et. al, Petroleum Engineer, 48, 1, (1976), 33, 36, 40, 42, 44, 46.
(9) Collins, A. G., Lithium Abundances in Oilfield Waters, U.S. Geol. Sur. Prof. Pap. 1005, (1976), 116-123.
(10) Collins, A. G. and Egleson, G. C., Iodide Abundance in Oilfield Brines in Oklahoma, Science, 156, (1967), 934-935.
(11) Collins, A. G. Chemistry of Some Anadarko Basin Brines Containing High Concentrations of Iodide, Chem. Geol., 4, (1969), 169-187.
(12) Collins, A. G., et. al, Iodine and Algae in Sedimentary Rocks Associated with Iodine-Rich Brines, Geol. Soc. Am. Bull., 82, (1971), 2607-2610.
(13) Environmental Studies Board, Water Quality Criteria 1972, U.S. Gov't. Printing Office, Washington, D.C., Stock No. 5501-00520, 1972.
(14) McKee, J. E. and Wolf, H. W., Water Quality Criteria: California State Water Quality Control Board Publ. 3-A, 1963.
(15) U.S. Public Health Service, Drinking Water Standards, U.S. Public Health Service Pub. 956, 1962.

(16)   Office of Saline Water, <u>1973-1974 Saline Water Conversion Summary Report</u>, U.S. Gov't Printing Office, Washington, D.C., Stock No. 2400-00796, 1974.

EVAPORATION CONTROL FOR
INCREASING WATER SUPPLY

C. Brent Cluff*

ABSTRACT/SUMMARY

A summary of the leading methods of evaporation control is presented.
Eight categories of evaporation control were discussed. The three leading
categories of evaporation control discussed were the monolayer, the reduction of the surface-area-to volume method and floating vapor barriers.
These methods are less expensive and appear to have a wider range of
application than destratification, wind barriers, shading the water
and floating reflective barriers. The other method of evaporation
control discussed was the use of sand or rock-filled reservoirs. This
method was found to be effective but limited to smaller size reservoirs.
The use of fatty alcohol to form monolayers will work on larger reservoirs
using either the airplane or a pipeline carrying the alcohol in slurry
or an emulsified form. An airplane was used to distribute alcohol
on the 12,000 hectare Vaal Dam in Africa during an extreme drought.
The estimated cost was $0.021/m^3. The method is less cost-effective
on smaller reservoirs due to rapid removal of the material by the wind.

The reduction of surface-to-volume ratio can be accomplished
through the use of proper site selection or if in flat terrain through
the use of the compartmented reservoir. The concept of the compartmented reservoir can be used on existing reservoirs but more easily
on new ones since it involves the use of construction equipment. The
proper use of the compartmented reservoir concept should result in a lower
unit cost of water saved than any other presently known evaporation
control method.

The use of floating vapor barriers of foamed rubber or wax-
impregnated expanded polystyrene seemed to have a wider range of use
than other floating vapor barriers. The recent development of the
wax-impregnated expanded polystyrene for evaporation control is
described. The paper describes it to be one of the most promising
floating vapor barriers in terms of cost effectiveness and
weatherability. Developed at the University of Arizona, this material
can be used in large 1.2 x 2.4 m sheets connected together by couplers
or rubber straps or as smaller floating squares that would be less
prone to vandalism.

*
C. Brent Cluff is an Associate Hydrologist at the Water Resources
Research Center, University of Arizona.

There is considerable merit to the use of the compartmented system with a floating vapor barrier in the "last" compartment to have water in it. This should increase the dependable water from a compartmented system at a relatively low cost.

INTRODUCTION

Reducing surface evaporation may be the most economical way of increasing water supplies in arid lands. The water thus saved is essentially distilled and is normally at the head of existing distribution systems and can be easily utilized. Evaporation loss is perhaps the major deterrent to fully utilize erratic flood flows in arid lands. The other water loss, seepage, can more easily be controlled using several available methods. In some cases, seepage loss can be recovered through wells located in the vicinity of the reservoir.

The amount of research funding that has gone into evaporation control is negligible compared with its potential. There was considerable activity following W. W. Mansfield's successful field test of a monolayer formed by long chain alcohol in 1953. For over a decade following this test various researchers tested the material in several parts of the world. Now over two decades have passed since the monolayer was first field tested. As far as the author knows, there are no commercial applications of long chain alcohol for evaporation control at the present time.

The author knows of only three research centers in the United States who have continued research in evaporation control since the 1960's. This research began with the use of monolayers but has since been centered primarily on floating reflective and/or vapor barriers. These three centers are the Agricultural Engineering Dept., Oklahoma State University; the U. S. Water Conservation Laboratory, Agricultural Research Service; and the Water Resources Research Center, University of Arizona. The findings of this research, along with others who have periodically looked into the problem, will be summarized below.

Methods of Evaporation Control

There are eight leading methods that have been tested for evaporation control. These are (1) minimizing the surface-to-volume ratio, (2) destratification, (3) surface films, (4) wind barriers, (5) sand or rock-filled reservoirs, (6) shading the water surface, (7) floating reflective covers, and (8) floating vapor barriers.

(1) Minimizing the Surface-Area to Volume Ratios

One method of minimizing the surface to volume ratio is to select a natural site where this is accomplished at low cost. This is difficult in flat terrain but under these conditions the concept of the compartmented reservoir can be used (Cluff 1977a, 1977b). Using this concept the reservoir is divided into separate compartments with low dikes and a portable high-capacity pump is repeatedly used to keep the water concentrated and thus minimize its exposure to the atmosphere. This method, when feasible, requires the use of construction equipment. It is much easier to compartmentalize a reservoir at the time of initial construction than later, however, there are many reservoirs where retrofitting would be beneficial.

(2) Destratification

One method of destratification consists of pumping air from the bottom of the reservoir to form air bubbles which carry the colder water to the surface. This reduces the evaporation loss. This works best on deeper reservoirs where the water in the deeper stratum would be cold most of the year. A test in California on a 130 acre lake resulted in a 15 percent reduction in May, June and July and a 9 percent increase in September, October and November. The net savings was only 6 percent (Frenkiel, 1965). Another method that might be used is making releases by drawing the surface water off, leaving the colder water in storage. This, however, would increase water loss downstream.

(3) Monolayers

The use of long-chain alcohol to form monolayers was first field tested in 1955 in Australia where W. W. Mansfield reported a reduction of evaporation up to 30 percent. Theoretically, it requires approximately 0.023 kg of alcohol to cover a hectare of water. Practically, it has been found to require closer to 0.057 kg, but even at that rate, one kilogram of alcohol will cover about 17.5 hectares of water surface.

The biggest detriment to the use of a monolayer is wind. Even a slight breeze will blow the monolayer off the pond. When the wind dies down the monolayer will tend to reform but evaporation is highest when the wind is blowing. This reduces the overall savings to less than 10 percent.

The effect of wind on monolayers suggests that the larger the reservoir the more economical the system of evaporation control becomes. The bigger the lake the longer is the residual time that a given amount of alcohol remains on the water. Various types of dispensers using alcohol in powder, solution and emulsified form have been tried in order to apply material to the upwind shore in proportion to wind speed.

Various types of raft dispensers have been tried for both powdered and semi-solid emulsified materials. The raft dispensers that were developed were simple in design but failed to supply sufficient material due to plugging. The best system that was developed at the University of Arizona was a propeller-driven dispenser, which applied alcohol in a fluid emulsion form proportional to wind speed. This wind-activated emulsion dispenser proved to be efficient in applying material as needed to replace that removed by wind. This resulted in savings up to 30 percent in tests on small ponds. The chief disadvantage of this system is that one dispenser would be required for each 9 m of shoreline. Thus a relatively large number of dispensers would be needed for each small reservoir.

For a square reservoir the cost of preventing water from evaporating, using the wind activated emulsion dispenser system, was estimated in 1966 to be more than $0.90 (U.S.) per cubic meter for reservoirs less than 0.4 hectare in size, dropping down to $0.70 (U.S.) per cubic meter for a 1.6 hectare reservoir. The cost was not less than $0.40 (U.S.) per cubic meter until the size of the reservoir was increased to greater than 6 hectares. These costs include material, maintenance and capital recovery costs for the dispensers (Cluff, 1966).

On larger reservoirs, pipelines feeding the alcohol from the upwind shore have been used in tests (Dressler, 1969; Reiser, 1969). Reiser estimated the cost at $0.04 (U.S.) per cubic meter for a 40 hectare reservoir dropping down to $0.002 per cubic meter for a 4,000 hectare reservoir. Another method that can be used on large reservoirs is the dispensing of alcohol by means of a crop-dusting airplane. This was initially investigated by Hansen and Skogerboe (1964). The method was used on a commercial scale on the 12,000 hectare Vaal Dam in South Africa during a period of severe drought (Bester, 1967). The estimated cost was $0.021 (U.S.) per cubic meter.

(4) Wind Barriers

Very little research has been done in this area. Crow (1967), in one test during a period of relatively high winds, achieved a 9 percent savings using wind baffles that were 0.27 m high and spaced on a 4.4 m spacing. Using the same baffles with a monolayer of fatty alcohol increased the savings to 31 percent. Cooley (1970) has done an energy budget analysis that indicates that reduction of wind velocity alone usually would not produce large savings. Vegetative shielding in some cases may utilize more water from the reservoir than it would save in the form of evaporation reduction. Some reduction may be achieved by placing excess excavated bank material on the upwind side of a small reservoir.

(5) Sand or Rock-Filled Reservoirs

This type of reservoir has been successfully utilized in many parts of the world. Most sand-filled reservoirs are built in stages with each stage or level being filled naturally with sand carried by the flowing water before the next stage is built. This reduces the amount of silt and clay trapped by the dam (National Academy of Sciences, 1974). Other sand-built reservoirs have not been planned but have developed naturally by the construction of dams across intermittant streams carrying large amounts of sand. The author knows of two reservoirs of this type on the Papago Reservation near Kitt Peak, west of Tucson, that are perhaps more efficient in their present sand-filled state than originally envisioned. In general, reservoirs that can be filled naturally with sand require a moderately steep gradient and a large supply of sand.

Colorado State University has extended the concept of the sand-filled tank by creating shallow plastic-lined storages in relatively flat terrain and then hauling the sand in to fill the small reservoirs (Smith and Corey, 1976). The major disadvantage noted in sand-filled reservoirs is the weed and phreatophyte growth that must be controlled.

Rock has been used instead of sand to fill three small plastic-lined reservoirs in Arizona. A commercial rock picker was used to collect the rocks around the reservoir site. Earth-filled, used-rubber tires were used to cushion the plastic prior to filling with rocks. It is important that the rocks be as uniform in size as possible to

maximize the porosity. The use of rock reduces the available storage by
50 to 60 percent. However, it provides an essentially vandal-proof, eva-
poration-free water storage. The resulting water is of excellent quality.
The limiting factors in the use of this method are the availability
of rock and the necessity of using sediment-free water (Cluff et al., 1972).

(6) Shading the Water Surface

Suspended shades have successfully reduced evaporation loss by an
amount close to the percent of shade cover (Crow, 1967; Drew, 1972). The
chief advantage to shading over floating methods is that the cover does
not impede the gas exchange primarily of oxygen and carbon dioxide between the water and the atmosphere. Also, it is relatively easy to
direct the runoff from a suspended shade into the water body (Cluff,
1967). The disadvantage is that a suspended shade is more difficult
to maintain than floating covers. Most available synthetic materials
have been found to oxidize at a faster rate when kept under the tension
required to maintain the suspension between supports. Natural materials
such as cornstalks, bamboo, etc. suspended above the water's surface
have been used to provide shade. The limitation with this method is
that the water is contaminated as the decaying organic cover drops
into the reservoir (Intermediate Technology Development Group Ltd., 1969).

(7) Floating Reflective Covers

This category, for the purpose of this paper, includes the use of
dyes in addition to floating reflective powders and particles that do
not form a vapor barrier. This type of cover reduces evaporation primarily by reducing the energy input into the system.

Yu and Brutsaert (1967), as reported by Cooley (1974), used shallow
evaporation pans to show the effect of color and reflectance on the
evaporation rate. White pans evaporated 35 to 50 percent less than
similar black pans. Cooley (1974) also reported that Gainer et al. (1969)
had savings of only 10 to 15 percent using dye or an oil and dye mixture.
The oil was used with the dye so that only the upper layer of water
would have to be treated.

Various types of floating powder and particles have been tried
including white spheres (Block and Weiss, 1959), and solid plastic
pellets (Salyer, 1964). Rogitsky and Kraus (1966) tested foamed
polystyrene powder, foamed polystyrene flakes, foamed polystyrene
spheres, and foamed polyurethane hemispheres, in evaporation pans. The
evaporation reduction ranged from 26 to 57 percent with the greatest
reduction coming from the 25 cm diameter polyurethane hemispheres. These
larger particles were obviously also serving as a vapor barrier in
addition to an energy barrier. Myers and Frasier (1970) found that polystyrene beads were more resistant to discoloration from dust and algae
than chopped styrofoam and perlite. The beads were much smoother and
the dust and algae did not accumulate. In a six month test on a small
insulated pan they achieved an efficiency of 39 percent using the beads.
The foam and perlite, although less expensive, became coated with dust
and algae and saturated with water, causing some of it to sink.
Cooley and Cluff (1972) obtained a 19 percent savings using perlite on
a 390 m^2 reservoir for a 9 month period. This was done at a cost of
$0.36 (U.S.) per cubic meter of water saved. During this test it was

necessary to periodically add perlite to the water to replace that which
had sunk. This requirement made this method uneconomical.

The author conducted a test beginning in February 1972, in which
styrofoam was run through a 0.6 cm screen in a feed mill and the
particles applied to the 390 m^2 pond used in the perlite test. For
one month the evaporation savings ran about fifty percent. Over one
weekend the pond encountered a heavy dust storm; the styrofoam lost
its water repellancy and became wetted and coated with dust. The
evaporation savings dropped to about 10 percent, making the method
impractical.

(8) Floating Vapor Barriers

Various types of vapor barriers have been used. The more successful ones have been foamed rubber, paraffin wax, which has been remelted
by the sun to form a continuous cover, and expanded polystyrene rafts
(Cooley, 1974).

Dedrick et al. (1973), reports installation of continuous foamed
rubber on a tank in Utah in 1971. More have been installed since that
time. After more than five years they are still in serviceable condition.
The floating cover is a low-density, closed-cell synthetic rubber
available in five foot rolls. Various thicknesses up to 1.2 cm are
available but the 0.6 cm has been the thickness most frequently used.
Small holes are drilled through the cover so that it is self draining.
Instruction for installation of the foamed rubber are available (Dedrick,
1976). Dedrick (1976) reports material costs of less than $0.30 (U.S.)
per ft^2 with cost of evaporation control at $0.25 (U.S.) per m^3
in a 2.3 meter evaporation zone. Both the continuous wax and
the rubber foam have been applied only to vertical sided tanks.

Most of the work in evaporation control at the Water Resources
Research Center, University of Arizona, has been centered around
maximizing evaporation savings through the use of floating vapor barriers.
Many different types have been tested over the last ten years. Rafts
made of polyethylene were first tried. First frames of styrofoam and
then wood were used to support the edges of the polyethylene. Frames
had to be more than three inches high to eliminate flooding but the
higher the frame the higher the wind resistance.

To reduce the height of the frame a "moat raft" design was tried.
This raft consists of a smaller frame inside a larger frame, connected
at the corners with open water in between. The water prevented waves
from reaching the inner frame which was attached to polyethylene or
other economical film. A small 0.6 to 1.2 m model successfully with-
stood flooding in winds up to 9 m/sec (Cluff, 1967).

A 7.8 x 7.8 m moat raft with a 0.6 m wide outer frame of 2.5 cm
styrofoam and an inner frame of 3.1 cm PVC pipe was constructed in April,
1972. It served as a support for an inner sheet of polyethylene, and
was built for a cost of $0.13 (U.S.) per m^2. This raft successfully
resisted flooding by waves on a 3600 m^2 reservoir. However, it fell
easy victim to turtles and frogs who punctured the polyethylene and the
raft was flooded from below. Floating concrete slabs, 60 x 60 x 5 cm,
constructed with a chopped expanded polystyrene aggregate such as were
used in tests in South Africa were tested (Engineering News Record,
1966). The specific gravity was about 0.8 percent which prevented the

Figure 1.   Coupled expanded polystyrene
asphalt-chipcoated (CEPAC) raft.

Above: Water Harvesting Agrisystem at Page
Experimental Ranch.
Below: One of three test ponds at the WRRC
Field Laboratory, University of Arizona.

slabs from piling up. The slabs are painted white to aid in reflection
of sunlight. The slabs were relatively expensive as compared to the
foam sheets. In addition, the slabs became water saturated which
reduced their efficiency.

Excerpt from "Delta Alternatives Review Status
Draft - February 1977"

Excerpt from "Delta Alternatives Review Status
Draft - February 1977"

ACTIONS NORTH OF THE DELTA

- Cottonwood Creek Project
- Glenn Reservoir

ACTIONS SOUTH OF THE DELTA

- Southern California Ground Water Storage
- San Joaquin Valley Ground Water Storage
- Los Vaqueros with Pumps
- Water Conservation
- Waste Water Reclamation
- Mid-Valley Canal

DELTA PROTECTION PROGRAM

- Environmental Monitoring
- Four - Agency Fish Agreement
- CVP - SWP Operating Agreement
- Limits on Delta Diversions
- Review of Standards
- Federal Legislation for Delta Protection
- South Delta Water Quality Improvement
- Relocate Contra Costa Canal Intake
- Delta Water Agency Contracts
- Suisun Marsh Protection
- Fish Screens on In-Delta Diversions

At this time research emphasis shifted back to the styrofoam raft which was first tested at the University of Arizona in 1966. Individual floating styrofoam rafts were first tried. The rafts were equipped with an extruded plastic edging which extended below the bottom of the raft. This caused a negative pressure to be formed thus preventing removal by wind. Later, due to increasing costs, the method was dropped in favor of coupling the rafts together to form a continuous blanket. Couplers were made of 5 cm sections of 3.75 cm PVC pipe split on one side. The sections of pipe can be used as "C" clamps and are opened up and placed in the middle on each side of the 1.2 x 1.2 m sheets. The raft is then tied together, one row at a time, before being pushed out on the water.

Various surface treatments have been tried and discarded. The one used predominately over the past five years has been an asphalt and gravel chip-coating. An assembly line was established to reduce the cost of coating. The total cost of this raft system was estimated to be $1.15 (U.S.) per m^2 (Cluff, 1972). It was found that this type of coating held up for two to three years. Its life could be extended if the asphalt chip-coating was painted with white acrylic paint. White acrylic vinyl paint has also been successfully used directly on the foam (see Figure 1).

Early in this testing program Dow styrofoam was used. This is a closed-cell, extruded polystyrene that does not absorb water. About 1972 a shift to molded expanded polystyrene was made because it was available in larger sheets and was less expensive. After a period of time it was found that the low-density expanded polystyrene did absorb water. At first this was thought to be a benefit since it would add weight to

the cover and reduce the chance of wind damage. Later it was found
that the absorption of water greatly weakened the foam. It also was
found to reduce the efficiency since eventually the water vapor began
passing through the foam. This loss could be lessened by using appropriate surface treatments but none were found to stop it completely.

In looking for a way to waterproof and weatherize expanded polystyrene the author, in the fall of 1975, began dipping the foam in
molten wax. Later, pressure was applied to the waxing process and
it was found that low-density expanded polystyrene could be completely
impregnated with wax increasing the specific density from 0.016 up to
0.224. The maximum attainable density is still much less than that
of water so there is ample flotation. This impregnation greatly increases the strength properties of the foam in addition to water-
proofing and weatherproofing the expanded polystyrene. A patent has
been applied for since the material has many potential uses in the
construction industry in addition to its use for evaporation control.
There are various types of waxes available so that several different
types of wax-impregnated expanded polystyrene can be made.

Small pieces of paraffin-wax-impregnated polystyrene have been
exposed for a year with little or no weathering effect. Cooley (1974)
has been using paraffin wax blocks for several years and reports
excellent weathering properties. Since the wax completely permeates
through the entire thickness of the foam the weathering properties of
the combined material should be similar to those of the wax. The
wax-impregnated foam is an excellent insulation material. Thus it serves
as both a vapor and energy barrier. An evaporation savings proportional
to the area covered will be obtained. The temperature of a partially
covered body of water has been observed to be close to the same as
the adjacent control reservoir using expanded polystyrene.

A pressurized system that will saturate 1.2 x 2.4 m sheets has
been successfully completed at the Water Resources Research Center at
the University of Arizona. Preliminary tests indicate that it has the
capability of wax saturating 200 to 400 m² of 2.5 cm thick foam per
hour. The wax-impregnated foam will cost about twice as much as the
parent material at the medium attainable densities but it should be
possible to use thinner sheets since the waxed material is much stronger.
The sheets can either be clamped together using sections of PVC pipe
described earlier or connected with straps of butyl or hypylon rubber.
The use of the straps will make it possible to accordian fold the waxed
sheets so that they can be quickly deployed in times of drought or
be easily stored during periods of excess rainfall. Both types of
connectors can be used on larger reservoirs provided some type of
wave energy dissipation is used to protect the outer perimeter of the system.

The use of smaller pieces should be possible since the wax-
impregnated foam is much heavier than the parent material. A test using
0.20 m squares on the 390 m² pond has been started and initial results
are encouraging. These smaller pieces should be used in areas subject
to vandalism. They are also recommended for smaller tanks or reservoirs
with circular or irregular water surfaces. It appears that this type
of treatment can also be used on larger reservoirs. The smaller
pieces can be easily cut using a standard table saw. This is another
advantage of adding the wax to the foam, it is much easier to cut
with a saw. However, the wax makes it difficult to cut the foam with
a hot wire. This is the method now used. Material costs could be as
low as $1.10 (U.S.) per m² for the 1.2 cm thickness.

DISCUSSION AND CONCLUSIONS

In the past there has been very little anyone could do to economically reduce evaporation loss. The fact that a suitable reservoir site was not available generally has ruled out the development of many projects due to the high evaporation loss.

The joint use of the compartmented reservoir with floating covers such as foamed rubber wax-impregnated-expanded polystyrene should make it possible to store water in flat terrain at a cost low enough for conventional agriculture.

The use of a floating cover without the concentrating effect of the compartmented reservoir would probably not be economically justified for conventional agricultural use except in times of drought. The cost of water saved would be low enough to use for domestic and industrial supplies or high valued crops.

The paper outlined many different methods that have been tried, most without success. Although many things have been tried, most tests have been done with modest budgets except perhaps for the work done in the area of monolayers. Both government and industry have not applied much effort to the economical solution of this problem. Lack of industrial interest is due primarily to the fact that there is no existing market. If any one method of evaporation control outlined here proves to be successful to the point that it is commercially viable, industry will probably become interested and come up with even better methods.

ACKNOWLEDGMENTS

Appreciation is hereby expressed to S. D. Resnick, Director of the Water Resources Research Center, University of Arizona, for his continued interest and support in the evaporation control research.

Funding for the development of the wax treating equipment has come from the State of Arizona, the Bureau of Indian Affairs, Argonne National Laboratories, and the U.S. Agency for International Development (AID).

REFERENCES

Bester, M. D. (1967), Aerial Spraying of the Vaal Dam to Control Evaporation, Technical Report No. 38, Department of Water Affairs, Pretoris, Republic of South Africa.

Block, M. R. and T. Weiss (1959), Evaporation Rate of Water from Open Surfaces Colored White, Letter to the Editor, Bulletin of the Research Council of Israel, Vol. 8A, pp. 188-189.

Cluff, C. B. (1966), Evaporation Reduction Investigation Relating to Small Reservoirs, Arizona Agriculture Experiment Station, Technical Bulletin 177, pp. 1-47.

Cluff, C. B. (1967a), Water Harvesting Plan for Livestock or Home, Progressive Agriculture, University of Arizona, Tucson, May-June.

Cluff, C. B. (1967b), Rafts: New Way to Control Evaporation, Crops and Soils Magazine, November.

Cluff, C. B. (1972), Patchwork Quilts Halt Pond Evaporation Loss, Arizona Cattlelog, July.

Cluff, C. B., G. R. Dutt, P. R. Ogden, and J. K. Kuykendall (1972), Development of Economic Water Harvest Systems for Increasing Water Supply, Phase II, OWRT Project No. B-015-ARIZ.

Cluff, C. B. (1977a), The Compartmented Reservoir: A Method of Efficient Water Storage, Dissertation, Civil Engineering, Colorado State University, Fort Collins, Colorado.

Cluff, C. B. (1977b), The Compartmented Reservoir: A Method of Efficient Water Storage on Flat Terrain, Presented at Conference on Alternative Strategies for Desert Development and Management, Sponsored by the United Nations, Sacramento, Calif., May 31-June 10.

Cooley, K. R. (1970), Energy Relationships in the Design of Floating Covers for Evaporation Reduction, Water Resources Research 6(3).

Cooley, K. R. (1974), Evaporation Suppression for Increasing Water Supplies, Proceedings of the Water Harvesting Symposium, Phoenix, Arizona, March 26-28, ARS W-22.

Cooley, K. R. and C. B. Cluff (1972), Reducing Pond Evaporation with Perlite Ore, American Society of Civil Engineers, Journal of Irrigation and Drainage Division, 98(IR2).

Corey, A. T. and G. L. Smith (1976), Soil and Rock Strata to Trap, Filter, and Store Water for Rural Domestic Use, Technical Completion Report, OWRT, December.

Crow, F. R. (1967), The Effect of Wind on Evaporation Suppressing Films and Methods of Modification, Extract of Publication No. 62 of the I.A.S.H. Committee for Evaporation, pp. 26-37.

Dedrick, A. R., T. D. Hansen, and W. R. Williamson (1973), Floating Sheets of Foam Rubber for Reducing Stock Tank Evaporation, Journal of Range Management, 26(6): 404-406.

Dedrick, A. R. (1976), Foam Rubber Covers for Controlling Water Storage Tank Evaporation, College of Agriculture, Extension Service, Miscellaneous Report #2, University of Arizona, Tucson.

Dressler, R. G. (1959), Method for Retarding Evaporation of Water from Large Bodies of Water, Patent No. 2,903,330, United States Patent Office, Patented September 8, 1959.

Drew, W. M. (1972), Evaporation Control -- A Comparative Study of Six Evaporation Restriction Media, Agua (January), pp. 23-26.

Frenkiel, J. (1965), Evaporation Reduction, Physical and Chemical Principles and Review of Experiments, United National Educational, Scientific and Cultural Organization, Paris, France.

Gainer, J. L., J. T. Beard, and R. R. Thomas (1969), Water Evaporation Suppression, Water Resources Research Center, Virginia Polytechnical Institute, Bulletin 27.

Hansen, V. E. and G. V. Skogerboe (1964), Equipment and Techniques for Aerial Application of Evaporation-Reducing Monolayer Forming Materials to Lakes and Reservoirs, Final Report of USBR Contract No. 14-06-D-4911, Utah Water Research Laboratory, Utah State University, Logan, Utah.

Intermediate Technology Development Group, Ltd. (1969), The Introduction of Rainwater Catchment Tanks and Micro-Irrigation to Botswana, London, England.

Myers, L. E. and G. W. Frasier (1970), Evaporation Reduction with Floating Granular Materials, American Society of Civil Engineers Journal Irrigation and Drainage Division, December.

National Academy of Sciences (1974), More Water for Arid Lands, Washington, D.C., U.S.A.

Reiser, C. O. (1969), A System for Controlling Water Evaporation, I&EC Process Design and Development, Vol. 8, January.

Rogitsky, M. and Y. Kraus (1966), Reduction of Evaporation from Water Surface by Reflective Layers, Preliminary Research Progress Report, Water Planning for Israel, Tel Aviv, Israel.

Salyer, I. W. (1964), Method of Reducing Water Evaporation Losses in Open Reservoirs, Patent No. 3,147,067, U. S. Patent Office, Patented September, 1964.

Yu, S. L. and W. Brutsaert (1967), Evaporation from Very Shallow Pans, Journal of Applied Meteorology 6(2), pp. 264-271

THE COMPARTMENTED RESERVOIR: A METHOD OF
EFFICIENT WATER STORAGE IN FLAT TERRAIN

C. Brent Cluff*

ABSTRACT/SUMMARY

The compartmented reservoir is presented as an efficient method of storing water in areas having a relatively flat terrain where there is a significant water loss through evaporation. The flat terrain makes it difficult to avoid large surface-area-to-water-volume ratios when using a conventional reservoir.

This paper demonstrates that large water losses through evaporation can be reduced by compartmentalizing shallow impervious reservoirs and in flat terrain concentrating water by pumping it from one compartment to another. Concentrating the water reduces the surface-area-to-water-volume ratio to a minimum, thus decreasing evaporation losses by reducing both the temperature and exposure of the water to the atmosphere. Portable, high-capacity pumps make the method economical for small reservoirs as well as for relatively large reservoirs. Further, the amount of water available for beneficial consumption is usually more than the amount of water pumped for concentration.

A Compartmented Reservoir Optimization Program (CROP-76) has been developed for selecting the optimal design configuration. The program was utilized in designing several systems. Through the use of the model, the interrelationship of the parameters have been elucidated. These parameters are volume, area, depth, and slope of the embankment around each compartment. These parameters interface with the parameters describing rainfall and hydrologic characteristics of the watershed.

The water-yield model used in CROP-76 requires inputs of watershed area, daily precipitation, daily and maximum depletion. In addition, three sets of seasonal modifying coefficients are required either through calibration or estimated by an experienced hydrologist. The model can determine runoff from two types of watersheds, a natural and/or treated catchment. Additional inputs of CROP-76 are the surface water evaporation rate and the amount and type of consumptive use.

*
C. Brent Cluff is an Associate Hydrologist at the Water Resources Research Center, University of Arizona.

Because of the large number of parameters it was found that repeated runs of the model are necessary to determine a near optimum design in a reasonable amount of time. The model computation time for the CDC 6400 computer for a 45 years length of record is less than ten seconds per run for the usual design. Usually no more than four or five computer runs were needed for design purposes. CROP-76 was used on several typical systems including a water harvesting agrisystem. The following general observations were made: (1) The rate of increase of efficiency of storage decreases as the number of compartments increase; (2) there was no significant difference in evaporation loss by varying the relative size of compartments provided the side slope, depth, total number of compartments and the total combined volume remained constant; (3) the increase in efficiency due to the use of the compartmented system decreases as the depth of the reservoir increases, becoming insignificant for depths of 20 or more meters; and (4) the use of a compartmented reservoir provides efficient storage for a water harvesting agrisystem.

INTRODUCTION

The need for a low-cost, efficient method of water storage in semi-arid and arid lands has long been recognized. The high evaporation loss coupled with flat terrain has prevented economical water storage except in rare instances where favorable reservoir sites are available. These favorable sites in most parts of the world have been utilized but the demand for water is far from satisfied and will continue to increase in the future.

IMPORTANCE OF IMPROVED STORAGE OF WATER

The importance of improved water storage can be verified easily by aerial flights over dry areas prior to the onset of the rainy season. These flights reveal that most small storage reservoirs are dry or close to it. An examination of many of these reservoirs by the author in Arizona and northern Mexico, as well as West Africa, has revealed that the average depth generally is less than the average annual water evaporation rate. This condition prevents withdrawal of water on a constant-rate basis and any chance of carry-over storage from one year to the next.

The importance of constructing deep reservoirs has long been known but there are several constraints which normally have prevented the construction of deep reservoirs. These are:

1. The grade of the bed of the contributing stream. Any conventional storage must be below the bed.

2. Shallow soils. These make excavation difficult.

3. Construction equipment. The equipment has constraints which restricts the depth.

4. Seepage control. This becomes more difficult in deeper reservoirs.

5. Erosion control on the steep banks. The problem increases with deeper reservoirs.

6. Safety constraint. Unless excavated, deeper reservoirs pose more danger to downstream occupants than shallow reservoirs.

7. Financial constraints. Deep reservoirs usually cost more money per unit-volume of storage than shallow reservoirs.

The dozer tractor commonly is used for constructing small reservoirs. The deeper a dozer excavates into the ground the greater is the unit cost. It has been the author's experience, when using a 1:2 embankment slope, that building a reservoir deeper than six meters is very expensive. This six-meter depth is usually accomplished with a three-meter cut combined with a three-meter embankment. Due to the constraint imposed by the grade of the stream the upper bank generally serves no useful purpose other than as a spoil area; hence, the effective depth is three meters or less.

The efficiency of storage is defined as the percent of water going into storage that is available for a desired beneficial use on a fixed demand basis. This efficiency can be increased by reducing evaporation loss. As indicated in the literature review in the next section, some research has been done on evaporation control which indicates that the costs of such control in general are prohibitive for use in some major applications such as conventional agriculture.

This study shows the advantages of controlling evaporation loss through the use of the compartmented reservoir. Using this system, the surface area to depth ratio is reduced by keeping the water concentrated. The increase in average depth reduces the amount of solar energy input into the reservoir as well as the exposure to the atmosphere thus reducing evaporation loss.

Figure 1 is a schematic drawing of a three-compartmented reservoir system. The reservoir consists of a receiving compartment which is called A. This compartment is located below the stream grade and therefore is usually shallow. Compartments B and C are shown as being smaller in surface area but deeper in depth. This reservoir is operated as follows: As runoff occurs during the rainy season, water is pumped from compartment A to fill compartments B and C. Water is first withdrawn for consumptive use from compartment A until the evaporation and seepage losses from B and C are equal to the remaining water in A. At this time, the pump is used to move the remaining water in A to fill the unused capacity of B and C. This eliminates further evaporation and seepage losses from A. Water is then withdrawn as needed for consumptive use from B until the water remaining in B is equal to the unused capacity in C. At this time, the pump is used again to move the remaining water from B into C. This eliminates further evaporation and seepage losses from B. At this point, C is filled and A and B are empty. A spillway would be needed from compartment A to protect the safety of the system. All inner dikes would have to be built higher than the maximum water level determined by the elevation of the spillway.

The compartmented reservoir concept can be applied to existing reservoirs or new ones. Since a pump will be used in flat terrain, all compartments other than the receiving compartment can be made deeper by building the embankments above the stream grade.

The recent development of portable, low-lift, high-capacity, tractor-operated pumps makes the compartmented reservoir system

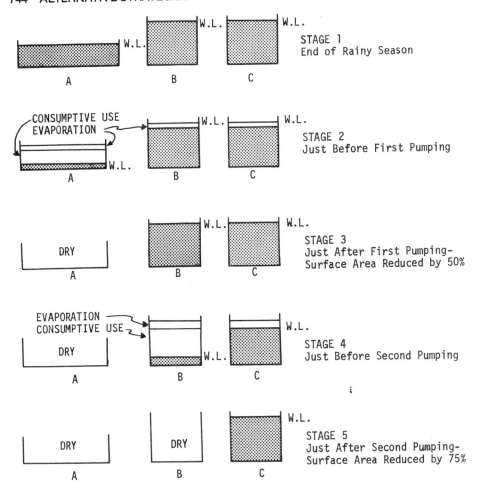

Figure 1.   Schematic cross-sectional diagram of a three-compartment
           reservoir showing water levels (W.L.) of various stages
           in the annual cycle of operation.

economically attractive. These pumps are powered by the power-take-off
(pto). They are available in capacities of up to 5000 cubic meters per
hour. One pump can service several small reservoirs. If tractors
are not available a suitable vehicle such as a British Land Rover could
be equipped with a pto and used to both transport and power the portable
pump.

     If the general slope of the topography is greater than three or
four percent, the concept of a gravity-fed compartmented reservoir can
be used. The compartments of this reservoir are separated by a sufficient distance to develop enough hydraulic head so that one compartment
can be completely drained by a gravity pipeline or an elevated canal

into the second and succeeding compartments. This reservoir system could be operated as before but without a pump.

Surface storage reservoirs in semiarid regions are usually fed by intermittent flood flows. However, in some cases, there may be a base flow going into compartment A. In this case, the base flow would be used to satisfy consumptive demands. The remainder, if any, could be pumped into storage on a continuous basis.

THE POTENTIAL OF THE COMPARTMENTED RESERVOIR

The potential of the compartmented reservoir is demonstrated in Figures 2 and 3 under idealized conditions.

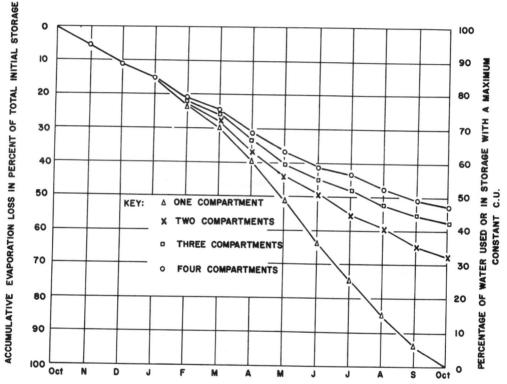

Figure 2. Evaporation Loss for Compartmented (but undeepened) Reservoirs with a Constant Volume and Area, a Maximum Constant Consumptive Use and a Depth Equal to Annual Evaporation Loss.

Figure 2 illustrates the use of compartments of equal size in a reservoir of depth equal to the evaporation loss. The reservoir is assumed to be filled by runoff only once a year, with no additional input. In this figure and Figure 3, an annual evaporation depth (ΣE) of 1.636 meter (m) is used. This is the evaporation for Parras, Coahuilla, Mexico and is close to the evaporation loss in Tucson, Arizona. It is less, however, than the evaporation in other parts of the world. A constant consumptive use that would be withdrawn each month

is selected so that there would be no water remaining in the reservoir
at the end of the year. This value is determined by trial and error.
It is called the maximum constant consumptive use. For the single
compartment (the typical reservoir) this consumptive use value is
zero. When the depth of the reservoir is equivalent to the annual
evaporation loss is is impossible to withdraw any water on a continuous
basis since all the water would be consumed by evaporation.

The efficiency can be increased beyond those shown in Figure 2 if
all compartments, other than the receiving compartments, are made deeper.
This is possible due to the use of the pump. This improvement in
efficiency due to deepening is shown in Figure 3 for a three-compartment

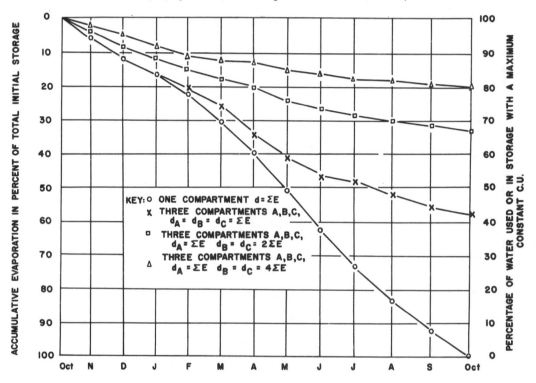

Figure 3.   Evaporation Loss in Percent of Total Initial Storage for
Compartmented Reservoirs with a Constant Volume, Varying
Depth and a Maximum Constant Consumptive Use.

system. If the depth of B and C are doubled and quadrupled the efficiency
goes up to 68 percent and 80 percent respectively. This is contrasted
with zero percent for the shallow single compartmented reservoir.

The above illustrations are idealized in that the reservoirs are
filled once a year with no additional input. If seepage is controlled,
the same savings would apply regardless of the total size of the reservoir system. It is readily apparent that the compartmented reservoir
concept can be applied to reservoirs of all sizes, from small, livestock
watering tanks to large reservoirs for agricultural use.

The amount of pumping required in a compartmented reservoir is relatively low compared to the water savings effected. For instance, the three-compartmented reservoir, with all compartments equal in depth to the evaporation loss, illustrated in Figures 2 and 3, requires the pumping of 20.8 percent of the initial storage to obtain a 42.5 percent efficiency when the water is used on a constant basis. This amounts to pumping 48.9 percent of the water beneficially utilized. This assumes that the water can be withdrawn by gravity flow for use from all compartments. The cost of pumping would be generally much less than pumping groundwater due to the low pumping lift.

The full concept of the compartmented reservoir was conceived by the author in the summer of 1975 while serving as a consultant for the Wunderman Foundation in the Sahel Region of Mali in West Africa. At this time an earlier concept of pumping from a shallow to a deeper compartment was expanded to include multiple compartments and repeated pumping. The discovery was made at this time that significant savings could be made just by dividing a reservoir into compartments and keeping the water concentrated. Deepening all compartments other than the receiving one was found to increase the efficiency.

The concept evolved as an attempt to help solve a critical water storage problem in Mali, Africa. Ten different sites were surveyed and recommended designs were made using a small programmable calculator (Cluff, 1975). Following his return from Africa, the author spent six months in Mexico working for the Food and Agricultural Organization (FAO) of the United Nations (Cluff, 1976) in support of the Fundo Candelillero, an action agency of the Mexican Government. Eleven compartmented reservoirs were built by the above agencies in the state of Coahuilla, Mexico, during the six-month period the author served as a consultant. More have been built since that time. These reservoirs range in size from a 8100 m^3 two-compartmented livestock reservoir dug by mules, to a 200,000 m^3 four-compartmented reservoir constructed using D-7 dozers. This largest reservoir which is used for agricultural purposes is shown in Figure 4. One small gravity-fed separated compartmented reservoir also was constructed.

The use of the compartmented reservoirs introduces additional design parameters for effectively using and storing water from any given watershed. The number of compartments and their depth and size relative to each other must be considered in order to maximize production of water from any given watershed. These parameters are a function of the seepage and evaporation losses. If needed, a floating cover can be used on the last compartments.

COMPARTMENTED RESERVOIR OPTIMIZATION PROGRAM (CROP-76)

A computer model has been developed to study the parameters involved in the compartmented system and their relationship to each other using historical data. This model is briefly described in this section. with examples of its use. A more complete description can be found in Cluff (1977).

The computer model converts daily historical rainfall data into runoff data from either a natural and/or a treated watershed. Runoff data is summarized and stored in a weekly array. The compartmented

Figure 4.   Four Compartmented Reservoir System near Parras, Coahuilla,
             Mexico, with a 200,000 m³ Capacity. -- Above: Under
             Construction. Below: Completed System.

reservoir is subjected to a domestic and/or agricultural demand as well
as evaporation losses. The design parameters of the compartmented
reservoir can be adjusted so that the "optimum" reservoir system would be
selected. The definition of an optimum reservoir is "the system that
would have the highest storage efficiency under the constraints imposed."
The definition of the storage efficiency is the percent of water that

passes into the storage system that is available for a desired beneficial use on a constant demand basis.

In the operation of the model the design parameters are usually adjusted so that the amount of overflow plus excess water is kept below a specified amount, usually 4 or 5 percent. An additional constraint is the requirement that the reservoir system is required to provide water for the desired beneficial use for a specified minimum, usually 95 or 96 percent of the time. The consumptive demand is reduced if necessary in order to fit the above constraints.

A water harvesting agrisystem option has been built into CROP-76. Under this option a soil moisture-accounting routine is used to account for storing water in the soil in addition to storing excess water in the compartmented reservoir system.

There are too many design parameters to obtain a satisfactory design in a single run of the computer, within a reasonable processing time. The design, however, can be obtained by repeated computer runs by a skilled operator who helps the computer in its selection of the parameters that will meet the constraints.

The 76 was added to the CROP acronym of the model since it was developed in 1976. It also serves as a reminder that there are additional improvements that can be made. However the model has been used to design compartmented reservoirs systems using a minimum amount of data. The model has been used on seven reservoir systems in Arizona, Coahuilla, Mexico and Mali, West Africa. These systems ranged in size from a 6,000 m^3 stock tank to a 90,000,000 m^3 reservoir system in Arizona designed to store excess floodwater in the Santa Cruz River near Tucson for proposed agricultural use. In the latter example, 23 years of daily runoff data was used in CROP-76. This computation showed that a reservoir system consisting of six 15,000,000 m^3 compartments only 5 meters deep could effectively re-regulate the erratic flood flows so that approximately 50 percent of the water could be beneficially used on a continuous basis.

A single compartment reservoir of the same depth would be dry approximately 10 percent of the time without any beneficial consumptive use.

CROP-76 has also been used to verify the design of a 20 hectare water harvesting agrisystem which has been constructed at the San Francisco Ejido near Parrus Coahuilla. The surface soil of the twenty hectares has been shaped and will be compacted this summer. Plantings of grapes and pistachios have been made in the artificially depressed drainage area. Excess water from both the artificial catchment and a natural watershed will be stored in a three compartmented system for use in the dry season. Ten years of daily precipitation data was used in CROP-76. This simulation indicated that the reservoir system would be dry only three weeks during the ten year period. There was, however, ample soil moisture during this period to maintain full production.

The use of CROP-76 also showed the advantages of using a floating cover in conjunction with the compartmented reservoir system. In one simulated example a floating cover was placed on the last compartment of a six compartmented system. The cover was placed over only 16 percent of the area but increased the dependable water supply from the system by 50 percent.

By making repeated runs with CROP-76 using different climatic regimes a better understanding of the inter-relation between parameters has been obtained. It was found that there is not much advantage in making the compartments different sizes. They can all be made the same size without reducing their efficiency.

The effect of the compartmented system on reservoir efficiency diminishes as the depth of the reservoir increases. The effect becomes negligible for depths greater than 20 meters. Also the rate of increase in efficiency diminishes as the number of compartments increase. Usually there is little to be gained by having much more than six compartments for the larger reservoirs and three or four compartments for smaller reservoirs.

It was also determined that the reservoir efficiency is relatively sensitive to the design of the system. In order to obtain the highest efficiency the system should be designed to match the amount and frequency of input to the consumptive demand. This is also true of shallow conventional reservoirs.

An extensive economic study of the compartmented system has not been made. The storage costs are site dependent, however a preliminary analysis shows that the cost of water from an intermediate size or large compartmented reservoir should be less than a conventional moderate depth irrigation well. This intermediate size reservoir would be between 200,000 to 400,000 cubic meters in capacity. The storage costs are significantly reduced if the reservoir is located in a soil type where seepage is negligible or where seepage control can be easily obtained by an inexpensive treatment such as sodium chloride.

DISCUSSION AND CONCLUSIONS

This study has shown that the use of a compartmented reservoir can provide efficient storage of water in areas of flat terrain and high evaporation loss. For intermediate and larger systems the unit cost of water should be low enough to permit its use for conventional agriculture. The system has been successfully used in conjunction with a water harvesting agrisystem. This use should make it possible to continuously cultivate large areas which could not be economically farmed any other way. Old projects that have been studied and then discarded due to poor storage efficiency should be re-evaluated. Existing reservoirs in high evaporation areas should be examined to see if storage efficiency can be increased. Any future studies of projects involving storage of water in flat terrain should include an analysis of a compartmented reservoir system.

ACKNOWLEDGMENTS

The help of Dr. G. R. Dutt, Soils, Water and Engineering, University of Arizona; Dr. E. V. Richardson, Civil Engineering, Colorado State University; Dr. H. Guggenheim of the Wunderman Foundation; Dr. N. H. Monteith of The Food and Agricultural Organization (FAO), United Nations;

Dr. J. Maltos R and Ing. F. Aguero M of the Fundo Candelillero, a Mexican Governmental Agency, is gratefully acknowledged.

Funding for the development and testing of CROP-76 was provided through the U.S. Agency for International Development (AID) under Contract No. AID/afr-C-1263.

REFERENCES

Cluff, C. B. (1975), Surface Water Storage Potential in the Nara and Songa Areas of Mali, Summer Project, 1975, Unpublished Consultants Report, Wunderman Foundation, New York, N.Y.

Cluff, C. B. (1976), Agro-Industrial Training, Research and Extension in Arid Zones, Report of the Expert Consultant in Hydrology, United Nations Development Programs, UNDP/MEX/74/003, Unpublished Consultants Report, Submitted to FAO, United Nations, Rome, Italy.

Cluff, C. B. (1977), The Compartmented Reservoir: A Method of Efficient Water Storage, Dissertation, Civil Engineering, Colorado State University, Fort Collins, Colorado.

ACCURACY OF CHANNEL MEASUREMENTS & THE IMPLICATIONS IN
ESTIMATING STREAMFLOW CHARACTERISTICS[1]

Kenneth L. Wahl
(U.S. Geological Survey, Menlo Park, Calif.)

ABSTRACT

Regional relations between flow characteristics and stream channel size offer a
promising alternative to available methods of estimating flow characteristics for
ungaged sites, particularly in semiarid regions. The reliability of such relations, and of flow estimates made from them are partly dependent on the user's
ability to recognize a suitable reach and the reference levels in that reach.

A test was made in northern Wyoming, U.S.A., to determine how consistently trained
individuals could measure channel size for three different reference levels.
Seven participants independently visited 22 sites and measured channel dimensions
in sections of their choosing. Assuming the functional relation between a discharge characteristic (Q) and channel width (W) is log Q = f(1.5 log W), and that
the average log W from seven measurements is the best estimate of log W at a site,
an average standard error for discharge of about 30 percent was attributed to
differences in width measurements alone.

INTRODUCTION

Hydrologists are frequently faced with the problem of estimating flow characteristics at ungaged sites. These estimates are usually made by transferring
information from gaged sites through regional relations between the flow characteristics and physical and climatic characteristics of the basins. Unfortunately, flows in arid or semiarid regions are often only poorly related to size
of drainage basin and other basin characteristics. Regional relations between
flow characteristics and stream channel size offer a promising alternative under
these conditions. Moore (1968) and Hedman (1970) describe such relations between
mean annual discharge and the width and mean depth of a channel section defined by
the tops of in-channel bars. Hedman, et al. (1972) also used width and average
depth of the section defined by in-channel bars but included relations for

---

[1]Paper approved for publication in the U.S. Geological Survey Journal of Research.

estimating floods of selected recurrence interval. Hedman, et al. (1974) used the width and average depth of a section defined by a feature of higher elevation termed the active-channel section. Riggs (1974) gave relations between floods of selected recurrence interval and the width of the main channel.

The reliability of flow estimates from such relations depends not only on the applicability of the regional relations but also on the ability of different individuals to recognize and measure the channel parameters used as independent variables. This paper reports the results of a test conducted to assess the magnitude of this personal error. The test was not concerned with defining a regional relation between flow characteristics and channel size; that such relations can be developed is demonstrated in the literature cited.

DESCRIPTION OF THE TEST

The purpose of the test was to determine how accurately trained individuals could independently measure the width and average depth of the channel as defined by three separate reference levels, and to determine the effect of variability in channel measurements on estimates of discharge characteristics. The three sections are the section defined by the lowest channel bars, the active-channel section, and the main-channel section. The seven participants were experienced in identifying at least one of the three reference levels and were generally familiar with all three sections.

Section Defined by Lowest Channel Bars

The section defined by within channel bars was described by Moore (1968), Hedman (1970), and Hedman, et al. (1972). The reference level is defined by the tops of the lowest prominent channel bars. In perennial streams the particles of the bars are moved annually, and the bars may be below the water surface for much of the year. In ephemeral streams, particles will be moved by significant flows but may not be moved annually.

Active Channel Section

The active channel section was described by Hedman, et al. (1974) and Riggs (1974) as the lower part of the channel entrenchment that is actively involved in transporting water and sediment during the normal regime of flow. Beyond the boundaries of the active channel the channel features are relatively permanent and usually are vegetated. The reference point for measuring the active channel section is the point at which the channel banks or tops of stabilized channel bars abruptly change to a flatter slope. In a straight reach devoid of channel bars the width of the section will be the width of the low water channel.

Main-Channel Section

The main-channel section was described by Riggs (1974) as that part of the stream channel bounded by the streamward edges of the flood plain, or by the lower edge of permanent vegetation. On perennial streams it is the same as the bankfull stage described by Leopold, et al. (1964), but is measured in a narrow section.

The Study Area

The test area is located in the Powder River and Bighorn River basins in northern Wyoming, U.S.A. This area was selected because a wide variety of hydrologic conditions and channel types exist in a relatively small area. The elevation ranges from about 1400 m in the plains on the east and west to about 3900 m in the Bighorn Mountains that bisect the study area. Mean annual precipitation ranges from roughly 170 mm at the lower elevation to about 1250 mm in the mountains. Streams flowing from the mountains derive most of their flow from snowmelt and are perennial; streams originating in the plains are ephemeral, with most flows resulting from thunderstorm activity. Streambed composition is quite variable, ranging from cobbles and boulders at the higher elevations, to gravel in the lower mountains, and to mixed silt, clay, and sand in the plains. The variation of stream type and channel size is indicated in table 1; the channel widths shown are the geometric means of the values determined by the test participants.

Table 1 - Average Channel Widths of Test Sites

Site	Stream type[1]	Geometric means of measured width, in meters		
		Low bar	Active channel	Main channel
1	E	1.03	2.21	4.11
2	E	1.56	3.27	4.72
3	P	10.1	13.6	15.6
4	P	5.81	7.15	8.20
5	P	4.61	5.55	7.31
6	P	5.30	5.94	7.48
7	P	6.82	8.40	12.1
8	E	2.59	5.18	9.86
9	E	2.37	5.55	9.20
10	E	2.65	6.37	10.3
11	E	1.56	2.72	6.08
12	P	16.7	60.8	65.2
13	E	.61	.92	2.11
14	E	1.10	2.06	3.58
15	P	6.52	8.40	10.1
16	I	1.39	2.78	4.21
17	P	1.30	2.11	3.50
18	P	17.1	18.4	21.6
19	P	11.6	12.1	13.9
20	P	2.54	2.98	4.02
21	P	2.98	4.31	5.67
22	E	5.68	9.20	12.1

[1]P = perennial, E = ephemeral, I = intermittent.

Test Procedure

The test was designed to simulate conditions that might exist in using regional relations to estimate flow characteristics at ungaged sites. To insure objectivity, the majority of the test sites were on ungaged streams. Furthermore, at the time of the test there were no regional relations based on channel size for the test area; thus, measurements at the gaged sites used could not be compared with any other estimates. The seven participants were given directions to the 22 sites, and they visited the sites independently. Only general reaches of each stream were identified so the specific cross sections at which individuals measured channel dimensions were of their own choosing. Thus, the variability of measurements by individuals reflects the combined effects of differences in cross section location within the test reach and differences in identification of the reference levels. This should be indicative of the true variability that would result if trained individuals measured channel size in an ungaged reach.

Weather was a factor in the test. Several participants were unable to visit some of the sites because of recent snowfall. Also, higher-than-normal precipitation caused increased flows, which inundated the lowest channel feature at a few sites. This was not felt to be detrimental to the test, however, as it added to the realism.

ANALYSIS OF DATA

The test was intended to define the variability of channel measurements by individuals and to give some insight into potential advantages and disadvantages of the three reference levels. There was no attempt to evaluate hydrologic considerations such as determining the reference level most closely related to a given flow variable.

The number of sites for which a given reference level could be identified is certainly one measure of the usefulness of the reference level. With 22 sites and seven participants there were potentially 154 measurements for each reference level. However, because of inclement weather, three participants did not visit site 10, one did not visit sites 12-15, and one did not visit sites 19-21. Thus the maximum sample for a given reference level would have been 144 measurements. A total of 109 measurements was made for the section defined by the within-channel bars, 141 were made for the active channel section, and 136 were made for the main channel section. These represent 76 percent 98 percent, and 94 percent, respectively, of the measurements that could have been made.

The low percentage of measurements for the low-bar section resulted in part from that reference level being submerged on perennial streams. Flows during the test period were higher than base flow, and two-thirds of the sites at which the low-bar section was not located by the participants were on perennial streams. This, however, would also be a factor in applying a regional relation so it must be recognized as a constraint on the utility of the low-bar section.

Agreement Between Participants

Two tests were conducted to assess the degree of consistency among channel measurements by different individuals. In one test the cross-correlation coefficients were defined for all possible pairs of individuals. Width and depth coefficients were considered separately for each reference level, and the mean and standard deviation computed for correlation coefficients from all the possible pairs. Results are shown in table 2, and indicate a high degree of consistency for measurements of width for all three reference levels. In contrast, the degree of consistency for depth measurements is relatively low. It appears from this test that different individuals can measure width more consistently than depth.

Table 2 - Summary of Cross Correlations Between
Measurements by Individuals

Section	Statistics of correlation coefficients		
	Mean	Standard deviation	Range
Low bar			
Width	0.95	0.055	0.74-0.99
Depth	.74	.128	.51- .93
Active channel			
Width	.97	.028	.91- .99
Depth	.59	.164	.27- .83
Main channel			
Width	.92	.067	.79- .99
Depth	.59	.193	.16- .89

Analysis of variance was used to test the hypothesis that there was no difference among individuals in the average value of a given channel size parameter. Widths and depths of the low-bar section for individuals 1, 4, and 6 and the depths of the active-channel and main-channel sections for individual 1 were not included in the analysis because of inadequate sample size. The hypothesis of no difference among means for individuals was accepted at the 95-percent level for the widths of all three reference levels and for the average depth of the low-bar section. The hypothesis was rejected for mean depth of both the active-channel section and the main-channel section; however, upon eliminating results for individual 1 and retesting, the hypothesis of no difference among means was accepted. Thus, while participant 1 apparently measured a section somewhat more shallow than that measured by other participants, the difference was not reflected in the measured widths.

Variability in Discharge Estimates

In analyzing the test data, the average of the logarithms of the seven measurements of channel width (W) at a given site was assumed to give the best estimate of log W, and departures of individual values of log W from the average for a site were examined. The mean and standard deviation of the departures are summarized in Table 3 for each participant and for the three reference levels. The mean of the absolute values of the seven average departures represents the average of individual bias in log W for the reference level without regard to the direction of bias. The average standard deviation for a particular section represents the mean standard error in log W resulting from the combined effects of individual differences in reach selection and identification of the reference level. The effect of this variability on computer discharge is defined as follows.

Relation between a discharge characteristic and channel width usually take the form

$$\log Q = \log a + b \log W,$$

where a and b are constants of regression. Given a relation of this form, the standard error in log Q produced by variation in estimates of W is b times the standard error of log W.

Table 3 - Statistics of Differences from The Mean Logarithm of Width at Each Site. Units are base 10 Logarithms

	Section					
Individual	Low bar[1]		Active channel		Main channel	
	Average departure	Standard deviation	Average departure	Standard deviation	Average departure	Standard deviation
1	−0.0459	0.1310	0.0001	0.0873	−0.0241	0.0956
2	.0322	.0812	−.0009	.0907	.0435	.1150
3	−.0418	.0876	.0418	.0554	−.0196	.0673
4	−.0127	.0737	−.0612	.0484	−.0557	.0904
5	.1241	.1068	.0622	.0859	.0153	.1140
6	−.0270	.1283	.0164	.1018	.0760	.1048
7	−.0805	.0770	−.0548	.0859	−.0239	.0524
Average	[2].0521	.0979	[2].0340	.0793	[2].0369	.0913

[1]Sites 8 and 10 were excluded as the low-bar feature was only measured by three individuals.

[2]Mean of the absolute values of average departure.

In relations developed to date, the regression coefficient, b, has averaged about 1.5. Using this value and the average standard deviations for log W in table 3, the corresponding standard errors in log Q are shown in table 4.

The average bias in estimates of log Q is also shown in table 4. This bias results from an individual consistently measuring either larger or smaller widths than the average. It must be emphasized, however, that the bias shown is only correct if the average of log W is the true value of log W.

The summary of errors in estimates of discharge characteristics shown in table 4 are the values that would result only from variability and bias in measuring channel width; they do not reflect the model error or the error of the streamflow characteristics used to develop the regional relation. The total error would include all of these components. Although these components cannot be separated at present, some insight can be gained by assuming that a regional relation between a discharge characteristic and width has a standard error of 0.13 log units (30.4 percent), and that the widths used to develop the relation were averages of measurements by a number of individuals. This standard error should approximate

Table 4 - Errors in Discharge Estimates Attributable
to Variability of Width Measurements

Section	Average standard error in Q		Average bias in Q	
	Log units	Percent	Log units	Percent
Low bar	0.147	33	0.078	18
Active channel	.119	27	.051	12
Main channel	.137	32	.055	13

the combination of model error and sampling error of the flow characteristics.
However, true error of applying the relation by one individual would include
components of error resulting from both variability and bias in measuring width.
The approximate magnitudes of these components would be 0.13 log units (30.4
percent) and 0.06 log units (13.9 percent), respectively, from table 4. If the
three errors are independent, the true standard error would be the square root of
the sums of squares of the components. Thus, the true standard error of discharge
would be about 0.193 log units or about 46 percent.

SUMMARY

Results of this test do not indicate a marked superiority of any one of the three
reference levels presently being used to develop regional relations for flow
characteristics. The variation in independent measurements of the three levels is
comparable. However, submergence of the low-bar section at medium and high stages
limits its usefulness.

As might be expected, the study indicated that trained individuals measure width
more consistently then depth. Cross-correlation coefficients and analysis of
variance failed to display any significant inconsistency between measures of width
by individuals. Cross-correlation coefficients for depth were lower than for
width, and analysis of variance indicated that at the 95-percent-confidence level,
one individual of the seven tested measured significantly smaller depths than the
average.

Given a regional relation between a flow characteristic and channel width, a
standard error of about 30 percent in estimated discharge could be expected from
the sampling error in width measurements by trained individuals, assuming a
perfect model and no sampling error in streamflow characteristics. Bias in
measuring width could produce about a 14-percent standard error in discharge. If
the errors are independent, a regional relation with model and streamflow sampling
error of 30 percent would have a true standard error of about 46 percent.

REFERENCES CITED

Hedman, E. R. (1970), Mean Annual Runoff as Related to Channel Geometry of Selected Streams in California, U.S. Geological Survey Water-Supply Paper 1999-E.

Hedman, E. R., et al. (1972), Selected Streamflow Characteristics as Related to Channel Geometry of Perennial Streams in Colorado, U.S. Geological Survey Open-file Report.

Hedman, E. R., et al. (1974), Selected Streamflow Characteristics as Related to Active-Channel Geometry of Streams in Kansas, Kansas Water Resources Board Tech. Rept. No. 10.

Leopold, L. B., et al. (1964), Fluvial Processes in Geomorphology, W. H. Freeman, San Francisco.

Moore, D. O. (1968), Estimating Mean Runoff in Ungaged Semiarid Areas, Bulletin, Internat. Assoc. Sci. Hydrology XIII, 1, 28-39.

Riggs, H. C. (1974), Flash Flood Potential From Channel Measurements, in Flash Floods Symposium, Internat. Assoc. Sci. Hydrology Pub., 112, 52-56.

Simulation of Regional Flood-Frequency Curves

Based on Peaks of Record

Kenneth L. Wahl (U.S. Geological Survey, Menlo Park, Calif.)

ABSTRACT

A method is described for defining a regional average flood-frequency curve from a set of concurrent short-term flood records. The method is based upon a modified station-year approach as suggested by P. H. Carrigan, Jr. Using a set of nine small-stream flood records, an average frequency curve is defined for a semiarid region of northeastern Wyoming, and the regional curve is shown to be in general agreement with the results from a more conventional regional regression estimate based upon a larger data base.

The regional relation is developed from simulated records using (1) the maximum event from each of a number of stations having concurrent record and (2) the cross correlation between stations. In order to use the method, however, records for the individual gages must be reduced to identical distributions by use of a scaling factor.

INTRODUCTION

Estimating the magnitude of design floods is one of the more pressing problems in hydrology. If records of peak flow are available at the site of interest, a flood-frequency relation can be developed by standard techniques and that relation used to estimate the magnitude of the design flood. The reliability of such an estimate is dependent on the length of record available. If the record length is less than about 25 years, estimates of high recurrence-interval floods, such as the 50-year or 100-year floods commonly used in design, are subject to considerable uncertainty. If the design site is ungaged or if only a short record is available, the usual practice is to transfer information from other gaged sites in the vicinity. One method of tranferring information is to relate flood characteristics (dependent variables) at the gaged sites to basin and climatic character-istics (independent variables) in a regression analysis. The desired flood characteristics can then be estimated at any site in the region by entering the relation with the values of the independent variables for the site. This assumes, however, that (1) the region is homogeneous so that the flow characteristics at the gages represent spatial sampling of the population of floods; (2) that a suitable mathematical model exists for transferring data from gaged to ungaged sites; and (3) that records at the gaged sites provide representative samples of the population of annual floods at the gaged sites.

Estimating design floods for small streams (less than 100 km^2) in semiarid areas often poses special problems. Few records of over 10-15 years are available in a given area. The problem is compounded by the large time variability inherent in floodflows from small basins. The combination of short records and large variability makes estimates of floods with high recurrence intervals (low probability of occurrence) of questionable reliability.

Conover and Benson (1963) suggested a method of estimating high recurrence-interval floods at a gaged site by combining independent records from a homogeneous area to form a long record for frequency analysis. Their approach was similar to the familiar station-year method (Fuller, 1914, and Clarke-Hafstad, 1942), in which records are added end-to-end, but Conover and Benson used only the maximum peak from each site. Carrigan (1971) suggested a modification of the Conover and Benson (1963) approach to allow for dependence between the records to be combined. He used a Monte Carlo technique to define exceedance probabilities and demonstrated the method using records for six highly dependent snowmelt stations in Idaho. The six stations had 46 years of concurrent record.

A method of combining short records is needed and should be especially useful for small streams in semiarid areas. Previous applications by Conover and Benson (1963) and Carrigan (1971) used long records and aimed at defining frequency characteristics at a gaged site. The purpose of this paper is (1) to report the results of an application of Carrigan's simulation model to small basins in the Powder and Cheyenne River basins of Wyoming and (2) to demonstrate potential of the technique for defining regional frequency curves for use at ungaged sites. Sensitivity of the relation to dependence between stations is considered. The sites used in the example are on ephemeral streams in the U.S. Geological Survey small-streams network.

PROBABILITIES FROM INDEPENDENT RECORDS

The station-year or composite-record method of frequency analysis was suggested for use on floods by Fuller (1914) and on rainfall by Clarke-Hafstad (1942). The method is based on the assumption that independent short-term records from a homogeneous region can be combined end-to-end to form one long composite record if the peaks of the individual records can be reduced to a common base. Flood records usually are reduced to a common base by dividing each annual peak for a particular site by the mean or the median of peaks for the site. The resulting standardized peaks are assumed to come from the same distribution.

In order to qualify for inclusion in the composite record, peaks must be random samples; they could not have been historical peaks that were measured simply because of their unusual nature. This precluded use of considerable information collected to document extreme events at ungaged sites. Conover and Benson (1963) proposed a modification of the station-year method to allow considering these historic peaks where the peaks were known to be the maximum for some interval of time. Their analysis allows determination of the exceedance probability of the maximum flood for all sites at which the maximum within a common period of time is known. The time periods at individual sites need not be concurrent but must be of equal length. The exceedance probabilities are based on the total length of the combined records.

In using the Conover-Benson approach all n-independent stations in the region that have k-years of peak-flow record, or for which the maximum peak in a k-year record is known, are ranked according to the maximum event. If x_{ij} ($1 \leq i \leq n$, $1 \leq j \leq k$) is the peak-flow event corresponding to the ith record and the jth rank within the record, the generalized array is of the form:

Rank of peaks at a site	Rank of peaks among sites				
	1	2	3	-----	n
1	x_{11}	x_{21}	x_{31}	-----	x_{n1}
2	x_{12}	x_{22}	x_{32}	-----	x_{n2}
3	x_{13}	x_{23}	x_{33}	-----	x_{n3}
---	---	---	---	-----	---
k	x_{1k}	x_{2k}	x_{3k}	-----	x_{nk}

Thus, x_{21} is the greatest peak at site 2 and is less than x_{11} but greater than x_{31}. Nothing is implied, however, about the relation among sites of the lesser peaks; for example, x_{21} may be less than x_{12}. The above matrix is intended to show only the relation between stations for the maximum event (x_{i1}). Nothing is implied about the time sequence of individual events (x_{ij}) within the k-year period.

Working only with the maximum event for each site and with order statistics, Conover and Benson (1963) determined that the probability, $p(x>x_{i1})$, of an additional event from the same continuous distribution exceeding x_{i1} was

$$p(x>x_{i1}) = \sum_{m=0}^{i-1} \frac{n!}{(n-m)! \ k[(n+\frac{1}{k})(n+\frac{1}{k} - 1)(n+\frac{1}{k} - 2)....(n+\frac{1}{k} - m)]}$$

They compared the theoretical probabilities from the above equation with results from sampling from known distributions using n=20 and k=50. One trial was made with an extreme values distribution and two trials used a normal distribution. Correspondence between the theoretical and sample results was close for all three trials, implying the formula to be valid regardless of the underlying distribution. It must be noted, however, that the approach is appropriate only if floods are independent between sites, are from the same distribution, and can be reduced to a common base.

PROBABILITIES FROM DEPENDENT RECORDS

The station-year method has not been widely used, partly because of the uncertainties introduced by the required assumptions of (1) independence among sites from a homogeneous region, and (2) that the standardized peaks from different sites are from the same distribution. While the Conover-Benson modification made it possible to include information on historical floods at independent ungaged sites, it did not allow adjusting for dependence between sites. Carrigan (1971) proposed an alternative method that eliminates the assumption of independence among sites. His approach uses Monte Carlo simulation techniques to generate sets of records

that preserve the dependence in the actual data. Use of Carrigan's model
requires that the cross-correlation matrix be defined for the records in the
region. Because the computer algorithm requires that the correlation matrix
be positive-semidefinite, the cross-correlation matrix must be based on
concurrent records.

Use of Carrigan's simulation approach requires that a geographic region
be defined within which the annual floods at different sites are identically
distributed. The steps in developing a regional flood-frequency relation
then proceed as follows (Carrigan, 1971, p. 2):

1. The records are reduced to a common base.
2. The stations are ranked in order of the standardized (reduced)
   maximum events.
3. The correlation matrix for the records is computed.
4. Computer routines are used to provide, through simulation
   algorithms that utilize the correlation matrix, estimates
   of exceedance probabilities associated with the n ordered
   maxima from the k-year records.
5. The frequency relation is defined graphically by relating
   exceedance probabilities to the ordered maxima. Reversing
   the transformation process that reduces records to an
   identical distribution produces regionalized frequency
   relations for annual floods at individual stations.

The simulated sets of records are generated using a normal multivariate
model

$$X = B\varepsilon$$

where X = matrix of n groups of k events,
      ε=n x k matrix of independent normal random numbers with zero mean and
        unit variance, and
      $B=E\lambda$=n x n principal-component transform matrix for the records of
        annual extremes, in which
      E=eigenvector matrix associated with the correlation matrix R obtained
        from n records of k hydrologic events, and
      λ=diagonal matrix whose n diagonal elements are the square roots of
        the eigenvalues for E.

This model was based on techniques of synthetic hydrology given by Matalas
(1967, p. 940). The maximum event x_{i1} for each simulated set is ordered so
that $x_{11} > x_{21} > x_{31} > .... > x_{n1}$, and their exceedance probabilities are determined.
Because the generated x_{ij} are from normal distributions with zero mean and
unit variance, the exceedance probabilities are given by

$$p(x > x_{i1}) = \frac{1}{\sqrt{2\pi}} \int_{x_{i1}}^{\infty} e^{-x^2/2} \, dx, \text{ for } i=1, 2, 3, ... n.$$

Carrigan's computer program currently repeats the simulation of n data sets
1,000 times; the exceedance probability for given x_{i1} is therefore the
average of 1,000 probabilities.

Carrigan (1971) tested his model for sensitivity to cross-correlation
coefficient ρ for n=3 and for k ranging from 10 to 50 years. He found the
average exceedance values to be relatively insensitive to values of ρ less
than about 0.6. Above ρ=0.6, however, the average exceedance probabilities
converged rapidly.

DEVELOPING A REGIONAL RELATION

Selection of Data

The area selected to test the application of Carrigan's simulation model on short records is in the Powder and Cheyenne River basins of northeastern Wyoming. The area is semiarid with a mean annual rainfall of about 300 mm, virtually all small drainages in the nonmountainous parts of this region are ephemeral. Peak-flow records are available from 1965-73 for about 20 basins of less than 65 km^2.

Because the present study was intended to test the potential of the method rather than to develop a frequency relation for general application, the analysis was limited to nine stations with concurrent records for the period 1965-73. The numerical identifiers, station names, and drainage areas are shown in the following list:

313020   Bobcat Creek near Edgerton, Wyo. (21.5 km^2)
313100   Coal Draw near Midwest, Wyo. (29.5 km^2)
313180   Dugout Creek tributary near Midwest, Wyo. (1.84 km^2)
316480   Headgate Draw at upper station, near Buffalo, Wyo. (2.85 km^2)
316700   Powder River tributary near Buffalo, Wyo. (3.89 km^2)
317050   Spotted Horse Creek tributary near Spotted Horse, Wyo. (11.1 km^2)
324900   Little Powder River tributary No. 2 near Gillette, Wyo. (10.2 km^2)
382200   Pritchard Draw near Lance Creek, Wyo. (13.2 km^2)
388800   Blacktail Creek tributary near Newcastle, Wyo. (0.65 km^2)

The simulation approach to flood-frequency analysis requires that annual flood records for stations in the area be reduced to a common base and that the reduced floods represent samples from the same distribution. Carrigan (1971) assumed that the annual floods were identically distributed if the confidence limits for the coefficient of variation, square of the coefficient of skew, and kurtosis were within the limits of the same moments for a base record. He reduced the records to a common base by dividing floods for each site by the average flood at the site. Record lengths in that study were 46 years, however, and the moments at each site were well defined. In the present analysis, records are only 9 years long; except for the mean, moments of the records are very poorly defined, and simplifying assumptions were necessary.

The Water Resources Council (1967) recommended that all Federal agencies use log-Pearson type III distributions to define flood-frequency relations. More recently the Water Resources Council (1976) recommended use of regional values of skew when records are short. Hardison (1974) found a regional skew of near zero for logarithms of annual floods for long records in northeastern Wyoming, and Lowham (1976) found that a regional skew of zero was applicable for small basins in the region. On this basis, logarithms of flood peaks at the nine sites used in this analysis were assumed to be samples from a log-normal distribution. A further assumption was that the logarithms of peaks at each site could be reduced to a common base by dividing each transformed peak by the average of the logarithms of the peaks for the site. The natural data are shown in table 1, and the standardized (reduced) data are shown in table 2; stations are ranked according to the extreme standardized event.

Table 1 - Annual peak discharges for the period 1965-73

Station number	Peak discharge, in cubic decimeters per second (dm^3/s)								
	1965	1966	1967	1968	1969	1970	1971	1972	1973
316480	155500	9910	28	2780	47300	32600	169	7500	962
313020	15200	28	5410	283	566	1700	2630	679	30000
316700	64900	22100	339	12700	46400	2750	2660	1920	283
313180	14000	1420	45000	2920	3740	14000	7050	8550	2720
382200	59500	34600	5300	114700	21200	1780	5610	24600	31200
317050	4810	3120	1700	877	594	169	1190	1640	5520
313100	9540	22800	28900	18100	679	46700	566	36800	8720
388800	2890	2350	1760	1130	1530	141	764	2260	1560
324900	7360	594	2150	3120	6090	3200	4110	2290	3340

Table 2 - Standardized (reduced) annual peak discharges. Stations ranked according to the extreme standardized event (underscored)

Station number	Logarithm of discharge divided by the average of logarithms for the site								
	1965	1966	1967	1968	1969	1970	1971	1972	1973
316480	1.4439	1.1114	0.4039	0.9577	1.3002	1.2551	0.6203	1.0778	0.8298
313020	1.3187	.4581	1.1776	.7735	.8685	1.0190	1.0790	.8934	1.4124
316700	1.3099	1.1826	.6891	1.1176	1.2705	.9362	.9324	.8942	.6675
313180	1.0861	.8257	1.2194	.9079	.9362	1.0861	1.0084	1.0303	.8999
382200	1.1216	1.0662	.8749	1.1887	1.0166	.7639	.8807	1.0317	1.0557
317050	1.1675	1.1076	1.0241	.9332	.8796	.7071	.9750	1.0195	1.1864
313100	1.0000	1.0950	1.1208	1.0700	.7117	1.1733	.6918	1.1473	.9902
388800	1.1165	1.0876	1.0467	.9853	1.0274	.6940	.9302	1.0824	1.0299
324900	1.1135	.7989	.9597	1.0059	1.0898	1.0093	1.0405	.9677	1.0147

Table 3 - Cross-correlation matrix for standardized annual peaks, based on 1965-73 period

Station number	316480	313020	316700	313180	382200	317050	313100	388800	324900
316480	1.00								
313020	-.20	1.00							
316700	.75	-.45	1.00						
313180	-.27	.52	-.34	1.00					
382200	.30	-.18	.46	-.56	1.00				
317050	-.14	.27	-.07	-.19	.57	1.00			
313100	.04	-.11	-.27	.22	-.00	.00	1.00		
388800	-.02	-.03	.20	-.18	.70	.82	-.07	1.00	
324900	.24	.64	.18	.32	.00	-.15	-.49	-.12	1.00

Average correlation coefficient $\bar{\rho}$= 0.07

Table 4 - Summary of simulation results

Station number	Reduced variate x_{i1}	Exceedance probability						Recurrence interval, years					
		Simulation results					Conover and Benson	Simulation results					Conover and Benson
		Actual ρ	ρ=0	ρ=0.1	ρ=0.2	ρ=0.4		Actual ρ	ρ=0	ρ=0.1	ρ=0.2	ρ=0.4	
316480	1.44	0.0147	0.0129	0.0132	0.0146	0.0177	0.0122	68.0	77.3	75.8	68.6	56.5	82.0
313020	1.41	.0269	.0265	.0270	.0287	.0327	.0257	37.1	37.8	34.8	34.9	30.6	38.9
316700	1.31	.0409	.0410	.0418	.0440	.0479	.0410	24.5	24.4	23.9	22.7	20.9	24.4
313180	1.22	.0575	.0575	.0594	.0622	.0656	.0584	17.4	17.4	16.8	16.1	15.2	17.1
382200	1.19	.0773	.0767	.0788	.0821	.0853	.0789	12.9	13.0	12.7	12.2	11.7	12.7
317050	1.19	.1006	.1011	.1017	.1066	.1084	.1038	9.99	9.89	9.83	9.38	9.23	9.64
313100	1.17	.1300	.1336	.1315	.1371	.1364	.1358	7.69	7.49	7.60	7.30	7.33	7.36
388800	1.12	.1732	.1760	.1752	.1797	.1735	.1813	5.77	5.68	5.71	5.57	5.76	5.52
324900	1.11	.2289	.2395	.2351	.2403	.2246	.2631	4.37	4.18	4.25	4.16	4.45	3.80

Cross Correlation

Cross-correlation coefficients were determined for the reduced peaks shown in table 2. The matrix of cross-correlation coefficients is shown in table 3, and figure 1 shows the cumulative probability distribution for ρ. The average correlation coefficient $\bar{\rho} = 0.07$ while the median (from figure 1) is about $\rho_{med} = -0.01$. Negative values of ρ for flood peaks would not be anticipated, thus the high incidence of negative correlation coefficients is evidence of sampling error. At the 99-percent confidence level and for nine pairs of data, absolute values of ρ must be greater than 0.8 in order to reject the hypothesis that $\rho = 0$ (Snedecor and Cochran, 1967, p. 557); only one correlation coefficient in table 3 exceeds 0.80. Therefore, at the 99-percent confidence level the hypothesis that the annual peaks are independent among stations would be rejected for only one pair of stations.

Simulation Results

The statistics of correlation coefficients based on the short records are by no means convincing evidence of dependence between sites. Because of the low $\bar{\rho}$ and Carrigan's observation that simulated exceedance probabilities were fairly insensitive to $\bar{\rho} < 0.6$, simulations were made for the observed correlation matrix (table 3) and for equicorrelation matrices with $\rho = 0$ (independent), $\rho = 0.1$, $\rho = 0.2$, and $\rho = 0.4$. The exceedance probabilities were also computed from the relation given by Conover and Benson (1963), which assumes independence between stations, i.e., $\rho = 0$.

The exceedance probabilities of the ranked extremes are summarized in table 4. Simulation results for the different assumed correlation matrices agree quite well; most of the influence of changing ρ is absorbed into the plotting position of the most extreme event. When loss of precision due to plotting is considered there is virtually no difference, except at the most extreme point, in simulated probabilities for the observed values of ρ and for a value of $\rho = 0$ (the mean value $\bar{\rho} = 0.07$). Furthermore, except for the highest event, Conover and Benson's plotting position agrees very well with the simulated value using the matrix of observed ρ.

The Conover and Benson (1963) approach is based on independence between stations ($\rho = 0$); therefore, the simulation results for Carrigan's model with $\rho = 0$ should be directly comparable to results using their relation. Comparison of Conover and Benson's values in table 4 with simulation results for $\rho = 0$, together with minor inconsistencies in simulated probabilities for different values of ρ, indicates that the present 1,000 interations in the simulation scheme should be increased when applied to records of less than 10 years.

The Regional Curve

The regional frequency curve is constructed from the data shown in table 4. The standardized extreme event for each site is plotted versus the recurrence interval (reciprocal of exceedance probability) as shown in figure 2, and the regional curve is defined graphically. Figure 2 also shows the relation of the regional flood-frequency curve developed in this study to a comparable relation developed by Lowham (1976) using multiple-linear regression. The curve developed by simulation gives consistently higher values for design floods than do the relations of Lowham (1976). His relations, however, were based on all data from the area whereas the present analysis used only nine stations.

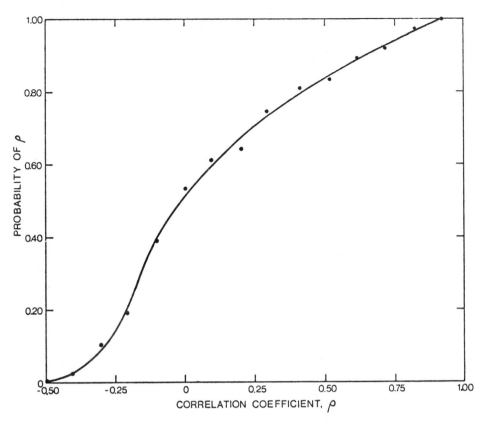

FIGURE 1.--Cumulative distribution function of the coefficient
of cross correlation.

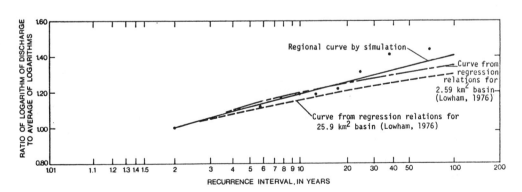

FIGURE 2.--Regional flood-frequency curve for small basins in the Powder and Cheyenne River basins, Wyoming.
Units of discharge are cubic decimeters per second (dm³/s).

The regional flood-frequency curve can be used in several ways. It may be used to extend an existing short-term record to include regional flood experience as demonstrated by Carrigan (1971). Or the curve can be used to estimate design floods for ungaged sites if the 2-year flood can be estimated from a regional regression relation. If records are short (10 years or less), a conventional regional analysis could be made, but the analysis would necessarily be confined to defining the 2-year flood. Because the logarithm of the 2-year flood is equal to the average of the logarithms of the annual events for a log normal distribution, figure 2 can be used to estimate higher recurrence-interval events at any site for which the 2-year flood can be estimated.

SUMMARY

Using the simulation approach suggested by Carrigan (1971), a regional flood-frequency curve was developed based on the maximum flood events from nine gages with 9 years of concurrent records. Size of the basins ranged from 0.65 to 29.5 km^2.

There are two points of concern regarding use of this technique with short-term records: (1) The cross-correlation coefficients that provide the basic input to the simulation phase are very unreliable for less than 10 years of concurrent record; and (2) a method is needed to reduce all flood records to a common base. This study indicates that error due to poor definition of the correlation matrix may be minimal in semiarid areas. Except for the maximum event (x_{11}), exceedance probabilities were shown to be quite insensitive to variations in ρ for $\bar{\rho} < 0.4$. The method of reducing records to a common base is an item of concern, however. In this study the logarithms of the events at a site were divided by the mean of the logarithms of the events to standardize the data. The effectiveness of this transformation is difficult to assess for short records. Because logarithms are used, all years must have non-zero peak discharges, but this is not a major problem; flows of at least 1 dm^3/s can easily be assumed to have occurred in any year. If, however, there are enough no-flow years that the average becomes a meaningless statistic of annual flows, the transformation used herein could not be used. The station-year approach might still be usable, but a different method of reducing the data to a common base would have to be conceived.

REFERENCES

Carrigan, P. H., Jr. (1971), *A Flood-Frequency Relation Based on Regional Record Maxima*, U.S. Geological Survey Professional Paper 434-F.

Clarke-Hafstad, Katherine (1942), Reliability of Station-Year Rainfall-Frequency Determinations, *Transactions*, American Society of Civil Engineers, 107, 2142, 633-652.

Conover, W. J. and M. A. Benson (1963), Long-Term Flood Frequencies Based on Extremes of Short-Term Records, in *Geological Survey Research, 1962*, U.S. Geological Survey Professional Paper 450-E, E159-E160.

Fuller, W. E. (1914), Flood Flows, *Transactions*, American Society of Civil Engineers, 77, 1293, 564-617.

Hardison, C. H. (1974), Generalized Skew Coefficients of Annual Floods in the United States and Their Application, *Water Resources Research*, 10, 5, 745-752.

Lowham, H. W. (1976), *Techniques for Estimating Flow Characteristics of Wyoming Streams*, U.S. Geological Survey Water-Resources Investigations 76-112.

Matalas, N. C. (1967), Mathematical Assessment of Synthetic Hydrology, *Water Resources Research*, 3, 4, 937-945.

Snedecor, G. W. and W. G. Cochran (1967), *Statistical Methods*, 6th ed., The Iowa University Press, Ames, Iowa.

Water Resources Council (1967), *A Uniform Technique for Determining Flood Flow Frequencies*, U.S. Water Resources Council, Hydrology Comm. Bull. no. 15, Washington, D.C.

Water Resources Council (1976), *Guidelines for Determining Flood-Flow Frequency*, U.S. Water Resources Council, Hydrology Comm. Bull. no. 17, Washington, D.C.

WATER SAVING DEVICES FOR SANITATION IN DESERTS.

A. Ortega - Architect - UN Adviser to the UAE and
B. LeFebvre - Architect - UAE Consultant.

THE PROBLEM

Using Water as a Transportation for Human Waste

The collection of human waste from house connections via an underground network of
sewers, and the disposal of the sewage in centralised locations, appeared in the
European cities in the middle of the nineteenth century. First in Hamburg (1842)
and later in London (1855) and Paris (1860), following disastrous cholera epi-
demics, underground sewers were begun. The first American example was in Brooklyn
(1857). However the problem of disposing of large amounts of sewage in spot
locations was not immediately appreciated, and it was estimated that in the
fifties, in the United States, more than a quarter of the systems discharged their
wastes without treatment.

The system of collection and neutralization of human wastes calls for high densities to justify the required network of water supply and sewer pipes, large
quantities of water, and facilities for waste disposal. When one or more of these
factors are absent from the equation, the network system approach cannot be
applied. This may be the case in rural areas where people live too far apart, in
poor countries which cannot afford investment required, in regions where water is
scarce.

Two factors deserve special mention; the use of large quantities of water
and the need to then purify this water, and the cost of the network approach.

The annual water consumption of the average family in a consumer society=400
cubic metres, and 40% of this water is used for toilet flushing. The "standard"
toilet requires 22-30 litres for each flushing. Obviously arid regions, or in
areas with dry seasons, or where water supply is critical, the conventional
water-borne system is wasteful and costly. Moreover, even where water supply is
not problem, it can be argued that the cost of purifying the water, (and sewage
is 99% water), is the cost of purifying the transporting medium, rather than the
waste, and so represents a misuse of energy and resources.

The use of drinkable water for the transport of relatively small quantities of body waste (120 grammes per person per day) does not make sense any longer. It is not reasonable to use 40 percent of the drinking water, now used by one person, only as a carrier for his body waste and also to give the chance to the disease causing bacteria to contaminate farther the slightly polluted water from baths, sinks and washing appliances.

It should be clear that water-borne waste disposal represents a (relatively recent) answer within a set of economic and physical conditions, and not clearly the least wasteful answer at that. Flush toilets should not be considered as "advanced", compared to the pit latrine. Under certain conditions the latter is ecologically sound, cheap and quite safe.

A SOLUTION. DECOMPOSITION.

Compost Privy.

Composting as a method for decomposition of animal and human waste to provide fertilizer has been in practice for a long time in India and China. Since the 1930's scientists have begun to study this phenomenon, and aerobic composting developed to the point where now in the Netherlands thousands of tons of compost are produced annually from municipal waste. Presently in India more than two million tons of compost are prepared annually.

The principles of composting have also been applied to small scale conversion of human and kitchen waste.

The compost privy consists of a tank with an air intake and a ventilation duct. The toilet is located directly above the tank, as is a garbage chute. Human waste, paper and organic kitchen refuse decompose together into a fertilized humus, whose volume is about 10% of the original. At the lowest point of the tank is an access door for removing the humus. It is important that a layer of straw, sawdust or leaves be first placed in the bottom of the tank. This absorbs the liquid waste and aids decomposition. No water, fuel or chemicals are used in the process, and valuable fertilizer results.

The impetus for the development of alternatives to the flush toilet occurred in Norway and Sweden in the mid-sixties when environmental protection laws virtually prohibited soakage pits and septic tanks in rural areas. One of the alternatives that proved popular was the composting method.

The original composting system was invented by a Swedish sanitary engineer, Rikard Lindstrom, in 1938. He patented his system in 1962 (U.S. Patent 3,136,608) and it was produced commercially in 1964, under the name Clivus-Mulltrum.

LOCALLY BUILT COMPOSTING TOILET

The objective is to build a composting toilet roughly of the same geometry and dimensions of the Swedish Clivus Mulltrum but at a much lower cost using local materials. The Swedish fiberglass unit has a cost F.O.B. US $1800.

Four prototypes were built in Canada and performed well. The next step was to see if they would function properly in hot-humid climates. So another three were built in the Philippines; one above ground because of high water table conditions and two other underground.

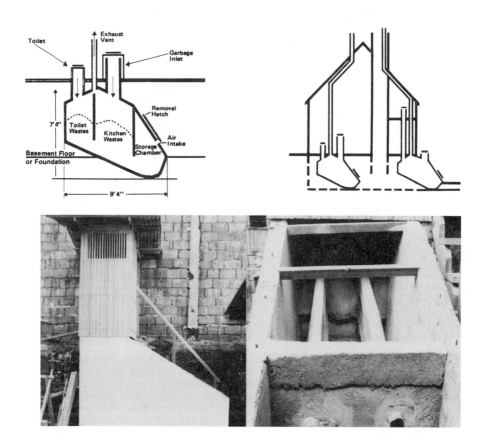

Above ground unit in the Philippines.

The demonstration in the U.A.E., presented in this paper, permits an evaluation of its performance in a hot desertic region. An excavation of 4m long by 2m wide by 1.5 deep was made to allow the container to be built underground.

Cement blocks of 20cm x 20cm x 40cm were laid in conventional fashion to a height of eight blocks. Then sand was put back inside and covered with 5cm of concrete to form a 30 degree slope bottom. This allows the compost pile to slide out slowly towards the collection chambers. Three V-shape ducts, shown in drawings, allow the air to pass from the collection chamber through the compost pile to aerate it and then escape out of the vent duct. Mosquito nettings were installed at all orifices of the system to keep insects out. The cover of the container was made out of plywood; it also could be reinforced concrete. For the garbage chute a plastic bin with the bottom cut out was used. For the toilet itself a new squatting plate, called "Watergate" was installed. This unit provides a water seal with only <u>one litre of water</u>. With this ingenious system of equilibrium the solids are evacuated by their own weight tilting a pan.

The <u>amount of water</u> added to the compost does not unbalance the decomposition process, evaporation being so rapid in desert regions.

OPERATING COMPOSTING TOILETS.

Temperature.   The temperature within composting toilets
                varies. A toilet such as the Clivus, builds
                up to a temperature of about 30°C which is
                below body temperature, consequently patho-
                genic organisms are not destroyed by heat,
                but rather by the long pasteurization
                period.

Volume.   It has been assumed that the volume output
                of organic waste and human excreta is in the
                order of 0.3 m^3 per person per year, of
                which only 20% is excreta. A long term
                composting toilet will require 5 m^3 for the
                three year period. The humus production is
                estimated to be about one or two buckets per
                person annually.

Starting.   In order to start a composting toilet it is
                necessary to build up a layer of rich soil
                on the bottom of the container in order to
                introduce micro-organisms to facilitate
                composting. A layer of peat moss or dry
                leaves is also required to absorb the urine
                until the mass of the pile is adequate.

Flies.   Most composting toilets seem to have a
                common problem with flies at the beginning
                of their operation. This may be due to the
                internal balance of the pile not yet having

HOW WATERGATE WORKS

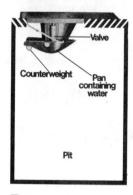

Valve

Counterweight Pan
containing
water

Pit

■ The water in the pan (approximately 2 litres) covers the bottom end of the shute. This completely seals off the pit so that flies cannot enter and bad smells cannot escape.

■ When Watergate is used, the weight of the waste matter causes the pan to tip, throwing the waste matter into the pit.

■ The counterbalanced pan closes again and the valve enables it to fill with water to the correct level.

been established, a situation which seems to last only a few weeks. The addition of sawdust to the pile helps at this point.

What goes in.	Urine, excrements, toilet paper, kleenex, tampax, kotex, paper diapers, paper towels, grease and fat, dust, vegetable and meat scraps, peelings, bones, and eggshells.
What doesn't go in.	Cans, glass, plastic, paints, toxic liquids, chemicals, pesticides, cardboard boxes, unshredded paper and especially any chemical sanitary agents.
Capacity.	The composting toilet is designed for regular use by eight to ten people. Heavier use for short intervals will not affect the toilet, but if more than the recommended number of people use the toilet for any length of time, urine will start to accumulate in the lower chamber.

This information is based on material from the Farallones Institute, Clivus Mulltrum

Underground unit in the U.A.E.

★ Materials and labour required.

  ◆ 120 cement blocks (20 x 20 x 40 cm)
  ◆ 10 bags of cement
  ◆ 3 air ducts (2 m long)
  ◆ 1 wood or concrete cover (1.60 x 3.60 m)
  ◆ 1 vent pipe (3 m long x 15 cm diametre)

  ┇ Approximately 36 man-hours are necessary to build one unit. ┇

U.S.A. and Bernard Lefebvre experience in
building composting toilets in Canada, in
the Philippines and in the United Arab
Emirates.

FINAL CONCLUSION

The use of composting toilets as a water saving device in desert areas should be
promoted as well as a way to reduce water pollution and produce fertilizer.

ADVANTAGES

1. Great water savings (40,000 litres per year for an average household.)
2. Eliminates the cost of installing and maintaining an underground sewer system.
3. No waste water treatment plant is necessary.
4. Produces a nutrient rich by-product that can be used as a fertilizer. The
    composting toilet produces about 30 Kgs of fertilizer per person per year.
5. Having no moving parts it requires no maintenance except to remove the compost
    produced once a year.

6.   It accommodates all household garbage thus eliminating garbage collection. The glass, metals and plastics which remain are separated more easily for recycling.

AN EXAMPLE OF SYSTEMATIC EVALUATION OF SURFACE WATER
RESOURCES SOUTH OF THE SAHARA

J.A. Rodier
Head Hydrological Service at ORSTOM*
Scientific Adviser at Electricité France

ABSTRACT

Through examples drawn from Africa, south of the Sahara, the author reviews the
problem of methodology with respect to the assessing of surface water resources of
rivers having their source in desert, subdesert, and Sahelian zones – these do not
include those rivers reaching subdesert zones but having their sources in areas
with much heavier rainfalls, such as the Nile, the Chari, the Senegal, and the
Niger Rivers.

The author explains that there does exist a methodology for assessing such resources. However, this methodology can be applied efficiently only after serious
attempts are made in the following recommended fields:

- Identification (quantitative, if possible) of physio-
  geographical parameters involved in the genesis of the
  flow; such parameters should be determined in a hydrological perspective;
- Better understanding of the action of the parameters
  on the hydrological cycle;
- Establishment and continuous monitoring of a few
  reference rain gauging stations, rainfall recorders,
  and hydrometrical gauging stations.

In many cases, however, the great irregularity of supply from one year to another
leads to serious deficiencies in the event of low probability drought, and whenever possible one should make maximum use of rivers flowing from zones with
heavier rainfalls down to the desert zones.

---

*ORSTOM: Office de la Recherche Scientifique et Technique Outre-mer (Office for
Overseas Technical and Scientific Research).

1.   INTRODUCTION

L'estimation des resources en eau suppose, en zone désertique, que l'on détermine essentiellement deux caractéristiques hydrologiques : le volume moyen annuel écoulé et le débit maximal de crue d'une fréquence donnée décennale, centennale, millennale, suivant la nature et l'importance des ouvrages. Ce second élément du régime doit être connu, car, souvent, il est si élevé qu'il rend très coûteux l'aménagement et peut en interdire pratiquement la réalisation.

Dans le cadre de la planification, les études hydrologiques ne sont pas effectuées cours d'eau par cours d'eau, mais elles intéressent de vastes ensembles présentant les mêmes conditions de ruissellement. C'est ainsi que les zones désertiques, subdésertiques et sahéliennes au Sud du SAHARA ont fait l'objet d'une étude générale, mais la plupart des résultats peuvent s'appliquer à beaucoup d'autres régions désertiques du monde.

On a considéré à la fois le régime subdésertique, le régime désertique qui concerne les régions recevant en moyenne 100 à 300 mm de pluie par an et le régime sahélien dont les limites sont fixées par les courbes de précipitations annuelles 300 mm et 750 mm.

On ne considère dans ce qui suit que les cours d'eau qui prennent naissance dans ces régions et non ceux qui parviennent aux zones subdésertiques après avoir pris leurs sources dans des régions beaucoup plus arrosées, comme le NIL, le CHARI, le SENEGAL et le NIGER.

2.   ELEMENTS DU REGIME HYDROLOGIQUE

2.1.   Généralités

Le régime d'un cours d'eau en Afrique topicale est déterminé principalement :

1°. par la quantité annuelle de pluie que reçoit son bassin versant, la répartition au cours de l'année de ses averses et leur intensité ;
2°. par les caractéristiques du bassin versant lui-même : relief, nature du sol, et parfois du sous-sol, réseau hydrographique, couverture végétale et utilisation des sols.

L'influence de la température sur le régime hydrologique est à peu près la même dans toute la zone étudiée, sauf pour les massifs montagneux pour lesquels l'altitude conduit à un régime climatique plus tempéré, donc à une baisse de l'intensité de l'évaporation, mais ceci n'est guère sensible qu'à partir de 800 - 1000 mètres d'altitude.

En zone sahélienne, la superficie cultivée n'occupe pas une partie très importante du sol, cependant, dans certaines vallées mises en valeur systématiquement, le ruissellement superficiel a augmenté de façon très importante depuis le remplacement de la couverture végétale naturelle par des cultures.

2.2.   Précipitations Annuelles

Elles jouent un rôle essentiel dans la formation des débits.

Elles sont irrégulières dans l'espace et dans le temps, c'est pourquoi il est
assez difficile de déterminer avec précision leur valeur moyenne en un point ; il
faudrait pour cela qu'elles aient été observées avec soin pendant au moins 100
ans. La comparaison des résultats de stations voisines permet de remédier, dans
une certaine mesure, à la brièveté des périodes d'observations. On trouvera
ci-contre la carte des précipitations annuelles (graphique n° 1) en zones
désertiques et sub-désertiques. Cette carte tient compte des observations faites
durant la dernière période de sécheresse. Les lignes d'égales précipitations
annuelles sont grosso modo légèrement inclinées sur les parallèles, la direction
étant la même que celle de la trace au sol du Front intertropical avec un certain
nombre d'irrégularités dues au relief ou à l'influence de zones marécageuses de
grande dimension, comme le delta central du NIGER. Si on considérait le régime
des averses à très grande échelle (au 1/20 000 ou 1/50 000) par exemple, on
trouverait un grand nombre d'irrégularités correspondant à l'influence d'une
colline ou d'un lac ; la carte correspond à une simplification de la répartition
réelle des pluies.

2.3. Répartition Saisonnière Des Précipitations

A part quelques averses de printemps (Mars à Mai) observées parfois dans les
régions subdésertiques, les pluies se produisent au passage de la mousson, pendant
une durée fonction de la latitude. En régime subdésertique, la saison des pluies
dure essentiellement du début de la mi-Juillet à la fin Août, début Septembre.

En régime sahélien, vers la limite Sud, elle dure de début Juin à début Octobre.
Il y a parfois quelques averses en Mai. La partie de la saison des pluies où les
averses se reproduisent assez fréquemment dure du 10 Juillet au 15-20 Septembre.

Le graphique n° 2 présente les variations des hauteurs mensuelles moyennes à
quatre stations caractéristiques. On notera que, pour les mois de saison sèche,
les moyennes représentent fort mal les faits ; il suffit d'une année sur cinq pour
laquelle le mois de Juin a été humide pour donner lieu à une moyenne non négligé-
able, alors que pour les quatre autres années, le même mois a été rigoureusement
sec.

On peut dire qu'en régime subdésertique, la saison sèche dure au moins dix mois,
et qu'en régime sahélien, elle dure à peu près neuf mois, un peu moins vers le
Sud. Cette saison sèche ne comporte, en général absolument aucune averse.

2.4. Irrégularité Interannuelle Des Précipitations

Les précipitations annuelles varient beaucoup d'une année à l'autre. On les
représente commodément par des diagrammes, tels que le graphique n° 3, avec une
échelle gaussique pour leurs fréquences et une échelle logarithmique pour la
hauteur correspondante. Dans le cas présent, au lieu des fréquences, on a reporté
leurs inverses: la période de retour pour les années sèches ou humides.

Le diagramme 3 a été établi en utilisant à la fois l'ensemble des résultats des
stations présentant la même hauteur de précipitations annuelles. L'étude statistique directe des relevés de stations prises une par une n'aurait pas permis
d'atteindre avec suffisamment de sûreté les valeurs des précipitations défici-
taires qui se reproduisent tous les 100 ans par exemple.

Le réseau de courbes représentées à la figure 3 correspond à des conditions
moyennes, et certaines stations correspondant à une situation particulière peuvent
avoir une courbe de distribution statistique différente de celle de la courbe du

diagramme 3 correspondant à la même valeur médiane de la hauteur de précipitations.

Pour les calculs sur ordinateur, le réseau de courbes du diagramme 3 a été mis en équation :

Soit F la fréquence au dépassement, PM la valeur médiane de la précipitation annuelle en mm :

$$F = \exp \left[- \left(\frac{x - x_o}{s} \right) \frac{1}{d} \right] \quad .$$

$\frac{1}{d} = 2,5 \quad x_o = 7,6 \times 10^{-4} \quad PM^2 - 52,4 \quad s = (PM - x_o)\, 1,1579$

PM : valeur médiane de la précipitation annuelle.

Ces formules sont valables pour PM > 500 mm.

Pour PM < 500 mm, les paramètres peuvent être déterminés par les équations suivantes :

$300 \text{ mm} \leq PM \leq 500 \text{ mm} \quad d = 0,4 \quad x_o = 7,6 \times 10^{-4} PM^2 - 10,7\, PM \times 10^{-2} + 1,1 \quad s = (PM - x_o)\, 1,1579$

$50 \text{ mm} \leq PM \leq 300 \text{ mm} \quad d = 0,4 + 0,009 \left(\frac{300 - PM}{100} \right) 3 \quad x_o = 7,6 \times 10^{-4} PM^2 - 10,7 \times 10^{-2} + 1,1$
$$s = (PM - x_o)/(0,69315)^d$$

Les courbes plongent vers la droite, d'où des valeurs très faibles pour les précipitations exceptionnellement sèches.

Pour une station en régime subdésertique qui reçoit 250 mm/an (année médiane), la hauteur de précipitation décennale sèche P 0,9 est de 125 mm, et la hauteur centennale sèche P 0,99 de 54 mm.

Pendant la sécheresse récente, presque tous les postes pluviométriques du SAHEL ont présenté au moins une valeur et souvent plusieurs dont les fréquences étaient comprises entre les fréquences décennale et centennale sèches, certaines régions étant particulièrement éprouvées, d'autres régions ont été relativement épargnées, avec un déficit désastreux, mais beaucoup moins sévère.

Les valeurs exceptionnellement humides sont élevées.

Les rapports entre P 0,1 et P 0,9 ou P 0,01 et P 0,99 donnent une idée de l'irrégularité interannuelle.

Pour l'isohyète 250 mm $\frac{P\ 0,1}{P\ 0,9} = 3,08 \quad \frac{P\ 0,01}{P\ 0,99} = 10$

Pour l'isohyète 500 mm $\frac{P\ 0,1}{P\ 0,9} = 2,32 \quad \frac{P\ 0,01}{P\ 0,99} = 4,55$

2.5. Etudes Des Séquences Des Précipitations Annuelles

En plus de la répartition statistique des précipitations annuelles, il est utile d'examiner quelle est la succession des années, en particulier d'analyser les séries d'années sèches ou humides. Un premier problème est à résoudre : le déficit pluviométrique observé une année a-t-il une influence sur la plus ou moins

grande abondance des précipitations l'année suivante ? C'est ce qu'on appelle le
problème de la persistance.

Après les travaux de Y. BRUNET-MORET, on peut répondre par l'affirmative.

L'équation générale donnant la hauteur de précipitation d'une année i + 1 par
rapport à l'année précédente i est la suivante :

$$x_{i + 1} = z_{i + 1} + Ax_i$$

z est une variable aléatoire et A le coefficient de persistance, on a trouvé A
voisin de 0,2 pour les régions sahéliennes d'Afrique.

On reconstitue ainsi des séries de 300 ans. On a représenté sur le graphique 4 la
série naturelle à la station de TOMBOUCTOU de 1897 à 1974 et une partie d'une
série de 300 ans reconstituée. Pour une hauteur de précipitation annuelle voisine
de 500 ans en moyenne, on retrouve bien les successions d'années sèches et humides
sur la période reconstituée. On en trouve même de plus sévères que celle de
1971-1973, ce qui est bien conforme à ce qu'enseignent à la fois les études
géomorphologiques et les traditions orales.

2.6. Précipitations Journalières

Leur étude présente un grand intérêt car le régime hydrologique d'un cours d'eau
en zone sahélienne résulte de série de crues de courte durée résultant d'averses
journalières et la seule façon de calculer les débits de ces cours d'eau à partir
des hauteurs de précipitations pour ce régime consiste à partir de ces averses.
Elles doivent être également bien connues pour calculer les crues.

On utilise souvent dans les projets d'aménagement l'averse décennale qui se
reproduit en moyenne tous les dix ans en un point. En fait, on peut observer à
une station en trois ans deux averses dépassant l'averse décennale et attendre
quinze ans sans qu'une seule averse atteigne la valeur décennale.

Au SAHEL, sa valeur P 10 est liée en première approximation à la hauteur de
précipitations annuelles. Le tableau ci-après représente les hauteurs de précipitations décennales de 24 heures en fonction des hauteurs de précipitations
annuelles.

TABLEAU I

Précipit. annuelle	P10 en 24 heures	Précipit. annuelle	P10 en 24 heures	Précipit. annuelle	P10 en 24 heures
100 mm	50 mm	400 mm	90 mm	700 mm	115 mm
200 mm	70 mm	500 mm	95 mm	750 mm	117 mm
300 mm	85 mm	600 mm	100 mm		

Bien entendu, les valeurs de ce tableau sont des moyennes, et des conditions
particulières peuvent conduire à des écarts notables par rapport à ces valeurs.

Il semble que l'altitude tende à réduire, par exemple, les hauteurs des précipitations décennales.

3.   REGIME DES COURS D'EAU AYANT LEUR ORIGINE DANS LE SAHEL

3.1.   Généralités

Les fortes averses orageuses de la saison des pluies donnent lieu à des phénomènes d'écoulement presque partout dans le SAHEL, tout au moins au Sud de la ligne isohyète annuelle 300 mm. Plus au Nord, l'ensablement des bassins est beaucoup plus fréquent, les pluies moins nombreuses et, dans de vastes régions, l'écoulement - assez rare - n'est pas organisé.

Ce qui caractérise la grande majorité des régions sahéliennes es ce qu'on appelle la dégradation hydrographique avec parfois tendance à l'endoréisme. Qu'est-ce que la dégradation hydrographique? Dans la plupart des cours d'eau normaux, le débit croît de l'amont vers l'aval sur le collecteur principal. Il n'en est pas du tout de même dans le régime sahélien : En tête des bassins, alors que la pente est relativement forte et si le sol n'est pas très perméable, on observe de fines rigoles qui donnent lieu à un ruisseau, à lit assez bien marqué, puis les débits se concentrent et, généralement, un petit cours d'eau se constitue avec parfois un lit sableux relativement large, la surface du bassin versant est alors de quelques km2. Très rapidement, des bras se détachent du cours d'eau principal, et celuici se perd assez vite dans une mare à fond argileux.

Mais le cours d'eau en question peut avoir la chance de déboucher dans une dépression collectrice. Celle-ci a souvent une pente générale faible. Chaque affluent remplit une mare à l'aval ou à l'amont par rapport à la pente générale du collecteur, qui peut être barrée ainsi par un certain nombre de seuils formés par les alluvions des divers affluents. Toutes les mares se remplissent, puis déversent les unes dans les autres, et il se produit alors un écoulement généralisé de l'amont vers l'aval, qui peut n'avoir lieu qu'en année exceptionnellement humide. Le cas le plus typique est celui de BELI à la frontière du MALI et de la HAUTE-VOLTA.

Si la pente est un peu plus forte, les affluents présentent l'allure d'oueds de l'AFRIQUE du NORD et la dépression principale couverte de végétation à l'état naturel est drainée, pas toujours très bien d'ailleur par une série de chenaux. Le tout se termine par une zone plus ou moins deltaïque, mais à ce moment le bassin versant atteint 2000 à 4000 km2.

En montagne, si la pente du collecteur est forte et continue, l'écoulement est torrentiel et il n'y a pas de dégradation. C'est par exemple le cas du Kori TELOUA au Nord d'AGADES, en région subdésertique.

Une série de hasards heureux peut conduire à l'existence de cours d'eau sahéliens importants. On citera deux exemples : le BA THA (Tchad) et le BAHR AZOUM (Soudan et Tchad).

La notion de bassin versant a peu de signification. En particulier, le débit spécifique , débit rapporté au km2, a peu de sens puis-qu'en quelques km le débit d'un cours d'eau peut passer de 50 m3/s à 0, et qu'une forte crue sur un "bassin" de 5 000 km2 résulte généralement d'un épisode pluvieux sur un bassin de 60 km2 situé immédiatement à l'amont de la station où on l'observe, le reste du bassin ne présentant aucun écoulement. Dans ce qui suit, on sera donc obligé de considérer plusieurs catégories de superficies de bassin.

Il a été précisé plus haut que tout le SAHEL ne présentait pas de phénomènes d'écoulement ; il existe, en effet, deux types de sol qui ne donnent pas lieu, en général, à un écoulement organisé : les régions sablonneuses et les régions argileuses très plates. Dans le premier cas, l'écoulement sur les pentes des dunes est presque nul, l'excédent des précipitations peut s'accumuler parfois au fond des creux de dunes ; ces mares s'assèchent généralement au début d'Octobre, mais, dans des cas tout à fait exceptionnels où le massif sablonneux repose sur un sous-sol imperméable à faible profondeur, il y a écoulement permanent, sauf en année très sèche.

En terrain imperméable et mal draîné, l'eau provenant des précipitations s'accumule sous forme de petites mares et s'évapore. On rappelle que, sous faible profondeur, l'évaporation annuelle dépasse largement 2 m par an au SAHEL. Les années très humides il peut y avoir un drainage assez efficace donnant lieu à des débits non négligeables s'il y a une dépression collectrice pas trop loin.

Ces phénomènes de dégradation hydrographique ont trois causes :

a)   une longue saison sèche au cours de laquelle la végétation herbacée disparaît
     et le sol nu est l'objet d'érosion intense;
b)   la durée de l'écoulement est faible et les cours d'eau n'ont pas le temps
     d'entretenir un lit continu dès que la pente est faible.
c)   on trouve dans ces régions, par suite de l'existence de périodes de très
     forte hydraulicité, il y a quelques milliers d'années, d'immenses étendues à
     très faibles pentes : cuvette tchadienne, marais du SUDD, etc.... où les
     crues s'étendent en nappes minces absorbées par l'évaporation.

On retrouve la dégradation hydrographique non seulement dans le SAHEL tropical africain, mais en bordure de nombreux déserts dans le monde.

3.2.   Bassins De Quelques Hectares A 1 km2

C'est la superficie des bassins versants susceptibles d'alimenter les citernes ou ce   ns lacs collinaires. Il n'y a pas de dégradation hydrographique et pratiquement pas de perte dans le lit du cours d'eau, le rapport entre le volume de pluie tombé sur le bassin et le volume annuel écoulé, ou coefficient
       coulement, est élevé. Mais, même sur un sol imperméable, les petites   ies de 1 à 2mm ne donnent pas lieu à écoulement, de sorte qu'en année m   le, le coefficient d'écoulement est compris entre 40 et 60 % ; dans ce cas la valeur la plus élevée ne correspond pas nécessairement aux zones les plus arides : l'absence de couverture vegétale favorise le ruisselement.

Pour des s   ssez perméables avec pente assez forte, le coefficient d'écoulement est nettement plus faible : 15 à 20 %.

Pour des perméables à pente plus faible et couverture végétale non négligeable,
     st compris entre 4 et 8 %.

En année traditionnellement humide, le coefficient d'écoulement sur ces petites surfaces atteint des valeurs beaucoup plus élevées, mais si l'on met à part les surfaces artificiellement imperméabilisées, et même pour des bassins de quelques hectares, il semble que dans les conditions les plus favorables, le coefficient d'écoulement annuel ne puisse pas dépasser 75 %. En terrain légèrement perméable, on peut considérer en zone désertique que la courbe de distribution statistique de l'écoulement annuel serait définie à partir des trois coefficients d'écoulement suivants : année médiane 40 % année centennale sèche 10%, année centennale humide 60 - 70 %.

En admettant que l'averse décennale comporte une pointe d'intensité de 140 mm/
heure pendant 5 minutes, on en déduit que le débit de crue décennale varie de à 35
m3/s pour un bassin de 1 km2 peu perméable ou imperméable. Pour un bassin perméable, il varie de 3 à 9 m3/s.

Tous ces c   es diminuent si la pente est modérée et encore plus si elle est
faible.

Pour ces très petits bassins, on n'a pas encore établi de gles bien définies
p   le calcul des volumes de crues.

3.3.   Bassins de 2 A 40 km2

3.3.1. Généralités

Sur les bassins perméables, ou à faible pente, il y a déjà dégradation, mais sur
les bassins imperméables, son rôle est encore négligeable. On subdivise souvent
cette catégorie en deux groupes : les bassins de 5 km2 et ceux de 25 km2. C'est
le genre de bassin qui a été le plus étudié. Ils sont souvent utilisés pour
l'alimentation de petits réservoirs et la traversée des cours d'eau correspondants
en crue a posé de sérieux problèmes aux constructeurs de routes. On a vu que,
malgré des hauteurs de précipitations annuelles modérées, les hauteurs des averses
journalières décennales sont voisines de 100 mm et présentent de fortes inten-
sités, la couverture végétale est peu dense et ne freine pas l'écoulement. C'est
dans ces régions qu'il est le plus facile d'observer le véritable ruissellement
superficiel, la majeure partie du bassin étant couverte d'une nappe d'eau.

Un assez grand nombre de bassins représentatifs a permis de connaître assez bien
le régime hydrologique qui varie très largement suivant les caractères physiques
du bassin. Les bassins versants de cette importance peuvent rarement être con-
sidérés comme étant homogènes. Si on veut les classer en fonction de leur
aptitude à l'écoulement, on doit considérer que chaque bassin comporte une
association type d'un certain nombre de sols de perméabilité généralement très
différente, comme on le montre ci-après.

3.3.2. Divers types de bassins versants

La liste qui suit, relative à la zone sahélienne, n'est pas limitative, certains
types de sols de bassins ne sont pas représentés, mais la plupart des cas courants
ont pu être analysés.

3.3.2.1. Sols sablex. Dunes mortes généralement, ou revêtement éolien, vers les
limites du bassin, on trouve des affleurements rocheux ou des lambeaux de
carapaces latéritiques et, dans le dépressions, des dépôts souvent argileux,
l'ensemble est très perméable. En général, l'écoulement correspond à moins de 1 %
des précipitations en année médiane.

3.3.2.2. Bassins sur granites ou granito-gneiss :   C'est le cas le plus difficile. On rencontre de l'amont à l'aval :

>   - des affleurements rocheux : perméables s'ils sont
>   décomposés en boules, moins perméables si le rocher est
>   en dalles ou en dômes, mais l'écoulement arrive dans les
>   zones suivantes où, souvent, il se perd ;
>   - des arênes granitiques très perméables ;
>   - des sols sableux plus ou moins profonds ; s'ils sont peu
>   profonds, le sol est vite saturé et ils peuvent avoir

une bonne aptitude au ruissellement. S'ils contiennent plus de 10 % d'argile, ce sont des sols battants et le résultat est le même ;
- des glacis ou pédiments à pente régulière : sols sablo-argileux compacts imperméables. C'est l'importance relative de ces deux catégories de sols qui définit l'aptitude à l'écoulement du bassin ;
- le lit des cours d'eau du bassin avec ses alluvions sableuses ;
- plus à l'aval, si la pente est faible, on trouve des sols hydromorphes, vertisols à larges fentes de retrait en saison sèche. Dès qu'ils ont reçu suffisamment de pluie, les fentes se referment et le sol devient imperméable. Mais la pente étant faible, le drainage est généralement mauvais;
- enfin, on peut trouver sur les parties amont du bassin, une couverture relativement homogène et épaisse de gravillons latéritiques, assez perméable.

Il est rare qu'un bassin de 25 km2 présente à la fois toutes ces catégories de sols. Mais l'écoulement sera d'autant plus important que la proportion de sol imperméable sera plus grande.

Pour en faciliter l'étude, on a défini en Afrique Occidentale trois bassins-types en zone sahélienne, auxquels on a donné le nom des "bassins représentatifs" les plus caractéristiques pour chaque type :

- le bassin-type ABOU GOULEM : sols perméables dans l'ensemble, avec pente notable, 12 à 14 m/km, pas de sol sablo-argileux compact, pas de sol hydromorphe, pas de sol sableux peu profond. Ces bassins correspondent à la limite de ce que l'on peut utiliser pour créer des réservoirs;
- le bassin-type BARLO : beaucoup moins perméable, a une pente plus forte, 20 à 30 m/km, avec au moins 25 % de sol sablo-argileux compact et de sol sableux peu profond et une proportion notable de massifs rocheux donnant lieu à un ruissellement ;
- le bassin-type CAGARA-Ouest : assez peu perméable, pente moyenne 4 m/km, constitué en majeure partie par des sols argilo-sableux, la masse d'alluvions sableuses dans le lit est négligeable.

Il existe des bassins un peu plus imperméables, constitués de regs homogènes dans le Nord, ou de sols plus argileux que CAGARA-Ouest, avec des pentes notables, dans certaines parties du Sud.

3.3.2.3. Bassins sur grès   (ADER DOUTCHI exclu) : ils comportent les formations suivantes :

- les grès : comme les granites, leur aptitude au ruissellement peut être très différente. S'ils sont très fracturés et sous forme d'éboulis plus ou moins envahis par les sables, ils sont très perméables. S'ils sont en couche subhorizontale, en bon état, et plus ou moins dénudés, ils ruissellent bien ;
- les carapaces latéritiques, si elles sont très démantelées, sont perméables, mais en bon état, elles donnent lieu à un écoulement superficiel important ;

- les sols sableux sur grès (voir sols sableux sur
  granites) ;
- les alluvions de fond de vallée sablo-argileuses ;
- les sols ferrugineux tropicaux lessivés.

Si la proportion de grès peu fissurés est importante, ces bassins ont un assez fort écoulement.

Le bassin-type est celui de KOUMBAKA (Mali) sur les grès dogons.

3.3.2.4. Bassins sur schistes.   Comme pour les granites, on retrouve de l'amont à l'aval :

- le schiste en place,
- les éboulis qui ruissellent assez mal,
- les glacis argileux plus imperméables en général que sur
  sous-sol granitique,
- les vallées avec les sols hydromorphes.

Les bassins les plus connus sont ceux de l'OUED GHORFA (Sud-Ouest de la Mauritanie). Ils présentent un fort ruissellement en raison de la proportion importante de glacis argileux.

Le bassin-type est celui de KADIEL.

3.3.2.5. Bassins de l'ADER DOUTCHI et de la MAGGIA.   On les trouve sur le continental terminal au NIGER. Ils comportent :

- des plateaux horizontaux de grès argileux démantelés qui
  ruissellent peu ;
- sur les pentes, des colluvions souvent peu perméables et
  des sols marnocalcaires imperméables ;
- des sols brun-rouge sur matériaux issus de grès, sols
  cultivés relativement perméables ;
- dans le lit des cours d'eau, des sables perméables ;
- plus à l'aval, des sols hydromorphes peu développés si
  le bassin ne couvre que 25 km2.

Vers le Nord, les sols des pentes plus argileux se présentent comme des regs.

L'ensemble donne lieu à un fort ruissellement, surtout si la superficie occupée par les grès est faible.

Les bassins-types sont ceux de :

- KOUNTKOUZOUT, qui ne comporte que 15 % de grès, mais
  près de 40 % de sols brun-rouge, ou de zones d'éboulis
  assez perméables,
- et de GALMI qui, avec 80 % à 90 % de colluvions argileuses et de sols marno-calcaires, semble correspondre
  aux conditions optimales pour le ruissellement.

3.3.2.6. Bassins des zones subdésertiques.   En régions subdésertiques, pour lesquelles il a été plus difficile de constituer une classification en règle, comme pour les régions sahéliennes, on a défini 3 catégories de bassins :

1° - bassins à très bonne aptitude au ruissellement
       (catégorie I), massif montagneux à très forte pente,

pratiquement pas de zone d'éboulis ni de recouvrement
éolien ;

2° - bassins à bonne aptitude au ruissellement : quelques
zones d'éboulis, une partie du bassin est recouverte
de sable (catégorie II) ;

3° - bassins à aptitude au ruissellement médiocre :
ensablement assez important, dégradation hydrographique. Bassins sur granite assez dégradé mais
avec une proportion de regs non négligeable
(catégorie III)

La plus ou moins grande dégradation du réseau hydrographique et l'extension des
zones endoréiques permet de classer, non sans difficultés, les bassins dans l'une
des trois catégories.

3.3.3. Ecoulement annuel en année médiane

En zone subdésertique, on peut caractériser l'écoulement annuel par les valeurs du
coefficient d'écoulement K_e en année médiane, et la lame d'eau écoulée annuelle E
en mm sous l'isohyète 200 mm.

Le volume disponible se déduit de E par la relation :

$$V \text{ en m3} = E \text{ mm} \times S \text{ km2} \times 10^3$$

S est la surface du bassin versant.

On trouve pour :

- Catégorie   I   Ke = 28,5 %   E = 57 mm

- Catégorie   II   Ke = 13,5 %   E = 27 mm

- Catégorie III   Ke = 5 %   E = 10 mm

On peut résumer, en zone sahélienne, l'aptitude à l'écoulement également par Ke et
E en mm pour l'isohyète 500 mm, pour les divers groupes de bassins et les bassins-
types suivants et pour l'année médiane :

- bassins sableux   Ke ≠ 0,45 %   E = 2,2 mm

- bassins sur granite ou
        granito-gneiss
    type ABOU-GOULEM   Ke = 3,5 %   E = 17,5 mm
    type BARLO   Ke = 7 %   E = 35   mm
    type CAGARA-Ouest   Ke = 14 %   E = 70   mm

- bassins sur grès ·
  (ADER DOUTCHI exclu)
    grès en bon état   Ke = 14 %   E = 70   mm

- bassins sur schistes
  (KADIEL)   Ke = 19,5 %   E = 98   mm

- bassins de l'ADER-DOUTCHI
  et de la MAGGIA
    type KOUNTKOUZOUT   Ke = 14 %   E = 70   mm
    type GALMI   Ke = 35 %   E =175   mm

Entre les deux cas extrêmes, l'écoulement annuel passe de 1 à 80. Si on ne considère que les bassins sur granite, il est possible que le rapport des valeurs extrêmes soit de l'ordre de 20.

Ces données concernent des bassins de 25 km2. Pour 5 km2, on trouverait un peu plus, surtout dans le cas des bassins perméables : dans les mêmes conditions, le coefficient d'écoulement varierait entre 3 et 40 %. Par exemple, ce coefficient pour un bassin type ABOU-GOULEM serait de 7 %, soit le double de la valeur admise pour un bassin de 25 km2.

En zone sahélienne, un coefficient d'écoulement médian de 5 % correspond à des possibilités d'aménagement assez peu intéressantes pour un bassin de 25 km2. Ceci élimine beaucoup plus de bassins qu'on ne pourrait le croire. Les deux premiers types de bassins présentés plus haut correspondent en effet à des cas beaucoup plus fréquents que le cas des bassins type GALMI. Encore convient-il de tenir compte des variations de ce volume annuel suivant les années, et, en particulier, lors des périodes sèches.

3.3.4. Distribution statistique de l'écoulement annuel

Le nombre d'années d'observation sur chacun des bassins représentatifs est insuffisant pour permettre une étude statistique des lames d'eau annuelles. Aussi, on a utilisé pour certains bassins-types, un modèle mathématique simplifié, mis au point par G. GIRARD, sur ces bassins, pour transformer chaque averse dans la crue correspondante. On peut donc, à partir de 50 ans d'observations de hauteur de précipitations journalières à une station du réseau météorologique, obtenir pour chacune des années, la série de crues correspondantes (un bon nombre d'averses ne donnant d'ailleurs pas lieu à écoulement), et, par suite, le volume annuel et la lame d'eau écoulée annuelle. On obtient donc une série de lames écoulées pour 50 ans. On a procédé par comparaison des conditions d'écoulement pour les bassins pour lesquels il n'y a pas eu d'étude sur modèle.

Dans ces conditions, on a établi pour la zone sahélienne, le graphique 5 qui donne la distribution des lames d'eau écoulées en fonction des périodes de retour (inverse des fréquences cumulées). Chaque bassin – type est représenté par deux courbes extrêmes : l'une correspond à la distribution des lames d'eau écoulées annuelles, pour le cas où le bassin-type reçoit en année médiane 300 mm, l'autre à la distribution des lames d'eau écoulées pour le cas où le bassin-type reçoit 750 mm. La courbe 500 mm vient se placer dans l'intervalle des courbes.

Une véritable étude d'exploitation doit, non seulement considérer les valeurs extrêmes, mais aussi une suite chronologique de lames d'eau écoulées annuelles ; on utilise alors la relation citée au point 2.5.

Pour la zone subdésertique, on a mis au point, en utilisant des procédés beaucoup plus sommaires, le graphique 6, sur lequel n'ont été représentés que les diagrammes correspondant aux catégories extrêmes I et III évoquées au paragraphe précédent.

En zone subdésertique, l'écoulement annuel est nul beaucoup plus souvent qu'en zone sahélienne.

3.3.5. Crues exceptionnelles

Pour l'étude des petits ouvrages, on considère, en général, la crue décennale ou centennale.

Il est important de connaître son volume et le débit maximal.

Les crues décennales ont fait l'objet d'études approfondies et l'ORSTOM a donné des règles de calcul en fonction de la surface, de la pente et de la perméabilité globale du bassin. Cette dernière a été définie par 5 catégories de P_1 à P_5, P_1 représentant un bassin rigoureusement imperméable, P_5 un bassin très perméable. Pour montrer les variations du débit des crues décennales suivant les catégories de bassins, on a repris les bassins présentés au paragraphe 3.3.3 et on a calculé : le volume de la crue décennale en milliers de m3, le débit de crue en m3/s et le débit spécifique rapporté à 1 km2 en l/s.km2, en supposant que les bassins subdésertiques sont situés sous l'isohyète 200 mm (précipitation décennale journalière ponctuelle 70 mm) et que les bassins sahéliens sont situés sous l'isohyète 500 mm (précipitation décennale journalière ponctuelle 95 mm). Les valeurs ainsi obtenues ont été reportées sur le Tableau II ci-après.

Chaque bassin correspond à des conditions géomorphologiques assez bien définies, en particulier, en ce qui concerne la pente et la perméabilité globale. La différence de comportement entre les différents bassins est très grande puisque les deux valeurs extrêmes du débit spécifique sont : 100 l/s.km2 et 12 000 l/s.km2. La perméabilité globale semble être, dans bien des cas, le facteur essentiel. On notera également que la valeur de la hauteur de précipitation annuelle n'a qu'une faible influence sur le débit de crue décennale. On pourrait placer la catégorie III du régime subdésertique entre les bassins types ABOU-GOULEM et CAGARA-Ouest. La catégorie I comprend des bassins de montagnes où les conditions de ruissellement sont plus favorables que celles des bassins type GALMI.

Pour les bassins sahéliens, on trouve très fréquemment pour les débits spécifiques de crues décennales des valeurs comprises entre 1500 et 3000 l/s.km2.

La conclusion pratique est la suivante : un bassin qui ruisselle bien, tel GALMI ou KADIEL, permet de garantir le remplissage d'un réservoir de volume appréciable, tous les ans, mais, malheureusement, il est nécessaire d'évacuer des crues de fort débit, ce qui nécessite de réaliser des déversoirs coûteux.

En zone sahélienne, le diagramme annuel des débits se présente comme une série de pointes de crues : les premières généralement au début de Juillet, les dernières en Septembre ou en Octobre, plus rarement en Mai. Après chaque crue, l'écoulement cesse. En zone subdésertique, se produisent quelques crues pendant 2 mois du 15 Juillet au 20 Septembre, sans écoulement entre les crues, sauf en année exceptionnellement humide où les rivières peuvent couler pendant 1 mois de façon continue.

3.3.6. Transports solides et qualité des eaux

Dans la partie du SAHEL au Sud du SAHARA, s'ils ne présentent pas l'importance qu'ils ont en Afrique du Nord, ils ne sont pas négligeables, surtout en zone subdésertique : on a mesuré au Sud d'AGADES des concentrations en sédiments de 30 g/l en crue. En zone sahélienne, elles sont un peu moins fortes à l'échelle de 25 km2.

Sur le bassin de GALMI I, l'érosion spécifique annuelle est de 18 tonnes/an/ha. La hauteur de précipitation annuelle médiane est de 500 mm.

Les transports solides croissent avec la pente, ils décroissent avec la densité de la couverture végétale, ils sont très sensibles au type de culture.

Au Sud du SAHARA, les eaux sont extrêmement douces tout au moins en zone sahélienne, c'est là une compensation à toutes les difficultés d'aménagement que nous venons de présenter.

Lorsque la superficie diminue, jusqu'à quelques km2, la pente est en général plus forte ; il n'y a plus du tout de dégradation hydrographique et l'écoulement annuel est plus élevé. L'augmentation des lames d'eau écoulées, très sensible pour les bassins perméables, l'est nettement moins pour les bassins imperméables.

Pour les bassins types ABOU-GOULEM ou BARLO, lorsque la superficie passe de 25 à 5 km2, toutes choses restant égales par ailleurs, l'écoulement annuel double en année médiane, il triple ou quadruple en année centennale sèche. Pour les bassins très perméables qui n'ont pas d'écoulement en année centennale sèche pour 25 km2, il y a écoulement pour des bassins de 5 km2 dans les mêmes conditions. Les crues exceptionnelles sont également majorées.

TABLEAU II
Crues Décennales Pour Des Bassins De 25 Km2
En Zones Subdésertique et Sahélienne

	V en 10^3 m3	Q_M en m3/s	qm en 1/s.km2
Zone subdésertique Catégorie I P = 200 mm	1 050	240	9 500
Zone subdésertique Catégorie II P = 200 mm	700	120	4 700
Zone subdésertique Catégorie III P = 200 mm	500	50	2 000
Bassin sableux P = 500 mm	95	2,5	100
Bassin type ABOU-GOULEM P = 500 mm	500	55	2 200
Bassin type BARLO P = 500 mm	650	115	4 600
Bassin type CAGARA-Ouest P = 500 mm	1 500	80	3 200
Bassin sur grès P = 500 mm	1 800	160	6 500
Bassin type KADIEL P = 500 mm	750	75	3 000
Bassin type KOUNTKOUZOUT	660	145	5 800
Bassin type GALMI	2 100	300	12 000

3.4.   Bassins Versants De 40 À 500 Km2

3.4.1.   Généralités

C'est le genre de bassin dont l'hydrologie est la plus difficile à étudier : la pente générale du bassin et celle du lit diminuent : en plus, la dégradation s'accentue, surtout dans les régions subdésertiques où le réseau hydrographique actuel s'enchevêtre plus ou moins avec un réseau fossile, relique de la dernière période humide, et tout ceci conduit à un régime hydrologique fort complexe avec une variété extrême de comportement des bassins dans l'espace et dans le temps, sauf lorsque l'ensemble de la région étudiée est peu perméable.

Ce sont les bassins sur sous-sol de granite, fort nombreux dans cette partie de l'Afrique, qui présentent les caractères les plus capricieux. L'exemple de l'Ouadi ENNE au Nord du OUADDAI (Tchad) est particulièrement typique. A BILTINE, son bassin versant couvre une superficie de 527 km2. Il prend naissance dans le massif du OUADDAI, où il est nourri par des tributaires dont certains ont des crues assez violentes, mais il se dégrade assez vite vers la zone de piémont à l'Ouest du OUADDAI, son lit se termine par une zone deltaïque, un bras se dégage sur sa rive droite et reçoit l'Ouadi AMBAR peu dégradé qui, au confluent, draine un bassin de 83 km2. Vers ce confluent, et jusqu'à la station de BILTINE, des zones de regs argileux assez imperméables ruissellent facilement, de sorte qu'en ce qui concerne l'écoulement à BILTINE, la situation est la suivante :

- En année très sèche, seules les zones de regs présentent un écoulement et, même en 1972, une pluie tombant sur le reg a donné lieu à la seule crue de l'année : l'écoulement total n'a peut-etre pas dépassé 20 000 m3, correspondant à une lame d'eau écoulée dérisoire : 0,05 mm.
- En année médiane, la hauteur de précipitation est de 330 mm : à peu près tout l'ouadi AMBAR concourt à l'écoulement, mais l'ouadi ENNE au confluent n'apporte rien. Répartie sur tout le bassin, 527 km2, la lame d'eau écoulée est de 6,6 mm.
- En année centennale humide (1961), l'ouadi ENNE rejoint AMBAR et y apporte un volume d'eau important. Pour une hauteur de précipitation voisine de 700 mm, la lame d'eau écoulée est de 40 mm, V = 21 000 000 m3.

Sur le graphique 7, on verra le diagramme de distribution qui montre le caractère très irrégulier de cet ouadi, lequel constitue un cas assez général.

En terrain imperméable, et avec des précipitations dépassant 500 mm, la dégradation est moins forte et le comportement du bassin est nettement plus homogène de l'amont à l'aval. Il en est de même dans les massifs montagneux où les lits gardent une forte pente.

On comprendra que, dans de telles conditions, on ne puisse pas fournir des données hydrologiques aussi précises que dans le cas précédent : chaque ouadi est souvent un cas d'espèce, il faut bien étudier la géomorphologie du bassin et du lit avant de le rattacher à un cas connu et surtout examiner de très près les tributaires situés le plus à l'aval.

3.4.2.   Ecoulement annuel sur des bassins de 40 à 500 km2

On distingue le cas des régions subdésertiques des régions sahéliennes.

3.4.2.1.   Bassins versants de 40 à 500 km2 en zone subdésertique.   La dégradation
hydrographique se produit dès l'arrivée en plaine. Seuls donnent lieu à écoule-

ment les bassins de montagne ou les bassins sur sols assez peu perméables avec
pente notable.

En montagne, en particulier dans l'AIR, et à condition que la pente du lit principal reste notable, en année voisine de la médiane, le coefficient de ruissellement est compris entre 10 et 25 %, correspondant à des valeurs de la lame écoulée
comprises entre 20 et 50 mm sous l'isohyète 200 mm. On peut retrouver les mêmes
chiffres pour des surfaces allant jusqu'à 1 000 km s'il n'y a pas dégradation
hydrographique. Même en montagne, l'écoulement décroît rapidement dès que l'année
devient déficitaire. En année décennale sèche, par exemple, le coefficient
d'écoulement total pourrait être compris entre 3 % et 10 %, ce qui correspond à 3
mm et 10 mm. Mais on doit noter qu'en 1971, 1972 et 1973, la plupart des cours
d'eau de l'AIR ont présenté un certain écoulement à chaque saison des pluies.
Cependant, en année cinquantennale ou centennale, pour de nombreux bassins de
montagne, l'écoulement annuel est nul.

Il existe des bassins à pente moins forte avec des sols assez imperméables, mais
également des zones de recouvrement sableux ou d'éboulis rocheux. Si le lit
principal garde une certaine pente, le coefficient d'écoulement reste compris
entre 3 % et 10 % en année médiane, la seconde valeur correspondant à l'isohyète
300 mm. En annéedécennale sèche l'écoulement est souvent nul, mais dans les cas
favorables, il atteint 5 %, soit 5 mm sous l'isohyète 200 mm. En année décennale
humide, on peut s'attendre à des coefficients d'écoulement de 15 à 20 %, soit 48 à
66 mm. Seuls les bassins les plus propices au ruissellement de cette catégorie
peuvent donner lieu à des aménagements dans des conditions relativement con-
fortables, mais on doit retenir qu'entre les années décennales sèches et humides,
l'écoulement varie de 1 à 50.

Ces deux catégories de bassins correspondent à une minorité. Il y en a beaucoup
dont les conditions d'écoulement sont beaucoup plus mauvaises.

3.4.2.2. <u>Bassins versants de 40 à 500 km2 en zone sahélienne</u>. Les bassins sur
sous-sol granite ou granito-gneiss couvrent au total de grandes surfaces et se
présentent sous différents aspects.

Ils peuvent être situés dans des massifs peu élevés à pente pas très forte et avec
sols assez perméables ; c'est le cas de l'Ouadi ENNE sous l'isohyète 300 mm à la
limite Nord de la zone sahélienne. On peut citer l'exemple du bassin versant de
TAYA (GUERA-Tchad) : avec une pente assez forte et un sol peu perméable, le
coefficient d'écoulement pour l'année médiane est un peu inférieur à 1 %, correspondant à un écoulement de 5 mm (isohyète 750 mm), mais ce bassin ne présente
pas à l'aval de regs argileux imperméables comme l'Ouadi ENNE.

La courbe de distribution des écoulements annuels est très raide, c'est pratiquement une droite (distribution gausso-logarithmique). En année exceptionnellement sèche, l'écoulement est nul, en année centennale humide, au contraire,
au bout de 1 mois ou 1 mois 1/2, les sols sont saturés et ils ruissellent assez
bien, compte tenu de la pente, d'où des écoulements très supérieurs à ceux de
l'année médiane.

Si un bassin du même type, en plus du sol perméable, a une faible pente, la courbe
est beaucoup moins raide et l'écoulement médian plus faible. On en a une idée par
la courbe du tributaire du Lac de BAM, qui garde encore un écoulement pas trop
faible en année médiane et non nul en année sèche grâce à quelques zones un peu
argileuses à l'aval.

Enfin, les granites ou granito-gneiss sont parfois recouverts de sols vertiques
argileux comme en HAUTE-VOLTA la FELLEOL. La pente reste toujours faible. Le

coefficient d'écoulement est alors plus élevé : 4 à 6 % en année médiane, soit une lame écoulée de 20 à 30 mm sous l'isohyète 500 mm.

En année décennale sèche, le coefficient d'écoulement ne décroît pas beaucoup grâce à la nature du sol, il reste compris entre 3 et 5%, la courbe de distribution est peu inclinée.

Les bassins sur schistes de l'Oued GHORFA conservent un ruissellement notable quand ils ne comportent pas de recouvrements éoliens ni de zone d'éboulis perméables. L'Oued DJAJIBINE, qui présente à peu près les conditions optimales de ruissellement, pour cette région, a, en année médiane, un coefficient de ruissellement de 19 %, soit une lame écoulée de 86 mm, mais à côté, l'Oued ECHKATA ne présente dans les mêmes conditions qu'une lame d'eau de 38 mm. En année sèche, l'écoulement à DJAJIBINE est encore très acceptable : 22 mm pour l'année cinquantennale. En année centennale humide, l'écoulement est de 365 mm, ce qui correspond comme on le verra à des crues très violentes.

Les bassins types ADER DOUTCHI - MAGGIA au NIGER, avec leurs pentes argileuses et les sols marno-calcaires ruissellent assez bien, cependant les grès argileux des plateaux assez démantelés absorbent beaucoup d'eau.

Le bassin de TAMBAS, en année médiane, présente un coefficient d'écoulement de 10,5 %, ce qui correspond à une lame écoulée de 49 mm ; en année décennale sèche, l'écoulement annuel tombe à 18 mm, en année centennale humide, il atteint 95 mm. Cette région, qui pour les bassins de 25 km2 présentait les records d'écoulement passe loin derrière la région de l'Oued GHORFA qui, elle, est dépourvue de plateaux de grès, lorsqu'il s'agit de bassin dépassant 100 km2.

Ces exemples ne couvrent pas tous les cas que l'on peut rencontrer en zone sahélienne. Il convient de bien étudier la morphologie de la région et surtout des collecteurs principaux et des tributaires aval avant de chercher à se rattacher à un exemple. Il faut essayer de reconstituer l'écoulement de chaque sous-bassin de 25 km2 et voir ce que peut devenir l'écoulement une fois arrivé dans la dépression principale, car, généralement, il s'y perd.

3.4.3. Crues exceptionnelles

Une des données les plus importantes à connaître est la crue décennale ou centennale. La crue de fréquence décennale n'est pas trop difficile à estimer et son débit maximal ou le volume qu'elle produit peut être utile pour comparer des bassins de divers types. On peut même passer de la crue décennale à la crue centennale par un coefficient, mais en zone sahélienne, ce coefficient varie très largement d'un bassin à un autre, surtout lorsqu'une proportion importante de la surface est perméable et que la pente générale est faible.

Pour des bassins de 100 à 500 km2, on ne dispose pas d'études générales comme pour les bassins plus petits, et, d'ailleurs, ce qui a été exposé au paragraphe 3.4.1 montre que toute généralisation systématique des résultats est pratiquement impossible. Nous ne pourrons donner que quelques indications pouvant servir de jalons, en précisant que le cas de bassins sur granite avec lit principal à faible pente et sous 200 à 500 mm, et le cas de bassins avec recouvrements éoliens partiels sont les plus difficiles.

Dans le tableau III, on donnera quelques exemples :

> - dans la première colonne, on a porté le nom du bassin et
>   de très brèves indications sur sa nature ;

  - dans la seconde colonne, la hauteur annuelle des pré-
    cipitations (valeur médiane),
  - dans la troisième colonne, la surface
  - dans la quatrième colonne, le volume de la crue décen-
    nale,
  - dans la cinquième colonne, le débit de pointe en m3/s,
  - et dans la sixième, le débit spécifique en l/s.km2.

Le tableau III ne présente que quelques exemples, et il est très délicat dans ces régions de tenter de donner des résultats généraux, car chaque cours d'eau est un cas d'espèce. Cependant, un certain nombre de constatations s'imposent :

  1°. On retrouve, en région subdésertique, les catégories
      I, II et III, l'Ouadi SOFAYA correspondant à peu près
      à la catégorie III, le Kori EL MEKI à la catégorie I.
      C'est un vrai cours d'eau de montagne, il ne présente
      pas de dégradation hydrographique. On trouve un
      rapport de 1 à 30 entre la crue décennale du cours
      d'eau de catégorie III et celui de la catégorie I.
      Un bon nombre de cours d'eau présentent des débits
      spécifiques de crue beaucoup plus faibles que l'Ouadi
      SOFAYA, ils sont généralement sans intérêt pour les
      aménagements.
  2°. Sur les bassins sahéliens, on rencontre d'abord la
      grande famille des bassins sur granite ou granito-
      gneiss. Là aussi les débits spécifiques sont très
      variables : dans l'ordre des débits croissants, le BA
      ADA présente une hauteur de précipitation annuelle
      voisine de 600 mm, c'est un bassin à la fois plat et
      perméable, aucune surface imperméable ne se trouve
      vers l'aval. La crue décennale est très faible. Le
      TAYA, dans le massif du GUERA au Tchad, est favorisé
      par la pente, mais une partie du bassin est perméable
      de sorte que le débit de crue décennale est très
      inférieur à celui des bassins de la catégorie I du
      régime subdésertique, mais il existe plus au Sud, au
      Nord-CAMEROUN, des bassins à forte pente très cul-
      tivés qui contiennent non seulement des arênes mais
      des sols argileux, et on retrouve alors des débits
      spécifiques dépassant 1 000 l/s.km2.
  3°. On trouve ensuite trois bassins voisins et tribu-
      taires de l'Oued GHORFA dans le Sud-Est de la
      MAURITANIE, caractérisés par des glacis argileux sur
      schistes. Le maximum de ruissellement correspond à
      l'Oued DJAJIBINE avec 2 200 l/s.km2, le minimum par
      l'Oued ECHKATA qui présente des éboulis, des zones
      perméables et un lit principal à faible pente encombré
      par le végétation.
  4°. Le Kori TAMBAS, dans l'ADER DOUTCHI, a un bassin
      couvert en grande partie par des pentes argileuses à
      fort ruissellement que l'on a déjà rencontrées sur le
      bassin de GALMI.

Les crues de fréquences plus faibles, centennales par exemple, sont très mal connues, elles présentent des débits très supérieurs à ceux des crues décennales lorsque le bassin ruisselle mal, plus particulièrement vers les zones subdéser-tiques.

TABLEAU III
Crues Décennales Pour Des Bassins De 100 À 500 Km2
En Zones Subdésertique Et Sahélienne

Bassins	P méd. mm	S km2	V 103 m3	Q10 m3/s	q10 l/s.km2
Ouadi SOFAYA (Tchad) pente modérée	120–130	345	750[1]	12	35
Kori EL MEKI (Niger) pente forte sous-sol cristallin	140–160	165	3 500	200	1 200
Oued DIONABA (Mauritanie) pente assez forte	280	111	3 400	60	540
Ouadi ENNE (Tchad) pente modérée sous-sol cristallin reg à l'aval	330–350	527	2 000– 2 500	70–80	130–150
TAYA (Tchad) forte pente sous-sol cristallin	850	167	3 800	100	600
FELLEOL (Haute-Volta) assez faible pente sous-sol cristallin couvert d'argiles vertiques	550	400	5 000	(20)	(36)
BA ADA (Haute-Volta) pente faible sous-sol cristallin sol perméable	610	500	(1 200)	(10)	(16)
Oued DJAJIBINE (Mauritanie) pente notable sous-sol schistes sol imperméable	450–475	148	4 800	325	2 200
Oued BOUDAME à ECHKATA (Mauritanie) pente assez faible sous-sol schistes sol perméable à l'amont	440–460	149	2 200	35	230
Oued BOUDAME à BOUDAMA pente modérée sous-sol schistes perméabilité variable	450–475	564	12 000	66	120
TAMBAS (Niger) pente modérée grès et argilo-calcaire	460	284	(6 000)	400	870

(1) Nota : 1 500 000 m3 si on considère le volume décennal, avec débit maximal 7 m3/s.

La concentration en sédiments, pendant les crues, varie encore plus d'un bassin à l'autre que dans le cas de bassin de 25 km2.

3.5. Bassins Versants De Plus De 1 000 Km2

3.5.1. Généralités

En régime subdésertique, de tels bassins sont rares, sauf en montagne. L'écoulement est souvent arrêté par une zone de dunes. En régime sahélien, de tels bassins sont beaucoup plus fréquents, surtout vers le Sud, mais si la pente du collecteur principal est faible et s'il est situé en terrain perméable, même sans recouvrement éolien, les apports du cours d'eau sont dérisoires. C'est le cas par exemple des cours d'eau du Sud du OUADDAI ou de la zone au Nord du Lac de BAM (Haute-Volta).

Presque toujours le régime hydrologique est commandé par le tributaire le plus proche de la station où on étudie le cours d'eau.

Le seul massif montagneux bien développé avec vallées principales à pente assez forte est le DJEBEL MARRA, dont les cours d'eau ne doivent subir que peu de dégradation hydrographique.

On retrouvera les mêmes types de bassins que plus haut.

3.5.2. Ecoulement annuel sur des bassins couvrant plus de 1 000 km2

3.5.2.1. Bassin versant de plus de 1000 km2 en zone subdésertique.
On peut se reporter à ce qui a été dit au paragraphe 3.4.2.1. Il y a tout d'abord les cours d'eau de montagne dont le Kori TELOUA dans l'AIR, dont on peut esquisser la courbe de distribution des écoulements annuels à partir des 3 points suivants :

- année médiane, lame écoulée 25 mm   Ke = 15 %
- année décennale sèche, lame écoulée
  8 mm   Ke = 8 %
- année décennale humide, lame écoulée
  50 mm   Ke = 19 %

Quelques cours d'eau, en MAURITANIE, présentent probablement des conditions aussi favorables.

Mais, dès l'arrivée en plaine, il y a dégradation hydrographique. Le Kori TELOUA peut perdre 75 % de ses apports en moins de 30 km.

En sol assez peu perméable, avec pente notable, lorsqu'une partie importante à l'aval du bassin est active, le coefficient d'écoulement peut être compris entre 2 à 5 % en année médiane. Mais si la pente est faible et le sol assez perméable, le coefficient d'écoulement descend en dessous de 1 %.

3.5.2.2. Bassins versants de plus de 1 000 km2 en zone sahélienne.
Sur sous-sol granitique et sur granito-gneiss, les bassins à très faible pente et au sol perméable ne donnent pas lieu à écoulement en année médiane.

Ce cas mis à part, on peut schématiser l'écoulement annuel par trois cas-types :

- le tributaire du Lac de BAM en Haute-Volta, avec sol
  assez perméable et pente générale assez faible ;

- le BAM BAM, cours d'eau assez torrentiel, avec une pente assez forte mais une forte proportion de sols perméables;
- le GOROUOL à DOLBEL (Haute-Volta) a un bassin de pente modérée comportant une proportion assez forte de sols assez argileux recouvrant le granite.

On trouvera les trois courbes de distribution des écoulements annuels sur le graphique 8.

En année médiane, les lames d'eau et volumes écoulés sont les suivants :

- Tributaires du Lac de BAM : S = 2 600 km2 E =   6 mm V =   15 000 000 m3

- BAM BAM   : S = 1 200 km2 E =   24 mm V =   29 000 000 m3

- GOROUOL   : S = 7 500 km2 E = 34,5 mm V = 260 000 000 m3

On constate sur le graphique 8 que les deux bassins perméables présentent de très faibles écoulements en année sèche. On manque d'exemples de bassins granitiques à forte pente qui présenteraient des coefficients d'écoulement variant de 10 à 20 %.

Les bassins argileux sur schistes ou sur les formations du continental intercalaire donnent lieu à des écoulements très acceptables tant que le cours d'eau principal présente une pente notable.

Le graphique 8 présente trois exemples de bassins de ce type :

- L'Oued BOUDAME à OULED ADDET (Mauritanie) avec un bassin versant de 1 125 km2.
- le Kori de BADEGUICHERI (Niger) à l'issue d'un bassin de 825 km2.
- la MAGGIA au pont de TSERNAOUA (Niger) : bassin versant de 2 525 km2.

L'Oued BOUDAME correspond à des conditions de ruissellement favorables, tout au moins dans la partie aval de son bassin. Les lames d'eau et volumes écoulés sont les suivants en années médianes :

- Oued BOUDAME   E = 65 mm   V = 73 000 000 m3
- Kori de BADEGUICHERI   E = 31 mm   V = 25 600 000 m3
- MAGGIA   E = 15 mm   V = 38 000 000 m3

De façon générale, ces bassins ruissellent mieux que les bassins sur granite, sauf lorsque ceux-ci sont recouverts de produits d'altération argileux (cas du GOROUOL).

En année sèche, les écoulements sont encore acceptables pour l'Oued BOUDAME et le Kori de BADEGUICHERI (20 mm en année décennale pour l'un, 10 mm pour l'autre).

La MAGGIA présente une lame écoulée beaucoup plus faible, les plateaux de grès démantelés qui recouvrent une partie des bassins ne donnent lieu à aucun écoulement et les pertes dans la vallée du collecteur principal sont importantes.

Notons qu'à l'aval des stations qui ont servi à ces études, le pente diminue, il y a une dégradation importante surtout pour les deux derniers bassins.

De façon générale, d'ailleurs, sauf pour l'Ouadi FERA, les chiffres que nous donnons ont été déterminés à l'amont de la partie du cours d'eau pour laquelle la dégradation hydrographique prend de très importantes proportions.

3.5.3. Crues exceptionnelles

Pour faciliter les comparaisons, on considèrera les crues de fréquence décennale comme pour les bassins de 100 à 500 km2. Comme dans ce cas, on ne peut donner que quelques indications. Sur le tableau IV sont reproduites les données des crues décennales pour les bassins déjà cités plus haut ; les colonnes correspondent aux mêmes caractéristiques que dans le tableau III.

Si on compare les chiffres du tableau IV à ceux du tableau III, les débits spécifiques pour des bassins de même catégorie : Kori EL MEKI et TELOUA, TAYA et BAM BAM, FELLEOL et GOROUOL, TAMBAS et BADEGUICHERI, on constate une diminution très forte, elle serait encore plus importante si les stations du tableau IV correspondaient à des cours d'eau déjà assez dégradés ! C'est-à-dire si on les avait implantées un peu plus à l'aval.

Il n'y a pas de rapport entre le débit spécifique de crue et la hauteur de précipitation annuelle. L'imperméabilité du sol et surtout la pente du collecteur principal jouent un très grand rôle . La valeur la plus forte correspond au Kori TELOUA qui est le moins arrosé mais qui est le seul vrai cours d'eau de montagne. On trouvera des chiffres nettement plus élevés au Sud de la limite méridionale du SAHEL dans les montagnes du Nord-CAMEROUN.

Le tableau IV montre l'extrême variété de types de crues décennales d'un bassin à l'autre.

Il y a souvent un rapport entre la lame d'eau annuelle écoulée et le débit spécifique de crue, mais il faut tenir compte de l'origine de la crue. Si elle est provoquée uniquement, ou à peu près, par une pointe de débit d'un affluent situé à l'aval, la montée des eaux peut être très brutale, le maximum relativement élevé, mais le volume total plus faible qu'une crue affectant tout le bassin avec un maximum beaucoup plus faible.

Les crues de plus faible fréquence, centennale, par exemple, peuvent présenter des débits très supérieurs à ceux des crues décennales, surtout pour les bassins qui en temps ordinaire ruissellent mal. A la limite, en zone sahélienne, certaines dépressions peuvent rester 30 ans sans aucun écoulement et être submergées par une crue. Il est impossible, bien entendu de donner des règles pour passer de la crue décennale à la crue centennale en zone sahélienne. Même le calcul de la crue décennale est difficile pour des bassins de plus de 1 000 km2.

3.5.4. Bassins versants sahéliens de superficie supérieure à 10 000 km2

De l'Océan Atlantique au NIL il y en a très peu qui parviennent à draîner plus ou moins bien une superficie pareille sans dégradation totale : ce sont de l'Ouest à l'Est : le KORAKORO (Mauritanie), la KOLIMBINE (Mauritanie) tous deux affluents du SENEGAL, la VOLTA BLANCHE (Haute-Volta) dans son cours supérieur, le GOROUOL (Haute-Volta, Niger) et la SIRBA (Haute-Volta, Niger) affluent du Niger Moyen, le GOULBI de MARADI (Nigéria, Niger), la KOMADOUGOU YOBE (Nigéria, Niger) tributaire du Lac TCHAD, le BA THA (Tchad) qui se jette dans le Lac FITRI et le BAHR AZOUM (Soudan, Tchad) qui, en principe, est un affluent du CHARI. Ils proviennent de zones qui ruissellent bien ou de massifs montagneux, ou de la limite Sud de la zone sahélienne ou même de la zone tropicale. En général, lorsque ces cours d'eau coulent du Nord au Sud, l'influence de la partie septentrionale du bassin est presque nulle, ceux qui viennent du Sud, comme la KOMADOUGOU perdent une bonne partie de leurs apports en progressant vers le Nord.

La valeur médiane de la lame d'eau écoulée annuelle est en général deux à trois fois plus faible que dans le cas des bassins de 100 à 500 km2.

TABLEAU IV
Crues Décennales Pour Des Bassins De Plus De 1 000 Km2
En Zones Subdésertique Et Sahélienne

Bassins	P méd. mm	S km2	10^3 V m3	Q max 10 m3/s	q10 1/s.km2
Kori TELOUA (Niger) forte pente, sous-sol cristallin pas de dégradation hydrograque	170	1 170	15 000	450	380
Ouadi FERA (Tchad) (1) cristallin, faible pente, endoréïsme	400-450	5 600	1 300	120	21
BAM-BAM (Tchad) cristallin assez perméable, pente assez forte, dégradation	800-835	1 200	30 000	350	290
GOROUOL à DOLBEL (Hte.Volta) cristallin recouvert d'argile peu perméable, pente faible	520	7 500	200 000	117	15,5
Oued BOUDAME (Mauritanie) argile sur schistes pente modérée	450-475	1 125	40 000	200	180
BADEGUICHERI (Niger) grès et marnes calcaires, pentes transversales assez fortes	470	825	15 000	300	360
MAGGIA à TSERNAOUA (Niger) grès et marnes calcaires, pente transversales assez fortes	475-500	2 525		140	55

(1)ruissellement partiel à l'aval du bassin.

On distingue deux groupes :

1°. la VOLTA BLANCHE, le GOROUOL, la KOMADOUGOU dont
l'écoulement annuel est voisin de 5 mm (valeur
médiane), soit 0,7 à 1,2 % des précipitations. Ces
bassins à faible pente comportent une partie importante de leur superficie à très faible écoulement,
ou présentent des zones de pertes très importantes
comme la KOMADOUGOU ;

2°. le BA THA, le BAHR AZOUM et la SIRBA dont les bassins
ruissellent mieux : les premiers parce que la pente
générale du bassin est notable, le troisième parce
que le sol est argileux. La lame d'eau écoulée
annuelle médiane est voisine de 15 mm correspondant à
un coefficient d'écoulement compris entre 2 et 3 %.

Bien entendu, l'écoulement diminue quand la superficie augmente. Par exemple,
pour le BA THA à OUM HADJER (S = 32 950 km2) la lame d'eau écoulée E est égale à
14 mm, plus à l'aval, à ATI, pour 45 290 km2, E est égale à 9,7 mm.

Les courbes de variation de la lame d'eau écoulée E, en fonction de la fréquence, ont été tracées pour les deux groupes : en année cinquantennale sèche toutes les courbes tendent à converger vers E = 2 mm, (1,5 mm peut-être pour la SIRBA, 2,5 mm pour le BAHR AZOUM).

En année humide, l'écoulement croît rapidement pour les cours d'eau du second groupe, sauf pour le BAHR AZOUM. En année centennale humide E = 66 mm pour le BA THA à ATI , E = 100 mm pour la SIRBA (S = 38 750km2).

Pour les cours d'eau du premier groupe, les pertes deviennent très importantes en année humide pour la KOMADOUGOU et le GOROUOL, de sorte que la valeur de E en année centennale humide reste comprise entre 5 et 10 mm, la VOLTA BLANCHE fait exception car elle coule du Nord au Sud et ses affluents méridionaux ont un régime un peu comparable à celui de la SIRBA, en année centennale humide E est peut-être de l'ordre de 20 mm à la station de WAYEN (S = 20 000 km2).

Le régime saisonnier est différent de celui des cours d'eau précédents. On ne se trouve plus en présence de pointes de crues isolées, mais d'une période annuelle d'écoulement continu. Sur le BA THA ou le BAHR AZOUM, le lit reste à sec pendant 9 mois. Le flot de crue arrive au début du mois d'Août, en quelques heures le débit atteint 100 m3/s et l'écoulement cesse fin Octobre ou fin Novembre.

On peut avoir une idée de l'importance des crues par le tableau V suivant qui présente les débits de crues de fréquence décennale.

TABLEAU V
Crues Décennales Pour Des Bassins De Plus De 10 000 Km2
En Zone Sahélienne

Stations	Surface km2	Débit maximal de crue m3/s	Débit spécifique l/s.km2
KOMADOUGOU à BAGARA	115 000	90	0,8
KOMADOUGOU à GUESKEROU	120 000	39	0,3
GOROUOL à ALCONGUI	44 850	100	2,2
VOLTA BLANCHE à WAYEN	20 000	250	12,5
BAHR AZOUM à AM TIMAN	80 000	360	4,5
BA THA à OUM HADJER	32 950	550	17
BA THA à ATI	45 290	500	11
SIRBA à GARBEY KOUROU	38 750	500	13

4. CONCLUSION :

Ce qui précède montre qu'il existe bien une méthodologie pour l'estimation des ressources en eaux superficielles des cours d'eau nés dans les zones désertiques, subdésertiques et sahéliennes, mais pour que cette méthodologie soit efficace on recommande qu'un gros effort soit fait dans les domaines suivants :

- caractérisation, quantitative si possible, des paramètres physiogéographiques intervenant dans la genèse de
  l'écoulement, ces paramètres étant déterminés dans une
  optique hydrologique ;
- amélioration de nos connaissances sur l'action de ces
  paramètres sur le cycle hydrologique ;
- création et observation permanente dans d'excellentes
  conditions d'un petit nombre de stations de références
  pluviométriques, pluviographiques et hydrométriques.

Mais dans de très nombreux cas, l'extrême irrégularité interannuelle des apports conduit à de graves défaillances en cas de sécheresse de faible fréquence et chaque fois que cela est possible il y a intérêt à utiliser au maximum, quand il en existe, des cours d'eau venant de zones mieux arrosées et parvenant jusqu'au désert.

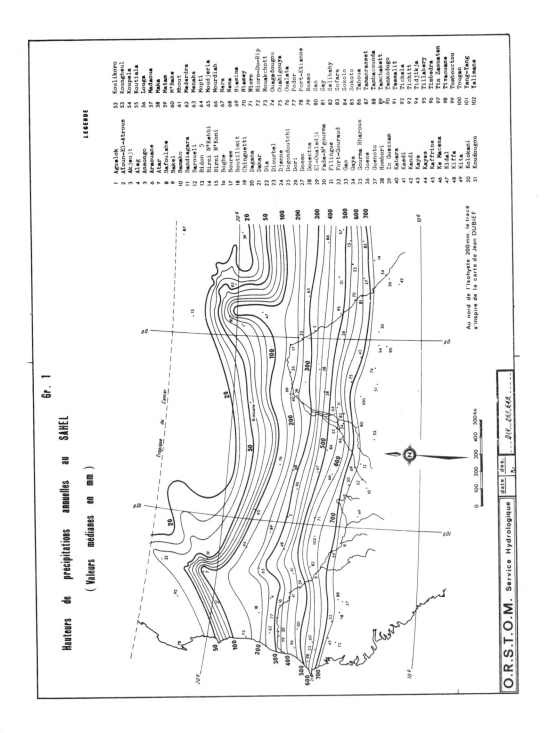

Gr. 1

Hauteurs de précipitations annuelles au SAHEL

(Valeurs médianes en mm)

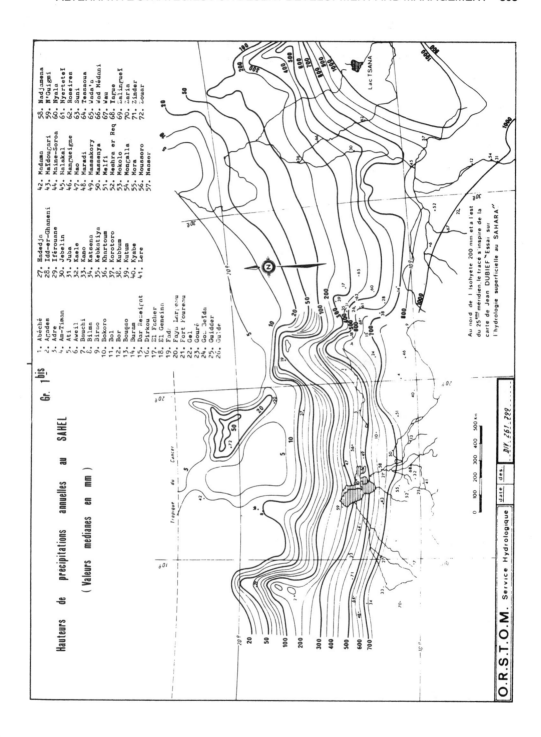

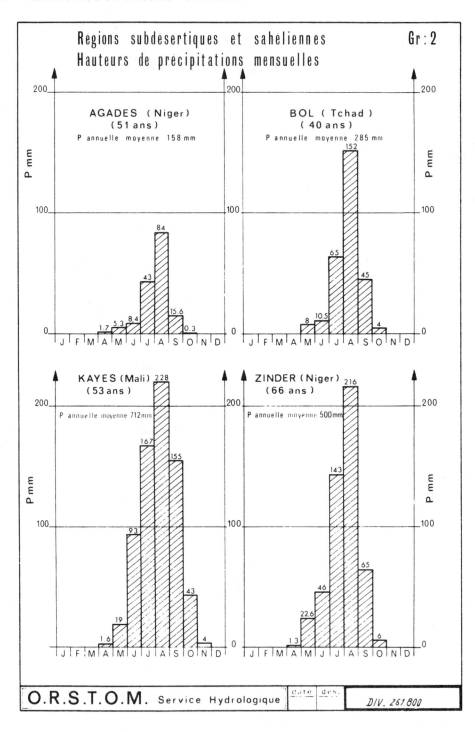

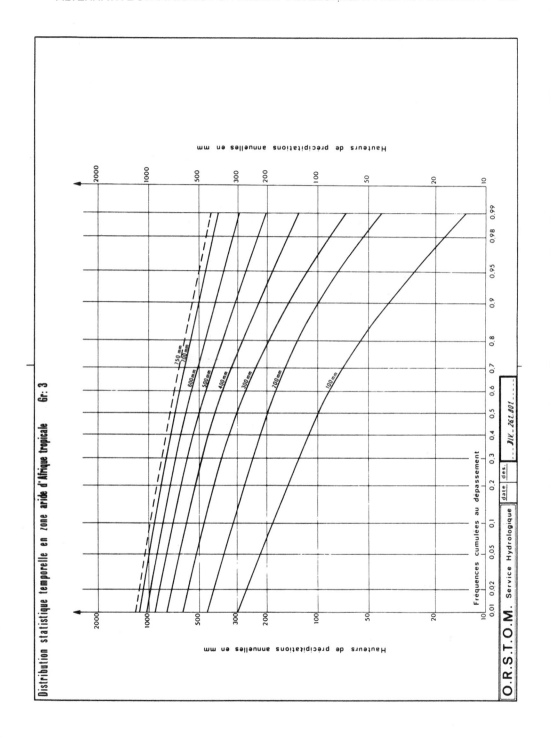

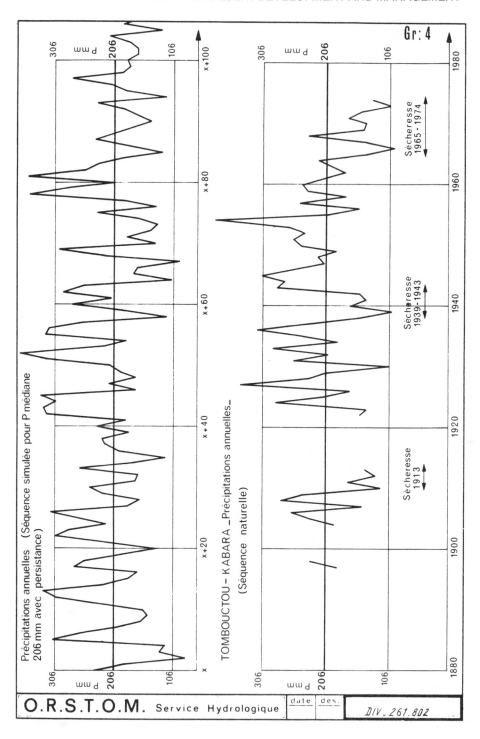

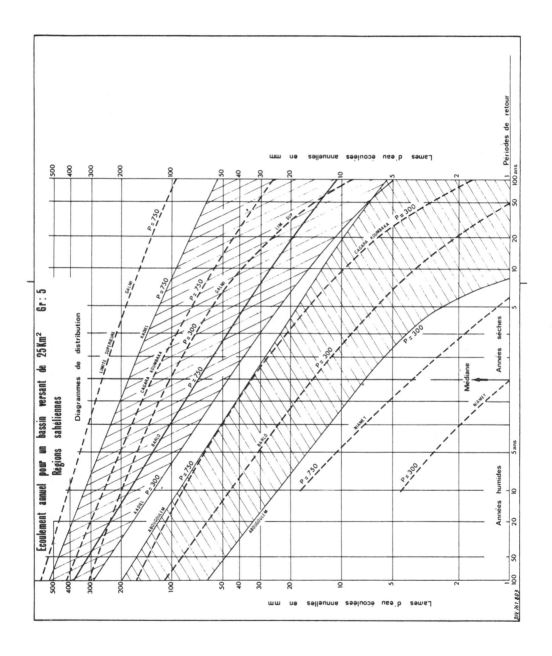

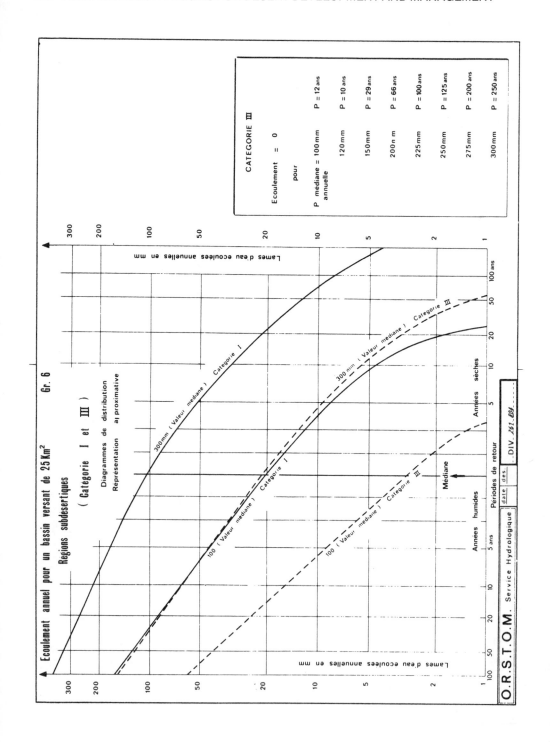

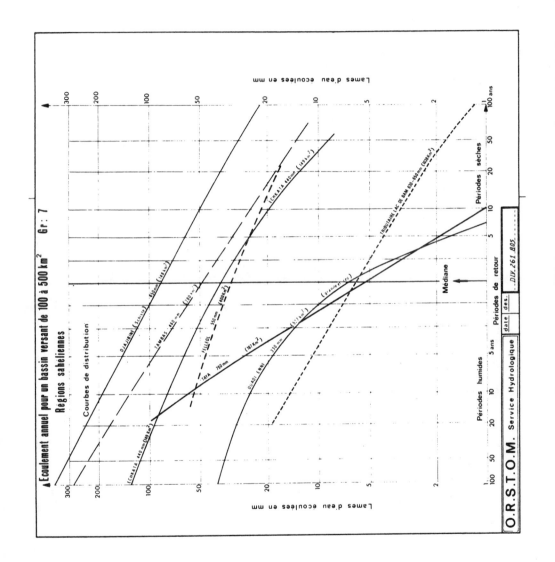

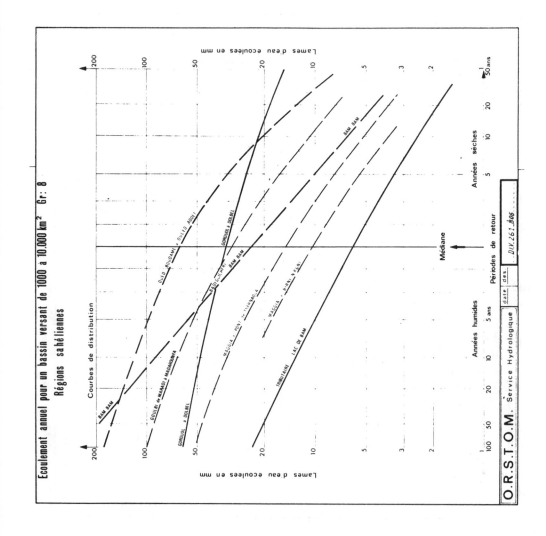

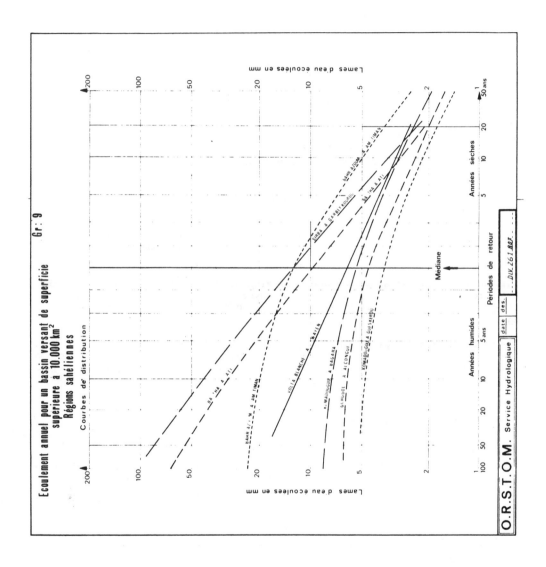

THE DEVELOPMENT OF THE EGYPTIAN WESTERN DESERT
FIFTY YEAR PLAN: 1975 - 2025

Mohamed Aly Ezzat
Undersecretary for Desert Irrigation
Ministry of Irrigation, Cairo, Egypt

ABSTRACT

The present paper deals with a suggested development program for the Western
Desert to the year 2025, in an attempt to solve the food and housing problems of
Egypt.

The Western Desert in Egypt, comprises an area of about 750,000 square kilometers,
which represents 75% of the whole area of Egypt. This area supports a population
of about 200,000 people, which represents only 0.5% of the total population of
Egypt.

The total present irrigated area using groundwater is about 32,200 feddans (13,644
hectares). The areas irrigated by rainfall along the coastal zone is about 20,000
feddans (8,474 hectares). The area irrigated by rainfall, supplemented by ground-
water is about 15,000 feddans (6,355 hectares).

With the present proposed program, it will be possible to increase the cultivated
areas in the following manner:

	Proposed Reclamation Program[1]				
Area	Rainfall	Rainfall + Ground water	Rainfall+ Winter Watering +Ground water	Ground water	Surface water
(feddans)	$\times 10^3$	$\times 10^3$	$\times 10^3$	$\times 10^3$	$\times 10^3$
Coastal Zone	20.0^2	15.0^2	4.750	--	42.740
Wadi El-Gedid	--	--	--	210	--
South-Kharga	--	--	--	100	--

[1]One crop per year

812

Thus total area with perennial crops will increase from 31,600 feddans (12,640 hectares), to 352,740 feddans (141,096 hectares). Of this area there is proven groundwater supply for an area of about 310,000 feddans (124,000 hectares). The rest, 42,740 feddans (17,096 hectares), will depend on surface water supply from the Nile.

With the proposed development program till year 2025, it will be possible to establish new towns and to accommodate about 11.45 millions persons in the Western Desert, with the following distribution:

Area	Population	
	Present	Future
Coastal Zone	80,000	6,000,000
Wadi El-Gedid	120,000	4,750,000
Qattara	-	700,000
TOTAL	200,000	11,450,000

After the construction of the proposed Qattara project 650 M^3/sec. water will flow from the sea to the depression. A reservoir of 50 million cubic meter capacity will produce a head of 262.5 giving a peak load of about 2400 million watts. This new power supply will help greatly in industrializing the Western Desert and in reducing the cost of development programs.

A network of asphalt desert roads is essential for the proposed development program. To accelerate communications in the Western Desert, it is proposed to establish airports in the Siwa, Bahariya, Farafra and South Kharga depressions.

I. INTRODUCTION

1.1 The Problem

In 1800, the population of Egypt was about 2 million, in 1900 about 5 million, and in 1947 jumped to 19 million. In 1975, according to the official statistics announced by the Government, the population on 15 March 1975 reached 37 million. The total area of Egypt, including the deserts is about 1 million km^2. Only 3% of this area, i.e, 30,000 km^2 is settled. The population density in 1975 was about 1267 person/km^2. For almost 500 years, no new towns have been built. The population now is increasing in the same towns and 4000 villages. For this reason, the population density was increasing at a severe rate, bringing with it many difficulties. The rate of population increase is almost about 2.5%. There are attempts to decrease this rate to 2%. If the rate of increase in population remains at 2% and taking the population density in 1975 as the datum (although 1300 persons/km^2 is still high) then it is clear that the area required for settling 63 million new Egyptians during the next fifty years should be increased by about 11 million feddans (46,000 km^2) to keep the population density of 1975 the same in 2025.

The present cultivated area is about 6 million feddans, i.e. the share per capita is 0.16 feddan. Keeping the same percentage of cultivated land, by 2025, the cultivated land should reach 16 million.

The only way to solve the problem of agriculture and settlement is by developing the land and water resources of the Egyptian Desert in the following regions:

    a. Western Desert
    b. Red Sea Coast
    c. Sinai and Suez Canal Zone

The present paper deals with a suggested development program for the Western Desert in an attempt to solve the food and settlement problems of Egypt.

2.   PRESENT SITUATION

2.1 Physiographic Provinces of the Western Desert

The Libyan Plateau is a large part of the Western Desert. It is a rolling, and in places, a very rugged upland surface with dry wadis. Dendritic drainage patterns are well established throughout the area. The regional slope of the plateau is eastward and its general altitude ranges from 200 to 400 meters above mean sea level. In part of the surface of the plateau, bedrock crops out or is covered by rock rubble, and part is covered by great swaths of south-southeastward trending sand dunes.

The Western Desert may be divided into three principal physiographic provinces - the Southern Plateau, the Central Plateau and the Northern Plateau (Figure 1).

2.1.1. The Southern Plateau   (Nubian Sandstone Series)

The Southern Plateau covers more than half of the Libyan Desert extending from the Nile in the east to the Tibesti and Ennedi highlands in the west, and from the highlands of the Kordofan in the south to the Eocene-Cretaceous escarpment in the north. This extensive plateau consists mainly of the Nubian series of Paleozoic and Mesozoic age and islands of the Precambrian crystalline rocks developed mostly in the south. The highest elevation is attained in the extreme southwest corner of Egypt, where the Gebel El-Oweinat crystalline massif attains a height of 1907 meters above sea level. From approximately the Gebe El-Oweinat highlands, this Nubian plain falls steadily northward and eastward towards the Nile Valley until it dips under the Eocene-Cretaceous escarpment.

2.1.2 The Central Plateau   (Cretaceous-Eocene Limestone)

This region covers an extensive area west of the Nile. It is a plateau consisting mainly of compact Eocene and Cretaceous limestone underlain by softer sediments of Mesozoic and Paleozoic age. In the south, it is well defined by the old es-carpment overlooking the sandstone lowlands that form the floor of Dakhla and Kharga; in the east, it is interrupted by the valley of the Nile where it forms towering cliffs. The northern boundary is the steep scarp on the north side of Siwa and the Qattara Depressions; on the east it is overlain by the Great Sand Sea.

The plateau is broad and attains a maximum height in the extreme south, where it rises to over 500 meters above sea level, from there it slopes gently northward. The area is one of vast uniformity and monotony, except for the large natural depressions of Bahariya and Farafra and some prominent siefs, such as the Abu Mobarik dune belt.

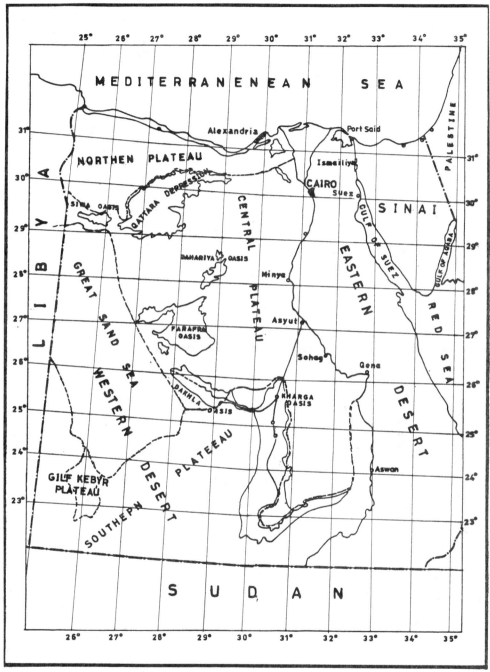

FIG 1 : PHYSIOGRAPHIC PROVINCES
WESTERN DESERT , EGYPT

The Great Sand Sea, one of the largest sand areas in the world, lies on the extreme western frontier between Egypt and Libya, and is usually considered a part of the Central Plateau. It stretches from Siwa in the north to the Gilf Kebir Plateau in the south, presenting virtually an impassable barrier for about 600 kilometers and separating Egypt from the Western part of the Libyan Desert. This sand covered area contains nearly 150,000 square kilometers. The rocks underlying the northern part of the Great Sand Sea consists of limestones of Miocene and Eocene age, while those in the southern part belong to the Nubian Sandstone series.

2.1.3   The Northern Plateau   (Miocene Limestones and Sandstones)

The Northern Plateau extends from the bold escarpment at Siwa and the Qattara Depression northward to the shore of the Mediterranean and from the Nile Delta Westward to Cyrenaica. This plateau slopes gradually to the north from the 200 escarpments overlooking Siwa to the flat-lying coastal zone at sea level. For the most part this landscape is fairly flat; occasionally it is interrupted by the isolated hills so characteristic of the Libyan Desert. The underlying rocks of the plateau are limestone and sandstones of Miocene age.

2.2   Population

The Western Desert in Egypt comprises an area of about 750,000 square kilometers, which represents 75% of the whole of Egypt. This area supports a population of about 200,000, which represents only 0.5% of the total population of Egypt. The population of the Western Desert is distributed as follows:

Coastal Zone	80,000
Siwa Oases	15,000
Bahariya Oases	10,000
Farafra Oases	2,000
Dakhla Oases	40,000
Kharga Oases	50,000

2.3   Mineral Resources

A large iron deposit exists in the northern part of the Bahariya oasis. The iron deposits are now under exploitation. It is transported by a railway line to the steel plant at Helwan near Cairo.

In the Abu-Tartour area, about 40 kilometers west of the Khargo oasis, a phosphate deposit is represented by three beds, separated by a shale layer, the lowest of which is 1.62 meters thick and contains 61.8% tricalcium phosphate. The quarry is located favourably near the main Kharga-Dakhla asphalt road. A preliminary estimation of the amount which is expected in this area was made and it was found to be in the range of three quarters of a million tons. At present, detailed studies are in progress to design the phosphate plant to be put in operation by 1980.

2.4   Agriculture

The total cultivated area in the Western Desert is about 67,000 feddans. Their geographical distribution is as follows:

2.4.1. Coastal Zone

Cultivated area rainfed and supplemented by groundwater is about 15,000 feddans.
Fruit trees occupy more than 13,000 feddans, of which 7,000 are olive trees and
over 5,000 fig trees. Vegetables, mostly tomatoes, onions, broad beans and water
melons, are cultivated for the most part in small plots and occupy in all some
1,500-2,000 feddans.

Of the field crops, barley is the most important and, in good rainfall years, may
occupy as many as 135,000 feddans of the 150,000 feddans put under crops. In poor
rainfall years, barley may occupy only 80,000 feddans, and the area harvested may
amount to only 20,000 feddans.

2.4.2 Siwa Oasis

The presently developed areas are chiefly located in the vicinity of Siwa and
Aghrumi, where about 900 feddans are maintained under permanent crops and about
200 feddans are farmed only in the winter. The outlying farm areas near Zeitun,
Maragi and Khamisa are believed to bring the total presently developed agricultural land to nearly 2000 feddans. The principal crops are dates with about
150,000 trees in production and olives which are cultivated on about 200 feddans.

2.4.3 Bahariya Oasis

There is considerable evidence of former cultivation on land that is not presently
being farmed, usually because of the lack of available water for irrigation. Some
of this land was abandoned centuries ago.

Bahariya contains about 85,000 date palm trees, compared to 150,000 date trees in
Siwa. Dates are the principal cash crop as well as an important local item of
food. There are more than 1400 olive trees in Bahariya, many of them interplanted
with the date palms. There are approximately 4,000 apricot trees, many of these
also being interplanted with the date palms. The total cultivated area is not
more than 5,000 feddans.

2.4.4 Farafra Oasis

Farafra contains about 1,7000 date palm trees. Dates are the most important crop,
since they produce most of the cash income of the trees. Interplanted with the
date palms are approximately 300 olive and 200 apricot trees. The olives and
apricots are used both locally and for cash income crops. There are some 350
citrus trees.

2.4.5 Dakhla-Kharga Oasis

The area under perennial irrigation is about 25,000 feddans (9,756 hectares)
distributed mainly between Kharga (10,000 feddans) and Dakhla (15,000 feddans).
Irrigation is by artesian wells.

The most promising crops experiments are:

> - Winter crops (six months November-February): onions,
>   carrots, broad beans, and other vege-
>   tables, medicinal crops, clover, wheat and
>   barley.

- Summer crops: rice, maize, sorghum, also dried fruits, raisins, figs, apples, apricots, oranges, etc.

- Round-the-year-crops: date plam, alfalfa.

Double cropping is a usual practice of perennial irrigation in the area.   The average rate of cropping is about 2.1 per year.

From the above discussions we can conclude that the area under irrigation in the Western Desert is 67,200 feddans.   About 20,000 feddans in the coastal zone depend on annual rain and the same area 15,000 feddans depend on rainfall supplemented by groundwater.   The rest, 32,200 feddans, depend entirely on groundwater.

2.5 Groundwater Use

2.5.1 Coastal Zone

Groundwater could be exploited by means of dug wells, equipped with windmills, drilled wells equipped with pumps, or by means of collecting galleries, equipped with pumps.

- Dug Wells   There are about 1040 dug wells equipped with windmills; at present it is estimated that about 30% of them are in operation.   The windmills generally have a capacity of 4-6 m^3/day.   Some of the wells are equipped with shadoofs and buckets in which case the capacity may be estimated at 1.5 m^3/day, total groundwater developed from dug wells, to be about 275,000 m^3 per year.
- Drilled Wells   At Fuka there are six drilled wells by the rotary method, equipped with turbine pumps.   Each well is capable of producing 200 m^3/day.   Annual groundwater withdrawn by drilled wells is about 360,000 m^3.
- Collecting Galleries   Groundwater stored in coastal sand dunes are exploited by collecting galleries.   The total annual withdrawal of water from collecting galleries near Nersa Matruh will be about 300,000 m^3.

Thus, total groundwater utilized in the area is about 935,000 m^3/year.

2.5.2 Siwa Oasis

In the Siwa Depression the source of the groundwater discharging from the Miocene bedrock appears to be from the deep underlying Nubian Sandstone reservoir.   Nubian water rises from a considerable artesian pressure to maintain substantial flows from various discharge points (ains) at the ground surface, and from many smaller seeps in the soil zone.

Parsons (1963) estimated total spring and well discharges in Siwa Oasis to be about 120,000 m^3/day.

2.5.3 Bahariya - Farafra Oases

Total discharge of all wells and springs in Bahariya and Farafra Oases is 120,000 m^3/day and 8,000 m^3/day respectively.

2.5.4 Dakhla-Kharga Area

Groundwater reaches the surface under artesian pressure. Total groundwater under exploitation is about 741,870 m^3/day in the Dakhla oasis and 215,710 m^3/day in the Kharga oasis.

2.5.5. Summary

The following table gives a summary of groundwater usefully utilized in the area.

Table 1

Groundwater Usefully Utilized

Area	Groundwater Exploited	
	m^3/day	m^3/year (millions)
Coastal Zone	935,000	341.275
Siwa	120,000	43.800
Bahariya	120,000	43.800
Farafra	8,000	2.920
Dakhla	526,160	192.048
Kharga	215,710	78.734
	1,924,870	702.577

2.6 Roads

There exist some asphalt roads 6 meters wide. (Figure 2) Other sand roads are present. Existing asphaltic roads are shown in Table 2.

Table 2

Desert Asphaltic Roads

Roads	Length, km.
Alexandria - Sallum (Sea Coast Road)	512
Cairo - Alexandria (Desert Road)	221
Matrouh - Siwa (Incomplete)	100
Cairo - Baweti (Bahariya)	300
Cairo - Fayoum	105
Fayoum - Beni Suef	45
Assiut - Kharga	226
Kharga - Beris	100
Kharga - Mut (Dakhla)	219
Mut - Farafra	60
TOTAL	1,888 km.

Table 3

Western Desert - Areas of Arable Land (Feddans)

Soil Classification

ZONE	I	1 & 2	2	2 & 3	3	3 & 4	4	4 & 5	5	TOTAL
Khanga	-			150,000	-	237,200	-	456,600	1,004,900	1,848,700
Dakhla	-			127,300	103,200	117,600	-	166,100	683,200	1,197,400
Farafra	-			200,000	-	675,000	-		2,000,000	2,875,000
Bahariya	-		11,350		24,150	-		2,150	-	37,650
Siwa	-		7,200		9,900	-	2,300	-	1,100	20,500
TOTAL			18,550	477,300	137,250	1,029,800	2,300	624,850	3,689,200	5,979,250
North Kharga		621,380		1,442,090			1,191,860		4,744,670	8,000,000
Coastal Zone		16,000	3,274	327,285		2,415		278,990	1,887,620	2,496,310
		637,380	21,824	2,246,675	137,250	1,032,215	1,194,160	903,840	10,321,490	16,475,560

3. LAND AND WATER RESOURCES

3.1 Land and Soils

As has been mentioned before, the present population of Egypt (1975) is about 37
million. The resulting population pressure on cultivated land (6 million feddans)
is 0.16 feddan per person, or 6.16 persons per feddan. In this situation, a main
concern of the Egyptian Government is to extend the area under cultivation and the
promotion of land reclamation, land development and resettlement has been its
policy for some years.

In accordance with this policy, the Government carried out an intensive program to
develop the land and soil resources in the Western Desert.

3.1.1. Coastal Zone

A reconnaissance soil survey was carried out which covered the area from Burg
El-Arab to Sidi Barrani. On the basis of this survey, five soil maps and five
potentiality maps on a scale of 1:100,000 were prepared. The total cultivable
area is about 2,496,310 feddans of different classes. Total soil area of classes
1, 2 and 3 is about 346,559 feddans (Table 3). At present only 67,200 feddans are
under cultivation, i.e., 19.4% of the soil classes 1, 2 and 3.

3.1.2. Wadi El-Gedid Area   (New Valley)

That part of the Western Desert which is designated as the Wadi El-Gedid area,
which means the New Valley, extends diagonally southeastward from the Siwa Oasis,
in the north, through Bahariya, Farafra, Dakhla and Kharga oases to a point in the
southern boundary of Egypt where the Nile crosses the Sudan boundary.

A reconnaissance soil survey of a scale of 1:1,000,000 was carried out over
6,250,000 feddans of which 3,689,200 feddans (Table 3) were classified from "good"
to "fair" soils to be subdivided into 2,100,000 feddans of which 200,000 feddans
were further surveyed on a scale of 1:10,000 to identify the project area to be
reclaimed and to give recommendations as to how to reclaim the soils.

The irrigable area under fair engineering and economic conditions is about 2
million feddans of good soils.

3.1.3. South Kharga Area

At a distance of 40 kms northwest of Nasser Lake lies the depression of South
Kharga. The area has been covered by aerial photographs on a scale of 1:40,000.
Thereafter controlled mosaic photographs on a scale of 1:50,000 were made for the
whole area. With the aid of these controlled photographs, a reconnaissance survey
on a scale of 1:50,000 was carried out over 8,000,000 feddans of which 2,063,470
feddans were classified from class 1 and 2 (very good) to class 2 and 3 (good)
soils. The rest, 5,936,530 feddans are of class 4 and 5 (fair).

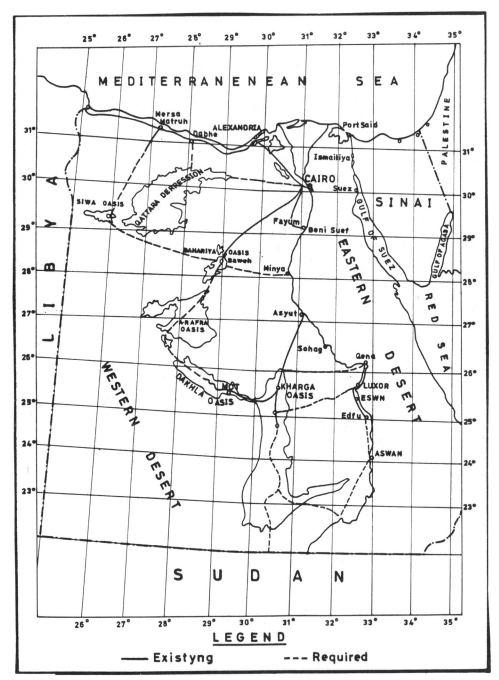

LEGEND

——— Existyng   - - - Required

FIG 2 : EXISTING AND REQUIRED ROADS
WESTERN DESERT EGYPT

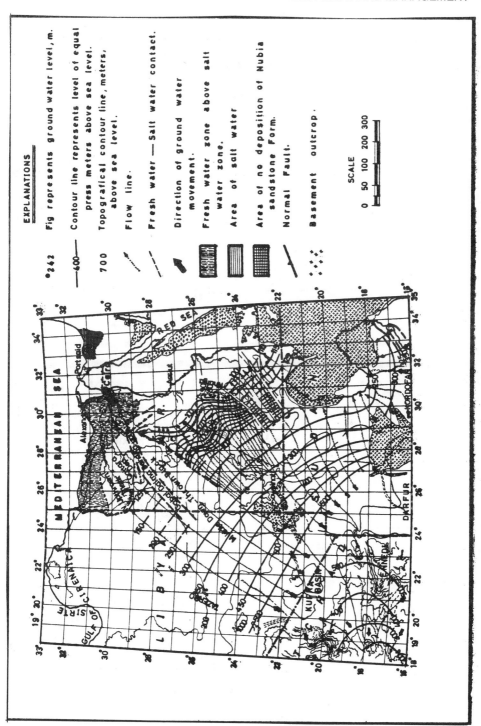

FIG. 4: REGIONAL GROUND WATER SETTING, NUBIA SANDSTONE FORMATION, LIBYAN DESERT.

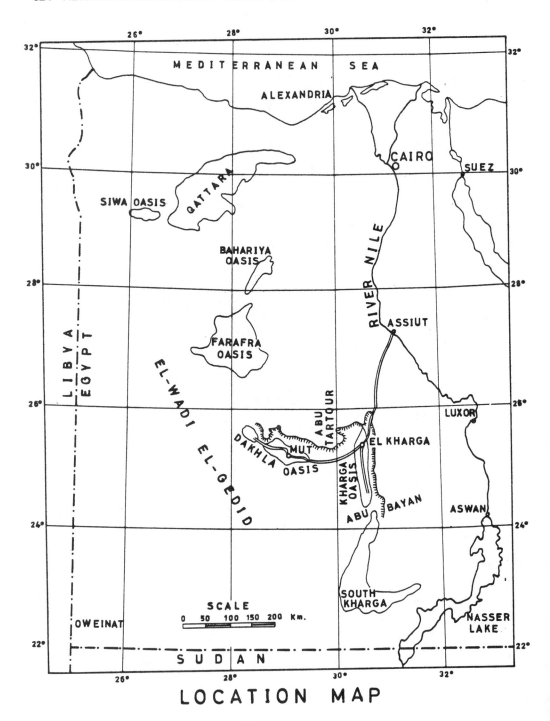

LOCATION MAP

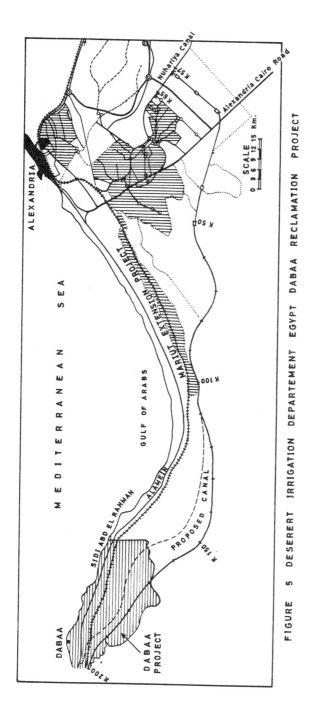

FIGURE 5   DESERERT IRRIGATION DEPARTEMENT EGYPT DABAA RECLAMATION PROJECT

COASTAL ZONE WESTERN DESERT HYDROGEOLOGIC REGIONS

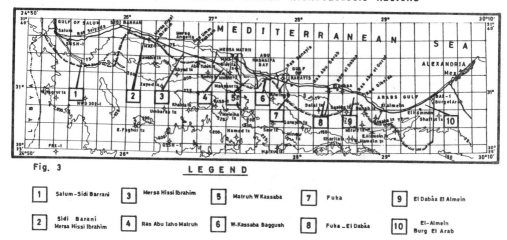

Fig. 3

LEGEND

1	Salum–Sidi Barrani	
2	Sidi Barani Mersa Hissi Ibrahim	
3	Mersa Hissi Ibrahim	
4	Ras Abu laho Matruh	
5	Matruh W Kassaba	
6	W-Kassaba Baggush	
7	Fuka	
8	Fuka – El Dabâa	
9	El Dabâa El Almein	
10	El-Almein Burg El Arab	

3.2. Surface Water

3.2.1. Rainfall

Rainfall exists only on the coastal zone area. From Burg El-Arab to Sallum, average amounts of rainfall range from 100-150 mm. Such a low amount places the zone at the driest limit of the arid climate. Only a small part of the coast, from Alexandria (181 mm) to Burgh El-Arab (157 mm), is a little more favored.

The rainy season begins during the second half of October. Three quarters of the total amount falls from November to February. December and January are the rainiest months with an average of 35 mm per month. Some showers are still observed in March, but Spring is dry and receives only 10% of the total.

According to a report published by the Ministry of Housing and Reconstruction (1975), the total catchment area is about 10,324 square kilometers (Table 4). With an average annual rainfall of 102 mm, the total annual rainfall is about 1,002 million cubic meters.

3.2.2 Wadi Runoff

Wadi runoff is that part of the precipitation which concentrates in wadis. 208 wadis were recognized in the area. Wadi runoff may eventually be carried into the sea or may just disappear, where the wadis peter out in some flat plain where no distinct channels are distinguishable any longer.

From the agricultural point of view, wadi runoff is an important item. It represents concentrated flows of water which might be utilized for agricultural purposes. Also it is an important constituent in the water budget of some hydrologic units, aiming at the determination of ground water flows.

The average wadi runoff coefficient is 1-2% of the rainfall. Therefore, total annual rainfall has been estimated to be 20.49 million cu. m.

3.2.3 Surface Reservoirs

The nearest sources of surface water is the Nile Valley and mainly the Lake Nasser reservoir of the high dam of Aswan, located 40 kilometers east of the south Kharga Depression. A volume of 164,000 million cubic meters is now in storage in the lake, of which 74,000 million cubic meters will be used annually for irrigation in the Nile Valley (55.5 Egypt; 18.5 Sudan) and over 10 or 15 billion lost through evaporation and seepage into the groundwater reservoir of the Nubian sandstone.

3.3 Groundwater

3.3.1. Coastal Zone

In the northwestern coastal zone, strata of hydrogeologic interest are found both in the Quaternary (Beach sediments, alluvial deposits, Alexandria Formation of Pleistocene age) and the Neogene (Early Miocene).

The groundwater flows present in this region depend on recharge through infiltration, sheet runoff and wadi runoff.

3.3.11.   Infiltration

Infiltration is a term used to designate the process of penetration of water through the surface of the soil downward to the water-table. At first, water replenishes the soil moisture deficiency, then excess water moves downward and may form a groundwater reservoir.

Average percentage value for infiltration is 20% of total rainfall. Thus, with total rainfall of 1,002.875 million cu.m., annual infiltration amounts to 203.11 million cu. m.

3.3.12.   Recharge to Groundwater

As mentioned above (3.3.11), rain water, when penetrating through the surface of the soil downward, first replenishes the soil moisture deficiency. Excess water then moves downward to the level of the water table. This amount of replenishment is assumed to be 50% of total infiltration. Thus, annual recharge to groundwater in the coastal zone is about 101.57 million cubic m. However, in order to prevent salinization of groundwater, exploitable groundwater is 50% of the annual recharge. Thus, annual exploitable water is estimated to be about 49.84 million cubic m.

3.3.2.   Wadi El-Gedid (New Valley)

3.3.2.1 Nubian Sandstone Aquifers

The Western Desert, including Wadi El-Gedid in Egypt, is the largest groundwater basin of the Sahara with an area of about 1,800,000 square kilometers, and a water storage capacity of 6,000,000 million cu. m., i.e., 36.58 times the capacity of Lake Nasser. It has an artesian flowing area of 57,000 square kilometers, including the Qattara Depression.

Below the arid surface of the Western Desert, and extending eastward till the mountain ranges of the Red Sea, is a huge natural reservoir of groundwater which could sustain human settlement and productive agriculture. The water bearing formation is the Nubian Sandstone Formation. In geologic age, it ranges from Precambrian till middle Cretaceous. This sandstone, which in many places is interbedded with shales, is 400-800 meters thick in the Kharga depression, 1400 meters in Dakhla depression and over 2,000 meters in Farafra depression. The sandstone series rest on Precambrian rocks of the basement. The French name equivalent to "Nubian Sandstone" is the "Continental Intercalaire." The groundwater reservoir is a confined aquifer overlain by a semi-impermeable stratum; it is likely to be under artesian pressure that will cause the water to rise above the top of the aquifer when the aquifer is penetrated by a well, whether or not the water rises enough to flow at ground level according to the ground elevation. The top of the aquifer is within 50 meters below ground level. Generally speaking all deep wells are flowing. Wells, with shallow depth, less than 100 meters, are non-flowing. Along with the development, flowing wells will be transformed into non-flowing wells. Under the present conditions, the average discharge of a flowing well is 200-350 m^3/h (5000-8000 m^3/day = 1.8 to 3 million cubic meters per year) and 200 m^3/h (with 10 meters drawdown) for a pumped well.

3.3.2.2 Annual Recharge

Annual recharge is at least equivalent to the natural losses through the outlets
before any important development of groundwater took place. The outlets are
visible as free flow in the depressions or invisible as evaporation directly from
the artesian aquifer or through transpiration from plants within the artesian
overflowing area and leaking upwards through the overlying upper Cretaceous and
Tertiary formations. Such a condition in groundwater reservoirs is called the
"Steady State", i.e., recharge is equal to discharge.

By the use of a computerized digital regional groundwater program (Ezzat, 1975),
it was possible to determine the steady state of the Nubian Sandstone aquifers in
the Western Desert of Egypt. Steady state results as of 1960, indicated that
total recharge to Kharga, Dakhla, Farafra, and Bahariya Oases, is in the following
order:

Eastern Desert	18.92	million cu. m/day
Sudan Border	193.70	" " " "
Gilf El-Kebir	449.59	" " " "
TOTAL	662.20	" " " (1.814 million cu. m/day)

For the Siwa upthrown block, Parsons (1963) estimated total spring and well
discharges in the Siwa Oasis to be about 120,000 m^3/day. The total area below
sea-level of nearby depressions such as Melfa, Areg, Bahrein and Sittra together
with Siwa Oasis is about 2500 square kilometers. Helstrom (1940) estimated
evaporation losses to be at the rate of 300,000 m^3/day in these depressions. This
brings total discharge from groundwater to be about 420,000 m^3/day (153.3 million
cu. m./year).

Thus, annual recharge to the Nubian groundwater reservoir is about 815.50 million
cu. m./year (2.234 million cu. m./day). This recharge is occurring at present
mainly at the south and southwestern edge of the Western Desert basin where the
rainfall intensity sufficiently increases on the Tibesti, Ennedi and Filf-El-Kebir
mountains to create a run-off water which percolates into the aquifer.

3.3.2.3. Long-Term Availability

Recharge and discharge of groundwater are usually in rather delicate balance under
desert conditions. The generally low rate of recharge may easily be overtaxed by
concentrated withdrawals within the depressions of Wadi El-Gedid. If and when
that happens, the groundwater is being mined. Any development based on the mining
of the aquifer is a rather delicate decision. However, in the case of the Nubian
Sandstone aquifer any reasonable mining could be envisaged for the next centuries
to come because in addition to a safe yield of about 815.51 million cu. m./year,
the aquifer has a capacity of 6,000 million cubic meters of dead storage accumulated during the past 50,000 years. Mining the first hundred meters of
aquifer could produce over 600,000 million cu. m. to be exploited within 200
years, i.e., about 300 million cu. m. per year, with 815 million cubic meters
yearly safe yield. Thus total exploitable groundwater for the coming 200 years is
about 3815 million cu. m./year. With 5000 m^3/year irrigation requirement per
feddan, it is possible to irrigate 763,000 feddans for 200 years.

It should be noted that we are assuming that present annual recharge will be
constant for the coming 200 years. This is not quite true. With the process of
mining the Nubian sandstone aquifer, the present steady state equilibrium will be

disturbed and the hydraulic gradient will be steeper, thus increasing annual recharge. However, in the suggested development program for the Western Desert (next chapter) we will assume that annual recharge will be kept constant as a factor of safety.

4.   PROPOSED DEVELOPMENT

The main concern of the Egyptian Government is to extend the area under cultivation and the promotion of land reclamation, land development and settlement has been its policy for the last 15 years. The proposed irrigation development plan is formulated on a long-term basis (1975-2025).

For development purposes, the Western Desert of Egypt could be divided into 4 development areas:

- Coastal Zone
- Wadi El-Gedid (New Valley) area
- South Kharga area
- Qattara Depression

We will deal with the development program for each area separately.

4.1.   Coastal Zone

This area is fortunate, as it has a good asphalt road and several exporting ports, such as Alexandria, Sidi Kreir, Mersa Matruh and Sallum.

4.1.1.   Surface Hydrology and Ground Water

The tentative programme as developed hereafter, aims at an optimal agricultural development. This means that the supply of water should be dependable. This could be achieved most efficiently if direct rainfall, wadi runoff and groundwater are utilized in an integrated way. Separately they may not be sufficient to sustain a crop, but in combination they may be. So combined utilization could be applied only where there is good soil available.

Wadi's have been identified which can be expected to have some discharge every year and which could support agriculture in selected areas. The total cultivable area, is 9,625 feddans (3,850 hectares). The assumption has been made that with appropriate measures, 80% of the Wadi runoff can be utilized effectively for the benefit of various crops. According to FAO (1970), at present only 20% of the total runoff is profitably used.

In the various cultivable areas, the total available amount of water varies from 130 to 245 mm. This certainly does not suffice to sustain a perennial crop. It is proposed to plant the greater part of the area with fruit trees. A minimum average water requirement of some 650 mm should be taken into account. If follows that the areas of fruit trees should be supplemented by winter watering (80% utilization of Wadi runoff) and by exploiting available groundwater.

4.1.2. Underground Reservoirs (Cisterns)

More than 3,000 cisterns dating back to the Roman period exist in the coastal
regions. Some of them (approximately 30%) have been cleaned and repaired by the
Government. They are the main drinking supply for the people and animals inland.
The cisterns have been excavated in the rock and their capacity varies from 100 to
3,000 m^3 with an average value of 440 m^3. Thus by cleaning all existing cisterns,
their total storage capacity is 1.3 million cubic meters.

4.1.3. Nile Water

At present Nile water is used in the Mariyut Extension irrigation project. The
main irrigation canal could be extended to the El-Dabaa area, to irrigate an area
of about 80,000 feddans of suitable soil there. The present existing canal,
El-Nahda is extended till Hamman town.

4.1.4. Desalination

The best method for desalination of seawater is by means of multi-stage flash
evaporators, similar to those presently in use in Mersa Matruh and Sallum. The
desalination plant at Matruh has a capacity of 0.35 million cubic meters annually.
For a well run plant the production costs of water would vary between L.E. 0.30 -
0.50 per m^3.

If groundwater would be available, with a maximum salinity of 4,500 ppm, reverse
osmosis would be a more profitable system. The price of the water in that case
would be about L.E. 0.25 per m^3. With these high prices for fresh water, it
follows that desalination is far more expensive than conveying Nile water into the
Western Desert.

4.1.5. Drinking Water

One of the most important items for the development of the Coastal zone, is to
provide fresh water for drinking purposes. At present, the new pipeline, Alex-
andria-Matruh, has a capacity of 7 million cubic meters annually or 0.22 m^3/sec.,
which is a small pipe line for the proposed development programme.

It is proposed, with the extension of the Nahda irrigation canal till El-Dabaa
town, to extend it, with a smaller section, till Sallum, mainly for drinking
purposes. The canal should have a section large enough for a supply of 5 m^3/sec
or 15.768 million cubic meters annually. According to the report of the Ministry
of Housing and Reconstruction (1975), estimated investment for the construction of
such an open canal till Sallum, is estimated to be about L.E. 78 million. Cost of
water is estimated to about L.E. 0.115 which is much less than the cost of fresh
water by desalination plants.

Such an open canal, with a capacity of 5 m^3/sec, can support 6.2 million persons
in the area. Moreover, with a continuous fresh water supply in the area, tourism
projects could be established, adding new capital income to the area.

4.1.6. Town Planning

With the development of 4750 feddans on rainfall and groundwater in the area
between Fuka till Sallu, and 80,000 feddans on Nile water at El-Dabaa, we propose
to establish the following towns, to support a population of 6 million:

Town	Population	Main Activity
Mersa Matruh	1,000,000	Capital, harbour and university
Dabaa	850,000	Agro-industrial center
Hammam	750,000	" " "
Burg-El-Arab	750,000	" " "
Sallum	350,000	Harbour & free zone market, tourism
Sidi Barrani	350,000	Argicultural Center
Mersa Angeila	250,000	" "
Mersa El-Asi	100,000	" "
Ras Abu-Laho	200,000	" "
El-Qasr	450,000	" "
Fuka	450,000	" "
Agiba	100,000	Tourism center
Baqqush	50,000	" "
Berbeita	50,000	" "
Ras El-Herkma	100,000	" "
Alamein	100,000	" "
Sidi Kreir	100,000	" "

4.2. Wadi El-Gedid Area (New Valley)

In Wadi El-Gedid Area, as mentioned before (3.1.2), the irrigable land under fair
engineering and economic conditions is about 2.4 million feddans of good soil, of
which 325 feddans could be irrigated entirely by groundwater and 1.1 million
feddans by combining groundwater in situ and surface water from Lake Nasser.

4.2.1. Strategy

Because water is the most important limiting factor, not capital, consideration
should be given first to potential production in relation to water consumption and
secondly to economic returns. The final target of irrigation development in the
New Valley could be 1.4 million feddans requiring 10.2 milliard cu.m. per year.
An area of 1.1 million feddans should be developed in South Kharga at a distance
of 40 kms from the lake and 0.7 milliard cu.m. from groundwater annual resources
and mining. Another area of 325,000 feddans should be developed by using only
groundwater in Kharga, Dakhla, Farafra, Bahariya and Siwa with first priority
given to the Dakhla-Kharga area where 85,000 feddans could be irrigated by using
620,500,000 cubic meters per year of groundwater.

4.2.2. Development

The present development program is based on the results of the soil survey and
the mathematical model made for the Western Desert.

The area to be developed should include the area of good soils in the five
depressions. The exploitation of groundwater should be based on freeflowing
wells, whenever possible, during a first temporary phase and sooner or later on
pumped wells. Also the groundwater extraction could start on a pumping basis in
the area of non-flowing wells, if proved economical.

The reclamation program is shown in Table 5.

Table 5

Proposed Reclamation Program

Location	Irrigated Area (feddans)	Proposed Additional Irrigation (feddans)	Total (feddans)	Total Water Requirements (Million cu. m./year)
South Kharga	--	100,000	100,000	730
Kharga	10,000	10,000	20,000	146
Zayat	--	5,000	5,000	36.5
Dakhla	15,000	45,000	60,000	438
Farafra	100	100,000	100,000	730
Bahariya	5,000	15,000	20,000	146
Siwa	1,500	18.500	20,000	146
TOTAL	31,600	293,500	325,000	2372.5

Such a plan should be implemented on a step-wise basis and readjusted accordingly. In the proposed plan, the water requirements for the Abu Tartour phosphate deposits between the Kharga and Dakhla Oases has been taken in consideration (30 million cu. m./annum).

REFERENCES

(1) Ayouty, M.K. and M.A. Ezzat, 1961 "Hydrogeological Observations in the Search for Underground Water in the Western Desert of Egypt, U.A.R." Symposium of Athens, Ground Water in Arid Zones, International Associations of Scientific Hydrology, Belgium, Publ. No. 56, pp. 114-118, 1961.
(2) Ball, J., 1927, "Problems of the Libyan Desert. The Artesian Water Supplies of the Libyan Desert: Geographical Journal, London, 70 pp. 1927.
(3) Ball, J., 1933, "The Qattara Depression of the Libyan Desert and the Possibilities of its Utilization for Power Production", Geographical Journal, London, Vol. 4, pp. 289-315, 1933.
(4) Bassler, F., 1975, "New Proposals to Develop Qattara Depression Hydroelectric, Waterpower and Dam Construction", June/July, 1975.
(5) Ezzat, M.A., H.M. El-Badry and M.M. Ibrahim, 1968, "Hydrogeology of the Wadi El-Gedid Project, Western Desert, U.A.R.," with special reference to the Kharga Oasis, Bull. of Faculty of Engineering, Cairo University, pp. 477-500, 8 Figs, 1965 - 1966.
(6) Ezzat, M.A., Abdel Azim Aboul Atta, 1974, "Regional Hydrogeological Conditions, El-Wadi El-Gedid Project Area", part 1 of Groundwater Series in the Arab Republic of Egypt, Ministry of Agriculture and Land Reclamation, Cairo, Egypt, 121 pp., 25 plates, 5 Figs, 1974.
(7) Ezzat, M.A., Abdel Azim Aboul Atta, 1974, "Hydrogeologic Conditions, Dakhla-Kharga Area, El-Wadi El-Gedid Project Area", Part II of Ground Water Series in the Arab Republic of Egypt, Ministry of Agriculture and Land Reclamation, Executive Agency for Desert Projects, 106 pp., 38 figs., 1974.
(8) Ezzat, M.A., 1975, "Exploitation of Ground Water-Dakhla Oasis, El-Wadi El-Gedid Project Area", Part III of Ground Water Series in the Arab Republic of Egypt, Ministry of Agriculture and Land Reclamation, General Organization for Rehabilitation Project and Agricultural Development, 149 pp., 26 plates, 37 figs, 1975.

(9) Ezzat, M.A., 1976, "Groundwater Regional Model, Groundwater Pilot Scheme, New Valley: Working Document. No. 4, AG. DP/EGY/71/561 presented to Ministry of Irrigation, Egypt, by: UNDP/FAO, Egypt, 20 pp., 15 figs, 1976.

(10) ILACO, 1975, "Regional Plan for the Coastal Zone of the Western Desert, Development Potentials", report presented to Ministry of Housing and Reconstruction, by: ILACO, Den Haag, Holland, 137 pp., 14 figs, 1975.

(11) INDUSTROPROJEKT, 1968, "Basis for the Analogue Model of Kharga and Dakhla Oases", report presented to General Desert Development Organization, Egypt, by: Industroprojekt Department for Exploration of Mineral Resources, Zagreb, Yugoslavia, 113 pp., 1968.

(12) Murray, G.W., 1952, "The Artesian Water of Egypt" Survey Department, Ministry of Finance and Economy, paper No. 52, 20 pp., 1952.

(13) Parson, 1962, "Bahariya and Farafra Areas, New Valley Projects, Western Desert of Egypt:, report submitted to the Egyptian General Desert Development Organization, by The Ralf M. Parsons Engineering Company, Los Angeles, California, U.S.A., 364 pp., plates 28, 1962.

(14) Parsons, 1963, "Siwa Oasis, New Valley Project, Western Desert of Egypt", report submitted to the Egyptian General Desert Development Organization, by The Ralf M. Parsons Engineering Company, Los Angeles, California, U.S.A., 218 pp. 22 maps, 1963.

(15) UNDP/FAO, 1970, "Pre-Investment Survey of the North Western Coastal Region, United Arab Republic, Comprehensive Account of the Project", Report prepared for the Government of the United Arab Republic by UNDP/FAO, technical report No. 1, ESE:SF/UAR 49, Rome, 109 pp., 5 maps, 1970.

WATER RESOURCES STUDY IN THE ATACAMA DESERT
(NORTHERN CHILE)

Uri Golani
United Nations Technical Adviser (Hydrogeology)

ABSTRACT

. water resources exploration project was carried out by the Government of Chile
ınd the United Nations in part of the Atacama Desert of Northern Chile. The area
is rich in copper and other minerals and the exploitation of these mineral deposits depend on water.

The present consumption of water in the region for the mining industry, agriculture and domestic supply is about 5 m^3/sec. Water supply would have to almost
double by the year 2005 to satisfy the demand of the growing mining industry.

The project has shown that the future water demand can be satisfied by development
of all surface and ground resources, by improving the water use practices in
agriculture and the mining industry and by re-use of water in the mining industry.

INTRODUCTION

Lack of water has long been the prime economic impediment to the development of
the Norte Grande* which extends over the northern quarter of Chile. This area
contains the driest desert on earth - the Atacama Desert. Precipitation along the
Pacific Coast and in the inner parts of the area is almost zero and in many places
even a sprinkle of rain is a very rare occurrence. Precipitation increases
towards the Cordillera in the east; it is about 100 mm at altitudes of 3000 meters
and 300-400 mm (mostly as snow) at the peaks of the volcanoes (altitude 5000 to
6000 meters).

_____

*The project work was concentrated in the area of the Second Region (formerly the
Province of Antofagasta) which occupies about one half of the Norte Grande. See
fig. 1.

835

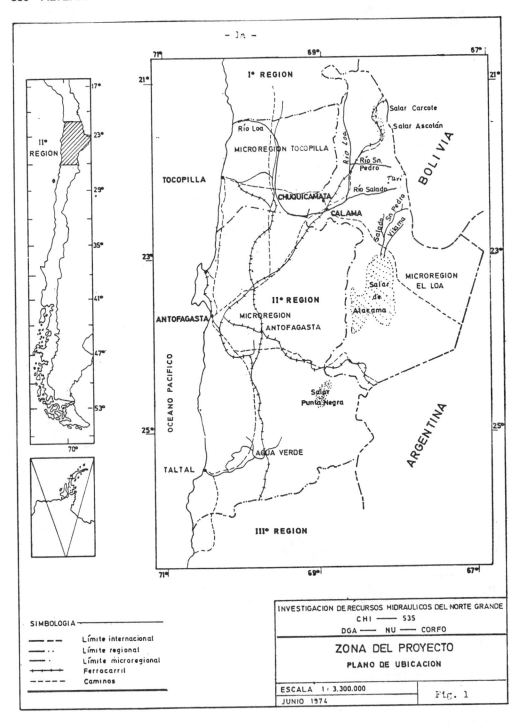

- Ja -

SIMBOLOGIA

— — —	Límite internacional
— ..	Límite regional
— .	Límite microregional
+—+—+	Ferrocarril
— — — —	Caminos

INVESTIGACION DE RECURSOS HIDRAULICOS DEL NORTE GRANDE

CHI —— S3S

DGA — NU — CORFO

ZONA DEL PROYECTO

PLANO DE UBICACION

ESCALA 1 : 3.300.000

JUNIO 1974

Fig. 1

The Norte Grande is a vast area of over 260,000 Km^2 with a predominantly urban population of 250,000 (1970 estimation) which is concentrated on the coast and in the inland copper production center of Chuquicamata-Calama.

The region's economic contribution in export earning is out of proportion to its share of less than 5 percent of the total population of Chile. About 40 percent of the Chilean exports come directly from the Norte Grande through exportation of copper, nitrates and other minerals, as well as fish meal.

Water for the coastal cities and towns and the copper mines has long been imported from the Cordillera by pipelines, which for the city of Antofagasta are 365 Km long. The quantities of known and easy to tap water resources of the region are smaller than the projected needs. The cost of water in the region is rather high due to the great delivery distances and the need for treatment (elimination of arsenic) of water for human consumption. Further development of the mineral resources, of industries and agriculture depend primarily on the availability of sufficient quantities of water of suitable quality and at an acceptable cost.

The Government of Chile has long been keenly aware of the water problems of the Norte Grande and their consequent limitation to development. Nevertheless, there has been no coherent policy for the allocation of water in this region. Government agencies such as the Direccion General de Agua (Ministerio de Obras Publicas) and the Corporacion de Fomento de la Produccion (CORFO) as well as the copper mining companies and other private entities have done some scattered exploratory work for water resources in the region. There has been no idea of the groundwater potential of the region and even the surface water resources were not adequately known and quantified. This situation has created a basic problem for making water policy and planning water development.

In 1967 the Government of Chile forwarded a request to the United Nations Development Programme (UNDP) for technical assistance in carrying out a water resources exploration, assessment and development project. The project started at the beginning of 1971 and is expected to terminate by the end of 1977. Some of the results presented in this paper are therefore not final.

The Occurrence of Water Resources in the Second Region

Hydrological and hydrogeological studies have been carried out in the Second Region by the CHI-69/535 project since 1971. Water exploration and studies as well as drilling of hundreds of shallow and deep wells have been carried out by Government agencies and private organizations in this area for many years previously. From the results of these studies and drillings the following conclusions have been drawn.

　　　　1. The greater part of the water resources in the Second
　　　　　　Region originates from the snow capped volcanoes
　　　　　　located in the eastern part of the region along and
　　　　　　beyond the border with Bolivia and Argentina.
　　　　2. Precipitation in the form of rain in the greater part
　　　　　　of the region is too low to sustain or, to contribute
　　　　　　to any permanent water resource. Occasional rain
　　　　　　storms in the eastern part of the area are of too
　　　　　　short a duration to produce any important recharge to
　　　　　　groundwater. A part of the waters of these rains
　　　　　　drain rapidly in the form of flash floods; the re-
　　　　　　mainder evaporates.

3. Snow melting is caused by solar radiation and in some places also by geothermal heat of the volcanoes. The water from the melting snow percolates into the very porous rocks which are found on the slopes of the volcanoes and flows as groundwater downward and away from there. The groundwater sooner or later re-appears on the surface in the form of springs, river flow (base flow), small saline lagoons and wet areas (vegas y salares). The quantity of water which re-appears on the surface diminishes rapidly with the distance from the recharge areas of the volcanoes. Water mineralization increases in the same direction.

4. In most cases the rivers' base flow and the water outflow from springs are steady throughout the year. The yearly variations are also very small. This indicates that either the groundwater storage is extremely large or that the rate of snow melting in the recharge areas is very uniform throughout the year, or that both of these factors work together to mitigate or even almost eliminate seasonal and annual climatic fluctuations.

5. The base flow in the rivers plus the water which evaporates in the salars and the vegas represent the bulk of the water involved in the hydrogeological cycle of the eastern part of the Second Region. A certain portion of this water can be utilized by tapping rivers and by pumping groundwater from wells. The total quantity of usable water resources of the area is quite limited.

WATER RESOURCES

Actual water consumption in the mining industry, agriculture and urban water supply is over 5 cubic meters per second. In order to carry out the planned development of copper mining and related industries the region will need by the year 2005 8-10 m^3/second of water.

The 5 m^3/sec. of water now consumed represent most of the easy to tap good quality surface water resources. The hydrological and hydrogeological studies showed that the addition required 3-5 m^3/sec. of water can be made available by exploiting the additional water resources of the area most of which are of poor quality and expensive to develop and transport and, by better water management, improved methods of use and recycle of the existing water resources. The following are the main ways to get the additional required quantities of water:

1. Tapping the yet untapped surface water. There is 1-1 1/2 m^3/sec. of brackish surface water that can be used in certain processes of the copper mining in-dustry instead of the present use of better quality water.

2. Storing of flood water in Rio Loa and Rio Salado can prove to be an important resource. However, more years of flood observation are needed to establish the reliability of this resource.

3. Improvement of irrigation practices and applying modern agricultural methods would increase agricul-

tural production and reduce the quantity of water presently consumed.

4. About one third to one half of the groundwater which flows to and evaporates in the salars can be intercepted by means of closely spaced small production wells. This has been proven in the Salars of Ascotan, Atacama and Punta Negra. Total quantity of water that can be obtained 1-2 m^3/sec.

5. Lowering water levels of groundwater by means of wells or drains would prevent evapotranspiration in the area of Calama and Vega de Turi. Estimated quantities of water that can be recovered 1-1 1/2 m^3/sec.

6. Extracting groundwater from one time storage may prove to be practical in the area of Pampas de Tamarugal.

7. Re-use of water in the copper industry could save up to 30-40 percent of the actual water consumption of that industry.

It is quite clear at this time that in spite of the high cost of developing the additional water resources of the region it is still more economical and more practical than to turn to non-conventional water resources such as desalination of sea water or iceberg towing.

WATER RESOURCES DEVELOPMENT IN SAUDI ARABIA

Dr. H.H. Hajrah
Head of Biology Dept.

INTRODUCTION

Saudi Arabia is a vast country covering approximately 1.5% of the global earth and 5% of arid and semiarid areas in the world. It is also covering more than 2 million square kilometers (2,253,000). It is one of the most thinly populated countries of the world, having on the average 2-3 persons per sq. km. At present the cultivated area comes roughly to 0.15% of the total land area of the kingdom. The average annual rainfall in most parts of the country is less than 100 millimeters while in the southwestern region (Asir) province it may get as high as 500 millimeters in a year. Only about 10 percent of the land gets 100 mm. or more. More than 80% of the cultivated area is under irrigation, rest is rainfed.

Agriculture in General:

Traditionally agriculture in Saudi Arabia has been limited to few areas with adequate water supplies in oasis, along wadi beds, in catchment basins and in relatively high rainfall regions in the southern parts of the country. In the past, dates used to be the only staple food crop but now the date gardens have declined in importance and the agricultural economy is gradually shifting to other forms of agriculture like growing of cereals, vegetables and fruit trees.

Traditional Water Resources:

Saudi Arabia is the largest country without a river. There are no perennial flows in the country but there are numerous streams which get once in a while during heavy rains. As already mentioned earlier that the average rainfall is low so the surface water resource is not enough for use in irrigation. Agriculture depends upon natural precipitation in the southern province to some extent, whereas the rest of the country is dependent upon ground-water, which may be taken out and used for irrigation purposes, either through shallow dug well, artisan wells or from drilled wells taking water from deeper aquifers.

Water Present Conditions:

Ground water meets the major needs of agricultural, industrial and urban while surface water satisfies a small share and limited amounts of water have been supplied by desalination.

Hydrological studies proved the existence of further usable quantities of ground water. Brackish water is abundant in several areas of the country; and sea water is available for desalination in unlimited quantities. Ground water and surface water are unable to provide an adequate and safe supply.

The government policy for development of water resources is to subsidize well drilling and water extraction machineries. This helped the people and encouraged them to use vast areas for cultivation. However, the extravagant use of water resources created a problem for the domestic water demands. The increasing number of population of the main cities aggravates the problem. So, it is intended to control water exploitation for agricultural purposes to keep an adequate supply for cities.

Rainfall and Surface Water:

The relative absence of vegetation and lack of settled population over most of the Arabian peninsula constitute reliable indicators of the lack of rainfall. Examination of long period records of rainfall from 57 stations scattered over the Kingdom reveals the extremes. Ten of 57 stations have average annual rainfall of less than 50 mm and 3 had averages over 500 mm.

The effectiveness of precipitation is reduced by the high temperature, bright sunshine and rather constant breezes that increase the rate of evaporation. Light showers may be largely dissipated under desert conditions.

Surface water as streams exists only after rains of sufficient intensity and duration to produce runoff. Rainstorms of enough intensity and duration to produce runoff fall on many water sheds with insufficient frequency to justify constructions of storage dams. In addition, the runoff waters usually carry much silt, and clay so that the useful life of most dams would be shortened by sedimentation. Runoff waters are the source of irrigation in south-west such as Jizan area with man-made water spreading.

It is the intention of the government of Saudi Arabia to increase the number of small dams in different parts of the country particularly in regions to recharge ground water aquifers and to supply the villages with domestic water. Over 20 small dams have been constructed or approved for construction. Additional small dams are required throughout the Kingdom for flood protection, aquifer recharge, and storage of surface runoff.

The constructed dams and their characteristics are given in Table 1. The largest dam in the Kingdom is that of Wadi Jizan with a storage capacity of 71 million cubic metres. The dam is located in the southwestern part of the country, where rainfall is relatively higher than in other parts of the country. The Abha dam, with a capacity of 2.4 million cubic meters has been constructed for the sake of water storage for domestic use of Abha city. Other dams in different parts of the country serve mainly for recharging the ground water aquifers. Perhaps dams will be of use in recharging groundwater to support water supplies for villages.

Table 1 - List of Dams and Their Characteristics

Name	Location	Catchment Km²	Type	Storage Capacity m³	Length m	Height m	Purpose	Completed date
Wadi Jizan		1100	concrete gravity	71 million	316	41.6	flood control and irrigation	1970
Abha		58.5	concrete gravity	2.4 million	350	33	water supply	1974
Batham	Medina	20	concrete and earth	800,000	266	12.5	recharge aquifers and flood control	1964
Majam'a		100	earth	1.5 million	360	11.0	flood control and recharge aquifers	1969
Hereimlah		350	earth	1.5 million	1250	6	diversion dam	69
Melham		20	concrete gravity	200,000	100	4	storage and recharge	68
Ananyyah	Qasim		earth		180	8	division and infiltration	64
Al Ainyyah	Riyad		masonry	1 million	400	5	storage and recharge	66
Safar	Diriyah	54	earth	300,000	325	4	recharge aquifers	74
Ghbeirah	" "	21	earth	90,000	170	6	recharge aquifers	74
Hriqua	" "	19	earth	80,000	190	5	recharge aquifers	74
Hanifa	Riyahd		Concrete	1.5 million	650		recharge aquifers	69
Ekrema	Taif						flood control and irrigation	56
Namer	Riyadh			1.5 million	400	4	recharge aq.	
Laban	Riyadh			2 million	500	12	recharge aq.	68
Galagel	Galagel			30	630	12	recharge aq.	74

Ground Water:

The first flowing artesian well was drilled in the Eastern province more than 30 years ago. Before that time the ground water potential of the country was completely unknown. Since that time, with the introduction of powerful drilling rings and deep well pumps, the total meterage of exploratory and producing wells drilled by the Ministry of Agriculture and Water and consultants exceeds 130,000 meters between 1961-1968.

In 1965 this Ministry divided the Kingdom into eight areas to facilitate the inventory and assessment of the water and agricultural resources by the International firms. Exploratory holes have been drilled with depths between 150-3000 meters.

Ground water is related to geology. Arabian shield area which is composed of Igneous and Metamorphic Rocks cover one third of the country, representing mostly western province. This area is the poorer region of the Kingdom in ground water potentials. The water in alluvial fill along drainage channel (Wadis) depends completely on rainfall for replenishment. Sedimentary deposits which are composed of sandstones and limestones cover the other two thirds of the country and contain considerable amount of ground water. 28 sedimentary formations are known and mostly water producing.

The recent investigations by consultant firms during 1964-1970 have located many aquifers with large amounts of stored water. It is to be noted that the cost budgeted for these surveys, including agricultural surveys, at that time was in the order of the equivalent of $28 million US dollars. Before the discoveries of the large amounts of ground water in many locaties, the already present artesian wells were overexploited. Such uncontrolled use of artesian water caused salinization of the soil in the surroundings of the well (Photo No. 1). The discovery of new aquifers resulted in the establishment of many wells from which water is pumped (Photo No. 2).

Uncontrolled flow of artesian water at Tabauk

Photo No. 1.

It is to be noted that warning is made regarding the depletion of many shallow wells where the extraction of water is more than the recharge to the aquifer.

Modern pump extracting water from a shallow
well at Al-Joaf

Photo No. 2.

Water Desalination in Saudi Arabia:

The distillation machine was imported in 1907 and started to distillate water from
the Red Sea.

The Arabian Peninsula is surrounded by boundless sources of saline water. Also
the abundant natural fuel resources in the Kingdom, pointed to the economic
advantages of using the process of desalination of sea water to meet the future
needs of the population.

In 1965, an ambitious sea water desalination programme was embarked upon in Saudi
Arabia by the Saline Water Conversion Corporation.

In the second 5-year development plan, which began in 1975, desalination of sea
water is planned to cover the major water needs of both the existing cities and
the new urban-industrial-agricultural centres along the Saudi coasts of the Red
Sea and The Gulf as well as to meet demands in selected locations away from the
coasts. By the end of the plan, in 1980 more than 212 million gallons/day
(800,000 cu.m./day) of fresh water will be produced by the desalination plants in
the Kingdom. This makes Saudi Arabia the largest producer of desalinated water in
the world.

The estimated cost of the 5 year plan is 18 billion US dollars during the period
1975-1980. The desalination plants have a dual purpose: producing desalinated
water and power, from a single fuel source. The distribution of the plants and
their producing capacity is shown on map 1. The plants in different localities
along the Red Sea coast and the Arabian Gulf, their capacity, as well as the time
of execution are given in Table 2. Production of desalinated water is so important to Saudi Arabia to meet the rapidly increasing water demand in the cities
along the coasts. Moreover, it will save the ground water supplies for other
purposes as the over exploitation of ground water in wadis adjacent to the coastal
cities has affected seriously the agriculture in these wadis.

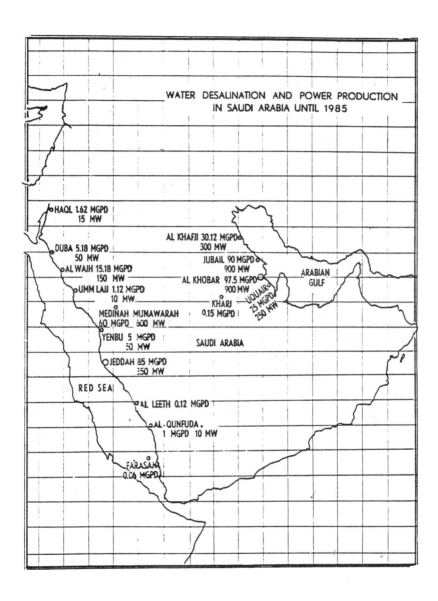

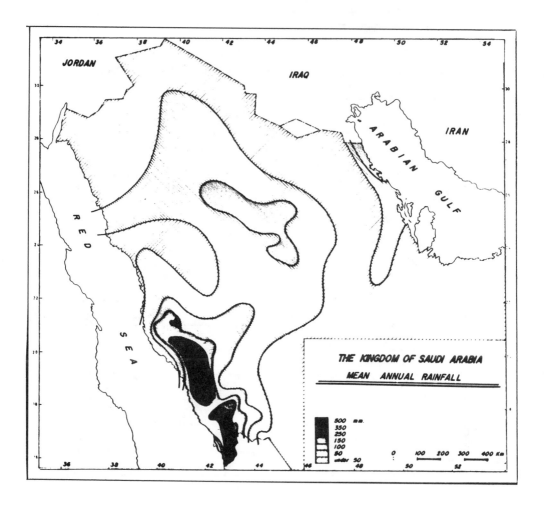

Table 2

Summary of Current Operations and Projects
Saline Water Conversion Up to 1985

Desalination		Water thousand gallons per day	Power Mega- Watts	Type of Fuel	Estimated Date for Operation	Region Benefitting From Project
	1	5,000	50	OIL	1970	Jeddah & Surroundings
JEDDAH	2	10,000	80	OIL	1977	a.a.
	3	20,000	200		1980	a.a.
	4	50,000	500		1983	a.a.
YANBU'	1	5,000	50	OIL	1979	Yanbu' & Surroundings
MEDINA	1	20,000	200		1980	Medina & Surr Surroundings
	2	40,000	400		1984	a.a.
RABIGH	1	240	-	OIL	1977	Rabigh & Surroundings
UMM LUJJ	1	120	-	OIL	1975	Umm Lugg & Surroundings
	2	1,000	10		1982	a.a.
AL-WAJH	1	60	-	OIL	1979	Al-WAJH & Surroundings
	2	120	-		1977	a.a.
	3	15,000	150		1983	AL-WAJH & Industrial Area
DUBA	1	60	-	OIL	1969	The Town of Duba
	2	120	-	OIL	1977	Duba & Surroundings
	3	5,000	50		1979	Duba & Industrial Area
HAQL	1	120	-	OIL	1977	The Town of Haql
	2	1,500	15		1979	Haql & Surroundings
AL-LITH	1	120	-	OIL	1979	Al-Lith
AL-QUNFU- DHAH	1	1,000	-		1980	AL-QUNFUDHAH & Surroundings
FARASAN	1	60	-	OIL	1979	ISLAND OF FARASAN
Total		174,020	1705			

CONCLUSIONS

The scarcity of water represents the main constraint to the development of agriculture in Saudi Arabia. Recent economic and social development of the country resulted in a great need for water, both for agriculture and domestic uses. Since 1964, the government embarked on agricultural and hydrological surveys. It has been shown that there is a reasonable potential of water resources. These include: (1) Surface water resources, particularly in the southwestern highlands and wadies in the central plateau, (2) Ground water resources, especially in the eastern part of the country in the sedimentary basin, and (3) Desalination of water along the Red Sea and the Arabian Gulf coasts for domestic uses. The de-

Table 2 (continued)

Desalination		Water Thousand gallons per day	Power Mega- Watts	Type	Estimated Date for Operation	Region Benefitting From Project
AL-KHOBAR	1	7,500	-	NATURAL GAS	OPERATING SINCE 1974	Al-Khober Dammam, Qatif Shihat, Salwa
	2	50,000	500	a.a.	1980	Towns Industrial Areas in the Eastern Region
	3	40,000	400	a.a.	1982	Eastern Region & Industrial Areas
JUBAIL	1	2,000	25	NATURAL GAS	1978	Jubail
	2	20,000	200	a.a.	1980	Jubail & Industrial Area of Petromin
	3	30,000	300	a.a.	1982	Jubail & Industrial Area
AL-KHAFJI	1	120	-	NATURAL GAS	OPERATING SINCE 1974	AL-KHAFJI & Al-Zurgani Border Station
	2	5,000	50	a.a.	1980	Al-Khafji & Surroundings, Industrial Area
	3	25,000	250	a.a.	1983	a.a.
AL-'UQUAIR	1	25,000	250	NATURAL GAS	1983	Al-'UQAIR & Industrial Area
AL-KHARJ	1	150	-		1980 ¿	Ater for brakish water
RIYAD		Under Consideration				

Page 1	204,500	1975	
	174,020	1705	
Total	378,520	3680	

velopment of these resources may remove the water deficiency problem, but not completely. It is hoped that modern technology, using the high solar energy and other fuel resources, will help to supply the country with additional water resources.

REFERENCES

Central Planning Organization Kingdom of Saudi Arabia, Economic Report, Riyad, 1970-1971.

Hajrah, H.H., Public land distribution in Saudi Arabia, Ph.D. University of
    Durham, England 1974.

Ministry of Agriculture and Water: Water Development Department, Water Devel-
    opment and Management, Riyad (Arabic).

Ministry of Agriculture and Water: Kingdom of Saudi Arabia, Seven Green Spikes,
    1974.

Ministry of Planning, Kingdom of Saudi Arabia, Second Development Plan, 1975-1980
    Riyad.

Saline Water Conversion Corporation, Kingdom of Saudi Arabia, Water Unlimited,
    Jeddah.

MANAGING A FINITE GROUND-WATER SUPPLY IN
AN ARID AREA: THE SANTA CRUZ BASIN EXAMPLE IN ARIZONA

by

Kennith E. Foster   *

INTRODUCTION

Abundant sunshine and the scarcity of water are two factors common
to all of Arizona. Today, it is estimated that people in Arizona use
about eight million acre-feet of water per year. In an average year
Arizona receives about 80 million acre-feet of water from rain and snow,
of which two million are captured. The balance, about six million acre-
feet, is obtained from underground water reserves for use by consumers.

SOURCES OF WATER IN ARIZONA

Arizona can be divided into three principal water provinces: 1) the
Plateau Uplands including the northern part of the state, 2) the Central
Highlands, a mountainous area extending diagonally across the state, and
3) the Basin and Range Lowlands, or desert, which includes the heavily
populated southern portion of Arizona (Figure 1).

The Basin and Range Lowlands, which contain more than 80 percent of
the state's population, consist of isolated mountain blocks jutting from
alluvial sediments that form the broad desert basins. The arid basins range
in altitude from about 100 feet above sea level at Yuma, to about 4,000 feet
in the southeastern part of the state. The climate in the lowlands is hot
and arid; the rainfall is light, with precipitation averaging less than ten
inches annually. But there are large desert areas which receive less than
five inches annually. No large surface water reservoirs are in this province
because the runoff is minimal. Average runoff from the driest 15,000 square
miles in the lowlands is less than 0.1 inch. The headwaters of the San
Pedro, Santa Cruz, and other large tributaries yield only about 0.5 inch
per year. Streams draining several of the higher mountain ranges yield
moderate amounts of runoff along the base of the mountains; but when the
water reaches the desert basin, it evaporates quickly leaving only a small
part to recharge ground-water reservoirs.

---

*Associate Director, Office of Arid Lands Studies, University of Arizona,
Tucson, Arizona 85719. This paper was prepared for the Conference on
Alternative Strategies for Desert Development, May 16-26, 1977, Sacramento, Calif.

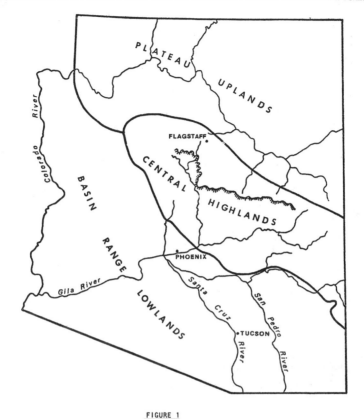

FIGURE 1

WATER PROVINCES OF ARIZONA

Streamflow in the lowlands occurs mostly as flash floods following thunderstorms. Storms originating in the Gulf of Mexico extend into the southeast corner of the state bringing precipitation to the mountainous areas. As flood waters move downstream from the mountainous areas, the flood volume decreases rapidly. Factors contributing to this depletion are infiltration, evaporation, channel storage, channel retention, and stream bank runoff.

The sediments in the alluvial desert basins constitute large storage areas for ground water. The occurrence of these large water reservoirs has made it possible for man to live prosperously and comfortably in the desert.

SURFICIAL WATER SUPPLIES

As in all arid or semiarid regions of the world, the dominant features of the rainfall in most of Arizona are scantiness and extremely variable from one year to the next. On the desert floor in the vicinity of Tucson, at a mean altitude of about 2,500 feet, the average precipitation is only about ten inches a year. Forty percent of the precipitation occurs in July and August (Figure 2). A large portion of the remaining 60 percent

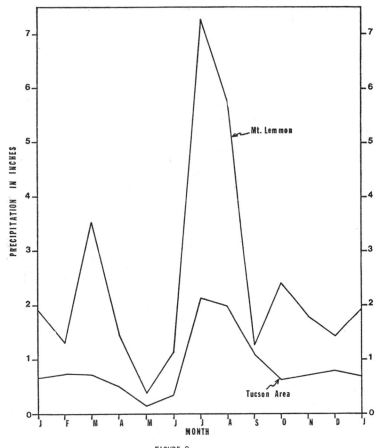

FIGURE 2

MEAN MONTHLY PRECIPITATION IN THE TUCSON AREA AND
AT MOUNT LEMMON IN THE CATALINA MOUNTAINS

falls as heavy showers which are scattered almost randomly through the rest
of the year.

Although the same rainfall regime is present at higher altitudes, amounts
are characteristically greater. For example, in the Catalina Mountains,
the Mount Lemmon rain gage, at slightly over 7,500 feet, receives an average
of 30 inches of precipitation each year - almost exactly three times that
falling on the Basin floor. However, monthly amounts (Figure 2) are extremely
variable from one year to the next, perhaps even more so than at lower altitudes.

Most of the rainfall in southern Arizona can be attributed to one of
four sources, depending mainly on the season of the year. A large percentage
of the summer thundershowers is associated with very warm, moist, and unstable
air which sweeps around the southern margins of the Atlantic Ocean high
pressure cell and advances into Arizona from the Gulf of Mexico. This air,
in passing over the highly heated land masses, is made even more unstable.
When it is forced to ascend over the numerous mountain ranges of southern
Arizona rain showers result. These showers have a very marked diurnal vari-

ation, being most intense over the mountains during the midafternoon when surface heating and the general convergence of air associated with the upslope mountain winds are maximal. In the valleys the heaviest summer rains usually do not occur until the later afternoon or early evening, when the desert floor is considerably warmer than the surrounding cloud-covered mountains.

Not all the warm-season rainfall is the result of simple convective activity. A small, but important, part is associated with tropical disturbances which form off the west coast of Mexico at about 150°N. latitude. These storms usually dissipate as they move northward into middle latitudes, but they are normally still intense and extensive enough when they reach the 30th parallel to produce heavy rainfall in southern Arizona. This type of rainfall differs from the normal convective type in several respects. It is more widespread, has lesser intensity but longer duration, and is only rarely associated with thunder and lightning. Some of the heaviest rainfalls on record, particularly in September, are associated with these tropical disturbances.

Winter, or cool-season, rains are generally less intense but more widespread than those of summer. They also vary less with ground elevation, sometimes being heavier on the desert floor than in the mountains. Part of this precipitation is associated with the middle latitude stormbelt, which occasionally moves far enough towards the equator in winter for its southern margins to affect Arizona. It is only when these cyclonic storms move in directly from the Pacific Ocean across the northern and central parts of the country that measurable amounts of rain occur. When the path of a storm is more nearly north to south, east of the 105th meridian, about all southern Arizona can expect is wind and subnormal temperatures.

Probably the heaviest rains of winter are associated with the so-called "cold lows" of the subtropical Pacific Ocean. These intense disturbances form in the vicinity of the Hawaiian Islands and move very slowly eastward to the coast of Southern California. In this region or slightly inland they often remain stationary for several days. But once caught in the strong upper-level westerlies, they move rapidly northeastward across the United States. These storms normally pass directly over Arizona, frequently advancing very slowly, retaining most of their moisture supply while moving in from the Pacific and can produce several days of moderate to heavy precipitation.

THE SETTING

The Santa Cruz Basin as shown in Figure 3 consists of several groundwater districts. The Basin is bounded by the crests of the Tortolita and Santa Catalina Mountains on the north, the Rincon, Santa Rita and Patagonia Mountains on the east, and the Atascosa, Tumacacori, Sierrita and Tucson Mountains on the west. The total drainage area as described contains about 2,240 square miles.

The Santa Cruz River has its source in the San Raphael Valley where it drains the east slopes of the Patagonia Mountains, the south slopes of Canelo Hills and part of the west slope of the higher Huachuca Mountains on the east. It flows south across the International Boundary into Mexico as a 35-mile loop before re-entering Arizona about six miles east of Nogales, and continues in a northerly direction to Tucson. The Santa Cruz then flows northwest and finally, about 12 miles southwest of Phoenix, joins the Gila River at a point 225 miles from its source in the San Raphael Valley.

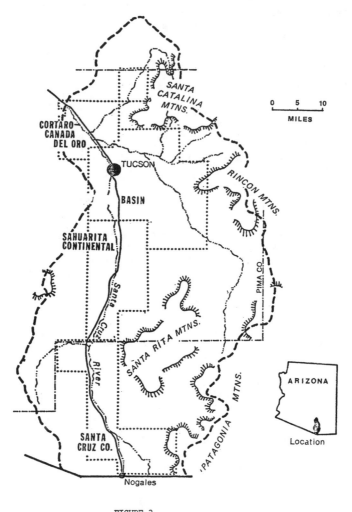

FIGURE 3

SANTA CRUZ BASIN

Structurally, the Santa Cruz Valley is a typical example of the Basin and Range Province of the Southwest United States. Northward trending mountain ranges border the broad, flat alluvium-filled valley (Figure 4).

Rillito Creek and its tributaries drain the eastern parts of the Basin and the adjacent SantaCatalina, Tanque Verde and Rincon Mountains. Rillito Creek is a tributary to the Santa Cruz River.

HYDROLOGIC CHARACTERISTICS

Runoff

Surficial runoff in stream channels is the most important source of re-

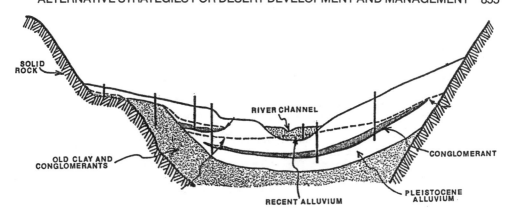

FIGURE 4

CROSS SECTION OF TYPICAL ARIZONA VALLEY

(FROM SCHWALEN AND SHAW, 1957)

charge to the ground-water reservoir of the area. The comparatively silt-
free runoff from snow melting in the early spring provides maximum recharge
through seepage into the Rillito Creek stream bed. The flows are more uniform
than summer floods, and may be continuous over a period of several months at
a season of the year when recharge is most effective with minimum losses to
evaporation or transpiration.

The short, intense summer storms of small areal extent result in short-
lived, silt-laden flows which are relatively ineffective as a source of
recharge to the ground water.

Natural Recharge

Matlock and Davis (1972) calculated annual ground-water recharge for the
Tucson-Sahuarita districts over a ten-year period, 1960-1969. Their study indicated an average annual recharge rate for the period of about 55,000 acre-feet.
Earlier studies (Matlock, Schwalen and Shaw, 1965) suggested a possible
range of 56,000 to 79,000 acre-feet for the average annual recharge.
The Arizona Water Commission (1975) estimated natural recharge in the
Upper Santa Cruz Basin to be 65,000 acre-feet per year. The recently completed Tucson Comprehensive Plan (1975) estimates recharge in the Basin to
be approximately 65,000 acre-feet per year. For discussion purposes, this
figure will be used.

Dependable Supply

The Arizona Water Commission (1975) estimates that 71,000 acre-feet
annually are available for a dependable supply. This consists of the
65,000 acre-feet of natural recharge plus 6,000 acre-feet of pumped ground-
water; however, since the aquifer is seriously overdrafted, the 6,000 acre-
feet cannot be considered dependable.

CURRENT WATER USE

Agriculture

Matlock and Davis (1972) estimated that the average agricultural consumptive use of water for approximately 15,000 acres in the Santa Cruz
Basin, excluding the Cortaro District, is approximately 46,000 acre-feet
annually. Demand for water is leading to retirement of agricultural land,
however, through purchase for water rights by mining companies. Since 1970,
over 8,000 acres (personal communication Jim Armstrong, Pima County Agent
Office) in the Santa Cruz Valley have been retired from production to use
the water for mining purposes. Less than 1,800 acres of irrigated cropland remain in the Basin adjacent to the city and south of Tucson
along the Santa Cruz River there is only approximately 6,800 acres of cropland, principally in pecan trees. Water use today in the Basin for agriculture is estimated to be 30,000 acre-feet per year, assuming a water use
of 3.5 acre-feet per acre.

Mining

Within the Santa Cruz Basin are five major open-pit copper mines plus
a sixth new mine under development. These mines are primarily in the
foothills of the Sierrita Mountain Range. They include the new San Xavier
Mine, operated by the American Smelting and Refining Company; the Mission
Mine, under the same management; the Pima Mine, operated by the Cyprus
Pima Mining Company; the Twin Buttes Mine, operated by the Anamax Company;
and the Esperanza and Sierrita Mines, operated by the Duval and the Duval-
Sierrita Corporation.

The daily tonnage of ore mined in this area is estimated to be 400,000
tons (personal communication Dr. David Rabb, Arizona Bureau of Mines).
The sulfide-copper ore is processed through the mills (300,000 tons
assumed) and the remainder is either overburden (50,000 tons assumed)
or stockpiled for oxide-copper ore leaching (50,000 tons assumed). The
ores, with copper content on the order of .5% or 100 pounds of copper
per ton of ore, are mined by open-pit methods.

The various kinds of water needs in the mining area are shown in
Table 1. In flotation sulfide copper is recovered in a froth from a
pulp containing suitable reagents and water in proper quantities. Optimum
density of the pulp ranges from 15 to 40 percent weight solids. A 15 percent pulp contains 5.67 tons (1,360 gallons) of water per ton of ore and
a 40 percent pulp has 1.5 tons (360 gallons) of water.

Leaching, another method of recovering copper, is particularly applicable to low-grade oxidized ores. Three types of leaching are commonly
used: (1) leaching in place, (2) heap or dump leaching, and (3) vat
leaching. Oxide-copper minerals are dissolved by a diluted solution of
sulfuric acid. Some sulfide-copper minerals are soluable in a diluted
ferric sulfate solution. Copper contained in the pregnant solution is precipitated by passing the solution over shredded scrap iron.

During leaching, evaporative losses and seepage must be replaced by
adding new water to the solvent. If ferric sulfate (formed by the action
of water on certain sulfides) is present in the solvent, some of the solvent
in the system may have to be removed to prevent an undesirable ferric ion
buildup.

TABLE 1

WATER USE IN SULFIDE AND OXIDE
COPPER ORE TREATMENT

Mining:	Bank sprays
	Wet down haulage roads
	Leach and precipitate copper from waste dumps
Crushing:	Cool crushing machine
	Collect and slurry dust
	Spray ore to minimize dusting
Concentrating:	Slurry the ore for grinding
	Particle size classification
	Manufacture milk of lime
	Flotation process
	Transport concentrates
	Transport tailings
Leaching:	Create diluted ferric sulfate solution

TABLE 2

AVERAGE AMOUNTS OF WATER REQUIRED AND CONSUMED
IN PRODUCING COPPER IN ARIZONA, 1960[1]

	Total Intake Gallons/pound Copper	%	Consumption Gallons/pound Copper	%
Mining	[2]1.1	1.9	[2]1.0	11.1
Flotation concentrating	[3]56.0	96.2	[3]7.5	83.3
TOTAL	57.1	98.10	8.5	94.40
Leaching	29.7	96.4	15.1	96.8
TOTAL	29.7	96.4	15.1	96.8

[1]Excludes water required for power plants, domestic and miscellaneous purposes

[2]Based on average grade of ore mined

[3]Based on average grade of feed at concentrators

(From Gilkey & Beckman, 1963)

Table 2 illustrates the relative water requirements for the mining flotation method and the leaching method of recovering copper. The additional water requirements for smelting is not treated here, since this copper operation is not performed in the Santa Cruz Basin.

Current water use by the mining district compares closely with the amounts given in Table 2. An estimated 90 gallons of water per ton of ore is consumed in the mining and flotation concentrating processes. Assuming a copper content of .5%, nine gallons of water per pound of copper would be consumed. On daily and annual bases, average water consumption would amount to 110 acre-feet per day or 40,150 acre-feet per year, assuming 400,000 tons per day of mined ore. The above consumption figures assume no recharge to the aquifer through the tailing dumps.

Municipal

The Water Resources Division of Tucson's Metropolitan Utilities Management Agency estimates current urban/recreation ground-water withdrawal to be 100,000 acre-feet per year. Of this amount 37,000 acre-feet returns to the sewage treatment plant where it is treated and finally recharged into the ground-water aquifer. Currently, the Tucson metropolitan area receives approximately 50 percent of its water from the Tucson district, 30 percent from the well fields in the Sahuarita area and 10-20 percent from Avra Valley. These percentages indicate that of the 100,000 acre-feet withdrawn, 80,000 acre-feet are pumped in the Santa Cruz Basin. Assuming a 37,000 acre-feet per year recharge volume, the Santa Cruz Basin depletion is 43,000 acre-feet. Water pumped from Avra Valley (20,000 acre-feet) represents an interbasin transfer.

The total per capita water use within the region is approximately 200 gallons per day. This includes residential, commercial, industrial and park usage. Current population is placed at 450,000 for the Tucson metropolitan area. Projected growth figures through 2000 place Tucson's population between 675,000 and 880,000. These population increases will produce an estimated consumptive water demand of 75,000 acre-feet and 97,000 acre-feet respectively, assuming a potential recharge of 50 percent.

SANTA CRUZ BASIN WATER BALANCE

The Santa Cruz Basin water balance is a function of inflow (natural recharge + under flow) and outflow (pumping).

Inflow

Inflow into the basin consists of two components: natural recharge and under flow into the Basin from the Santa Cruz River.

Values for these components are estimated to be:

1.	Natural recharge	65,000 acre-feet/yr.
2.	Under flow	2,000 acre-feet/yr.

Outflow

Current depletion resulting from pumping by the agricultural, mining, and municipal sectors in the Santa Cruz Basin is estimated to be:

1.	Agriculture	30,000 acre-feet/yr.
2.	Mining	40,000 acre-feet/yr.
3.	Municipal	43,000 acre-feet/yr.
	Total	113,000 acre-feet

The Dilemma

Current water use is exceeding natural replenishment of the aquifer by 46,000 acre-feet annually. As a result, the Tucson and Sahuarita-Continental districts have been experiencing significant declines in the water table. For the period 1947-1965, water levels in the Tucson district have declined over 100 feet. The maximum decline was 172 feet in one location. Over the 22-year period (1947-1969) wells in the Sahuarita-Continental district have declined one to three feet per year. For the six-year period (1969-1975) average annual declines of six to twelve feet per year are common.

MANAGEMENT ALTERNATIVES

Various alternatives for water-use patterns in the Santa Cruz Basin to reduce water level declines have been discussed by several researchers and governmental agencies over the past 20 years. Among alternatives most frequently discussed are: 1) importation of water from the Colorado River; 2) municipal exchange of sewage effluent with mines or farms for fresh water; 3) interbasin transfers of water; 4) retirement of farmlands for water rights; and 5) implementation of a combination of alternatives.

Alternative 1 - Importation of Water from the Colorado River

The history of large scale diversions from the Colorado River dates back to the early 1900s when farmland in the Imperial Valley of California was irrigated with river water. Frequent flooding of farmland resulted in today's regulation of the lower Colorado River through a system of dams and diversion canals. Water allocation was accomplished through the Colorado River Compact and later the Boulder Canyon Project Act which invited the states of the lower basin to come to an agreement on the final diversion of water. Nevada receives 300,000 acre-feet annually, Arizona, 2.8 million acre-feet and California 4.4 million acre-feet.

To move the water from the Colorado River to the central part of the state will require an aqueduct system, and as early as 1947 the first Central Arizona Project (CAP) proposal was submitted to Congress for such a project. The final version of the CAP was enacted in 1968.

CAP consists of an aqueduct to carry water from Lake Havasu on the Colorado River to the Phoenix and Tucson areas (see Figure 5). The initial water volume would be about 1.6 million acre-feet per year, declining to about 380,000 to 600,000 acre-feet in 2035.

Tucson's CAP allotment of 50,000 acre-feet will represent half of the current 100,000 acre-feet of ground water being pumped for municipal uses. By 1985, it will be less than half of the anticipated demand; however, it will provide an important means of reducing the rate of withdrawal.

In addition to the municipal allotment, mining and farming interests in the Basin have proposed purchasing 83,000 acre-feet of CAP water by 1990 (Tucson Comprehensive Plan, 1975). If the CAP were operational today, then ground-water pumping would not exceed natural replenishment, but by 1985 when the CAP does open,the demand picture may be markedly different. Water requirements for the mining industry could possibly reach 160,000 acre-feet per year by 2000. Farm requirements are projected to decline to about 20,000 acre-feet per year by 1985 and stabilize at that level. City of Tucson consumptive requirements could increase to at least 75,000 acre-feet per year

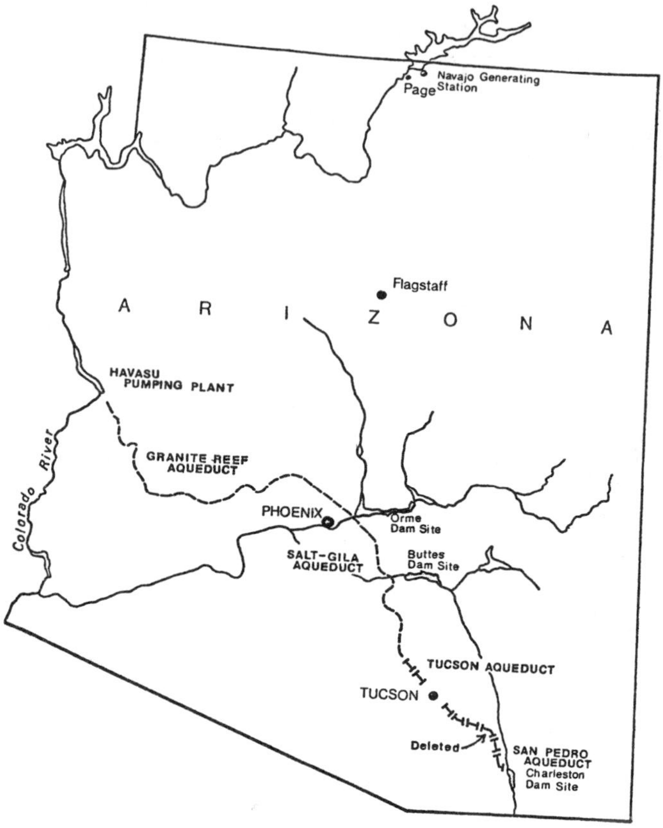

FIGURE 5

CENTRAL ARIZONA PROJECT MAJOR FACILITIES

by 1985 (The Central Arizona Project, 1974). Total annual requirements in the Basin could reach 225,000 acre-feet per year. If this growth occurs, the Basin overdraft will be 35,000 acre-feet per year, even though 133,000 acre-feet of CAP water will be supplied to the city, mines, and farms, and the 20,000 acre-feet of water importation from Avra Valley continues.

Alternative II - Municipal Exchange of Sewage Effluent with Mines
              or Farms for Fresh Water

Mines

As discussed in Alternative I, 37,000 acre-feet of Tucson's annual ground-water pumpage currently is recharged after passing through primary and secondary sewage treatment. This recharge is projected to reach 100,000 acre-feet per year by 2000. Ultimate reuse of this water, however, has been by agricultural interests in the lower Santa Cruz Basin, north of Tucson.

A possible alternative would be to pipe this municipal waste water to the mines as a substitute for ground water used in the milling process.

Tucson, in turn, could withdraw a like amount of ground water for municipal uses, and there would be a net reduction in withdrawals of ground water.

Fisher (1976) concluded that such a substitution was technically feasible, but that sewage effluent caused a measurable decrease in recovery of both copper (2.4%) and molybdenum (16.2%).

Farms

Most of Pima County's agricultural areas are west of Tucson in the Avra Valley, which is a separate ground-water basin.

Cluff and DeCook (1974) proposed the establishment of a metropolitan-operated district to provide wastewater for irrigation of farmland in the Avra-Marana area (Figure 6). Exchanging wastewater for ground water to be used in the City water system would be a viable alternative to Tucson's present practice of purchasing farmland and retiring it to acquire rights to the ground water. The City could pump up to 37,000 acre-feet of additional ground water in Avra Valley if the farms used the effluent and reduced their pumping by a similar amount.

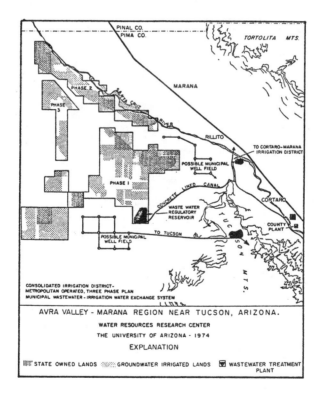

FIGURE 6

WASTEWATER FOR IRRIGATION OF FARM LAND

This alternative is not without its disadvantages because considerable acreage of lettuce is irrigated, for which secondary effluent cannot be used. The nitrate concentrate also would have to be diluted with fresh water before applying it to any agricultural crops. Such a solution would require up to 50 percent fresh water, thus decreasing the potential ground-water savings by a like amount.

Alternative III - Interbasin Transfer of Water

Since the early 1970s the City of Tucson and the mines have purchased and retired over 13,000 acres of irrigated farmland. About 5,000 acres of this land is located in Avra Valley, west of Tucson, in a separate ground-water basin. Once purchased, the City of Tucson established a well field and now transfers about 20,000 acre-feet per year (20 percent of Tucson's annual demand). Current plans call for Tucson to pump up to 40,000 acre-feet of ground water per year from Avra Valley by 1985. This represents a net gain for the Santa Cruz Basin, but the gain only decreases the over-draft from 46,000 acre-feet to 26,000 acre-feet annually.

To effectively balance current water use and replenishment in the Santa Cruz Basin, an additional 26,000 acre-feet of water importation per year is necessary. This amount is technically feasible to deliver, however, Avra Valley also is experiencing a serious ground-water overdraft because of pumpage for Tucson and agricultural uses; hence this approach would simply shift the overdraft burden from one basin to another unless other measures were taken.

Alternative IV - Retirement of Farmlands for Water Rights

The present policy of the City of Tucson and of the mines is to purchase and retire farmland to obtain its water rights. This policy has been in effect a number of years and 4,800 acres of farmland in Avra Valley have been retired by Tucson. Mining interests also have been purchasing farm-lands for water rights, especially south of Tucson in the Santa Cruz Basin. About 8,000 acres of farmland have been retired by mines purchases.

The Arizona Supreme Court is allowing the City to pump an average of 2.4 acre-feet of ground-water per acre of farmland retired. Assuming this to be an average ground-water pumping amount for the mines as well, almost 31,000 acre-feet per year is being diverted from agricultural use by the City and mines.

In 1973, 41,000 acres were still under cultivation in the Santa Cruz and Avra Valleys. Assuming a retirement of all but 6,000 acres of this acre-age over a ten-year period, an additional 8,400 acre-feet per year would be diverted for City and mining use. The total ground-water pumpage would be al-most 84,000 acre-feet per year after total farmland retirement.

As discussed in Alternative I, by 1985 mines and City requirements could total 235,000 acre-feet per year. If all but 6000 acres of farmland were retired, the Santa Cruz Basin would still be in an overdraft situation by 86,000 acre-feet per year.

Alternative V - Implementation of a Combination of Alternatives

The magnitude of potential water demand in the Santa Cruz Basin by 1985 precludes any one previously discussed alternative from meeting total water needs.

By applying all of the previously discussed alternatives, a water balance can be achieved for the Santa Cruz Basin by 1985.

Inputs and source are listed below:

Source	Alternative	Amount
Natural replenishment		67,000
CAP import	I	113,000
City effluent	II	50,000 (estimated)
Interbasin transfer	III	40,000
Farmland retirement	IV	84,000
	Total	354,000

Estimated demands are listed below:

Demand by 1985	Amount (acre-feet)
City of Tucson	150,000
Mines	160,000
Farms	20,000
Total	330,000

A positive balance can be maintained if Alternatives I - IV are implemented by 1985. The small surplus, however, will become a deficit if any one of the four alternatives is not implemented simultaneously with the other three alternatives.

CONCLUSION

The magnitude of the water demands in the Santa Cruz Basin precludes any single alternative from meeting those demands. This paper has attempted to deal with only the physical realities of water import, reuse, and transfer. No attempt has been made to justify the alternatives in terms of costs, political considerations, social attitudes or economic return to the community.

The amount of ground-water in storage, the length of time that the Basin can continue to be overdrafted and the consequences of the overdraft also were not discussed. Consistent water-level declines are a warning, however, that water demands are being fulfilled from a dwindling ground-water supply; use is exceeding natural replenishment.

This paper serves as a guide to the magnitude of the problem and emphasizes the need for immediate water management planning and implementation.

REFERENCES

Arizona Water Commission (1975), Phase I - Arizona State Water Plan, Inventory of Resources and Uses, Phoenix, Arizona

Cluff, C.B. and DeCook, K.J. (1974), Metropolitan Operated District for Sewage Effluent - Irrigation Water Exchange, Proceedings of the 1974 Meetings of the Arizona Section - Arizona Water Resources Association, and the Hydrology Section - Arizona Academy of Science, Vol. 4, 94-98.

Fisher, W.W. (1976), Utilization of Clear Water Sewage Effluent in Mineral
    Processing, Project Completion Report No. A-046-ARIZ, Water Resources
    Research Center, University of Arizona.

Gilkey, M.M. and Beckman, R.T. (1973), Water Requirements and Uses in
    Arizona Mineral Industries, U.S. Bureau of Mines Information Circular
    8162.

Matlock, W.G. and Davis, D.R. (1972), Groundwater in the Santa Cruz Valley,
    Agricultural Experiment Station. Technical Bulletin 194, University
    of Arizona.

Matlock, W.G., Schwalen, H.C. and Shaw, R.J. (1965), Progress Report on
    Study of Water in the Santa Cruz Valley, Agricultural Experiment
    Station Report 233.

Nebeker, J.S. (1973), Water Use In Arizona's Copper Industry, Paper
    Presented to the Arizona Water Pollution Control Association,
    Tucson, Arizona.

Schwalen, H.C. and Shaw, R.J. (1957), Water in the Santa Cruz Valley,
    Agricultural Experiment Station, Bulletin 288, University of
    Arizona.

The Central Arizona Project, as Staff Report to the Metropolitan Utilities
    Management Agency Board and the Mayor and Council of the City of
    Tucson (1974), Tucson, Arizona.

Tucson Comprehensive Plan (1975), Phase IV, Physical Development Guide:
    Water, City of Tucson, Arizona.

ECONOMICS OF DEVELOPING WATER SUPPLIES IN A SEMIARID REGION

Billy E. Martin

INTRODUCTION

Although California is not usually thought of as a desert, particularly by Californians, considerable portions of the State are semiarid. The climate of almost all of the State can be characterized as Mediterranean. This means that rain does not fall for six months in the summer. Accordingly, man has found it necessary to modify water supplies as they occurred in nature. This modification permits redistribution of water in a more timely sequence to areas where it has greater economic, social, and environmental benefits.

The Central Valley Project is one of the water systems which Californians and the United States Government have developed for that purpose. Conceived and built in response to the needs of the people of the State of California for food, flood protection, and jobs, the Central Valley Project has helped stabilize and enhance the economic and social development of the valley and the State. Irrigation from the project has turned both semiarid lands and marshlands into fertile fields. It should be possible to draw some lessons from experience with the Central Valley Project in the recent past to apply to other regions in the future. That is the purpose of this paper.

The basic factors which started Central Valley Project construction were a widely recognized need and a favorable social and political climate. Four further essentials for successful operation are water, fertile soil, money, and experienced technical and administrative people to employ these resources. Additionally, the prosperity of the project has been dependent upon the energy and "know-how" of its water users, and the advantages provided by a highly developed social and economic structure to market the crops and develop the industry and commerce of the cities.

To some extent, the four essentials of water, soil, money, and experienced staff are interchangeable. With enough money, water can be brought from distant places, or distilled from the ocean, soils treated, experienced staff hired, irrigators trained, and urban complexes developed. Conversely, water and fertile soils can constitute the foundation for a golden empire.

In its broader sense, water development economics goes beyond the concepts of classical economics. These concepts merely provide a system of accounting for part of the social and economic consequences of an actual or contemplated course of action. Usually they measure the resultant benefits primarily in monetary terms. Many project effects are nonmonetary, the so-called "intangibles." Thus classical economic methods applied to water development provide only a crude--although extremely useful--measure of project effects. This paper goes beyond the narrower concepts of classical economics to include consideration of the significant intangibles and "trade-offs" which the Central Valley Project has encountered since it went into operation 30 years ago.

865

Writing in the 1930's, when the project was still in the planning stage, the State Engineer of California emphasized the urgency of the need for the Central Valley Project in terms of the monetary losses which the Central Valley was experiencing from deficient water supplies, uncontrolled flooding, and seawater intrusion into the Delta. He provided a simple cost-revenue balance sheet to demonstrate the feasibility of the project. However, he also pointed out that unless something were done, a large area of land in the San Joaquin Valley would go out of production with a resultant loss to business, agriculture, and taxable wealth. As he was seeking financing, he emphasized those aspects of the project which were of interest to the entities which might be induced to provide it. For the Federal Government he mentioned its established policies relating to the support of navigation, flood control, and reclamation. For the State he cited as precedents appropriations for flood control and water development studies. For private entities, he noted advances of funds by San Francisco and Los Angeles commercial interests for studies of areas of distress in the San Joaquin Valley.

Since financing was the primary problem at the time (the beginning of the Great Depression of the 1930's), first priorities dealt with the economic impacts of the project and the social consequences of these impacts: the provision of jobs, the maintenance of agriculture to provide food and fiber, and the improvement of business.

Having successfully met the first priorities, and because these were being met, and taken for granted, additional social and environmental objectives have become incorporated into the project. These include recreation, fish and wildlife propagation, water quality control, and a variety of environmental considerations.

The preceding discussion points up two of the lessons to be learned from the experience of the Central Valley Project: the need for a sound economic and financial base, and the need to be prepared to respond to changing priorities in water development. Further details upon the method by which the Central Valley Project has been able to respond to changing priorities are contained in the final section of the paper: Changing Concepts.

Not only do social and political priorities change, but experience has been a great teacher in educating us about the physical impacts of the project. Some highlights of our experience are the unexpected growth of irrigation from the Delta-Mendota Canal and change of use from agricultural to urban for the Contra Costa Canal. In the environmental field has come additional understanding of the effects of water temperatures upon fish and upon agriculture, to cite but two examples of the changes wrought in stream regimes by large-scale reservoir storage.

This experience reinforces the need for building flexibility into the original project plans and providing contingency allowances to deal with the constant stream of operational changes which the passage of time and the pressure of events has entailed. This is the third great lesson to be gained from our operational experience with the Central Valley Project.

The following sections of this paper summarize project accomplishments in the economics field. They describe the physical features of the Central Valley which governed the engineering plans for the project, the project itself, environmental impacts, and the changing concepts which have and are influencing its development.

PROJECT ECONOMIC ACCOMPLISHMENTS

In 1975 the Central Valley Project delivered 6.8 million acre-feet
(8.4×10^9 m^3) of water. Most of this was used for agriculture, municipal
and industrial, and fish and wildlife purposes; with 2.6 million acre-feet
(3.2×10^9 m^3) used for the satisfaction of prior water rights. Project
powerplants produced nearly 6.1 billion kilowatt-hours of energy for project
pumping (1.1×10^9 kwh) and for commercial sale (5.0×10^9 kwh).

Twenty recreation areas, and 10 fishing access sites, and other areas
are now provided at various project facilities. The areas are used for
camping, fishing, picnicking, swimming, boating, water skiing, hiking, and
sightseeing, with attendance during 1975 estimated at 13 million visitor-
days.

The eight counties of the San Joaquin Valley are the heartland of
California's agriculture. The area contains 1.7 million people with an
employment base in excess of 600,000 and total personal income amounting to
$7.5 billion. Fresno County, located in the center of the San Joaquin Valley,
since 1951 has maintained its prominence as the Nation's most productive farm
county. Tulare and Kern Counties rank second and third in the State.

The Sacramento Valley's 10 counties include the governmental and technological center of the Central Valley. They have a total population of 1.2
million, employment of almost 400,000, and total personal income approaching
$5.0 billion.

Supplemental irrigation water is provided by the project to most of the
irrigated lands in the Central Valley. Gross income from farm operations for
the crop year 1972 in the Sacramento Valley is estimated at $547 million. In
the San Joaquin Valley, gross returns are estimated at $2.42 billion, or 45
percent of the State total, and more than 80 percent of the Central Valley's
gross farm receipts.

Fruits, nuts, grains, vegetables, sugar beets, alfalfa, rice--a few of
the crops grown--help to supply the increasing agricultural demands of the
State, the Nation, and the world.

REGIONAL GEOGRAPHY

The geography of the Central Valley and the nature of its climate
favored the expansion of irrigated agriculture and established the primary
basis for planning. The Mediterranean-type climate and long growing season
favored irrigated agriculture with its abundant and great variety of crops.

Physiography

The Central Valley Basin (figure 1) comprises two major river basins,
that of the Sacramento River on the north and the San Joaquin River on the
south. These two rivers join in the Sacramento-San Joaquin Delta. The combined basin extends nearly 500 miles (800 km) in a northwest-southeast direction and averages about 120 miles (190 km) in width. It includes more than
one-third of California. The basin is entirely surrounded by mountains except
for a narrow gap on its western edge through which the combined Sacramento
and San Joaquin Rivers flow to the Pacific Ocean through the San Francisco
Bay. The valley floor is a gently sloping, practically unbroken alluvial
plain which comprises nearly one-third of the basin area. The surrounding

two-thirds of the basin is mountainous. The southernmost portion of the San Joaquin Valley is a closed basin.

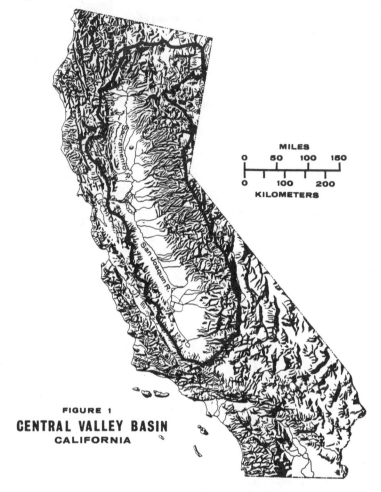

FIGURE 1

CENTRAL VALLEY BASIN
CALIFORNIA

To the east of the main valley is the massive Sierra Nevada with its rugged peaks, some of which are more than 14,000 feet (4,200 m) in elevation. The coast ranges, to the west of the main valley, are less rugged and lower in altitude, separating the interior Central Valley from the Pacific Ocean.

Population, agriculture, and industry are principally on the main valley floor. Heavier precipitation in the mountains provides the main source of the water supply essential for the successful development of the valley. The mountain regions, in addition, have considerable areas of land suitable for crop production and grazing purposes.

Climatic Characteristics

A controlling feature of the Central Valley development is the general climate of the basin. The main valley floor has a climate characterized by warm, dry summers with an almost complete absence of rainfall during the midsummer months, and mild winters with relatively light rainfall. The surrounding mountains are also generally warm and dry in summer, but winter temperatures, particularly in the Sierra Nevada, frequently drop below freezing. Precipitation in the mountains is much greater than on the valley floor, and a large portion of it falls in the form of snow.

The summer drought is caused largely by a subtropical high air pressure belt located off the coast.

The mild winter climate in the main valley is the result of two influences, the moderating effect of the Pacific Ocean, and the high mountain barrier of the Sierra Nevada. The valley floor is free from frost during the normal growing season, with the average warm period between freezing temperatures more than 7-1/2 months. The moderate winter climate of the Central Valley enables the production of citrus fruits, the less hardy deciduous fruits, and other specialized crops which required mild winters and long growing seasons. In certain areas the production of winter vegetables is important.

Precipitation and Runoff

The absence of rainfall during the summer months makes irrigation imperative for production of most agricultural crops. Winter rainfall in the coast ranges and lower foothills of the Sierra Nevada causes immediate stream runoff, often in flood proportions, which flows to the ocean. In the higher mountains the winter precipitation occurs principally as snow. Snowmelt in the spring and summer months results in a relatively large stream flow from rivers which rise in the high mountains. Because of the low natural flow in the rivers in the middle and late summer months, irrigation from unregulated runoff is limited and unsatisfactory in most areas.

The precipitation and runoff in the Central Valley vary from winter to summer of each year, and from year to year. In most of the valley the dependable supply of water which can be made available for irrigation or other uses is limited by the quantity of water available in a series of successive dry years.

Pre-Project Development

The Central Valley in 1933 had a population of about 880,000. About 3 million acres (1.2×10^6 ha) were irrigated. There were many large irrigation systems, some of them embracing 100,000 to 200,000 acres (40,000 to 80,000 ha) with only limited storage facilities. As a result, ground-water overdraft began to occur in water-deficient areas of the San Joaquin Valley. Hydroelectric power had been extensively developed in the Sierra tributaries of the Sacramento and San Joaquin Rivers. The agricultural products of the valley were sold nationally and internationally through a highly developed marketing system. Rights to water were governed by a large and complex body of water rights law which embraced both the riparian and appropriative doctrines.

Plans for the Central Valley Project had to be fitted into this highly developed social and economic system. Storage reservoirs and waterways were designed to be compatible with existing systems, and to minimize interference with established rights.

THE CENTRAL VALLEY PROJECT

Initial Planning and Construction

The concept of the Central Valley Project (figure 2) as a means to correct the imbalance of the water supplies in the Sacramento and San Joaquin Valleys was developed from the Marshall Plan by the State of California and the Reclamation Service, which later became the Bureau of Reclamation. Difficulties of State financing led to adoption of the project by the United States, and construction by the Bureau of Reclamation.

FIGURE 2
CENTRAL VALLEY PROJECT
FEATURES IN OPERATION

MAY 1977

By the 1930's the Central Valley faced growing problems of ground-water overdraft in the San Joaquin Valley, increasing seriousness of flood damage to developed areas, and in some cases the threat of ocean salinity intrusion. The initial features of the Central Valley Project were designed to meet these problems by storage of water at Shasta Dam on the Upper Sacramento River and its release down that river into the Sacramento-San Joaquin Delta. From the Delta, the water is pumped up into the San Joaquin Valley via Tracy Pumping Plant and the Delta-Mendota Canal to replace San Joaquin River flows stored at Friant Dam. Water from Friant Dam is in turn transported in the Friant-Kern

Canal to the southern San Joaquin Valley and in the Madera Canal to the
northern San Joaquin Valley. Another canal from the Delta, the Contra Costa,
transfers water toward the San Francisco Bay area.

Continuing Development

In 1937 construction began on the initial features, planned primarily for
development of irrigation supplies. During the 1940's while agriculture continued to expand, the valley experienced a rapid urban and industrial growth
which increased the demands for water and power. Different kinds of water
uses--municipal and industrial, recreation--became important and competitive.
Responding to this growth, Congress authorized additional features as they
were needed. These include Folsom and Nimbus Dams and Powerplants on the
American River in 1949, the Sacramento Canals in 1950, the Trinity River
Division in 1955, the San Luis Unit in 1965, Auburn-Folsom South Unit in 1965,
Tehama-Colusa enlargement and San Felipe in 1967, and Allen Camp Unit in 1976.

Each addition to the project is analyzed to determine if it meets the
tests of economic justification and financial feasibility. The economic
justification test measures and compares benefits and costs, resulting in a
benefit-cost ratio. The financial feasibility test determines whether the
proposed project can meet repayment obligations imposed by law and policy.

Besides the monetary benefits and costs measured in determining economic
justification, other aspects considered include:

1. Unmeasured and intangible effects such as providing jobs for new
workers entering the labor force,

2. Effectiveness of the plans in meeting overall objectives which might
include preservation of established economies in areas of rapidly declining
ground water,

3. Relationships of the plan to comprehensive development of large areas,
such as providing canal capacity for use by a separate development which
would move water through the project area, and

4. Consistency with national and State objectives for conservation and
development of natural resources.

The financial feasibility test determines the project's ability to meet
its repayment obligations, most commonly using the separable cost-remaining
benefits method of allocating costs. Rules and standards of repayment of
costs allocated to the various project purposes are established by a series
of Federal laws, dating back to the Reclamation Act of 1902.

Objectives

The main multiple functions of the Central Valley Project are to provide
a dependable and ample supply of water for irrigation in the Central Valley,
and for municipal and industrial uses. The facilities which store and deliver
this water provide many additional benefits, including power, flood control,
navigation, fish and wildlife enhancement, recreation, water quality improvement, and environmental protection and preservation.

Project service facilities are authorized, designed, and constructed
for the life of the project--conservatively estimated as 100 years--to meet

the increasing project water needs of a service area over a contract period
extending 40 to 50 years in the future. During the interim, from completion
of the project until water use in that area reaches full design capacity of
the project, additional water is available for other purposes either within
that area or in other service areas of the project.

Physical Features

Shasta Dam, north of Redding on the Sacramento River, is 602 feet (183 m)
high and creates a reservoir of 4.5 million acre-feet (5.6×10^9 m^3) of water.
Water released from Shasta Dam drops through a powerplant with an installed
capacity of 456,000 kilowatts.

Keswick Dam, on the Sacramento River below Shasta Dam, catches the
fluctuating flow from Shasta Dam and Trinity Division and controls the releases
through a 75,000-kilowatt powerplant into the Sacramento River.

Trinity Dam, an earthfill structure 538 feet (164 m) high and 2,450 feet
(747 m) long at the crest, conserves the flow of the Trinity River. From
Clair Engle Lake, the 2.4 million-acre-foot (3.0×10^9 m^3) reservoir behind
Trinity Dam, water flows through a 106,000-kilowatt powerplant into Lewiston
Lake.

Lewiston Dam releases some of the flow into the Trinity River; some it
diverts into a tunnel which runs about 11 miles (18 km) through the mountains,
to the Sacramento River Valley. During this transmountain diversion, water
is dropped through a 141,000-kilowatt powerhouse into Whiskeytown Lake on
Clear Creek.

Whiskeytown Dam, 282 feet (86 m) high, releases water to maintain a live
stream in the creek downstream from the dam, and diverts the majority of the
flow through a tunnel into a 150,000-kilowatt powerplant, which empties into
Keswick Reservoir.

Between Redding and Sacramento, some of the water released into the
Sacramento River from Keswick is diverted directly by more than 140 individuals and districts to irrigate half a million acres (200,000 ha).

A diversion dam at Red Bluff provides water for delivery along the west
side of the Sacramento Valley through the 21-mile-long (34 km) Corning Canal
and the uncompleted 122-mile (196 km) Tehama-Colusa Canal.

Near the city of Sacramento, Sacramento River flows are augmented by
water from the American River. Folsom Dam, 340 feet (104 m) high and 10,200
feet (3,110 m) long, stores the water in its 1-million-acre-foot (1.2×10^9
m^3) reservoir. Releases are made through a 199,000-kilowatt powerplant into
Lake Natoma behind Nimbus Dam. Nimbus Dam, with its 13,500-kilowatt powerplant, provides regulatory service on the American River.

Below the city of Sacramento, the Delta Cross Channel carries water from
the Sacramento River into natural channels of the Sacramento-San Joaquin Delta
leading to the Tracy and Contra Costa Pumping Plants. The Contra Costa Canal,
including a series of pumping plants, delivers water to the Antioch-Pittsburg-
Martinez area of San Francisco Bay, primarily for municipal and industrial use.

The Delta-Mendota Canal diverts water through an intake channel and the
six 22,500-horsepower (16,800 kW) pumps at Tracy lift water 200 feet (61 m).
The canal has a water depth of 18 feet (5.5 m) and is 48 feet (15 m) wide at

the bottom of its largest section. It supplies irrigation water to farms along its 116-mile (187 km) length, but delivers most of its water into the pool on the San Joaquin River formed by Mendota Dam. From Mendota Pool, the water is diverted principally into private canals providing an exchange source of water for the lands formerly irrigated by San Joaquin River flows and to wildlife refuges.

Friant Dam, a concrete structure 319 feet (97 m) high, extends 3,488 feet (1,063 m) across the San Joaquin River. Friant Dam regulates the San Joaquin River and diverts most of the regulated water 36 miles (58 km) to the north through the Madera Canal and 152 miles (245 km) to the south through the Friant-Kern Canal as supplemental irrigation supplies for more than 1 million acres (400,000 ha).

During the late fall, winter, and early spring, the Delta-Mendota Canal conveys surplus water from the Delta to the San Luis Unit for storage behind San Luis Dam, 3-1/2 miles (5.6 km) long, in a 2,100,000 acre-foot (2.6 x 10^9 m^3) reservoir. The San Luis Unit's major facilities are used jointly and are financed 55 percent by the State of California and 45 percent by the Bureau of Reclamation. The State's water delivered through the joint San Luis Unit and State facilities is being used primarily in Kern County and in the metropolitan areas along the coast of southern California. The joint-use facilities also include the San Luis Pumping-Generating Plant; the 102-mile-long (164 km) San Luis Canal, and the Dos Amigos Pumping Plant. The joint-use facilities were substantially completed in 1968. The Federal-only portion of the unit, including the relift pumping plants for the distribution system and the San Luis Drain, are scheduled for completion in the 1980's.

Under construction are Auburn Dam and the Folsom-South Canal in the eastern Sacramento and San Joaquin Counties, and the Tehama-Colusa Canal in the western Sacramento Valley. New Melones Dam on the Stanislaus River is under construction by the U.S. Army Corps of Engineers and will be operated by the Bureau of Reclamation as part of the Central Valley Project. The San Felipe Division and Allen Camp Units are authorized for construction as additions to the project to serve areas in the central coast and northern portion of the Sacramento Basin. Other water storage and conveyance facilities are being planned to serve farms and communities in the Central Valley.

Administration and Financing

The Central Valley Project is administered by the Bureau of Reclamation, an agency of the Department of the Interior, in accordance with the laws enacted by the Congress of the United States. Funds have been appropriated as required by the Congress for construction, operation, and maintenance of the main project features. Some of the project features are operated and maintained by local water agencies. Operating revenues, principally from the sale of water and power, are returned to the Treasury of the United States. The Central Valley Project also includes the San Luis facilities jointly administered and financed with the State of California, and five projects built by the Army Corps of Engineers that are authorized for integration into the Central Valley Project.

The total cost of the presently authorized features of the Central Valley Project is approximately $4+ billion, with construction of currently authorized facilities being almost 50 percent complete. About 10 percent of the project cost is allocated to nonreimbursable functions such as flood control, recreation, and fish and wildlife, which are considered to be national responsibilities. The balance of the project cost is anticipated to be repaid by local agencies, mainly the irrigation, municipal and industrial, and power customers.

Operation

The multipurpose objectives of the project are met by storage of stream-
flows, primarily in the winter and early spring months; and controlled
releases and transport of water for agricultural and other uses, principally
during the late spring, summer, and fall months.

An operational year comprises two relatively distinct periods. The
first is primarily the flood season (November through April) when the reservoirs (figure 3) are operated to provide flood control, thereby minimizing
flood damage and concurrently, insofar as operationally permissible, storing
the excess winter flows for use during the remainder of the year.

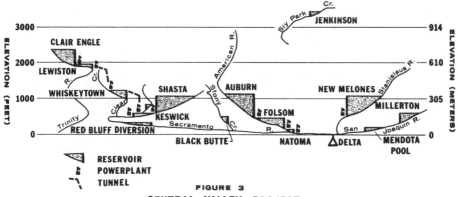

FIGURE 3

CENTRAL VALLEY PROJECT
SCHEMATIC PROFILE OF RESERVOIRS

The second period (March through September) is principally the water use
season when attention is shifted to the Sacramento-San Joaquin Delta and the
Central Valley Project reservoirs are operated on an integrated basis to meet
irrigation, municipal and industrial, navigation, fishery mitigation and
enhancement, and recreation demands. Substantial salinity control benefits
result from operation for other purposes. During both periods, hydroelectri-
cal power production is an important byproduct of project operations.

Water is pumped at Tracy Pumping Plant from the Delta on a year-round
basis. From late spring to early fall, the water is used primarily to meet
irrigation demands along the Delta-Mendota Canal and from the Mendota Pool.
From late fall, through winter and early spring, excess water in the Delta is
pumped at Tracy into the Delta-Mendota Canal for delivery and storage in
San Luis Reservoir. The water is released from storage as needed to meet the
water requirements in the Federal service area along the San Luis Canal.

Maintenance and modernization of the Central Valley Project facilities
is a continuing job, requiring the efforts of approximately 500 employees.

IMPACT OF DEVELOPMENT

Incidental Effects on the Natural Environment

Storage of winter floodwaters in Shasta Lake, and the use of the
Sacramento River as a channel for summer water releases altered the regimen

of the river and the activities of the people living and farming along its banks. Reductions in winter flooding along the upper river encouraged greater human encroachment on the flood plain.

Releases of the colder waters of Shasta Reservoir lower the summer water temperatures in the upper reaches of the Sacramento River. The colder waters made possible a trout fishery, but also slow the maturing of rice irrigated directly from the river.

The upper strata of water in the Central Valley Project reservoirs are warm in summer and do not mix with the lower, colder waters. At Folsom Dam on the American River, the outlets were modified so that the warmer water would be released first so that the lower and colder waters could be released to provide controlled temperatures for the fall salmon run.

It is difficult to prove what effect, if any, changes in Sacramento River stages due to project operations have had on the seepage and erosion problems in the alluvial valley. Stages are lowered in the winter but increased in the summer, and the net effect is further masked by tributary floodflows.

On the Trinity River, measures to alleviate the effects of siltation upon fish spawning gravels have been undertaken as part of an effort by an Interagency Task Force.

To compensate or enhance fishery resources, fish hatcheries have been constructed and water releases made. A special type of louvered fish screen was developed at Tracy Pumping Plant. Special provisions for spawning channels were incorporated in one of the Sacramento Valley canals. To enchance fishery, project reservoirs are stocked with sport fish, and reservoir releases are made during October through December to the Sacramento and American Rivers to provide suitable riverflows during the period of fishery migration and propagation.

The major irrigation deliveries from Mendota Pool are made to the irrigation districts with prior water rights on the San Joaquin River. Their demands are met by the Central Valley Project by exchange through the Delta-Mendota Canal and Mendota Pool. As an essential part of that exchange, the San Joaquin River water stored by Friant Reservoir is delivered to districts between Fresno and Bakersfield by the Friant-Kern Canal and north of Friant by the Madera Canal. These surface water deliveries are integrated with ground-water pumping by the irrigation districts to create one of the most productive agricultural areas in the Nation.

San Joaquin Valley dryland grain and hay acreages are being replaced with permanent irrigated orchard and vineyard crops at the rate of 70,000 acres (28,000 ha) per year.

The protection of the unique blend of environmental and natural resources is of vital concern to the Bureau of Reclamation as well as the millions of Central Valley residents.

Changing Concepts

Changes in the economic and social climate have conditioned the human response to the project's operation.

The increase in the population of California has been dramatic, rising from 6 million in 1930 to 21.5 million at present. The economy of the State has moved through a depression into a war, a period of growth and expansion, and then into a period of inflation.

In earlier periods, employment opportunities and economic accomplishments of the project were emphasized. Since then recreation, fish and wildlife propagation, water quality control, and a variety of environmental considerations have been incorporated in project objectives.

Change in emphasis can be attributed in part to the success of the project in meeting its early economic objectives of providing food, fiber, and energy. The accomplishments are now taken for granted as part of a familiar environment.

Experience with these changes has demonstrated the need for not only providing the generous contingency allowance in constructing project features, but also building as much flexibility into project plans as possible.

This has been shown by experience with Contra Costa and Delta-Mendota Canals. World War II accelerated the growth of California's population. Agrarian areas are being replaced by urban and industrial developments, and choice farmlands are giving way to suburban housing. The Contra Costa Canal, designed before World War II, mainly to provide for agriculture, now operates mainly to supply water for municipalities and industries. The Delta-Mendota Canal was originally conceived as a transfer canal with no water deliveries en route, but a demand soon developed to divert from it to irrigate adjacent lands.

Stage construction provides a large measure of the flexibility which enables the Central Valley Project to respond to changing needs. General plans for ultimate basin development, prepared from time to time by State and Federal agencies, have furnished a framework into which specific new features can be fitted.

The Bureau of Reclamation is working with the California Department of Water Resources and the Corps of Engineers to develop needed water for augmentation of the Central Valley Project and State Water Project supply. Cottonwood and Marysville developments on Cottonwood Creek and the Yuba River are already authorized. In addition, under study are several off-stream reservoir sites, including the Glenn Reservoir complex and Sites Reservoir in the Sacramento Valley, and the Los Vaqueros and Los Banos Reservoirs south of the Delta. Water conservation, waste water reclamation, and ground-water management are also under study. These developments could produce as much as 3.4 million acre-feet of new water supply; enough to go a long way toward meeting the anticipated 4 million acre-feet of needed new supplies.

In addition to the basic storage facilities, additional conveyance facilities would be required. Principally, these include a Peripheral Canal; a Hood-Clay Connector; South Delta Facilities; Contra Costa Canal intake relocation; Delta-Mendota Canal enlargement; added State Water Project pumps, and a Mid-Valley or an East Side Canal system.

During the authorization process prior to construction, each proposed feature is reviewed to assess its ability to update and imporve the project in the light of current needs. Examples of this process are the inclusion of specific authorization for recreation and water quality releases at newer reservoirs such as New Melones. Similarly, the additional features meet increasing needs for flood control, water supply, power, and fish and wildlife resources.

An environmental assessment is made of each new feature before construction starts. If the impact is major, an environmental impact statement is also required, and positive steps are implemented to insure minimal degradation to the environment.

Further analyses of the project's ability to meet today's needs are being made in a Total Water Management Study now in progress. This study considers water and related resources along with the legal and physical structures which manipulate these resources as an integrated system for man's benefit.

Much of the valley's economic development has been achieved because of the availability of the right amount of water, at the right place, and at the right time. Both agriculture and industry have benefited, while the threat of flooding has been practically removed from most areas.

MANAGEMENT OF A GROUNDWATER DEPOSIT*

James E. Osborn**

INTRODUCTION

Management of a groundwater deposit is dependent on the social, economic,
and political environment which was dominant during the planning phases for a
region with the natural resource. Ownership of the groundwater has the determining force on the use of the resource. However, political actions by regional, state, and national units have, in some instances, developed policies to
control the use of groundwater which has ownership rights vested with land-
owners.

Numerous social, political, and economic situations can be cited for
specific management schemes for groundwater deposits. This paper will emphasize the situation in Texas and more specifically in the Texas High Plains.

The Texas Legislature was very specific when it noted, "The ownership and
rights of the owner of land, his lessees and assigns, in underground water are
hereby recognized,...". This unequivocal recognition of private ownership of
groundwater is embodied in House Bill 162, 51st Legislature, 1949. It is commonly referred to as the Underground Water Conservation Districts' Law that
makes possible the creation of local groundwater conservation districts. There
are three active conservation districts in the Texas High Plains.

The Texas Water Law provides for conservation districts to:

1. Make and enforce rules to provide for conserving, preserving,
    protecting, recharging and preventing waste of the underground
    water.

2. Employ registered professional engineers to make surveys of
    the underground water reservoir.

3. Develop comprehensive plans for the most efficient use of
    the underground water.

4. Carry out research projects, develop information, and deter-
    mine limitations which should be made on withdrawing under-
    ground water.

*Paper presented at the Conference on Alternative Strategies for Desert
Development and Management sponsored by the California Department of Water Re-
sources and the Programme of Future Studies of the United Nations Institute for
Training and Research, Sacramento, California, May 31 to June 10, 1977.
**Professor of Agricultural Economics, Texas Tech University, Lubbock, TX.

878

5. Collect information regarding the use of underground water
    and the practicability of recharging the reservoir.

6. Publish its plans and the information it develops.

7. Acquire land to erect dams or to drain lakes, draws, and
    depressions, draws, and creeks; and install pumps and other
    equipment necessary to recharge the underground water
    reservoir.

8. Require that records be kept and reports be made of the
    drilling, equipping, and completing of water wells and of
    the production and use of underground water.

9. Require permits for the drilling, equipping, or completing
    of wells, or for substantially altering the size of wells
    or well pumps.

10. Require that accurate drillers' logs be kept of water wells
    and that copies of drillers' logs and electric logs be filed
    with the district.

11. Provide for the spacing of water wells and may regulate the
    production of wells.

12. Enforce its rules by injunction, mandatory injunction, or
    other appropriate remedy in a court of competent jurisdiction.

Water Conservation Districts are generally funded with advalorem taxes
within the district. The remainder of this paper will concentrate on the
economic effects from the Ogallala on the Texas High Plains economy as the
natural resource is mined.

TEXAS HIGH PLAINS

The Texas High Plains is a major cotton, grain sorghum and wheat producing
area. One-fifth of the total U.S. and three-fourths of the total for Texas
cotton was produced in a 56 county area in the Texas High Plains in 1971 (5).
Of the total wheat production in Texas for 1971, 81 percent was grown in the
56 county area of the Texas High Plains (6).

Groundwater, which is exhaustible and non-renewable, is a major limiting
resource in producing crops in the semi-arid area. The estimated total value
of irrigated crops was $633 million (82 percent of the value of total crop
production) in the 56 counties in 1967. The number of irrigated acres increased from 3.0 million acres in 1954; to 4.1 million in 1959; to 4.5 million
in 1965; and to 4.8 million in 1969 (8:9).

Groundwater is the primary source of irrigation water. The source of
groundwater is an aquifer called the Ogallala. The Ogallala does not receive
a significant volume of recharge when compared to the withdrawal by pumping
for irrigation. The water table has declined at a fairly rapid rate (3 to 4
feet per year) in the areas where irrigation has been developed. Natural recharge has been estimated to range from 10 to 15 percent of the annual withdrawal. The formation can be recharged from surface sources, but technological
and environmental problems have not made artificial recharge feasible on a
large scale. The quantity of water that is available for recharge is less than

the amount pumped at the present time. Although the aquifer underlies large portions of the Texas High Plains, it is relatively thin (75 to 100 feet) in many areas.

Studies have indicated that development of irrigation and associated bene-fits will increase until approximately 1980 and then decline as the groundwater supplies become exhausted. Grubb has estimated that benefits from irrigation will increase to $517 million by 1980 and decline to $195 million by 2020 (1, p. 21).[1]/ The production of crops will revert to dryland production as the supply of groundwater is exhausted. Hughes and Harman have estimated that cotton and grain sorghum output in the Southern High Plains will decrease by 65 and 90 percent, respectively, by 2015, from the 1966 levels for the most de-veloped portion (2.8 million acres) of the Ogallala formation (2, p. 23).

Growth of the cattle feeding industry in recent years has been a major contribution to the Texas High Plains economy. Cattle feedlots with capacity of one thousand head or more, have increased from 105 feedlots in 1968 to 118 in 1971 for this region (3).

The swine feeding industry is experiencing a rapid growth in the Texas High Plains. There was an estimated 72 percent increase in the number of hogs located in a 56 county area of the Texas High Plains from 1968 to 1971. Most of the growth occurred in large confinement feeding operations in the region.

A major factor which is contributing to the growth of the industry for feeding livestock in the High Plains of Texas is the supply of feed grains produced in the region. The production of a large quantity of feed grains is providing an economical and stable source of feed for cattle and swine feeding operations. The feed grains produced are vital to the growth of the feedlot industry.

The depletion of the supply of groundwater in the Texas High Plains will reduce the amount of feed grain that can be produced. The total effect of a reduced supply of feed grain for the region's feedlot industries is not known, but the expected decrease will partially eliminate an economical source of feed for the feedlots. The continued growth of the feedlot industries will require sources of additional feed grains.

Description of Texas High Plains

The study area included 56 contiguous counties in west and northwest Texas which was called the Texas High Plains. The average size farm in the study area was 1,342 acres in 1969 (10). Farms in the intensive irrigated areas averaged 708 acres in 1969. The total value of crop and livestock production in the study area was $1.3 billion in 1970 which included 59.3 percent from crops (8). Agricultural production in predominant throughout the study area. However, rainfall varies from eight inches in the southern counties to 21 inches in some of the northern counties in the study area. In addition, the amount of rainfall varies widely from year to year.

Population in the study area was 960,479 in 1970 which indicated a decline of 3.4 percent from 1960 and comprised 8.2 percent of the population of the state of Texas in 1970 (11). There are four Standard Metropolitan Statistical Areas which included over 50 percent of the population in the study area. Al-

---

1/ The benefits were based on net effects from irrigation which include primary, secondary, and tertiary benefits.

though the population decreased by 3.4 percent from 1960 to 1970, employment increased by one percent in the same period.

GENERAL PROCEDURES

An input-output model with 98 processing sectors, five final payment sectors, and six final demand sectors was used. Data were assembled on a producer's price FOB for the shipper for 1967. An establishment (individual firm) basis for collection of data with a stratified (by employment) random sample was used for all sectors except some of the agriculture and construction sectors. Data were collected for major purchases by sector and location as well as taxes, salaries, wages, depreciation, total sales, inventories, and employment. The predominant method of data collection was personal interviews with the establishments.

Technology was assumed to be constant for the study. Particularly, water saving technologies in irrigation of crops were not considered in the analysis. Although several water saving techniques are being developed, it is not known how significant the developments will be in terms of water use efficiency. The relationships of production inputs (column vector of direct requirements per acre) for each agricultural sector were assumed to be constant.

Irrigated acres for the 56 county area were projected for each decade from 1970 through 2010 as well as 2015. The projected acres were based on linear programing and information about availability of groundwater. The information about groundwater was based on results of work by the Texas Water Development Board. The projected acres of dryland and irrigated production were used in the model to determine the effects of the declining groundwater supply on the regional economy. In turn, the growth in the livestock feeding sectors was estimated so that the associated economic effects could be determined.

The economic effects of each of the processing sectors of the input-output model that were generated by the crop, feedlot, and meat processing sectors were estimated for each of the seven study periods with interindustry coefficients. The effects resulting from estimated adjustments by the eight crop sectors from declining supplies of groundwater were calculated for all processing sectors. The effects resulting from changes in estimated growth in the feedlot sectors as well as the beef and pork processing sectors were calculated without double counting of interindustry effects. Although the purchase of feed grain by feedlots was in interindustry impact from feedlots, this estimate was included in the crop effects. The combined effects were then determined to estimate the total economic effects for each of the processing sectors for each of the study periods 1967, 1970, 1980, 1990, 2000, 2010, and 2015.

ECONOMIC PROJECTIONS

Total output for irrigated crop sectors was estimated to increase from $633.1 million in 1967 to $639.7 million in 1970 (Table 1). The output for irrigated crop sectors was estimated to decrease to $366.0 million in 2015 for a 42.4 percent decrease from 1967. The output of dryland crop sectors increased from an estimated $142.7 million in 1967 to $240.0 million in 2015 or an increase of 68.2 percent. The total output of the crop sectors declined by 21.9 percent from 1967 to 2015 or $169.8 million.

The economic effects from the output of the crop sectors was used to estimate the impact of the sectors on the regional economy (Table 2). The

Table 1.   Total Output of Irrigated and Dryland Crop Sectors, Texas High
Plains, 1967-2015

Year	Irrigated		Dryland		Total	
	Output ($1,000,000)	Percent of 1967	Output ($1,000,000)	Percent of 1967	Output ($1,000,000)	Percent of 1967
1967	633.1	100.0	142.7	100.0	775.8	100.7
1970	639.7	101.0	141.5	99.2	781.2	100.7
1980	559.0	88.3	176.7	123.8	735.7	94.8
1990	506.6	80.0	199.2	139.6	705.8	91.0
2000	457.9	72.3	212.1	148.6	670.0	86.4
2010	418.0	66.0	221.7	155.4	639.7	82.4
2015	366.0	57.8	240.0	168.2	606.0	78.1

Table 2.   Economic Effects from Crop Sectors in the Texas
High Plains, 1967-2015

Year	Sectors		Total Economic Effects
	Chemical	Banking & Credit	
	- $1,000,000 -		
1967	52.9	43.4	2,045.2
1970	51.7	43.9	2,189.8
1980	44.2	40.9	2,031.2
1990	38.4	37.4	1,996.7
2000	34.2	33.8	1,895.8
2010	31.6	32.5	1,819.8
2015	27.2	29.5	1,718.5

economic effects include direct, indirect, and induced effects from the production of crops. The effects were estimated to increase from $2,045.2 million
in 1967 to $2,189.8 million in 1970. However, total effects were estimated to
decline to $1,718.5 million by 2015 for a 16.0 percent decline from 1967.

Economic activity in the Chemical Sector that was generated from the output
of the eight crop sectors was $52.9 million in 1967 (Table 2). The output as-
sociated with crop production was estimated to decrease by 16.4 percent in 1980.
The total output for 2015 was estimated to decrease to $27.2 million, a 48.6
percent reduction from the 1967 total value. The 48.5 percent decrease in
economic activity is attributed to a smaller volume of fertilizers, pesticides,
and herbicides required by the agricultural industry. As irrigated acreage was
transferred to dryland acreage in each consecutive study period, the amount of

chemicals used in the production of the eight field crops decreased. The output of the Chemical Sector in 1967 was $337.4 million. The amount which was associated with the output of the crop sectors was $52.9 million or 15.7 percent in 1967.

Output of the Banking and Credit Sector that was generated from the eight crop sectors was $43.4 million in 1967, decreasing to $40.9 million by 1980 (Table 2). A 32.0 percent decrease in output from 1967 to 2015 was estimated with the total value for 2015 being $29.5 million. The decrease in irrigated acreage in the Texas High Plains was the primary factor for the estimated decrease in activity within the Banking and Credit Sector. The investment required for crop production was reduced when irrigated acreage was transferred to dryland acreage as the supply of groundwater declined in each consecutive study period of the study. The Banking and Credit Sector received $43.4 million in 1967 of its $170.1 million from economic activity associated with the output of the crops sectors.

The output of the Cattle Feedlot Sector was estimated to increase by 98.6 percent from 1967 to 1970 and by 49.5 percent from 1970 to 1980 (Table 3). A decrease in output from 1980 to 2015 was projected. The decrease is directly related to the projected decrease in feed grain output in the Texas High Plains. The output of the Cattle Feedlot Sector for 1967 was estimated to be $267.6 million and to increase to $794.4 million by 1980 (Table 3). The number of cattle that were projected to be fed was increased by 30 percent per year. This was the annual growth rate for recent years. The number was increased until the available feedgrains were exhausted. Feedgrains were not assumed to be imported into the area for feeding. As the available supplies of feedgrains declined, the total output of the Cattle Feedlot Sector was assumed to decline. The projected value of feedlot cattle in 2015 was estimated to be $479.8 million, a decrease of 39.6 percent from 1980.

Table 3. Output of Feedlot Livestock Sectors, Texas High Plains, 1967-2015

Year	Cattle Feedlot		Swine Feedlot		Total	
	Output ($1,000,000)	Percent of 1967	Output ($1,000,000)	Percent of 1967	Output ($1,000,000)	Percent of 1967
1967	267.6	100.0	9.4	100.0	277.0	100.0
1970	531.4	198.6	23.7	252.0	555.1	200.4
1980	794.4	296.9	33.8	359.3	828.2	298.9
1990	699.2	261.3	27.7	294.7	726.9	262.4
2000	594.3	222.1	21.8	231.9	616.1	222.4
2010	564.5	211.0	21.8	231.9	586.3	211.6
2015	479.8	179.3	15.7	167.0	495.5	178.8

The output of the Swine Feedlot Sector in the study area increased an estimated 152 percent from 1967 to 1970 and was projected to increase by 42.6 percent from 1970 to 1980 (Table 3). The output of the Swine Feedlot Sector was assumed to expand by 25 percent per year when local feedgrains were available. This rate is based on recent annual rates of growth. Output was an estimated $9.4 million in 1967 and $33.8 million in 1980. From a peak in output in 1980, the output of the Swine Feedlot Sector was projected to decrease

53.6 percent by 2015. The projected output of the Swine Feedlot Sector in 2015 was $15.7 million. The projected output for 2015 indicated a 67.0 percent increase over the estimated output in 1967 for the Texas High Plains.

The output of other economic sectors that could be attribubed to the feedlot and meat processing sectors in 1967 for the study area was estimated to be $827.4 million (Table 4). Total value of economic activity increased by 100.2 percent from 1967 to 1970. The economic effects associated with the growth in the feedlot and meat processing sectors in the Texas High Plains continued until 1980 when the total estimated value was estimated to be $2,433.1 million. For each consecutive study period after 1980, the total value decreased. The value of economic activity generated from the feedlot and meat processing sectors in 2015 was estimated to be $1,477.9 million, a decrease of 39.5 percent from 1980.

Table 4. Economic Effects from the Feedlot and Meat Processing Sectors, Texas High Plains, 1967-2015

Year	Sectors		Total Economic Effects
	Chemical	Banking & Credit	
	- $1,000,000 -		
1967	5.4	18.0	827.4
1970	10.7	36.2	1,656.8
1980	15.5	53.7	2,443.1
1990	12.9	46.1	2,172.9
2000	10.8	38.3	1,841.6
2010	10.2	36.4	1,754.1
2015	8.2	30.4	1,477.9

Economic activity in the Chemical Sector that could be associated with the livestock feedlot industry increased by 187.0 percent from 1967 to 1980. Although the level of economic activity that was generated by the livestock feedlot sectors was estimated to decrease from $15.5 million in 1980 to $8.2 million in 2015, the amount in 2015 was 51.8 percent greater than the 1967 value.

Capital expenditures which were required for the feedlot and meat processing industries increased the output in the Banking and Credit Sector in the Texas High Plains economy. The economic activity within the Banking and Credit Sector that was attributed to the feedlot and meat processing sectors in 1967 was estimated to be $18.0 million (Table 4). The rapid growth of the feedlot and meat processing industries in the study area from 1967 to 1980 increased the value of economic activity in the Banking and Credit Sector to $53.7 million, or an increase of 198.3 percent. The value of economic activity for 2015 was estimated at $30.4 million, a 47.5 percent decrease from the 1980 value. Although there was an estimated 43.5 percent reduction from 1980 to 2015, the value of economic activity for 2015 indicated a 68.9 percent increase over the estimated value for 1967.

The estimated economic activity for 1967 in the study area that was generated from the crop, feedlot, and meat processing sectors was $2,872.6 million (Table 5). Of the estimated value for 1967, 71.2 percent was generated by requirements of the crop sectors. The crop sectors generated $2,045.2 million in 1967 and the feedlot and meat processing sectors generated $827.4 million.

The study periods 1990, 2000, 2010, and 2015 indicated consecutive decreases of economic activity in the Texas High Plains that was generated from the eight crop sectors, the two feedlot sectors, and the two meat processing sectors (Table 5). The estimated economic activity for the study area in 2015 generated by the crop, feedlot, and meat processing sectors was $3,196.4 million, a decrease from the estimated value of $4,474.3 million in 1980. Although the

Table 5. Economic Effects from the Regional Crop, Feedlot, and Meat Processing Sectors, Texas High Plains, 1967-2015

Year	Crop Sectors	Feedlot Processing Sectors	Total
	- $1,000,000 -		
1967	2,045.2	827.4	2,872.6
1970	2,189.8	1,656.8	3,846.6
1980	2,031.2	2,443.1	4,474.3
1990	1,996.7	2,172.9	4,169.6
2000	1,895.8	1,841.6	3,737.4
2010	1,819.8	1,754.1	3,573.9
2015	1,718.5	1,477.9	3,196.4

projected value for 2015 indicated a 28.6 percent decrease from the value of 1980, the value in 2015 was 11.2 percent higher than the value for 1967. The economic activity generated from the crop sectors was $1,718.5 million in 2015 compared to a value of $1,479.9 million generated from the feedlot and meat processing sectors.

Although the economic effects from crop production reached a maximum in 1970, the impact of the feedlot livestock sectors did not attain a maximum until 1980. The effects from crop sectors were 147.2 percent greater than the effects from the feedlot livestock sectors in 1967. However, the effects from feedlot livestock sectors exceeded the effects from the crop sectors from 1980 to 1990. The economic effects from the feedlot livestock sectors were sufficient in 2015 to compensate for the decline in economic activity from crop sectors as groundwater supplies declined. That is, total economic activity that can be associated with the crop and feedlot sectors was greater in 2015 than in 1967.

ALTERNATIVE SOURCES OF WATER

The Bureau of Reclamation has conducted a study for diverting excess water from the Mississippi River to West Texas and Eastern New Mexico (12). A system of canals, reservoirs, and pumping plants would convey the water to terminal reservoirs in the High Plains. The system would total over 1,400 miles of canals and about 71 pumping plants. It would require the equivalent of 50-2/3 billion kilowatt hours of energy. The estimated construction costs of the system to deliver 5.8 million acrefeet annually was $16.6 billion in January 1972 prices. Estimated costs per acre foot was $125 for the Texas High Plains.

Recently, a consulting firm was employed by the Texas Water Development Board to assess the feasibility of diverting surplus water from Arkansas (7). The results of the study indicated that six million acre feet are available for diversion. The report included proposed systems for moving surplus water to Texas. However, estimated costs for the diversion as well as movement from East Texas to West Texas were not included.

At the present time, the Texas Water Development Board is developing a Water Plan for Texas. This document, when completed, will replace a prior Texas Water Plan which was published in 1968.

SUMMARY

Various management schemes can be used for utilizing water from ground-water deposits. In the Texas High Plains, groundwater conservation districts have been developed since groundwater rights to the Ogallala are vested with the land owner.

Since the Ogallala is a finite supply of groundwater, it is being ex-hausted. As the supply of groundwater is exhausted, economic effects will occur throughout the area. The estimated economic effects were projected through 2015 in this paper.

Selected Bibliography

1. Grubb, Herbert W. Importance of Irrigation Water to the Economy of the Texas High Plains. Texas Water Development Board Report No. 11, January, 1966.

2. Hughes, William F. and Wyatte L. Harman. Projected Economic Life of Water Resources, Subdivision Number 1, High Plains Underground Water Reservoir, Texas Agricultural Experiment Station Technical Monograph 6, December, 1969.

3. Texas Department of Agriculture and U.S. Department of Agriculture Texas Livestock Statistics, 1968-1971. Texas Crop and Livestock Reporting Service, Bulletin 77, May, 1971.

4. Texas Department of Agriculture and U.S. Department of Agriculture Texas Cotton Statistics. Texas Crop and Livestock Reporting Service, Bulletin 91, June, 1972.

5. Texas Department of Agriculture and U.S. Department of Agriculture Texas Field Crop Statistics. Texas Crop and Livestock Reporting Service, Bulletin 82, October, 1971.

6. Osborn, James E. and William C. McCray. An Interindustry Analysis of the Texas High Plains: Part I. Division of Planning Coordination, Office of the Governor (Texas Tech University, College of Agricultural Sciences Technical Publication T-1-101). April, 1972.

7. Stephens Consultant Services, Inc. An Assessment of Surface Water Supplies of Arkansas. Little Rock, Arkansas, 1976.

8. U.S. Department of Agriculture. 1970 Texas County Statistics. Statistical Reporting Service, Austin, 1971.

9. U.S. Department of Commerce. Census of Agriculture. Bureau of the Census, Washington, D. C., 1959.

10. U.S. Department of Commerce. Census of Agriculture. Bureau of the Census, Washington, D. C., 1969.

11. U.S. Department of Commerce. Census of Population. Bureau of the Census, Washington, D. C., 1970.

12. U.S. Department of the Interior. West Texas and Eastern New Mexico Import Project. Bureau of Reclamation. Washington, D. C., 1973.

URBAN WATER USE IN A DESERT ENVIRONMENT--
CALIFORNIA AND SOUTHWESTERN UNITED STATES
By
Glenn B. Sawyer
Supervising Land and Water Use Analyst
Department of Water Resources
The Resources Agency

INTRODUCTION

The growth and nature of urban water use in the arid and semiarid portion of
the southwestern United States, to a large extent, has been a function of
the unique climate conditions of this area. The long frost-free crop growing
season and the attractiveness of the dry climate to people of the eastern part
of the country have combined to encourage development of the very limited
water supplies to sustain a large and very diverse economic development.

From the earliest colonization by the Spanish to the present day, water supply
development has been a principal effort of the people. This development has
ranged from minor stream diversions and well installations by individuals to
vast water storage and delivery facility construction by public districts and
by local, State, and Federal governments. Population growth has continued
to the point where, today, the options for satisfying water demands have become quite limited. One of these options, conservation, i.e., using less
water to accomplish the same purposes, is receiving a great deal of attention.
The following discusses the climate of the area and outlines some of the
history of economic growth and water development, the nature of current
water use, and the possible accomplishments of water conservation.

CLIMATE

Figure 1 depicts the range of precipitation zones found in the southwestern
United States. Table 1 presents the long-term annual and monthly distribution
of precipitation at various major cities in the area.

A major characteristic of precipitation found throughout the area which has
been one of the most critical factors to deal with in securing sufficient
water for economic growth is that very little precipitation occurs for nearly
half the year in most areas. This period is the same period that water use
rates for most purposes are at the maximum. This has required either development of surface water storage facilities, sometimes at locations far removed
from the place of water use, or reliance on ground water.

TABLE I: NORMAL MONTHLY AND ANNUAL PRECIPITATION
AT LOCATIONS WITHIN SOUTHWESTERN UNITED STATES

State	Weather Station Location	Normal Annual Precipitation (Inches)	Normal Monthly Precipitation (inches)											
			Jan.	Feb.	Mar.	Apr.	May	June	July	Aug.	Sept.	Oct.	Nov.	Dec.
Arizona	Phoenix	7.05	0.71	0.60	0.76	0.32	0.14	0.12	0.75	1.22	0.69	0.46	0.46	0.82
	Tucson	11.00	0.82	0.84	0.53	0.27	0.13	0.29	2.06	2.88	1.00	0.64	0.62	0.92
California	Red Bluff	22.05	4.29	3.31	2.70	1.83	1.13	0.45	0.04	0.06	0.37	1.37	2.29	4.21
	Sacramento	17.22	3.73	2.68	2.17	1.54	0.51	0.10	0.01	0.05	0.19	0.99	2.13	3.12
	Fresno	11.14	2.03	2.19	1.96	1.13	0.30	0.07	-- *	0.01	0.10	0.43	0.95	1.97
	Bakersfield	6.36	1.17	1.14	1.06	0.81	0.22	0.09	0.01	-- *	0.08	0.32	0.49	0.97
	Los Angeles	14.05	3.00	2.77	2.19	1.27	0.13	0.03	0.00	0.04	0.17	0.27	2.02	2.16
Colorado	Denver	12.95	0.52	0.54	1.00	1.55	2.24	1.62	1.53	1.02	0.94	0.97	0.65	0.37
	Pueblo	11.84	0.31	0.48	0.52	1.18	1.80	1.22	1.82	1.85	0.84	0.99	0.53	0.30
New Mexico	Albuquerque	7.77	0.30	0.39	0.47	0.48	0.53	0.50	1.39	1.34	0.77	0.79	0.29	0.52
	Roswell	11.62	0.48	0.42	0.50	0.73	1.28	1.05	1.77	1.62	1.82	1.07	0.34	0.54
Nevada	Las Vegas	3.90	0.53	0.44	0.35	0.23	0.08	0.04	0.50	0.48	0.34	0.20	0.31	0.40
	Reno	7.20	1.21	0.86	0.70	0.47	0.66	0.40	0.26	0.22	0.23	0.42	0.68	1.09
Texas	El Paso	7.77	0.39	0.42	0.39	0.24	0.32	0.60	1.53	1.12	1.16	0.78	0.32	0.50
	Amarillo	19.67	0.65	0.62	0.82	1.32	3.37	2.89	2.34	2.58	1.89	1.76	0.66	0.77
Utah	Salt Lake	15.17	1.27	1.19	1.63	2.12	1.49	1.30	0.70	0.93	0.68	1.16	1.31	1.39

*Trace

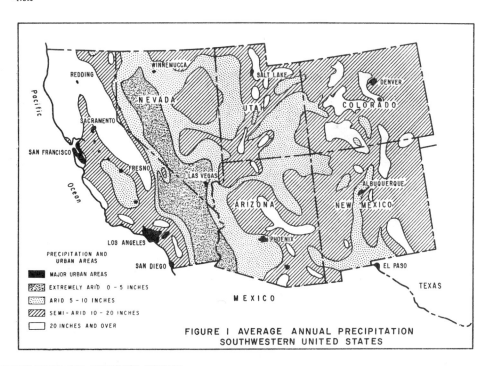

FIGURE I   AVERAGE   ANNUAL   PRECIPITATION
SOUTHWESTERN   UNITED   STATES

POPULATION AND ECONOMIC GROWTH

Figure 1 shows the location of principal urban development throughout the
arid and semiarid southwest area. Table 2 presents the 1920 and 1975
population of the area by the several states. In 1920 only about 5 percent of the people of the United States were located here. Over the last

TABLE 2: POPULATION OF ARID AND SEMIARID
SOUTHWESTERN UNITED STATES

| States | POPULATION | | Population Increase 1920-1975 | Percent Increase |
	1975	1920		
Arizona	1,963,000	334,000	1,629,000	488
California	21,000,000	3,427,000	17,573,000	513
Colorado	2,364,000	940,000	1,424,000	151
New Mexico	1,076,000	360,000	716,000	199
Nevada	533,000	77,000	456,000	592
Texas (West. 1/3)	1,120,000	311,000	889,000	285
Utah	1,127,000	449,000	678,000	151
TOTAL	29,263,000	5,898,000	23,365,000	396
United States	210,000,000	106.022,000	103,978,000	98
Percent of U. S.	14%	5.5%	---	---

55 years the national population has more than doubled, but the area now accounts for 14 percent of the United States total.

What has been the stimulus for such growth in areas that, under natural conditions, are critically short of one of the most basic resources required to sustain human life and economic activity?

Generally, early growth was associated with mineral extraction. Soon after the initial flurry of this activity the relative importance of mineral extraction gave way to agricultural crop production. As the potentials for a wide variety of crop production due to the long warm growing seasons became apparent, efforts to develop irrigation water supplies were undertaken. These efforts provided water throughout the normally parched dry season not only for irrigation, but also for the food and fiber processing industries and other related community requirements.

As time passed, word of the development potential of the arid and semiarid portions of the West spread throughout the country. In addition, land developers were instrumental in promoting the desirability of living in some of these areas. In response, there was a great increase in westward migration. This created an expanded economic development. The great expansion of manufacturing in the West during World War II stimulated increased migration. With the resulting large work force available, manufacturers continued to locate facilities in this area following the War--and to the present day, this growth has continued.

Table 3 presents the estimated 1975 industrial output and employment by type of activity in California. This illustrates the wide diversification of the economic development which has occurred.

TABLE 3: CALIFORNIA EMPLOYMENT
AND GROSS STATE OUTPUT 1975

	Employment (thousands)	Gross State Output (millions dollars)
Agriculture, forestry, fishery products	338	9,775
Mining	34	2,797
Construction	303	11,728
Manufacturing	1,587	41,415
Wholesale and retail trade	1,788	28,078
Transportation and utilities	459	16,984
Finance, insurance, real estate	447	24,497
Services	1,543	30,754
Government	1,668	
Total	8,167	166,028

THE HISTORY OF WATER DEVELOPMENT

From the earliest time, a major use of water was for crop production. Initially, farms were located along rivers where water could be readily diverted as needed. As agricultural activities expanded in response to market opportunities and the communities grew in support of the activities, small dams for storing winter flows and canals to transport the water by gravity to the place of use were constructed by private companies and local districts.

The history of development of facilities for storage and transport of Colorado River to Southern California is an example of the course of water development which has been followed when local water supplies were completely utilized. In 1902 the California Development Company began construction of the International Canal. By 1904 this canal was transporting water for irrigation to 30,000 hectares (75,000 acres) in the Imperial Valley in southeastern California. By the 1920s it was increasingly apparent that the rapidly growing Southern California population, located mainly along the Pacific Ocean coast, would soon require additional sources of water; the local surface supplies and developed imports, notably from Owens Valley, would have to be augmented. Therefore, in 1923 a plan for an aqueduct to carry Colorado River water to the metropolitan area of Los Angeles was prepared by the City of Los Angeles. In 1929 the Colorado River Compact was ratified by the United States Congress. It provided for the diversion of available water of the Colorado River system by the seven southwestern

states of the river's basin. In 1928 the Congress authorized construction
of the Boulder Canyon Project. By 1935 Boulder Dam had been constructed.
Since that time, up to 1,500 cubic hectometres (1.2 million acre-feet)
annually have been delivered to The Metropolitan Water District of Southern
California, which serves much of California's south coastal area. California's
allocation of Colorado River supply will be reduced in future years, so with
continued growth in water demands additional supplies had to be developed.
This was one of the factors that resulted in the construction of the
California State Water Project, which is projected to satisfy Southern
California demands beyond the year 2000.

This pattern of water development is typical of the history throughout the
arid West--first the individual diversions from natural streamflow to
adjacent lands; then private company or community development of small
storage and transport facilities; and finally, construction of large-scale
dams and canals with pumps to transport the water very long distances.

However, the options for obtaining additional supplies in the future are
much fewer now than in the past. The best dam sites in terms of water yield
per dollar cost of construction have already been utilized. Large increases
in labor and equipment costs have greatly reduced the number of economically
feasible water development possibilities. In addition, concerns for the
environmental impacts of facilities and their operation further constrain
future water supply options.

Therefore, those people concerned with how we may satisfy future demands
are taking a hard look at how we currently use water to determine what
changes could be made to stretch the use of the existing developed and less
costly water supplies.

CHARACTERISTICS OF CURRENT URBAN WATER USE

Figure 2 presents the quantity of water used for urban purposes in California
in 1972 by hydrologic regions. Over half of the State's total is used in the
driest portion of the State (the South Coastal, South Lahontan, and Colorado
Desert regions). Figure 3 shows the 1972 water use in California by type
of use, i.e., residential, industrial, commercial, and governmental.
Over two-thirds of the use was for residential purposes. Less than 20 percent
went to industry. Table 4 presents the average annual rate of water use on
a per capita basis by the various regions of the State for the period 1966
to 1970.

There are many factors that influence the amount of water used, resulting
in a wide range of per capita values within each region. Figure 4 shows
examples of the range of per capita water use rates for 1970 of cities within
the south coastal region.[1]

The unit per capita values are a function of the climate, type of residential
development, affluency of the population, price of water, existence of meters
on individual connections, condition and design of the water delivery system,
amount of water-using industry within the area, and other factors. Per capita
water use for any community is the net result of many variables. Each area
has a unique combination of these variables.

----

[1]Department of Water Resources Bulletin No. 166-2, "Urban Water
  Use in California", presents water delivery and per capita use
  data for communities throughout the State.

URBAN WATER USE – 1972

	1,000 ac–ft	cubic hectometres	%
NORTH COASTAL	93	110	1.8
SAN FRANCISCO BAY	990	1,220	19.6
CENTRAL COASTAL	181	223	3.6
SOUTH COASTAL	2,370	2,920	47.0
SACRAMENTO BASIN	470	580	9.3
DELTA–CENTRAL SIERRA	173	213	3.4
SAN JOAQUIN BASIN	192	237	3.8
TULARE BASIN	363	448	7.2
NORTH LAHONTAN	23	28	0.5
SOUTH LAHONTAN	89	110	1.8
COLORADO DESERT	99	120	2.0
STATE TOTAL	5,040	6,210	100.0

FIGURE 2. URBAN WATER USE BY HYDROLOGIC STUDY AREAS, CALIFORNIA, 1972

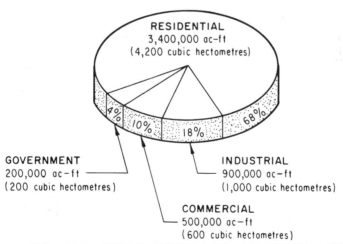

RESIDENTIAL
3,400,000 ac–ft
(4,200 cubic hectometres)

4%   10%   18%   68%

GOVERNMENT
200,000 ac–ft
(200 cubic hectometres)

INDUSTRIAL
900,000 ac–ft
(1,000 cubic hectometres)

COMMERCIAL
500,000 ac–ft
(600 cubic hectometres)

FIGURE 3. URBAN WATER USE, CALIFORNIA, 1972

TABLE 4. AVERAGE ANNUAL URBAN UNIT WATER USE
FROM ALL SOURCES BY HYDROLOGIC STUDY AREAS,
CALIFORNIA

Hydrologic Study Area 1966 through 1970	TOTAL Agency and Private Industry–Produced Water			
	Gallons per capita daily	Litres per capita daily	Acre-feet per capita annually	Cubic metres per capita annually
NORTH COASTAL	521	1,970	0.584	720
SAN FRANCISCO BAY	179	678	0.200	247
CENTRAL COASTAL	194	734	0.217	268
SOUTH COASTAL	179	678	0.200	247
SACRAMENTO BASIN	351	1,300	0.393	485
DELTA–CENTRAL SIERRA	315	1,190	0.353	435
SAN JOAQUIN BASIN	436	1,650	0.488	602
TULARE BASIN	363	1,370	0.407	502
NORTH LAHONTAN	492*	1,860	0.551	680
SOUTH LAHONTAN	305	1,150	0.342	422
COLORADO DESERT	378	1,430	0.423	522

*Includes tourist use.

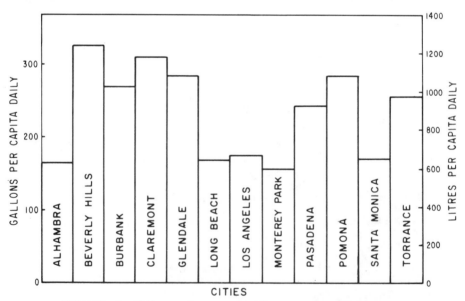

FIGURE 4. EXAMPLES OF VARIATION IN RATES OF
PER CAPITA WATER USE FOR CITIES WITHIN
SOUTH COAST REGION, CALIFORNIA 1970

Not unexpectedly, the climate of a given area usually has the greatest influence on the rate of water use. Climate also influences the kinds of water use in any given area or the proportion of the water use going to specific purposes. The principal influence of climate on urban water use is in respect to landscape irrigation. On the average in California, about half of the residential water use is for irrigation. Water use by vegetation, including lawns and ornamental plantings, is directly affected by temperature, humidity, and wind movement. Precipitation is important to the extent it helps meet the landscape water needs. The difference in average daily per capita water use rate between the City of Long Beach (605 litres or 160 gallons) and the City of Pomona (1,060 litres or 280 gallons) is in part due to a difference in climate. Long Beach is situated on the ocean where it receives cool breezes and summer fog. Pomona, on the other hand, is located inland where it experiences warmer temperatures.

The type of residential development is often reflected by the water use rate. Areas with large lots usually use significantly more water due to increased landscape irrigation. On the other hand, communities with a high proportion of apartment houses and similar multifamily dwellings show significantly lower per capita water use rates due to less area, on a per capita basis, being irrigated.

Income level of the population may also affect water use rates. High income families often have large residential lots and swimming pools, and also may have more household use. The City of Beverly Hills, with an average daily per capita water use rate of 1,250 litres (330 gallons) is an example of this.

Water price appears to have had little effect, with only a few exceptions, on a rate of urban water use in California because of the relatively low prices found throughout the State. In addition, by far the most common water pricing structure is one which charges less for each unit of water in excess of a basic quantity (declining block rate). However, comparisons of per capita water use rates of cities having meters on individual connections with those that do not, indicate that water bills based on the quantity used rather than on a flat rate result in more frugal use. Figure 5, based on 1970 data, presents such a comparison of cities within the Central Valley of California where most of the unmetered service is found.

The characteristics of the water delivery system significantly impacts the total quantity of water used in some communities. In some cases, old systems have developed serious leaks which may go undetected for considerable time. In some service areas, differences in elevation result in high water pressures at some locations which contributes to excessive water use.

Although industry uses less than 20 percent of the total urban fresh water supply in California, the impact of such use on per capita rates in specific communities is great. As indicated in Figure 6, food processing, paper production, and petroleum processing each account for a large portion of the total fresh water use by manufacturing industries in California. Paper production occurs in Northern California at locations where water is plentiful. Food and petroleum processing occur at various locations throughout the State, including the south coastal region. In the southeastern desert regions, food processing accounts for most of the industrial use of water, although agricultural chemical production and mineral processing use significant quantities of water. More detailed information concerning industrial use is presented in the Department of Water Resources Bulletin No. 124-2, "Water Use by Manufacturing Industries in California--1970".

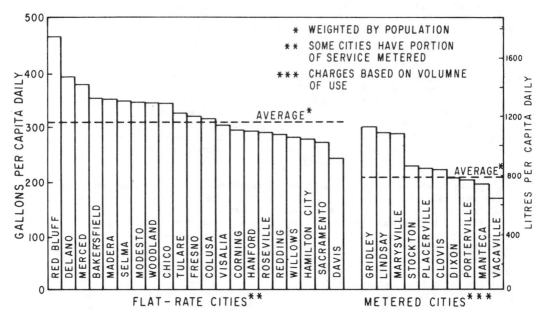

FIGURE 5: COMPARISON OF AVERAGE PER CAPITA WATER USE
IN METERED CITIES AND FLAT RATE CITIES
IN THE CENTRAL VALLEY OF CALIFORNIA — 1970

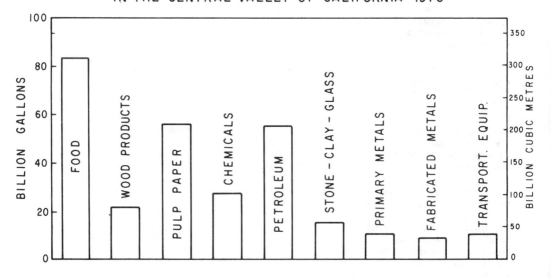

FIGURE 6. ESTIMATED TOTAL ANNUAL WATER USE BY MAJOR
WATER-USING MANUFACTURING INDUSTRIES, CALIFORNIA

Another kind of variation in per capita water use that is important to
consider in water planning is change over time. Figure 7 shows the change
in urban per capita water use from 1940 to 1975 in the south coastal area.
The annual variation (from one year to the next) reflects temperature
variations for a large part; however, there also are numerous other factors
involved--specific industries and commercial enterprises going into or
out of operation, population shifts to higher or lower water using resi-
dences, and other factors.

The general trend upward is characteristic of urban per capita water use
values in most areas of California that have experienced significant population growth. The newer residential developments have been principally
of the suburban type with larger lots and more exterior water uses. Along
with this has been greater use of high water using household appliances
and increases in the number of swimming pools.

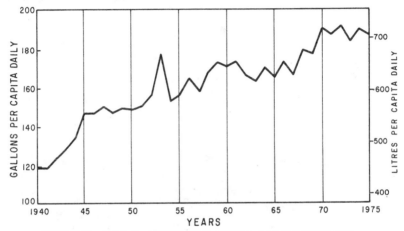

FIGURE 7. ANNUAL URBAN PER CAPITA DAILY WATER USE
SOUTH COAST REGION, CALIFORNIA

WATER CONSERVATION

As pointed out earlier, the traditional action for satisfying the continually
increasing demand for water in California has been to construct major water
storage and transportation facilities.
Due to greatly increased cost of facility construction; cost of energy required for transporting the water; and concerns for the preservation of the
native values of our rivers, fisheries, and recreation opportunities; there
has been increasing pressure that every effort possible be made to increase
the efficiency of water use. Water conservation, i.e., using less water to
accomplish the same beneficial purpose, has been the subject of considerable
study and action by the Department of Water Resources. In May 1976, the
Department published Bulletin No. 198, "Water Conservation in California",
which presents an assessment of the measures that can be taken to reduce
water use, estimates of the potential water savings, and actions which the
Department and others can take to accomplish water conservation. Since
that time, the drought of 1976 and 1977 has given impetus to implementation
actions throughout the State.

The findings of Bulletin No. 198 indicated there are opportunities for
achieving water savings and other benefits, such as energy savings, to

various degrees statewide. Because about 70 percent of the State's urban water use is in coastal metropolitan areas, which discharge a large portion of the excess applied water into the ocean after only once-through use, considerable attention has been focused on urban water conservation potentials.

With nearly 70 percent of the State's urban water use going to residential purposes, a close examination has been made of the specific manner of use by this sector. Figure 8 depicts the average proportion of residential use for specific purposes. Over 50 percent goes to interior uses, of which the toilet and bath account for about 75 percent. Table 5 presents the standard and possible "improved" levels of water use for fixtures, appliances, and other features of the interior water system. Table 6 indicates that by year 2000 as much as 1,500 cubic hectometres (1.2 million acre-feet) savings in interior residential water use might result from installation of lower water-using devices and other actions. This would be a 40 percent reduction of total statewide interior residential water use estimated for that state.

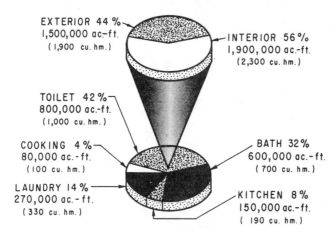

EXTERIOR 44%
1,500,000 ac-ft.
(1,900 cu.hm.)

INTERIOR 56%
1,900,000 ac.-ft.
(2,300 cu.hm.)

TOILET 42%
800,000 ac.-ft.
(1,000 cu.hm.)

COOKING 4%
80,000 ac.-ft.
(100 cu.hm.)

BATH 32%
600,000 ac.-ft.
(700 cu.hm.)

LAUNDRY 14%
270,000 ac.-ft.
(330 cu.hm.)

KITCHEN 8%
150,000 ac.-ft.
(190 cu.hm.)

FIGURE 8: RESIDENTIAL WATER USE IN CALIFORNIA-1972

By far, the greatest saving would result from modification of existing toilets and showers and installation of low-water using models in new construction. This fact has led to efforts by many local water agencies, as well as the Department of Water Resources, to promote installation of water dams and other devices in toilet reservoirs and installation of shower flow restrictors as measures to cope with the drought-imposed water shortages.

According to Bulletin No. 198, exterior residential water use also could be reduced significantly. Just a 10 percent reduction statewide would amount to nearly 170 cubic hectometres (140,000 acre-feet) at the current rate of use and over 250 cubic hectometres (200,000 acre-feet) at year 2000. How much of this would be actually saved depends on how much of the excess normally reenters a useable water supply source; either surface streams or ground water basins. The experiences of the current drought situation indicates that in areas where people have been impressed with the critical need to conserve water, significantly more than 10 percent reductions in landscape irrigation have been obtained simply by more attention being given to using the water.

TABLE 5. POTENTIAL WATER SAVINGS FROM RESIDENTIAL INTERIOR FIXTURES

FIXTURE /ACTION	WATER USE		PERCENT SAVINGS TOTAL INTERIOR		INCREMENTAL COST		IN-HOUSE
	STANDARD	IMPROVED	NEW	RETRO	NEW	RETRO	ENERGY SAVINGS
TANK TOILET	5-7 gals. (19 to 26 litres) per flush	3.5 gals.(13 litres) per flush	18	10-18	$0-$10	$0-$6	No
SHOWER	Up to 12 gals. (45 litres) per minute	3.0 gals.(11 litres) per. minute	9-12	9-12	$0-$5	$1-$5	Yes
KITCHEN AND LAVATORY FAUCETS	Up to 5 gals. (20 litres)per minute	1.5 gals. (5.7 litres) per minute [1]	2^2	$0-2^2$	$0-$5	$1-$5	Yes
PRESSURE REDUCING VALVE	80 lbs. per sq. inch (550 kilopascals)	50 lbs. per sq. inch (340 kilopascals)	0-10	0-10	$0-$25	$25	Yes
HOT WATER PIPE INSULATION[6]	Not insulated	Insulated	$1-4^2$	$0-1^2$	$0.50-$1.00 per foot	$0.50 per foot	Yes
AUTOMATIC CLOTHES WASHER[7]	27-54 gals. (100-200 litres)per load	16-19 gals.(61-72 litres)per load	0-5	—	$20-$30	Not practical	Yes
AUTOMATIC DISHWASHER[7]	7.5-16 gals.(28-61 litres) per load	7.5 gals. (28 litres) per load	$0-4^4$	—	0	Not practical	Yes
TOTAL			$30-55^2$	$19-43^5$			

1. ATTACHMENTS MARKETED WITH 0.5 GALS. (2 LITRE) PER MIN. FLOW. RESIDENTIAL ACCEPTANCE UNKNOWN BUT COMMERCIALLY PROVEN.
2. NO FIELD QUANTIFICATION.
3. RETROFITTING MAY NOT ALWAYS BE PRACTICAL.
4. BASED ON ONE LOAD PER DAY.
5. EDUCATE TO ONLY WASH FULL LOADS, TURN OFF WATER FAUCETS UNLESS ACTUALLY USED, ETC., COULD ADD ANOTHER PERCENT OR TWO TO THE TOTALS.
6. INSULATION OF CERTAIN CONTINUOUSLY CIRCULATING HOT WATER PIPING IS ALREADY REQUIRED.
7. 59% OF THE HOUSEHOLDS IN LOS ANGELES AREA HAVE WASHING MACHINES AND 24% HAVE DISHWASHERS.

TABLE 6. POTENTIAL RESIDENTIAL INTERIOR WATER SAVINGS, CALIFORNIA

FEATURE	ADDED COST PER UNIT ($)	WATER SAVINGS AS A % OF INTERIOR USE	POTENTIAL YEAR 2000 STATEWIDE WATER SAVINGS	
			(1,000 ACRE-FEET)	(CUBIC HECTOMETRES)
NEW CONSTRUCTION:				
• LOW-FLUSH TOILETS	0-10	18	185	230
• LOW-FLOW SHOWERHEADS	0-5	12	125	155
• LOW-FLOW KITCHEN & LAVATORY FAUCETS	0-5	2	20	25
• PRESSURE REDUCING VALVES	0-25	5	50	60
• INSULATED HOT WATER LINES	0.50-1.00 [1/]	4	40	50
• LOW-WATER USING CLOTHES WASHERS	20-30	5	50	60
• LOW-WATER USING DISH WASHERS	0	4	40	50
		SUB TOTAL	510	630
EXISTING HOUSING:				
• PLASTIC BOTTLES OR WATER DAMS IN TOILET RESERVOIR	0-6	18	345	425
• REPLACE SHOWERHEADS WITH LOW-FLOW VARIETY OR INSTALL FLOW RESTRICTORS	1-5	12	230	285
• PLACE LOW-FLOW AERATORS ON KITCHEN & LAVATORY FAUCETS OR REPLACE ENTIRE UNIT	1-5	2	40	50
• PRESSURE REDUCING VALVES	25	5	95	115
• INSULATED HOT WATER LINES	0.50 [2/]	1	20	25
		SUB TOTAL	730	900
		TOTAL	1240	1530

1/ PER FOOT OF LINE (1.65-3.00 PER METRE)
2/ 0.51 OR MORE PER FOOT OF LINE (1.65 + PER METRE)

In addition to giving greater care to irrigation, exterior residential water use can be reduced over time by more use of drought resistant landscape vegetation. The type of landscaping found in the West commonly includes broad areas of lawns and high water using shrubs and trees that are not native to the semiarid and arid environments. More use of Mediterranean types of vegetation would greatly reduce the need for irrigation.

In the industrial sector, some decline in unit water intake is expected because of governmental regulations on disposal of waste water. In many instances, the stringent treatment requirements for waste water disposal provide incentives to industries to reduce unit water intake by either reusing waste water or changing production processes, or both. Some California industrial concerns have already made such changes because of strict waste discharge requirements. Bulletin No. 198 and Bulletin No. 124-2 ("Water Use by Manufacturing Industries in California--1970") contain descriptions of actions taken by some concerns to date. They include primary metal, petroleum processing, food processing, and other plants. Recycling of waste water has been the principal action, with water intake savings of 30 percent.

A public awareness of the need to conserve water and a general commitment to use water more prudently is essential for accomplishing water savings in urban areas. Reactions to the current drought emergency situation have demonstrated that people generally will readily cooperate with water conservation programs if the need is adequately presented. This is the task that lays before the California Department of Water Resources and water managers throughout the West. When the current critical drought period has run its course and water supplies return to "normal", the importance of continuing water conservation practices for the survival of the economies of the arid and semiarid western regions of this country must be communicated to the general populace. Because it is true--our water resource is finite and we are approaching the limit of feasible water development. Lavish use of water supplies and disposal of large quantities of still useable water to salt sinks must come to an end.

Groundwater and Its Overuse in Arizona

Wesley E. Steiner

The State of Arizona lies in the arid southwestern portion of the
United States. Approximately half of the State receives less than 25 centimeters of precipitation per year. The low relative humidities and high
temperatures that are characteristic of Arizona's climate result in high
rates of potential evaporation that drastically limit the amount of precipitation that becomes available as stream flow or as additions to ground-
water reserves. Of an estimated average of 100 billion cubic meters (80
million acre-feet) of precipitation falling in Arizona each year, more than
95 percent is lost to evaporation and plant transpiration.

The total average dependable or renewable supply available for
consumption in Arizona, including the developed portion of Arizona's
entitlement to waters of the Colorado River, which River flows through
northwestern Arizona and forms the state's western boundary, is
estimated to aggregate only 3.45 billion cubic meters (2,800,000 acre-feet).
Of this amount only 11 percent, or slightly more than 370 million cubic
meters (300,000 acre-feet), represents the annual rate of recharge or
dependable groundwater supply. The remaining 3.1 billion cubic meters
(2,500,000 acre-feet) or 89 percent of dependable supply represents that
available from surface water flows. With overdrafting, however, approximately 60 percent of the waters diverted or pumped for application to
beneficial use represents a withdrawal from groundwater reserves.

Arizona's growth has been rapid during the last three decades in
spite of limited water resources. Today the state's population growth rate
is the highest in the nation. This is not to say, however, that water supply
limitations have not constrained development in Arizona, e.g. of the state's
13.4 million hectares (33,000,000 acres) of arable land and only 526
thousand hectares (1,300,000 acres) are being irrigated, largely because
of an inadequate water supply.

Arizona's economy would have been much more severely constrained, however, had it not been for the ready availability of extensive
groundwater reserves. In many areas, groundwater is the only supply
and in areas with a surface water supply, groundwater is often a supplement which assures a continued supply in times of low surface flows or

901

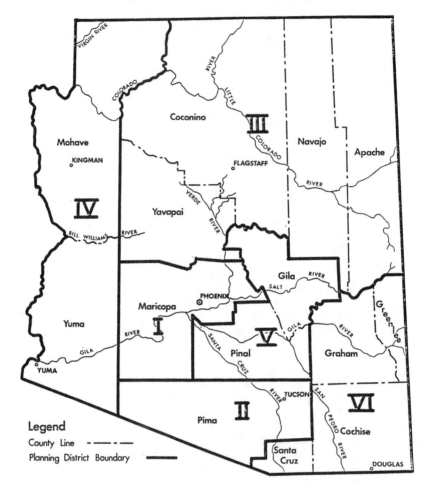

ARIZONA WATER COMMISSION

State Planning Districts

Figure 1

which extends the dependable water supply of an area through coordinated use of both sources.

The development of groundwater supplies in Arizona began about the turn of the century and increased gradually through the 1930's. Pumpage increased rapidly during the 1940's and 1950's to a level of 6.2 billion cubic meters (5 million acre-feet) annually by the late 1950's. This general level of development has continued to date.

Table 1

Estimated Annual Groundwater Pumpage, Overdraft and Storage
Arizona - Normalized 1970 Conditions

Unit: Million Cubic Meters

Planning District and County	Groundwater Pumpage	Groundwater Overdraft	Groundwater in Storage 1/
Planning District I			
Maricopa Co.	2,527	1,110	359,000
Planning District II			
Pima Co.	508	329	267,000
Planning District III			
Apache Co.	14	0	-2/
Coconino Co.	7	0	-2/
Navajo Co.	49	0	-2/
Yavapai Co.	28	15	65,500
Total	98	15	65,500
Planning District IV			
Mohave Co.	42	6	58,100
Yuma Co.	667 4/	97	322,000
Total	709	103	380,100
Planning District V			
Gila Co.	20	0	-3/
Pinal Co.	1,375	765	148,000
Total	1,395	765	148,000
Planning District VI			
Cochise Co.	624	331	176,000
Graham Co.	207	33	41,000
Greenlee Co.	33	0	23,200
Santa Cruz Co.	23	10	10,400
Total	887	374	250,600

1/ Groundwater in storage in alluvial aquifers to 366 meters below land
surface. Data not available for all areas.

2/ Although small amounts of water may be developed from localized
alluvial aquifers, the important groundwater resource is found in "C",
"D", and "N" multiple-aquifer systems. About 6.5 million hectares
is underlain by the three aquifer systems and about 308 billion cubic
meters of groundwater is stored in a 30.5 meter thick section of
aquifer.

3/ Gila County undoubtedly contains some groundwater in storage in
alluvial aquifers, but the amount has never been quantified.

4/ Groundwater pumpage includes approximately 445 million cubic meters
pumped for drainage purposes only.

As indicated in the last column in Table 1, the groundwater
reserves available in Arizona's alluvial aquifers are quite substantial,
aggregating approximately 1.5 trillion cubic meters (1.2 billion acre-feet)
to a depth of 366 meters (1,200 feet) below land surface. Many factors,
however, limit the utilization of this important resource. Most of the
water is stored in separate basins which are spread from the Mexican
border on the South to Lake Mead on the North and from the Colorado River

on the West, to New Mexico on the East. As only some 5 percent of the state's land resources are developed, full utilization of the stored ground-water would require periodic movement of groundwater consumers onto undeveloped basins or the construction of a vast groundwater collection and distribution system. Full utilization of the state's groundwater resources is further limited by the fact that only 40 percent underlies lands in private ownership and is available for development by private interests. It is further estimated that at least 10 percent of the water included in the fore-going estimate of the available groundwater resource is unusable due to quality considerations. Large depths to water and/or low well yields further reduce the potential for developing viable groundwater supplies in many areas.

In the developed areas of the state, annual water table declines of up to 6.1 meters (20 feet) are being experienced. Typical declines in groundwater levels as a result of irrigation pumpage are shown on Figure 2.

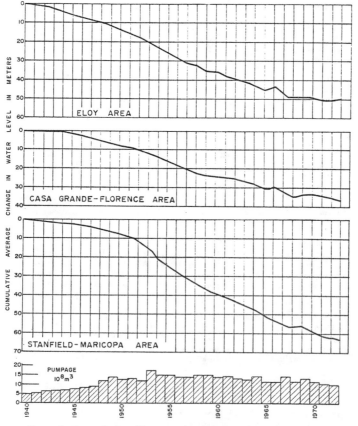

Cumulative Avg. Change in Water Level by Areas &
Est. Annual Pumpage in the Lower Santa Cruz Basin

SOURCE : U.S. Geological Survey

FIGURE 2

It has become obvious to Arizonans that the annual rate of
recharge of the groundwater resource is very limited and that most of the
water stored in the ground is available only for one time use. With few
exceptions, groundwater resources are being mined, that is they are being
taken out faster than the replenishment rate. Rates of consumption
approach 100 times the magnitude of recharge in some of the smaller
hydrologic basins. In Maricopa County, site of a very large metropolitan
area and the nation's 15th largest city, consumption of groundwater
reserves is occuring at over 30 times the rate of natural recharge.
Similarly in the other two central Arizona counties consumption is occuring
at rates of 12 (Pinal County) and almost 5 (Pima County) times replenishment.

The resulting overdrafts under 1970 conditions are shown in the
3rd column of Table 1. The three heavily developed central Arizona
counties of Maricopa, Pinal and Pima and the southeastern agricultural
county of Cochise account for 2.5 billion cubic meters (2,057,000 acre-
feet) or 94 percent of the current annual overdraft. With the exception
of Yuma County, which is supplied largely by diversion of surface waters
from the Colorado River, consumption in the state's other 10 counties is
relatively small and overdrafts either do not exist or are negligible.

As is the case in other western states, most of the water consumed in Arizona represents use by irrigated agriculture. Agriculture
currently consumes 89 percent of all water used, municipal and industrial
uses, 10 percent; and single purpose fish and wildlife uses, 1 percent.

The quality of groundwater supplies varies from acceptable to
unusable. In the northeastern part of the state, groundwater in the
Coconino aquifer contains as much as 100,000 milligrams per liter of total
dissolved solids or about three times the level of sea water. In central
Arizona, groundwaters in the Verde River Basin often contain toxic
amounts of arsenic in addition to excessive amounts of sulphate and
chloride. In portions of the developed area of central Arizona, water from
the alluvial aquifers is often unacceptably high in fluoride.

Quality of groundwater in central Arizona, where 72 percent of
current pumping takes place, varies by area and with depth depending
primarily upon the mineralogical makeup of the unit of the aquifer from
which the water is pumped. Wells in the area typically produce water
ranging from 300 milligrams per liter to 4,500 milligrams per liter of
total dissolved solids. In a large part of the area, the dissolved solids
concentration exceeds the U.S. public health service recommended maximum for domestic supplies of 500 milligrams per liter. Most of the major
irrigation supplies constitute a medium to high salinity hazard to irrigated
crops according to the U.S. salinity laboratory classification system. The
weighted average quality of all groundwaters pumped in central Arizona in
1965, the last period for which comprehensive data are available, was 955
milligrams per liter. In some areas of continual pumping of groundwater,
the as pumped quality of groundwater has deteriorated. In other areas as
pumped quality has improved as depths to the water table increased.

LAND SUBSIDENCE

Arizona is experiencing subsidence and earth cracking as a direct result of the mining of groundwaters. Up to 3.7 meters (12 feet) of subsidence have been realized since 1948 in the agricultural area of Pinal County lying between the cities of Phoenix and Tucson. Unlike California, subsidence in Arizona is accompanied by earth fissuring. Several mechanisms have been postulated for this phenomenon. Normally, the center of a groundwater basin subsides more than the margin of the basin as a result of larger groundwater level declines near the center. This produces a bending of the land surface and a tensional stress on the sediments at the edges of the basin and of nonsubsiding areas. This stress is eventually relieved by earth fissuring. Variations in aquifer compressibility due to lithographic differences or buried fault scarps adjacent to mountain masses along the edge of the basin represent probable zones of weakness and/or sites where maximum tensional stresses develop. Fissures are most likely to occur in these locations.

Subsidence doesn't always accompany overdraft in Arizona, but earth fissures always seem to accompany subsidence. To date large scale subsidence and cracking have been limited, with one minor exception, to areas of irrigated agriculture. Subsidence and earth fissures have adversely impacted on farming operations, highway and railroad maintenance. The prospects of subsidence and fissuring in an urban area and the havoc that it would cause has prompted the City of Tucson to undertake a program to transfer pumping from within the urbanized area to outlying areas.

GROUNDWATER LAW

Dating from early territorial days, Arizona courts repeatedly found that percolating groundwaters were not appropriable and belonged to the surface landowner, subject only to reasonable use. These opinions established groundwater as a property right in Arizona. In 1948 the Arizona legislature enacted the first groundwater code. This code reaffirmed the body of court opinion with respect to the ownership of groundwater and the rules of reasonable use. The most notable accomplishment of the code was establishment of procedures for the designation of critical groundwater areas. A "critical groundwater area" is defined by statute as any groundwater basin or any designated subdivision thereof not having sufficient groundwater to provide a reasonably safe supply for irrigation of cultivated lands in the basin at the then current rates of withdrawal. The code prohibits the drilling of irrigation wells within a groundwater area designated as critical for the irrigation of lands which on the date the area was declared critical were not irrigated or had not been cultivated within the five years prior thereto. The code does not provide, however, for control of extent of pumpage or the apportionment of pumpage among the land owners, nor does it prohibit the development of new wells within the designated critical groundwater areas for purposes other than irrigation. Most of the state's irrigated land, except for that lying along the lower Colorado River, is located within the 10 areas currently designated as critical and is closed to the drilling of wells to irrigate additional acreage.

Several decisions by the Arizona Supreme Court are of extraordinary importance in establishing the legal framework of the state's groundwater law. Bristor vs. Cheatham in 1953 reaffirmed that Arizona was committed to the American Doctrine of reasonable use and that the court announced rule had become a "rule of property". The holding on rehearing stated "...when a decision does become a rule of property, the rights acquired thereunder are entitled to protection under the laws as declared". The Bristor decision and the decision in Jarvis vs. State Land Department in 1969 were consistent in holding that owners of land overlying groundwaters may not concentrate the groundwaters and convey them off the land if other users of the same supply are damaged. However, in 1970 the Court modified its decision in Jarvis and allowed the City of Tucson to export water from the critical groundwater area to the extent the city purchased and retired land under cultivation and exported no more than the amount of water previously consumptively used for irrigation. In doing so, the Court specifically cited legislative policy that assigned a priority to municipal needs over those of agriculture.

In 1976 in Farmers Investment Company vs. Andrew L. Bettwy the State Supreme Court ruled that the mines could not transfer groundwater from the critical groundwater area that lies south of the City of Tucson from the parcels on which the wells are located to other parcels removed therefrom because of the resulting damage which would accrue to the agricultural land owners cultivating lands adjacent to the mines' wells. In the same decision the Court ruled that the City of Tucson could not pump and transfer groundwaters from City wells constructed subsequent to April of 1972.

This decision threatened approximately 30 percent of the water supply of the City of Tucson and all of the supply of copper mines that produce 25 percent of the nation's copper. The decision also threatened water supplies for many of the state's other cities and major industries. The pressure for accommodation was immediate and enormous and in the current session of the Arizona State Legislature the legislature after much travail enacted legislation which permits transfers under certain conditions without threat of injunction and establishes a joint legislative-public member groundwater management study commission charged with developing comprehensive groundwater management legislation by December 1979. In the event that the legislature fails to adopt groundwater management legislation by September of 1981, the Commission's legislative proposal automatically becomes law.

While it will be an extremely difficult undertaking, the chances appear excellent that the demands of the people of the State of Arizona for meaningful groundwater management statutes will be heard and that a much needed groundwater management program will be implemented in Arizona within the next four years.

PROSPECTS FOR THE FUTURE

       The State of Arizona is involved in the development of a State
Water Plan. As the first two steps in this process, current uses and
supplies have been inventoried and alternative levels of water use through
year 2020 have been projected for each of the state's fourteen counties.
It is believed that the alternatives investigated bracket the range of possible
future water use by the agricultural, municipal, mining and energy
development sectors of the state's economy.

       The State of Arizona is committed to completion of the Central
Arizona Project, a federal project which will bring into central Arizona,
the area of greatest imbalance between supply and demand, most of
Arizona's remaining entitlement to waters of the Colorado River. The
supply made available by the Central Arizona Project will relieve current
overdraft in central Arizona by about two-thirds. The remaining one-third
plus the supplies necessary for continued growth must come from one or a
combination of the following: (1) additional importations of waters from
areas of surplus such as the Columbia River or from desalted sea or
geothermal water resources, (2) continued mining of groundwater resources
or (3) a combination of improved efficiency of use by all users and a
decrease in agricultural use of Arizona's limited supply. The economic
and political constraints on desalinization or importation from distant
sources of surplus would seem to clearly dictate that Arizona resort to
additional mining or to maximum conservation and a retirement of less
economic uses in favor of those with a higher return. In our state water
planning investigations we have determined county by county the amount of
agricultural acreage that would have to be retired to effect and maintain a
balance between water supply and use as growth of other sectors of the
economy continues. In some areas of the state we find that even with
importation of Arizona's remaining entitlement in the Colorado River via
the Central Arizona Project it will be necessary to retire all or a significant portion of the existing irrigated acreage if balance is to be achieved
without additional importation. In a number of the mountain counties
where there is very little agriculture and most of the prospective use is
associated with mining we find that even with the lowest level of projected
development it would be impossible to achieve a balance through the
medium of reducing agricultural acreage.

       By way of example, the evaluation for Pima County is shown
on Figure 3. The City of Tucson, Arizona's second largest city with a
population of 300,000 is located here along with five mines that produce
25 percent of the nation's copper and 21,900 hectares (54,000 acres) of
irrigated agriculture, consisting primarily of high return pecan orchards.
In 1970, the cities were consuming 85.1 million cubic meters (69,000
acre-feet) per year, the mines and other major industrial users 72.8
million cubic meters (59,000 acre-feet) per year, and agriculture 260
million cubic meters (211,000 acre-feet) per year for a total of 418
million cubic meters (339,000 acre-feet) per year.

       On Figure 3, the upper line marked Alternative I represents the
water requirements associated with population projections of the Arizona

FIGURE 3

PIMA   COUNTY

ALTERNATIVE   FUTURES

PROJECTED .ALTERNATIVE WATER DEPLETIONS
AND   DEPENDABLE   SUPPLY

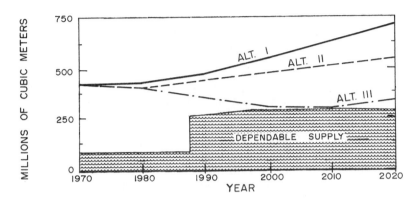

State Office of Economic Planning and Development.   This projection was
selected to represent the upper limit of what might be anticipated in the
way of growth.   Alternative II represents a medium level of growth as
projected by the Arizona Department of Economic Security.   Alternative
III represents medium growth with a reduction in agricultural acreage to
effect a balance between water supply and use.   In Alternative III for Pima
County, all agricultural acreage has been retired by year 2000.

The lowest line on the chart, represents the dependable water
supply.   The difference between this line and any of the others constitutes
the amount of overdraft or deficiency.

Currently all water consumption in Pima County is supported by
the groundwater resources of the area.   None of the needs are satisfied
from developed surface waters.   The average annual recharge or dependable supply of the groundwater basins underlying Pima County is estimated
to be only 89 million cubic meters (72,000 acre-feet) per year.   Hence,
as shown on Figure 3, consumption is taking place at a rate almost five
times the dependable supply and results currently in an annual overdraft
or mining of groundwater of 333 million cubic meters (270,000 acre-feet)
per year.   Were all agricultural retired by year 2000, and with a moderate
rate of growth of municipal and mining uses, consumption would still
occur at a rate 3.3 times the local dependable supply.   Fortunately,
Arizona has available, as previously cited, some 1.2 million acre-feet
per year of remaining entitlement in the Colorado River.   The state
proposes to dedicate 189 million cubic meters (153,000 acre-feet) per
year of that supply to support municipal and mining uses in Pima County.
Figure 3 shows that with the introduction in 1987 of that additional supply

via the Central Arizona Project and retirement of all agriculture and/or restrictions on municipal and industrial growth, it would be possible to effect a water balance well into the next century.

The third phase of the State Water Plan, now under preparation, will articulate plans for meeting the water requirements of alternative levels of growth and will assess the economic costs and environmental impacts of selected plans.

Depths to groundwater throughout most of the developed area of the state are from 91 to 152 meters (300 to 500 feet). As these levels continue to drop the energy required to lift the water to the surface increases and the longer Arizona waits to effect a balance between use and recharge, the greater will be the energy requirement. With energy resources also of diminishing magnitude and of increasing concern it would seem to behoove Arizona to balance supply and use of groundwater as soon as possible.

The need for groundwater management in Arizona is manifest. Ways must be found to bring water use and supply into balance even as rapid growth continues. Awareness of the state's water supply limitations by the public and at all levels of government is growing rapidly and prospects for adoption of a meaningful management program improve steadily.

References

Arizona Water Commission. Inventory of Resource and Uses. Phase I, Arizona State Water Plan. Phoenix, Arizona: Arizona Water Commission, July, 1975.

Arizona Water Commission. Alternative Futures. Phase II, Arizona State Water Plan. Phoenix, Arizona: Arizona Water Commission, February, 1977.

Diminishing Ground Water
in the San Joaquin Valley

Carl L. Stetson, Chief
San Joaquin District
Department of Water Resources
The Resources Agency
State of California

The purpose of this paper is to explore the past and present relationships between irrigation and ground water in California's San Joaquin Valley.
In the midst of a severe drought, the demand for water throughout the State is
higher than ever before; and in agricultural areas the combination of a lack
of precipitation and cutbacks in State and federal allotments of imported
surface water has prompted farmers to look to underground sources for needed
irrigation water. In the Valley, drilling companies are swamped with requests
to construct new wells or to deepen existing ones. Though drilling costs are
high (from $30 to $60 a foot, depending on the well's diameter and depth),
water is so valuable in the Valley that many farmers are willing to pay the
price.

They pay because the San Joaquin Valley is one of the world's largest and
richest agricultural areas. Though its climate is semiarid, and local precipitation is insufficient for intensive farming, the soil is so fertile, and the
days of the growing season so warm and sunny, that bumper harvests make the
effort and expense of irrigation worthwhile. For years, extensive ground
water pumpage has provided farmers with substantial irrigation supplies. Up
until the 1950's, much of the water pumped from the ground was replaced through
percolation; but in the last few decades, pumping has far exceeded recharge.
This condition, where a great deal of water is taken from the ground and relatively little is returned, results in lowered ground water levels and land
subsidence -- two significant agricultural problems in the Valley.

THE SAN JOAQUIN VALLEY

The San Joaquin Valley is walled by the Sierra Nevada on the east, the
Coast Range on the west, and the Tehachapi mountains on the south. The valley
floor is about 48 kilometres (300 miles) long and averages about 65 kilometres
(40 miles) wide. It contains about 3.2 million hectares (8.0 million acres),
almost all below an elevation of 150 metres (500 feet). The valley floor
slopes gently from the eastern and western foothills to the north-south-
trending trough located in the western third of the Valley. The valley trough
slopes gently from south to north and drops from an elevation of about 100
metres (300 feet) in Kern County to sea level at the Sacramento-San Joaquin
Delta.

Most streamflow enters the Valley from the Sierra Nevada on the east.
Between the Kings and San Joaquin Rivers, the gentle northward slope of the
valley floor is interrupted by a low divide that prevents draining of the
surface waters from the southern portion into the northern part.

The area south of the divide is called the Tulare Lake Basin, and that
north of the divide the San Joaquin River Basin. The nearly imperceptible
divide rises to only about 8 metres (25 feet) above the lowest point in the
Tulare Lake Basin, the Tulare Lakebed.

Before development, the valley floor was dominated by vast lakes and
swamps surrounded by tree-dotted grasslands. The huge lakes -- Kern, Buena
Vista, Goose, and Tulare -- were replenished annually by the flows of the
Kern, Tule, Kaweah, and Kings Rivers in the Tulare Lake Basin. Excess waters
of Tulare Lake Basin flowed over the divide at the north and joined the San
Joaquin River near the present City of Mendota. A major swamp covered the
central area of what now is Merced County. The waters of the San Joaquin,
Fresno, Chowchilla, Merced, Tuolumne, and Stanislaus Rivers flowed to the
trough of the Valley and then northward and westward via the San Joaquin River
through the Sacramento-San Joaquin Delta and San Francisco Bay to the ocean.

Early settlers found grizzly bears, pronghorn antelope, and Tule elk, as
well as numerous other animals, warmwater and anadromous fish, water birds,
and both resident and migratory waterfowl.

Resident valley Indians had established relatively permanent settlements
on the valley floor because of the abundance of wildlife and plants which
provided food and shelter. Some tribes apparently wintered in the Valley and
spent summers in the mountains, while others apparently lived year-round on
the valley floor and hunted in the foothill and mountain areas.

HISTORY OF IRRIGATION IN CALIFORNIA

The practice of supplying California land with water by means of ditches
or artificial channels is as old as the State itself. Initially, irrigation
grew up around each of the missions as they were established along the Pacific
Coast. Indians, under the direction of Spanish padres, built rather extensive
irrigation systems to bring water into the fields in and around the missions.
As long as the Church maintained administrative control, these irrigation
systems were successfully operated and maintained. But when the missions were
secularized about 1833, the irrigation systems, despite their demonstrated

value, soon fell into decay and the fields reverted to desert. Only the
ditches necessary to supply the forts with domestic water were maintained.

The first Anglo-Saxon settlers to successfully practice irrigating crops
in California were driven by the desire and the necessity to adapt to the
rigors of their new land. Although many settlers' irrigation attempts were
unsuccessful, those who did succeed were oriented generally toward some sort
of community betterment and managed to stimulate cooperation among their
neighbors to achieve it. With limited knowledge and equipment at their disposal, individual settlers were able to plow a furrow and guide water from a
stream onto small, shallow meadows. But when it came to building canals and
ditches of sufficient size to bring enough land under cultivation to support
a community, such minimal developments were inadequate. Only when a number of
farmers worked together in a cooperative project were extensive developments
possible.

In the San Joaquin Valley, the typically dry climate made irrigation
even more essential than in the wetter northern areas of the State. Early
irrigation systems diverted water from rivers and streams crisscrossing the
Valley. One trouble with this practice was that agricultural growth was
largely restricted to lands in the vicinity of surface water supplies.
Shortly before the turn of the century, farmers who had previously used ground
water (from shallow wells) for drinking and sanitary purposes started using
it to irrigate farmlands away from surface water sources. In the Valley, the
first ground water for irrigation was extracted by the windmill-operated
piston pump.

Extensive, large-scale pumping began a decade later when the centrifugal
pump, and later the deep-well turbine, arrived on the scene. Ground water
utilization then moved into a new phase, for instead of being limited to
extractions from depths of 9 metres (30 feet) or less, wells could be put
down to depths determined solely by the amount of money the owner was willing
to pay. By 1929, extractions were occurring at a rapid rate, with each
farmer trying to get his well level below that of his neighbor, to insure
an adequate supply of water for irrigated acreage. This lowering of well
depths was expensive, and in order to turn a profit, farmers began asking and
receiving more money for their crops. This increase in profit was yet another
incentive to pump water from greater and greater depths, and the combined
effects of deep-well turbine pumps, abundant power, excellent soil, favorable climate, a 300-day growing season, and a huge demand for farm products,
all contributed to an increasing reliance on ground water supplies.

Large pumps and widespread ground water reserves allowed farmers to
irrigate fields throughout the San Joaquin Valley, and until large-capacity
reservoirs were constructed in the 1920's on major rivers such as the
Stanislaus, Tuolumne, and Merced, ground water pumping continued at a swift
pace. By collecting surface runoff from the nearby Sierra Nevada, these
reservoirs provided farmers with surface water for irrigation (on a guaranteed
annual basis) and allowed many of them to buy water from governmental agencies
at a lower rate than it cost them to pump ground water. Other farmers, however -- especially those whose lands were not provided with low-cost surface
water -- remained dependent on ground water supplies to meet irrigation needs.
In the 1950's additional reservoirs were constructed, and many of the reservoirs constructed in the 1920's were replaced with significantly larger
reservoirs in the 1960's and 1970's. In dry years, when Sierra Nevada runoff
into the reservoirs was light, even those farmers supplied by the reservoirs
resorted to underground pumping to make up shortages. This is still true
today.

In 1968, the California State Water Project began importing vast quantities of surface water from Northern California into the San Joaquin Valley. Like the water provided by local reservoirs, State Water Project supplies were hailed as a reliable irrigation source that would dissuade many farmers from pumping water out of the ground. Unfortunately, in the past two years, drought conditions have forced the State to cut back on its State Water Project allotments, and many valley farmers have resumed large-scale ground water pumping. With little rain or surface runoff to replace the ground water farmers pump out, overdrafting is still increasing throughout much of the Valley.

HYDROLOGIC EQUATION

An important fact about ground water is that you cannot get more water out of a basin than is put into it on a long-term basis. A ground water basin is virtually a tank. The inflow consists of: (1) percolation from the surface flows of streams and from rainfall; (2) percolation losses from irrigation water-conveyance facilities and from water application on overlying lands; (3) direct artificial replenishment of ground water with imported water or excess local supplies; and (4) subsurface flows, if any, in the form of ground water movement from surrounding substrata. Outflow consists of: (1) evaporation and plant transpiration; (2) seepage and surface drainage flowing in surface streams and passing out of the area; (3) exportation of water pumped from the ground; and (4) subsurface ground water movement from the basin into surrounding substrata.

If there is no change in the basin's water table elevation over a long period of time, the outflow is equal to the inflow and the equation is balanced. Whenever a water table is lowered and not replenished year after year, it is apparent that more water is being extracted from the basin than is normally flowing into it, and the equation is thus out of balance.

It is a fact that any prolonged overdraft of ground water supplies disrupts the hydrologic balance. Before extensive settlement occurred in the San Joaquin Valley, nature had developed ground water conditions wherein the rate of replenishment to the underground aquifers equalled the rate of outflow. The Department of Water Resources has estimated that the annual overdraft in the San Joaquin Valley in 1972 was 1 800 cubic hectometres (1,500,000 acre-feet). During the drought of 1976 this overdraft increased to in excess of 3 600 cubic hectometres (3,000,000 acre-feet), and for 1977 the overdraft is predicted to be even greater.

LAND SUBSIDENCE

Competition for water in the San Joaquin Valley is expected to increase according to the demand for new irrigated land; and as long as surface water remains scarce, overdrafting of ground water supplies will likely continue. Besides depleting a valuable fresh water source, this practice of overdrafting also causes land subsidence in some areas. The mechanism and nature of this subsidence are complex, but in general, as the underground pressure decreases due to withdrawal of water, more load is placed on the grains of sand and silt that make up the sediments. Grain structures break down under the weight and are then rearranged more compactly (with fewer voids between them). This change in grain structure is irreversible, and the corresponding changes in the elevation of land surface are the results of overdrafting.

An outstanding example of subsidence and its environmental consequences is found along the last 48-to-64-kilometre (30-to-40-mile) stretch of the San Joaquin Valley's Delta-Mendota Canal. The canal was designed and constructed by the U. S. Bureau of Reclamation as a part of its extensive water resource development complex, the Central Valley Project. It extends approximately 182 kilometres (113 miles) from the Tracy Pumping Plant in the Sacramento-San Joaquin Delta to the Mendota Pool on the San Joaquin River.

The occurrence of land subsidence in the lower reaches of the Delta-Mendota Canal was not recognized during preconstruction and construction surveys. Puzzling discrepancies between old and new bench mark elevations and topographic maps were, however, frequently encountered. Originally, earthquakes were blamed for these variances, but by 1952 land subsidence was identified as the major contributing factor.

Due to subsidence, the top of the canal's concrete and earth linings were submerged below the water level; and the effects on numerous bridges and pipe crossings were similarly striking. Normal 1-metre (3-foot) clearances between the bottom of a bridge and the water surface gradually decreased to a few centimetres (inches) in the subsiding areas. In areas of maximum subsidence, bottoms of bridges actually became submerged. (Just a few miles away from the canal, near Mendota, subsidence on farmland has been measured at 8.5 metres, or 28 feet.)

OUTLOOK

In the San Joaquin Valley today, a scarcity of surface water resulting from a severe drought has led to a situation where many farmers are pumping water from the ground faster than it can be replenished.

Increases in farm-applied water use efficiency would produce a concurrent decrease in demand on surface and ground water supplies. Both ground water overdraft and surface drainage flows would be reduced. In the southern portion of the Valley, however, increased efficiency on a regional basis does not appear very promising, in that a high degree of efficiency is now being achieved. This southern area is where over 75 percent of the entire Valley's overdraft occurs.

To preserve ground water supplies, and to maintain a healthy agricultural environment, ways must be found to import more surface water into the Valley. Ambitious projects like the proposed Mid-Valley Canal and the Auburn Dam (already under construction in Northern California) are intended to augment existing supplies and reduce overdrafting. These new imported water supplies should be used primarily for the purpose of alleviating the ground water overdraft.

Even if these projects were constructed as planned, the additional supplies of imported water would only alleviate -- not eliminate -- the Valley's water shortage. Long-range planning should be dedicated to overcoming the shortage, for until this is accomplished, overdrafting of ground water in the San Joaquin Valley will undoubtedly continue.

SOME APPROACHES TO SALINITY CONTROL
IN THE COLORADO RIVER SYSTEM

Ernest M. Weber[1]

INTRODUCTION

The Colorado River, which begins in the snow-capped Rockies of Wyoming and Colorado, travels 1400 miles through lush forests, farm lands, and arid deserts of the seven basin states, finally emptying into the Gulf of California in Mexico. Along its route, the water, power, and recreational potential are used and re-used as the river forms the lifeline of 12 million people and two million acres of irrigated agriculture. It is the most controlled and used river in the world. By the time it reaches the gulf, it is thoroughly spent and its chemical quality greatly altered.

The drainage basin of the Colorado River covers an area of 242,000 square miles or about one-twelfth of the conterminous United States. It drains portions of seven states: Colorado, Wyoming, Utah, New Mexico, Arizona, Nevada, and California, as well as 200 square miles in Northern Mexico. (Figure 1.) The basin contains climatic extremes ranging from year-round snow cover in the high peaks of the Rocky Mountains, to desert conditions with hardly any precipitation in the lower portions of the Basin.

The Colorado River Basin has been divided into an Upper Basin and a Lower Basin for the purpose of apportioning surface flows. The Upper Basin is defined as that portion of the basin drainage above Lee Ferry, a point one mile below the mouth of the Paria River near the Arizona-Utah border (Figure 1). Annual virgin flow for the Colorado River for apportionment purposes is determined at Lee Ferry.

The surface flows of the Colorado River Basin are relied upon very heavily by the basin states and Mexico. Yet the flows are small when compared to other major drainage systems. The long-term average annual virgin flow of the Colorado River at Lee Ferry is 14.9 million acre-feet while the analogous flow of the Columbia is 180 million acre-feet and that of the Mississippi is 440 million acre-feet.

1/ Engineering Geologist - Colorado River Board of California

COLORADO   RIVER   BASIN

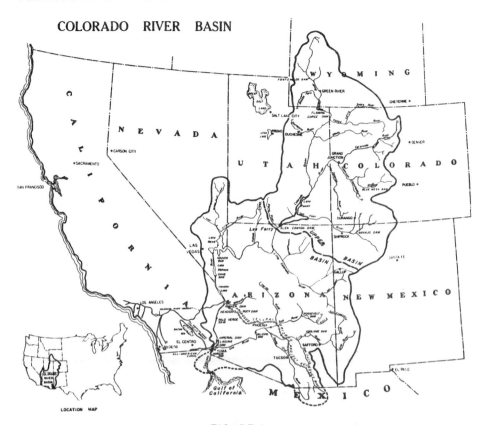

FIGURE I

Of major importance, too, are the large variations in annual flows which have ranged from 5.6 million acre-feet to 24 million acre-feet.

Total demands placed upon surface flows of the basin currently exceed 12 million acre-feet per year and depletions exceed 9 million acre-feet per year.

Aquifers capable of significant yields are quite limited in the basin except in central and southern Arizona. The reliance upon groundwater varies accordingly.

Groundwater consumed in the basin is about 67,000 acre-feet in the Upper Basin and 3.5 million acre-feet in the Lower Basin. Some three million acre-feet, or 85 percent of the total groundwater use, occurs in the Gila Subbasin in southern Arizona. However, groundwater management on a local basis plays an important role in salinity control.

The control of surface flows in the Colorado River Basin has been very extensive. The initial major storage feature on the Colorado main stem and still the keystone of the Lower

Basin control system is Hoover Dam, completed in 1936. Construction of other storage, diversion, and control features
in the Lower Basin during the first half of this century
resulted in virtually complete control of the Colorado below
Lee Ferry.

Major development in the Upper Basin began in the 1950's,
the most important feature being Glen Canyon Dam. Completed
in 1964, Glen Canyon Dam provided the Upper Basin the storage
needed to meet downstream obligations.

The waters of the Colorado are totally allocated through
interstate compacts, court decisions, international treaties,
and water supply contracts. All of the entitlements to water
are not yet fully developed, however, it is anticipated that
full utilization of each of the states portion of their compact
apportioned waters will be put to use early in the twenty-first
century. When this occurs, the demands on the river will exceed the estimated long-term supply.

The quantity of flow in the basin is not the only problem;
more critical, is the quality (salinity) of the water in the
Lower Basin.

THE SALINITY PROBLEM

As a pollutant, salinity increases in western rivers is not
a new or unique situation. Water quality problems in the Colorado River, for example, were recognized as early as 1903.
Today, the salinity conditions of the Colorado River are viewed
as the bell-weather of similar problems on other western rivers
such as the Rio Grande, and parts of the Arkansas and Platte,
which are also affected by increasing salinity levels.

The basic causes for the Colorado River's high salinity
are both natural and man-made. The primary natural causes are
the saline shales underlying much of the Upper Colorado River
Basin. Even though precipitation falling on the high peaks of
the Rocky Mountains runs off in mountain streams with a low
salinity, by the time these streams have joined to form the
major rivers draining the basin and have picked up runoff and
seepage from the underlying shale formations, the Colorado
River is carrying a heavy load of dissolved salt. It is estimated that the salt load at Lee Ferry was approximately five
million tons a year under natural conditions.

Man's activities have added to this problem, as the consumptive use of water in irrigation and other activities remove
water from the system but return all of the dissolved salts
back to the river, concentrating the salts in a smaller volume
of water. Also, when irrigators apply water to their crops on
lands that overlie the saline formations, the excess applied
water is flushed through the soil and underlying formations,
sending additional salts back to the river. Salinity also increases as a result of reservoir evaporation, trans-basin
exports of low salinity water, and municipal and industrial
uses. It is estimated that currently the river carries approxi-

mately 8.5 million tons of salt each year at Lee Ferry in a
flow less than that under natural conditions.

Without any salinity control measures and with continued
Upper Basin development, the average salinity at Lake Mead,
behind Hoover Dam, would increase from the present average of
around 730 mg/l to around 1,000 mg/l by the year 2000. The
salinity at Lake Mead approximates that at Lake Havasu, downstream from Lake Mead, where The Metropolitan Water District of
Southern California diverts for use on the coastal plain of
Southern California and where the Central Arizona Project, now
under construction, will also divert. Further downstream, at
Imperial Dam, where the bulk of diversions are made for agricultural use in California and Arizona, the comparable values
are 830 mg/l at present and over 1,200 mg/l by the turn of the
century if there are no salinity control measures. It should
be noted that the above values are annual averages and that
higher seasonal values will occur.

Harmful Impact of Salinity

The harmful impact of high salinity water is felt by both
irrigators and municipal and industrial users. Irrigation
farmers are affected through reduced crop yields, the inability
to grow high value, salt-sensitive crops, and increased costs
of irrigation and drainage systems. Municipal and industrial
water users experience detriments in a number of ways, including high soap consumptive, reduced service life of plumbing and
appliances, added water treatment and conditioning costs, and
increased process water costs. In addition, groundwater basins
are degraded by use of saline water.

Colorado River water users will experience increases in
detriments costing the users millions of dollars a year as a
result of projected salinity increases. Estimates are that the
total economic loss to the Lower Basin users will amount to
$425,000 per milligram per liter of future increase in salinity.
Current annual damages in this area attributable to salinity
are $89 million, and by the year 2000, will reach $160 million
unless appropriate salinity control action is taken.

EFFORTS TO CONTROL SALINITY

In the mid-1960's, concern over the quality of the
nation's water supply brought forth new water quality legislation. As international difficulties arose with Mexico over
the salinity of the water delivered to that country from the
Colorado River, agreement was reached between the two nations
to set a limit on the salinity of the water delivered to Mexico.

In order to provide for domestic salinity control and to
authorize the facilities required to fulfill the agreement with
Mexico, the U.S.-Congress passed in 1974 Public Law 93-320, the
Colorado River Salinity Control Act which authorized facilities
for enhancement and protection of the Colorado River water in
the U. S. and that which is delivered to Mexico. The major
facility for meeting the agreement with Mexico is a desalting
complex to reduce the salinity of irrigation return flow to the

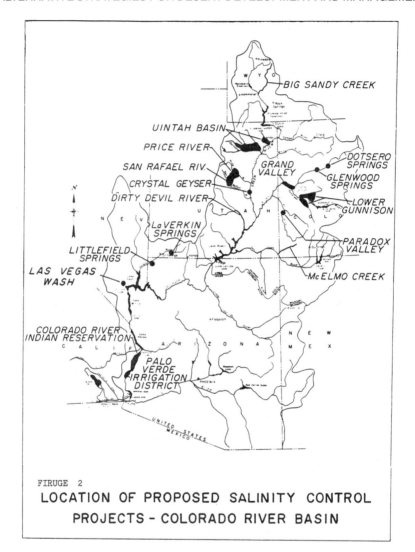

FIRUGE 2

LOCATION OF PROPOSED SALINITY CONTROL
PROJECTS - COLORADO RIVER BASIN

Colorado River from Wellton-Mohawk Valley in Arizona. The estimated cost of these facilities is $300,000,000 will an annual
O&M of about $15-20 million. The program for salinity control
to benefit the U. S. users consists of 16 salinity control
projects (Fig. 2) to control point, diffuse, and agricultural
sources.

    The seven Colorado River Basin states acting through an
interstate group have set salinity standards for the river and
developed a plan of implementation for salinity control to
maintain the standards while the states continue to develop
their compact-apportioned waters. The plan is designed to
reduce the salt load of the river and to minimize future increases in salt loading by the most cost-effective means

(environmentally, economically, and socially). The principal
components of the plan are:

1.   Prompt construction and operation of the four
salinity control units authorized by P. L. 93-320, the Colorado River Basin Salinity Control Act.

2.   Construction of 12 other units specified in P. L.
93-320 or their equivalents.

3.   The placing of effluent limitations, principally
under the NPDES permit program provided for in Section 402 of
P. L. 92-500 on industrial discharges.

4.   The reformulation of previously authorized, but un-
constructed, federal water projects to reduce the salt loading
effect.

5.   Use of saline water for industrial purposes whenever
practical, programs by water users to cope with the river's
high salinity, studies of means to minimize salinity in municipal discharges, and studies of future possible salinity control
programs.

SELECTED CONTROL UNITS

Three units involving the control of saline groundwater
discharges to the river system are examined.

Grand Valley, Colorado

Grand Valley is at the confluence of the Colorado and
Gunnison Rivers. Grand Junction, the principal City in the
Valley, is an industrial and commerical center of Northwest
Colorado.

About 76,000 acres are irrigated in the Valley by means of
200 miles of canals and 500 miles of laterals, all of which are
unlined. Estimates are that Grand Valley contributes by salt
pickup between 600,000 and 1,000,000 tons of salt annually. The
bulk of this salt load is derived from underlying saline marine
shales and is moved to the river by deep percolation of applied
irrigation water and water delivery systems losses.

Irrigation efficiency in Grand Valley is quite low, ranging between 30 and 40 percent. Much of the loss is in surface
runoff, the remainder returns to the river by means of deep
percolation and groundwater movement. About 8-9 feet of water
are applied to croplands, about double the amount required for
optional crop production. The delivery system is comprised
of unlined canals and laterals which have high seepage losses.

Water loss from the delivery system and excess water application passes through the very thin soil cover and leaches
salts from the underlying Mancos shale. The Mancos is a very
thick, 3000-5000 feet, sequence of marine shale which contains
a high percentage of salts: mainly calcium sulfate, with lesser
amounts of sodium chloride, sodium sulfate and magnesium sulfate. Gypsum is commonly found in open joints and fractures.

The shale is compact and has a low permeability in the deep, unweathered zone. However, in the shallow-weathered zone, the formation is rather highly fractured and posses good secondary permeability. Percolating water, flowing along the joints and fractures dissolves salts present in the shale. Since most of the soils were derived from the Mancos shale, they also are a source of salinity.

The proposed salinity control program to be implemented for the area consists of two two-parts: (1) lining of the delivery canals and laterals and irrigation management services, to be provided by the U. S. Bureau of Reclamation and (2) an on-farm improvement program consisting of such improvements as: lining of farmer's head and tailwater ditches, providing automatic water control gates, land contouring, precise land leveling, and improved application methods such as sprinkler or drip irrigation systems. The on-farm program will be conducted by the U. S. Department of Agriculture, Soil Conservation Service in a cost-sharing program with the local farmers.

The USBR has completed a study based on the three years, 1974, 1975, and 1976 to estimate the salt reduction that can be achieved by lining the main canals and laterals. Numerous ponding tests on canals and laterals performed by a number of governmental agencies indicate that, on the basis of soil type, the weighted average seepage loss from canals is 0.38 cfs per square foot per day (individual tests ranged from 0.15 to 1.3) and from laterals, 0.25 cfs per square foot per day. On the basis of these seepage rates, the Bureau has estimated that the seepage from the main canals and laterals is 100,000 acre-feet annually, 50,000 acre-feet of which occur in the canals, and the other 50,000 acre-feet in the laterals. Assuming a seepage rate of 0.1 cfs per square foot per day for concrete lining, the Bureau has estimated that the lining program will reduce the annual seepage by 70,000 acre-feet and salt pickup by 200,000 tons.

There is little hard data available regarding the on-farm improvements program or the irrigation management services program. It is roughly estimated that the former program could reduce salt pickup by about 150,000 tons and the latter by about 50,000 tons.

The estimated total salt reduction to the river system from the full program will be about 400,000 tons annually at a capital cost of $100 million. Impact at Imperial Dam will be 38 mg/l.

Paradox Valley, Colorado

Paradox Valley, located in Montrose County of Southwestern Colorado, has been identified as a significant natural contributor to salinity in the Colorado River Basin. Studies conducted over the last three years by the Bureau of Reclamation have indicated that the Dolores River picks up over 200,000 tons of salt annually in Paradox Valley from a natural source and discharges the salt into the Colorado River northeast of Moab, Utah.

Paradox Valley, about 24 miles in length and about 3 to 5 miles wide, is crossed near its midpoint by the Dolores River. The surrounding walls of sandstone and shale layers are quite steep and rugged, while the floor itself is relatively flat. Economic activity consists of about 3,600 acres of privately irrigated cropland, livestock grazing, mining, oil exploration, and a lumber mill in the town of Paradox.

Paradox Valley is one of five major collapsed salt anticlines (elongated swells) in southwestern Colorado and southeastern Utah. Paradox Valley, lying along the axis of one on the largest anticlines, has been formed by the erosion of faulted and uplifted sandstone and shale formations, exposing a residual gypsum cap which covers about 15,000 feet of nearly pure salt and salt-rich shale.

Uplift on the sides of the area have placed intense lateral pressures on the intervining sedimentary formations causing warping. The pressures and the weight of overlying formations has caused a deeply buried layer of salt to intrude into the fractured area of the folds. Erosion by the Dolores River has brought the salt relatively close to the surface.

Ground water comes into contact with the top of the salt formation, where it becomes nearly saturated, and surfaces as salt brine in the channel of the Dolores River near the middle of the valley, Fig. 3. The effect of the brine varies considerably, depending upon the amount of water in the river. High riverflows mix with and considerably dilute the brine; consequently, the salt content of the stream may increase from 250 milligrams per liter (mg/l) as it enters the valley to 450 mg/l as it exits the valley. With low riverflows, however, the salinity of the river increases from 1,000 mg/l to as high as 166,000 mg/l (approximately five times as saline as sea water) as it crosses Paradox Valley.

The Bureau has studied various control methods to prevent brine from entering the Dolores River in Paradox Valley. These can be divided into two major concepts: management of surface water and management of ground water.

The first major approach is through management and regulation of the surface-water elements of the system. This approach includes various ways of diverting the Dolores River away from its present channel which would prevent the brine from entering the river by separating the river from the area where the brine is entering. Alternatives based on this concept proved to be infeasible due to very high construction costs and extreme environmental impacts.

The second major concept, to manage and regulate ground water, includes two basic methods -- control of an unknown, diffuse source of recharge or control of the known discharge point source where the brine surfaces. The first method would be to intercept the recharge to the

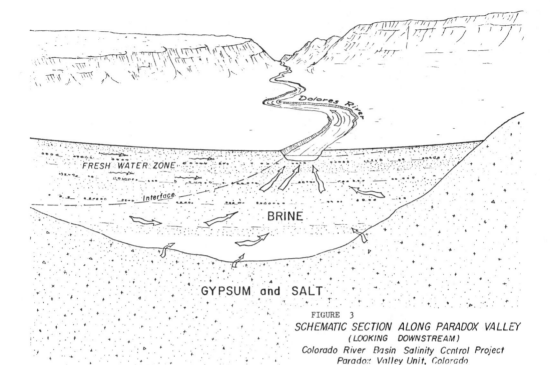

FIGURE 3

SCHEMATIC SECTION ALONG PARADOX VALLEY
(LOOKING DOWNSTREAM)
Colorado River Basin Salinity Control Project
Paradox Valley Unit, Colorado

aquifer, thereby minimizing flow through the aquifer, the
dissolution of salt, and the consequent discharge of brine.
A very detailed and accurate definition of the flow system
throughout the whole valley is a prerequisite for planning
this type of approach. Even if the data were available,
it is unlikely that recharge could be controlled sufficiently
to control brine inflow into the river.

The second method based on the ground water management
concept would control the brine at the point source where
it enters the river. Because the brine discharge is concentrated in a small area, management and regulation of the
groundwater can be achieved. This method would involve a
well field in the area of brine discharge, and evaporation
pond or storage area, and a pipeline from the well field
to the disposal area. This approach would intercept and
remove the brine from the aquifer before it discharged into
the river, Fig. 4.

The plan being implemented by the USBR is to drill a
series of wells on both sides of the river into the brine
zone and pump the saline ground water. This pumping would
lower the interface between the relatively fresh groundwater
and the underlying brine. The brine would be pumped from
the well field to an evaporation reservoir.

The planning and future development of project facilities is divided into two stages. The first stage involves

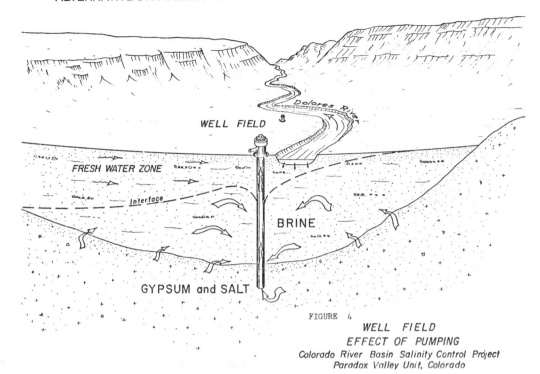

FIGURE 4
WELL FIELD
EFFECT OF PUMPING
Colorado River Basin Salinity Control Project
Paradox Valley Unit, Colorado

drilling and testing of wells and short-term disposal system.
The second stage would consist of converting the wells to
full production status, construction of a hydrogen sulfide
stripping plant, a long-term evaporation reservoir, and a
pipeline and pumping plants to the reservoir.

Preliminary sizing of the facilities included in Stage
2 has been based on a pumping rate of 5 cfs from the brine
well field. The development stage will permit final preconstruction sizing of the hydrogen sulfide stripping plant,
pipeline and pumping plants, and evaporation reservoir.

The evaporation reservoir would be located about 20
miles west of the well field in a shallow natural basin. The
site would be enclosed by the construction of a dam across
the stream channel on the northern edge of the area and a dike
across a low saddle on the eastern edge. Since the reservoir
site is underlain by impermiable shales, lining of the reservoir would not be required.

The project will remove about 180,000 tons of salt
annually at total capital cost of $21 million. The impact
at Imperial Dam will be 16 mg/l.

Las Vegas Wash, Nevada

The Las Vegas Wash is a natural drainage channel which
drains 2,200 square miles of Las Vegas Valley watershed into

the Colorado River. Prior to man's coming, Las Vegas Wash
was dry except for occasional floods when discharge into the
Colorado River occurred. About 1930, the first influx of
people came to the Las Vegas area, and accelerated in the
1940's with industrial development in the Henderson area.
Groundwater was the only supply until the 1940's when Colorado River water was imported to the industrial complex in
Henderson. With increased development in Las Vegas, along
with industrial development and limited farming in the neighboring areas, the character of the Wash changed. Ground and
imported waste waters were discharged into the Wash. Water
applied in the Valley and industrial wastes deep percolated
in the shallow aquifer increasing the groundwater return to
the Wash. Consequently, the Wash became a perennial stream
contributing some 200,000 tons of salt annually to the Colorado River system.

        Much of the salt comes from industrial and municipal
waste water and some saline regional groundwater. Pollution
controls on industrial sources will remove most of those salts
from the system. Municipal wastewater will be given advanced
treatment and a portion bypassed around the Wash and returned
to the Colorado River. A component of wastewater will be discharged
to the Wash to maintain it as a greenbelt. The saline ground-
water will be collected and removed from the system by solar
evaporation.

        The proposed salinity control program calls for a ground-
water barrier, collection system and evaporation ponds. The
groundwater barrier will consist of a trench excavated to bedrock across a natural geologic constriction in the Valley, and
back filled with impervious material to block the downstream
movement of groundwater. A number of large drains installed
upstream of the subsurface dam will collect the saline ground-
waters which will be pumped to lined evaporation ponds.

        A two-stage development is planned. The first stage
consists of constructing the barrier, collection system and
evaporation ponds. The second stage, consisting of a desalting unit, will be added when more water supply is needed for
the Las Vegas area. The first stage will remove about 3,600
a.f. of water annually and 56,000 tons of salt, the second
stage would remove an additional 1,500 a.f. of water and
30,000 tons of salt for a total salt reduction of 86,000 tons
annually, or 9 mg/l at Imperial Dam. The estimated cost for
the first stage is $32 million.

CONCLUSIONS

        The maintenance of current salinity levels in the Colo-
rado River can be achieved while the states continue to develop
their compact-apportioned waters through a basinwide salinity
control program. The techniques described, although site-
specific, can be applied to other point, diffuse and agricultural sources of salinity. These techniques plus additional
and inovative schemes on the identified sources will achieve
most of the control required.

The control of waste discharges and areawide water quality planning combined with control projects will achieve the objective in a cost effective manner. Thus, maximizing a limited water resource in an arid region.

INTEGRATION OF WATER SUPPLY STRATEGIES
IN A DESERT AREA
by
James L. Welsh, Chief
Statewide Planning Branch
Division of Planning
California Department of Water Resources

Integration of available water supply strategies can best be accomplished
through a continuing and evolving master plan process. The objective is not a
rigid plan for development, but rather a process to evaluate the relative
roles of development strategies to meet water-related issues of the area.
Experience with the California Water Plan may serve as a guide to development
in other arid and semiarid parts of the world. Even while local conditions
may be different, as long as the basic needs of the population and the
economy are considered, and the water resources, relative to both quantity
and quality, are treated as a complete system, the application will still be
pertinent.

The California Water Plan is a dynamic example of the role of an overall master
plan for water resources development and management in an arid region. The
success of the plan is primarily due to its implementation by many different
agencies--federal, State, and local--and to the continuing planning studies
carried out on the basis of changing conditions. It serves to illustrate the
advantage of broad regional planning, reflecting that full consideration of
water supply and requirements for water may transcend political boundaries.

The world is in a state of rapid change, with expanding populations and modern
technological developments. Natural resources are being rapidly depleted.
One fundamental need for any type of development is water. This becomes particularly significant as we consider the expanding population of the world,
especially since the present food supply is far from adequate. In the many
desert areas of the world additional badly needed crops can be produced only
through irrigation. This relationship is as true in the United States as in
some of the less developed areas of the world, where basic technological and
economic developments are in earlier stages.

The population of the United States increased four times during the first half
of the nineteenth century, and it more than tripled in the second half of that
century. By 1950, it had doubled again. In the decade 1950-60, the United
States population increased by 28,000,000, and at the beginning of the 1960s
our country was growing at a rate of nearly 3,000,000 persons a year.

Even with this rapid increase in population, the industrial and agricultural
production increased dramatically--continuing to challenge Malthus--and our
standard of living continued to increase at a rapid rate. Our agriculture has

become so productive that today one farm worker supplies not only himself and 29 other Americans with necessary farm products, but also he produces a surplus for export to other parts of the world.

The State of California has been a dynamic participant in the growth of the United States; however, much of California's finest agricultural land lies in semiarid zones, and during the major growing season practically no rainfall is available for crop growth. Approximately 95 percent of the agricultural crop tonnage harvested in California receives some irrigation.

In 1880, only 160,000 hectares were devoted to irrigated agriculture in California. By 1955, over 3,000,000 hectares were under irrigation. During the last 20 years, about 800,000 hectares of additional irrigated land have been brought into production, bringing the State's total to over 3,700,000 hectares. This may well increase by another 400,000 hectares by the year 2000. In 1972, approximately 39,000 cubic hectometres of water were used for irrigation, and about 6,200 cubic hectometres of water for municipal and industrial purposes.

The statewide long-term runoff in California averages about 87,600 cubic hectometres per year. Flows have varied, however, from as little as 22,200 cubic hectometres this year to more than 166,500 cubic hectometres in another. These flows are distributed in numerous rivers, creeks, and streams. Many have only intermittent flow.

The water supply in California is further complicated by the areal distribution. Seventy percent of the State's total precipitation--both snow and rain--occurs in the northern third of the State. The use of water is just the opposite: above 70 percent occurs in the southern two-thirds of the State. Throughout the State, the bulk of precipitation occurs in a few winter months while the summers are long and dry. In addition to these characteristic variations in natural water supply within the year, California is subject to extended wet and dry periods of several or many years' duration.

Of the 45,500 cubic hectometres of water applied in California in 1972, about 18,500 came from ground water basins. The remainder of the water came from direct diversion from surface rivers and streams and from reservoirs built to conserve the water supply from the rainy season and the years of high runoff.

Under the concept of reuse, a specific quantity of water can be made to serve a greater demand by the reapplication of a portion of the supply that can be recaptured after initial use. A net supply of 38,200 cubic hectometres is required to meet the total applied water demands of 45,600 cubic hectometres.

California's water problems are not confined to conservation and development alone. They also include the major problem of flood control. Many of California's nearly 22,000,000 people have settled in areas subject to natural flooding. Thus, flood control is an important purpose in the multiple-purpose development of the water resources of California.

With such wide variation in the water supply, careful, long-range planning must be maintained to ensure the availability of adequate water for the expanding population and the resulting increased demands for water for agriculture, industry, and municipal purposes. A master statewide plan has therefore been maintained to ensure that water development at each potential site will meet its optimum capability for the many project purposes, i.e., water supply, flood control, hydroelectric power, recreation, and the maintenance and enhancement of the fish and wildlife of the area.

In the early stages of development of the water supply of California, the least costly and most easily constructed projects with the most promising yield were developed first. Now, in the advanced stages of development in California, only the more expensive and more complex development projects remain, often at great distance from their major service areas. Other water supply strategies incorporating water conservation, waste water reclamation, desalting, and conjunctive operation of surface water projects with ground water basins are becoming much more important. The role of the master plan in California is to manage California's water resources to meet the social and economic needs of the people most efficiently.

Thirty years ago, the phenomenal growth of population in California, and the realization that water supplies then available in many parts of the State could not be expected to meet future needs, prompted the State Legislature to authorize the Statewide Water Resources Investigation. This authorization called for a comprehensive investigation of the water resources of the State and the formulation of a plan for orderly development to meet the State's ultimate water requirements. The Statewide Water Resources Investigation, which forms a basis for the State's present water development planning, was conducted in three phases.

The first phase consisted of a current, detailed inventory of all the water resources of the State. It included a concise compilation of data on precipitation, runoff of streams, flood magnitude and frequency, and quality of water throughout the State.

The second phase concerned the determination of present and ultimate requirements for water. This phase of the investigation developed the detailed information on the 1950 level of water use throughout the State for all consumptive purposes and projected estimates of ultimate water requirements, based on the capabilities of the land to support further development.

Finally, the water resources and requirements were equated in the formulation of an overall master plan, which was called "The California Water Plan", published in 1957.

An additional continuing phase is the appraisal of changing conditions, the emphasis on economic consideration, consideration of the timing and sequence of development, and full consideration of water quality factors.

The California Water Plan demonstrates that California has enough water so that, if prudently captured, controlled, conserved, and distributed, it will fulfill the foreseeable future water requirements in all areas of the State. This is a master plan that provides the framework within which management of California's water resources would fit. It is intended as a guide to all levels of government--local, State, and federal--in the further development of California's water resources.

The California Water Plan is designed to include or supplement, rather than supersede, existing water resources development. It provides for development by individuals and local governmental agencies to meet local needs, as well as federal and State development of major facilities for export of surplus water from areas of surplus to areas of deficiency. A basis is provided for management of the water resources of the State whereby the most applicable water supply strategy can be used. It contemplates the conjunctive operation of surface and ground water reservoirs as being essential to the regulation of the large amount of water required under full development. The Plan also gives

full consideration to associated benefits such as flood control, power generation, reclamation, fish and wildlife protection, recreation, and other beneficial purposes.

A key element of the plan is to solve critical problems and provide for the
specific needs as simply as possible as they arise within the framework of
comprehensive management. As the water system is stressed and the demands
broaden, both quantity and purpose and the supply facilities become more
complex, additional measures are undertaken in light of greater economic
ability, changed perceptions of values and increased knowledge of the resource.
This process provides for maximum economic efficiency.

Development of the California Water Plan required a team of technical experts
in many fields. Several branches of civil engineering--planning, design, construction, water quality, hydrology--were involved throughout the development
of the plan. Soil specialists, water use experts, and economists played a
key role in the inventory and evaluation of the land resources of the State
and the estimate of the future water requirements for agricultural, municipal,
and industrial purposes. Electrical and mechanical engineers carried out the
more detailed analysis of hydroelectric power and pumping plants. Geologists,
both ground water specialists and structural, played important roles in the
evaluation of the ground water resources of the State and in damsite exploration and the location of materials for dam construction.

Recreation planners determined the recreation potential of the various proposed
development areas, because water-associated sports--boating, swimming, fishing,
camping, and sight-seeing--have become popular with the people. If recreation
planning is not incorporated in the initial stages of development, access to
the recreational areas around a reservoir may be lost. And, because their
value quickly increases, future acquisition of such areas becomes quite expensive. A strip of publicly-owned land around each reservoir preserves public
access for recreation. Fishery biologists and wildlife management experts
were called in to plan for the protection and enhancement of the existing
fisheries and wildlife.

The services of specialists from other State agencies were obtained to carry
out certain aspects of the study, particularly when another agency had the
basic responsibility and the appropriate staff to carry out the work.

Thus, the California Water Plan represents a team effort of many technical
disciplines and State agencies. Federal and local agencies also cooperated
by furnishing much needed information and data.

Many of the features proposed in the California Water Plan have been built.
The State Water Project by the State of California is in operation. Major
extensions to the federal Central Valley Project, other federal reclamation

and flood control projects, and the continuing development by local water
agencies and public and private utilities have also been completed.

Much of the development by local agencies was for dams, reservoirs, and
power plants for the generation of hydroelectric power. Other significant
development was for municipal water supply. In addition to this, irrigation
districts developed water to supply the needed water for their expanding
irrigation development. Although less dramatic than the surface water
projects, ground water development has also continued. This water was used
for agricultural, municipal, and industrial purposes. In many locations
ground water development has reached the stage where the annual use exceeds

the replenishment of the basin. This has resulted in lowered water tables
and in sea water intrusion into some of those aquifers adjacent to the
coastline. Complex planning studies are underway to protect and maximize
the use of the ground water basins without detriment to their use by future
generations.

Development by the Federal Government has added major reservoirs, hydroelectric power plants, aqueducts, and flood control levee systems. In addition to conserving water, the major reservoirs have added flood control
protection for downstream areas and have supplied water for fish, wildlife,
and recreational purposes.

To meet the expanding future needs for water will require the combined effort
of all water management agencies at all levels of government. The federal
Central Valley Project and the State Water Project must be closely coordinated
because some facilities are actually joint projects. Together they constitute
the major water development system of the State.

The California Water Plan has demonstrated that one broad master plan can
accommodate and facilitate water development by several different governmental
and private agencies. In implementing the Plan, regional jealousies have had
to be overcome to secure development for the State as a whole. Through the
export of water from areas of surplus, it has been possible to carry out
development for water-deficient areas. Thus, areas of surplus and areas of
deficiency have all benefited from the California Water Plan.

The California Water Plan typifies a process that might be applied to studies
of other desert areas. Even though basic demands for water development may be
somewhat different in other areas--less emphasis on hydroelectric power and
more seasonal regulation of irrigation water, for example--the fundamental
planning studies required and the process of developing and maintaining an
overall master plan would be much the same.

The plan has also demonstrated sufficient flexibility to accommodate significant redirection of social goals and rapidly changing perception of resource
limitations. The conditions which spawned the California Water Plan were those
of rapid economic and population expansion. Following initial completion of
the plan, California continued a rapid growth and maintained surface water
development for nearly a decade. In the late 1960s, strong concern for
environmental values provided a redirection of water resource priorities.
These included a strong program of maintaining and enhancing water quality,
upgrading fishery and wildlife values and dedication to maintaining some
streams and land resources in natural conditions. In the intervening ten

years traditional water demands increased 3,700 cubic hectometres, environmental reservations increased 27,000 cubic hectometres.

These increased reservations and recognition that both water and energy
resources are limited, combined with greatly increased economic and environmental costs of surface water development, have fostered a conservation ethic
dedicated to more efficient use of the resources we have before developing
new conventional supplies.

In earlier stages of development, planning was a discrete process with a
beginning and end--usually resulting in or leading to a specific water
development project. But today the picture has changed. A simplistic
concept of planning is no longer applicable. The changing public values,
the diminishing available resources, and the rapid advance of water development technology demands that we plan in a manner which will keep open as many

options as possible and maintain maximum flexibility. Instead of the goal being a discrete plan with specific long-range developmental projects, we now look upon it as a dynamic iterative process carried on under almost continuous uncertainty.

The continuing planning process which results in periodic updating of the plan reflects actions which have been taken, identifies and addresses current specific issues, defines policies, and recommends actions to be taken within a time horizon required for implementation of those actions.

Current water management issues include:

1. Reallocation of water supply to a greater variety of in-stream needs.

2. Resolution of competing water supply and environmental concerns in the Sacramento-San Joaquin Delta.

3. Integration of water management and water quality planning.

4. Agricultural drainage and salt buildup in the San Joaquin Valley.

5. Annual ground water overdraft in the San Joaquin Valley of 1,900 cubic hectometres.

6. Risks associated with attaining planned water supply project yields considering the present extreme drought.

By 1990, net water demand for consumptive purposes in California is expected to increase as much as 4,000 cubic hectometres. Because of recent and continuing commitments of water for environmental purposes, there is a remaining available undeveloped supply of about 33,000 cubic hectometres, of which only about 13,600 cubic hectometres is considered potentially developable. Most of this yield will be extremely costly.

A number of alternatives to satisfy increasing demands are under consideration. These can be broadly grouped into four categories:

1. Reduction of demand through water conservation--simply using less water to accomplish the same beneficial purpose.

2. Conventional surface water storage.

3. New sources of water or augmentation of existing supplies through waste water reclamation, desalting, and weather modification or cloud seeding.

4. A combination of physical facilities and integration concepts for off-stream storage, both surface and underground.

Both the water issues and water demands are inextricably intertwined.

In integrating water supply alternatives into a master plan benefits, detriments, and trade-offs between economic and environmental values must be evaluated. This must include the effects of additional supply projects on existing development and the existing system of development.

For example, increased agricultural water application efficiencies can be attained through better irrigation methods, irrigation scheduling, and drainage control. However, over half of current ground water recharge is accomplished through deep percolation of excess applied water. A reduction in this recharge would also increase the ground salinity. Similarly, over-application of surplus surface water creating an increase in usable ground water recharge can result in more efficient operation of the total resource.

Conversely, conservation in an urban area adjacent to the ocean where water is used once, treated and discharged, will result in a net saving equivalent to the demand reduction. However, this same urban conservation will make reclamation of the resulting waste water less viable since the waste quantity will be reduced and the quality more degraded.

A new surface water development project may result in environmental damage which must be mitigated through project design and operation. A new project also offers opportunity for enhancement of in-stream values at downstream locations or it may even permit reoperation of an existing project to provide for enhancement on a different stream.

The current drought demonstrates the value of integrated surface and ground water supplies. Since it is uneconomic to plan for full surface supplies in all years, deficiencies are expected during drought conditions. Where ground water supplies are available, these deficiencies can be reduced by increased pumping. The current level of integration of the surface and ground water prevented the current drought from causing a major disaster. However means must be found to recharge this large cyclic overdraft created both by increased pumping and decreased recharge to prevent a continuing economic penalty for ground water users. The drought has also demonstrated that we have neither enough ground water extraction facilities nor surface water interconnections to provide adequate water system flexibility. Thus each new experience serves as valuable input toward improving the master plan on a practical basis.

These few examples illustrate the concepts of relating water management strategies in maintaining a current master water plan.

DESERT FLOOD AND EROSION CONTROL

by

W. E. Bullard, Consultant

INTRODUCTION

For the purposes of this discussion, a few definitions may be in order.
First, desert: a dry, barren region largely treeless and usually uninhabited;
an area so deficient in water as to support only a sparse, widely-spaced vegetation, or none at all; an area in which few forms of life can exist because of
lack of water, permanent frost, or absence of soil. Since we are concerned
primarily with the warm to hot deserts, we can omit permanent frost from consideration here. Some sciences go further than the dictionary and put numbers
with the adjectives; but even so, the limits are variable, from 8 down to 3
inches (200 to 80 millimeters) or less of precipitation received on the average
each year said to define desert.

Second, flood: a great flowing of water, an overflowing of water on land
not usually submerged. Too much water at one time in one place causes a flood,
even in deserts.

Third, erosion: the process by which the earth is worn away, the action of
all the forces of nature that wear away the earth's surface. These forces include precipitation and runoff, wind, gravity, diurnal temperature changes,
freezing of soil moisture, earthquakes, solution and leaching, activities of plants
and wild animals. In the recent history of the earth, the activities of man and
his domestic animals must be included.

We will consider in this paper the significance of floods and erosion in
desert areas, their influence on formation and maintenance of desert conditions
and on improvement or rehabilitation of desert areas. Each of the specifics in
the definitions above affect these matters.

FACTORS IN FLOODS AND EROSION

To cause a flood, an excess of water may be provided by rapid runoff or
drainage from another area, whether from rainfall or snowmelt or the breaking
of an impoundment, or from heavy precipitation on the area itself. Again, too
much water may result from the lack of infiltration capacity in the soil receiving the rainfall. And that, in turn, may be due to a number of causes,
separately or interacting.

Climate

A major factor in rapid runoff, flooding, and erosion is rainfall intensity; large amounts of rain in large drops have a beating action that puddles and seals the soil surface to inhibit infiltration. High rainfall intensities are fairly common in arid areas; sometimes the total annual rainfall may come in an August afternoon thunderstorm that drops 3 inches (80 millimeters) of rain in 40 minutes or less. In flat country this fills the playas; in hilly or mountainous terrain it causes flash floods.

Wind, at least seasonally, is a factor in erosion. Strong winds can lift and move fairly large soil particles, and may carry the fine silt particles for thousands of kilometers before they settle out. Where this wind erosion is common, and erosion pavement of pebbles forms because the finer and desirable fertile elements of the soil have blown away. Wind creates and builds and moves sand dunes; and may affect flooding and erosion by water as it blows sand and silt into channels to plug them.

Aridity is an expression of a couple of the climatic elements working together. It is a descriptive term generally applied to fairly large areas; we shall, however, consider it here in relation to the soil. Solar radiation heats the soil, causes evaporation of soil moisture to a depth of about 12 inches (30 centimeters), increases transpiration of plants growing on the soil and thus may exhaust soil moisture to a depth of several meters. Wind has similar effects. Together, sun and wind are highly significant in determining the soil moisture regime throughout the year, especially where rainfall is light and irregular.

Seasonal regularity of the climatic elements also plays a strong role. Solar radiation varies with the height of the sun as it changes from season to season. It may also vary according to the amount of cloud cover or fog, though these may not be important except along coasts in desert regions. Precipitation - rainfall, except in cold high deserts - varies seasonally and from regular to irregular even in the seasons of usual occurrence. Winds such as the chergui of North Africa and the Santa Ana of California are expected in certain seasons, but even so, without any definite regularity. Plant cover has to adapt to these conditions; and so must human occupation and development.

Hydrology might be considered a separate element; but it is closely related to climate. Water movement depends on water supply, on rainfall and snowmelt. Streams flow according to the water received on their watersheds, from whatever source. They also flow according to conditions of their watersheds; - topography, type and density of plant cover, depth and texture of the soil, permeability of the rock basement. Few desert areas have streams with permanent flow; some may have seasonal flow from snowmelt, but most have dry washes or oueds or wadis with flow only after rare high-intensity rainstorms. Groundwater, however, does feed some springs and oases; and often is available at depth. Sand dunes in coastal areas where they may receive moisture form night-time condensation and fog as well as from occasional rain can store considerable amounts of water not far below the surface, and may even lead to interdune swamps or ponds. These, however, are more of a desert fringe than true desert phenomena.

Geology and soils

As wide a range of rock types can be found in desert regions as in the rest of spaceship earth. While the alignment of mountain ranges relative to the circulation of the atmosphere may have considerable bearing on aridity, rock types have little or no relation to desert formation and maintenance. They do, however, affect hydrology; some being much more permeable than others. Also, they affect soil development and to some extent the type of soil produced. Granites break down to coarse crystalline sand particles, and from certain of their constituent minerals, fine clay particles. Basalts and limestones and other rocks break down to other percentages of other particle sizes and shapes. The soils formed are often red in color because of exidation of the iron component of the soil minerals.

Desert soils, because rarely subject to leaching through the entire soil profile, may show strong accumulation of calcium and other minerals. This is a result of continuing alternate up-and-down movement of soil moisture caused by evapotranspiration plus the lack of any through drainage of consequence. The final result is deposition of a hard and often impenetrable layer at 20 to 50 centimeters depth in the soil; the layer itself being perhaps at most some 10 centimeters thick, and usually less.

Again, if the rocks from which the soils are formed by weathering contain significant quantities of elements toxic to plants, so will the soils. Leaching and evaporation may concentrate them, and there is no drainage to carry them away.

Because plant cover is usually sparse and scattered, desert soils do not incorporate much organic matter recycled from the plant cover. Thus the soils show a color largely determined by the mineral composition of the parent rock rather than by any organic content. Desert soils are usually light-colored, reflecting rather than radiating heat.

Except where the caliche or hardpan layers have formed, desert soils do not develop much structure but tend to remain single-grained. Interior dune sands and silts are a prime example. Salt accumulation and the lack of organic content tend to maintain this condition. This characteristic makes the soils readily eroded both by wind and water, and does not provide good infiltration except in sands.

Desert soil may be too shallow to hold much water, of too tight and heavy a texture - like clay - to accept water rapidly, of too coarse a texture to hold water against rapid drainage, crusted over the surface and practically impervious as are some salt soils, or containing impervious lime deposits near the soil surface. Any of these may be natural conditions.

Plant cover

Excepting the surface of rapidly moving dunes or rapidly eroding areas or excessively saline areas, one may find plant life growing on nearly all exposed soil. Even on rock; - though tiny lichens may be the only examples there. Where vegetation has opportunity to develop and adapt to conditions of the site, it may develop into a dense cover. In desert regions, however, the cover of perennial plants - grasses, herbs, brush, and trees - is usually scattered; the roots of each plant must draw their necessary moisture from a large volume (and surface area) of soil. These plants, then, are spaced rather widely; - a meter or two apart at the closest. If the area includes an occasional - not necessarily frequent or regular - wet season, at that season the area will also have a considerable growth of ephemerals that will produce a dense plant carper if only for a week or so.

While the ephemerals are often tender and delicate, occupying a microclimate that exists only briefly and is atypical of desert regions throughtout most of the year, the perennial plants tend to be deep-rooted, armed with spines or thorns against heat and drought and foraging animals, tough, and with a waxy bark and small leaves to protect them against excessive evapotranspiration. Perennials

tend to be longlived, and to reproduce by coppice and root-shoots more often than
by seed. Though the ephemerals are not themselves robust and durable, their seeds
are; - they may endure many years while awaiting the rare rainfall and head combination that will bring them to germination. The exchange of soil nutrients
and organic matter by the ephemerals relatively insignificant in the desert
ecology, so that they have little effect on runoff and soil erosion.

The tufts of desert grasses, the thick clumps of desert shrubs, the coppice
growth of desert trees like the arganier in Morocco, all tend to catch windblown
soil particles and thus to create hummocks of fertility scattered across the
erosion pavement. These clumps and hummocks provide homes and hiding places and
food for much of the desert biota. In addition to holding some of the soil in
place, they also provide some recycling of soil nutrients. They are the only
natural deterrent to wind erosion, and tend to expand slowly if undisturbed.
Where loss of soil moisture by evaporation affects little more than the uppermost
30 centimeters of soil, transpiration by the deeprooted desert perennials may draw
water from a depth of 10 meters.

In the widely scattered oases that develop around springs and seeps and more
or less permanent ponds, as well as along channels where the ground water is not
too deep and subirrigation is available, a different class of vegetation develops.
A few of the true desert plants may hand on, and a few species from more humid
zones; but a special group has developed for this particular habitat, too. The
largest and best-known element is the date palm in Africa; others may include
a few willows and poplars, the carob, tamarisk, camelthorn, tree tobacco, etc.
Along watercourses, these trees stabilize sediment deposits and streambanks in
addition to providing some food and nutritious forage.

Land use

On desert fringes in Africa there may be dryfarmed cultivation of cereal
crops such as barley or millet;- the production usually low and chancy according
to the amount of rainfall received. Around oases and along channels where there
is some subirrigation, trees such as fig and olive and carob may grow, and the
native date palm be carefully tended. Small vegetable plots may be found where
irrigation is possible, often formed by a series of steps formed by dams in
channels where water flows infrequently that have accumulated silt to build
deep-soiled terraces that are subirrigated.

The most widespread use of the land, however, is as range for hardy livestock, principally goats and camels. These animals can thrive on thorny forage,
and even overuse it as they do the more succulent and tender species. Most of
the desert range use is uncontrolled, at least in the sense of being managed.
The range is open; livestock may be turned loose to roam within large areas, or
be herded in fairly close groups to wherever a bit of green shows. Livestock
numbers and intensity of use are rarely in any balance with available forage
under such conditions; forage use is extremely heavy, and most of the desirable
forage species have disappeared.

This heavy grazing use involves repeated trampling by millions of hooves
over the years, which increases soil density and reduces infiltration as well
as breaking down any primitive soil structure that may have begun to develop.
The process has been going on for several thousand years in North Africa, and
is believed to be a major factor in the presence and continuing expansion of the
Sahara Desert.

Land cultivated for grain is usually prepared by shallow plowing with a
stick plow, a practice that does not greatly disturb the soil. However, at
harvest, the practice of pulling the barley plants up by the roots (every bit

of it is used!) does disturb the soil and cause some loss. It also prevents any recycling of organic matter or replacement of the mineral nutrients drawn from the soil. The disturbed surface is that much more open to erosion by wind, and the productivity of the soil is gradually lessened.

Interior dunes, formed as much of silts as of sands, are often sufficiently fertile to be recovered by volunteer vegetation when grazing pressure is removed. Once a reasonably dense cover is reestablished - perhaps a matter of six to ten years - controlled grazing can make use of the forage resources without further damage to the land. However, the return of uncontrolled grazing is soon followed by uncontrolled wind erosion and loss of a forage-producing area.

EROSION

The two principal causes of erosion, wind and water, have already been described. There are others usually less important though occasionally and locally significant. Earth displacements, from small landslides to wide area tremors, may cause rapid downslope movement of masses of soil and rock. Earth movement damages include destruction of developments, loss of crops, death of people and livestock, creation of new channels to drain a slope, and plugging existing channels with debris. However, we will consider here only erosion by wind and water.

Water erosion

Water erosion requires rainfall or snowmelt at rates greater than the infiltration capacity of the soil can take up, so that water flows across the soil surface. The faster it moves, the more soil particles it can pick up and carry away. Thus water erosion tends to proceed most rapidly where there are large concentrations of water received on steep slopes, where the soil is compacted and structureless, and where it is unprotected by vegetation. As heavy runoff moves downslope it erodes and carries away much of the soil, as it converges into streamflow it enlarges old channels and may dig new ones. Below on the flatter slopes, the eroded sediment settles out to damage crops, roads, villages.

Under conditions of depleted cover and reduced infiltration capacity, rapid runoff from almost any rain that does occur is to be expected. It happens;- water that should be held where it falls and slowly percolate to feed into channels below is instead suddenly accumulated in rills and gullies and finally as flood waves rolling in once dry stream channels. The rapidly moving water carries erosion sediments until it reaches breaks in gradient, usually in the major stream channels, where the sediment load may be dumped to fill the channels and cause the water to spread over adjacent lands. The rills and gullies deepen and widen and lengthen with each storm; the volume of soil carried away per hectare per year even from almost level ground may reach as much as 2,400 metric tons during the period of most rapid gully growth. The measured average annual erosion loss for an area of 26,000 hectares in the Souss Valley of southern Morocco over a 19-year period was 1,200 tons per hectare under desert conditions; this land had suffered from both overgrazing and the dryfarmed culture of barley. It is not necessary that the slopes be steep for such erosion to proceed;- the erosion rates cited were measured on valley slopes of less than three percent. Part of the reason for this is gully head-cutting where there is a vertical step or drop that permits erosion to eat its way upstream whenever there is enough water to carry any erosion sediments. Erosion is in large part a gravity process.

Based on appearances there are two kinds of erosion;- sheet erosion and rill or gully erosion (gully erosion being considered just the enlargement of rill erosion). Most water erosion is of the rill or gully type because water tends

to concentrate in channels. This erosion begins with small rills that become
enlarged to gully size as the flow accumulates downslope. However, in a resistant soil only the fine particles may be eroded away, and these over the entire
soil surface where runoff occurs, to leave a coarse-textured surface over-all.
The loss to the soil surface may be almost invisible yet contribute large amounts
of sediment where several hectares of drainage area are concerned.

Erosion takes place wherever soil is exposed. Vegetation protects the soil
surface against the beating action of rain. Litter from dead leaves and twigs
blankets the soil surface and protects it in addition to providing organic material that my be incorporated into the soil. Plant roots bind the soil particles
together, and provide openings for infiltration. Desert soils often are easily
eroded and always susceptible to erosion because there is usually too sparse a
cover vegetation to provide such services. Particularly with a delicate balance
in the development and maintenance of even a sparse vegetation cover, the often
much too heavy grazing use leads to cover and soil deterioration and rapid erosion.

Wind erosion

The effect of strong constant wind is most often noticed along seacoasts and
lakeshores where it creates and moves dunes built of sand from the shorelines.
However, the same force can create and move dunes in interior areas; or pick up the
finer particles, lift them to considerable altitudes, and carry them hundreds of
kilometers in dust storms. The chergui of North Africa may carry a load of dust
far west over the Atlantic Ocean and drop it on ships at sea.

Materials in such loads come mostly from desert soils; they represent the finer
and more fertile soil particles. The heavier sand particles will ride the wind only
short distances though the fine silts may travel far. Gradually the source soils
become coarsened and impoverished, though continuing weathering of rock minerals
provides new fine elements to replace some of the losses.

Wind erosion is mainly sheet erosion, but it is a serious source of soil loss.
Strength of the wind and windborne sand is shown by sculpturing of the rocks in
desert regions.

Wind and deserts interact;- unprotected soil surfaces heat up and the heat is
transfered to the air to produce rising currents. These hot dry moving air masses
take up moisture from the soil and the vegetation cover to an extent that severely
limits growth of the cover. The soil remains without sufficient cover, and the
desert condition is maintained. The air moves rapidly across the surface with no
cover to slow it down, and removes soil particles that are unprotected.

Wind erosion is a cause of damage by sandblasting both crop plants and developments, by carrying away fertile soil elements, and by dropping sediments where they
bury crops and developments and plug up stream channels.

FLOODS

Flood occurrences may be regular and seasonal, or infrequent and unpre-
dictable. Both types occur in desert regions. Snowmelt floods from mountains
bordering desert regions may be expected predictably, as may occasional rainy
season floods. Rarer and more catastrophic because unprepared for are the unseason-
al floods that arise from sudden local thunderstorms in which the total average
annual rainfall expected over a five-year period may fall in a couple of hours.
It is floods such as these that build up huge deltas of boulders and gravels and
sands below canyon mouths in mountains of arid regions. With high-intensity rainfall the runoff builds up rapidly on unprotected soil surfaces and has nothing to

slow its movements as it accumulates. Collected into torrents in the canyons it
can carry huge boulders great distances, grinding them to produce finer sediments
as it goes.

These flows tear out developments, empty the channels of old sediments and re-
fill them as the flow decreases after the peak. They undercut channel sidebanks
and slopes to cause landslides, they dump coarse sterile sediments on croplands to
ruin any crop present and to lower productivity of the site. The floods also represent a loss of water, some being wasted to saline sinks or to the ocean instead
of improving soil moisture content on the watersheds where they originate. Floods
cause death to both people and livestock, destroy cropland, damage homes and other
buildings, and make roads impassable. The costs of restoration and rehabilitation
usually are high.

Both wind and water cause erosion; and both carry away the eroded soil particles. In flood, water can move tremendous boulders as well as all sizes of stones
and gravel and sand and silt. Both causes of erosion may be checked by a plant
cover;- the denser and more complete the cover, the greater the degree of protection
provided.

CONTROL METHODS

The natural factors in erosion and runoff and the floods have been described;
and we can see that there are numerous ways in which we may make use of them to
avoid or prevent or reduce erosion and runoff and to control floods. We will start
by considering protection of the soil surface where rain falls and where erosion
begins. If we have any kind of degree of cover vegetation we have some protection;-
how may the cover and its benefits be improved?

Controlled use of the plant cover

Two activities of man reduce the cover;- grazing his livestock, and the collection of fuel materials. Both forage and fuel are produced by vegetation, but there
are limits to the volume that can be produced on a given site. Once conditions
are established - or allowed to develop - that produce the greatest amounts of fuel
and forage from the current growth each year, management must be applied to hold
consumption and utilization of the resource at the level that permits maintenance
of that rate of production. Studies have shown that plant cover tends to be self-
maintaining if not damaged or overused. Both soil protection and runoff control
are best afforded by a healthy cover in undamaged condition. Inspection of the
area to be managed will show what species of those one would normally expect to
find have disappeared and others that are enroute to disappearing. Removing use
from the area and seeding in or planting desirable species (and "desirable" might
include palatable nourishing forage species, legumes to build the soil, woody-stemmed
species for use as fuel, etc.) may be sufficient to rebuild cover. Control of the
cover is based on control of land use.

Which sounds simple enough, and indeed is; though difficult to put into effect.
It first may be necessary to overcome some thousands of years of traditions, to
change attitudes toward the land, to teach proper use and management of the soil,
to introduce new forage crops and perhaps new types of livestock, to provide alternatives for uses that are stopped or temporarily suspended. Generations may be
needed to change attitudes toward the land, to get new crops and livestock accepted,
and to supplement the traditional land use with scientific management. Alternatives must be made available first, before restrictions can be made effective.

Plant cover improvement

    Plants that can be used to make windbreaks, to bind the soil surface, or to supplement the forage resource may be found among the species of the native flora, and these can be reintroduced the most rapidly because they are already adapted to the site. However, there are others, exotics, that have been tested under a wide variety of conditions over the world; these include several species of Eucalyptus for windbreaks and wood production, and of Acacia (often the only tree or shrub that can be grown in minture with the Eucalyptus), the spineless variety of the cactus Opuntia, the shrub Medicago arborescens, certain poplars, certain pines and cypresses, the grass Cynodon dactylon, the locust tree Robinia, shrubs of the genus Atriplex for saline soils, and some others.

    Windbreaks reduce the drying and sandblasting action of the wind around cultivated plots as well as furnishing fuelwood and small construction wood, some forage, some fruit, and shade for livestock. While they will draw heavily upon available soil moisture and take up some soil area, their benefits in increased crop production alone can more than compensate such losses. Windbreaks are effective for a distance of several times their height.

    Dune stabilization is usually begun with low slat or brush fences as windbreaks, plus a cut brush cover over the dune surface to protect the dune grasses with which the development of a permanent vegetation cover starts. Species of the reedgrasses Ammophila and Saccharum can be used for the initial plantings; both can adapt to soil piling up around their stems. As the sand movement becomes slowed, leguminous shrubs of several genera may be planted to begin the soil-building process. Sometimes trees may be planted in interdune troughs, subirrigated from the dune water table built up from rainfall and condensation. Grazing use of stabilized dunes must usually be prohibited, or where a dense cover and ground litter has developed, be limited and carefully controlled to avoid exposing bare soil to the wind. Even foot traffic by recreationists in coastal dunes can cause the cover reestablishment works to fail; a small sand exposure can soon become a blowout, and dune movement begins again.

    On depleted pasture areas, once grazing pressure is removed, palatable forage species can be reintroduced by seeding and planting. After establishment they should be limited to the annual growth or volume of good forage produced, and done at seasons when there is least damage caused to the soil and plant cover. Enough of the forage plants must be left to provide seed and growing stock for next year's forage crop. In situations of delicate balance, it may be necessary to harvest the forage by hand and feed it to the livestock offsite.

    In tree plantings, whether woodlots, greenbelts, or windbreaks, protection must be provided against grazing animals until the trees are tall enough to escape browsing and trampling. Even well-developed plantations may be damaged by trampling which compacts the soil and reduces aeration and infiltration. Taking fuelwood should be prohibited until the growing stock has reached full productivity; and then should be restricted to the annual volume increment. During this primary period dead wood and prunings and thinnings are all that should be used for fuel.

    Livestock droppings are better used returned to the soil as conditioners and fertilizers then as fuel. However, until an adequate fuel supply becomes available from the woodlot or other tree planting, it may be necessary to continue to use dried manures as fuel.

Engineering methods

Impounding water to reduce streamflow peaks and to save water provide a traditional and widely-used method of flood control and reduction of flood damages in
major river systems in humid climates. In semi-arid and arid regions where high-
intensity rainstorms and subsequent floods are scattered and infrequent and there
are no channels where floods tend to occur regularly, flood control impoundments
are rarely justified. The runoff must be controlled where it starts, and the
erosion be controlled where the soil particles are first separated and carried
away. This is on the watershed;- not in the channels downstream.

On the slopes, contour trenching to catch and hold immediate runoff is a
useful measure. It is also one that requires much work whether done by hand or
by machine. Trenches are usually built deep enough - about a meter - and with a
berm sufficiently solid to catch and hold at least 25 millimeters of rainfall in
a short period. Trenches must be built to have short segments with cross dams
every three or four meters to avoid creating heavy flow in case of a break.

Where slopes are less steep, terraces with berms can be built to provide both
stormwater storage and in the same place level cropland to benefit from the occasional irrigation. Again, storage space for at least 25 millimeters of rain should
be provided. Terraces should not be continuous, but be separated by cross berms
at intervals of about ten meters.

Gullies must have their banks plowed down and become part of the trench or
terrace system where not to be continued in use as channels. If needed for channels,
they should have rock or brush checkdams to retard the flow and to trap silt, and
paving or lining to prevent further erosion. Trapped silt will change the channel
gradients to about 70 percent of the original slope, which will help reduce the
erosive power of any flows. These channel deposits also provide a foothold for
vegetation which will bind them in place. At gully heads, drop-inlet structures
may be needed to prevent further erosion of the gullies upslope.

Engineering in dune control would include both construction of slat and brush
fences to cut down the force of the wind, and covering exposed surfaces with crude
oil to bind the sand and silt particles against erosion.

Biological control methods

Any of the engineering contols just described can be made safer and more effective by tree and shrub and herb and grass plantings. Berms and both inner and outer
slopes of contour trenches can be stabilized by seeding in herbs and grasses and by
planting shrubs. Terrace berms and slopes may be treated in similar fashion, or
have cactus or some unpalatable shrub planted along the berms to keep livestock
from trampling them down. Terrace areas may be used to grow forage plants which
should be harvested by hand and fed to livestock offsite, to grow fruit and nut
trees, or to grow truck crops. If water is available, bermed terraces may be
irrigated.

Use of various types of plants in dune stabilization has already been mentioned;
it should also be noted that certain hardy grasses and herbs will naturally invade
bare dune source areas such as blowouts and flood silts deposits if grazing
pressure is removed. The tough creeping grass Cynodone dactylon is one
of the first and best of the invaders for quickly binding the soil and stabilizing
an area. Prime need in dune control and stabilization is a carpet of plants to
cover and bind the dune surface.

Stream flood channels may similarly be first invaded by tamarisk (a small
tree), genus Tamarix, and the shrub oleander, genus Nerium. Tamarisk is a pro-

lific seeder and grows rapidly; it often forms a dense cover on channel bars and flood silt deposits. Oleander spreads by root suckers or coppice, is unpalatable to livestock, and makes an excellent soil binder for the coarser channel deposits. However, it requires subirrigation. In fact, both these plants will draw heavily on available water.

Smaller watercourses and gullies that are kept as watercourses may be paved with stone or lined with sod, and have their banks planted to shrubs that will serve to hedge livestock out. There is a thorny Acacia excellent for this purpose. Trampling down treated streambanks must not be permitted, nor direct use of forage from channel lining. Any forage available in channels should be cut by hand and taken offsite to feed to stock.

Windbreaks for small cultivated plots of low truck plants may be grown from thorny Acacia or Casuarina or the reedgrass <u>Saccharum roseum</u>. Taller barriers for orchards may be grown from Eucalyptus and Acacia, or from cypress and poplar if irrigation for them is available. The windbreaks prevent sandblasting and a certain amount of sunscald, plus reducing evapotranspiration from the crop plant, in return for the space and water they take up. They also provide barriers against entry by livestock or crop thieves, as well as furnishing wood for fuel and other purposes. Windbreak trees should be planted at close intervals to provide some barrier protection right from the start. As they grown they can be thinned to furnish poles and eventually some heavy timbers for construction, or wood for fuel or charcoal-making. Windbreaks have a significant effect in reduction of wind force out to a horizontal distance equal to five times their height; and generally should be planted across the prevailing wind direction.

<u>ALTERNATIVES IN LAND USE AND MANAGEMENT</u>

Before land can be taken out of use to be improved or reclaimed, some provision must be made to replace the lost uses. Because they take time to develop, some of the alternatives must be established before anything else is begun. Flood and erosion control measures apply to all land, and may consist of avoiding operations that damage the soil and its vegetation cover as much as of reclamation operations, tree-planting, and water and sediment storage structures.

<u>Forage reserves</u>

Since the bulk of the desert economy rests on livestock and grazing, alternative forage must be provided during the period of several years needed to improve existing pasture and range areas. This can be accomplished by planting strips of spineless cactus (which is highly acceptable to goats and camels) between the areas to be treated by planting clumps of native forage grasses and herbs and shrubs. Forage also may be collected from other areas and brought in to feed local livestock;- this might include grass from erosion control treatments, for example.

<u>Water development</u>

Additional sources of water must be developed so that more crops can be grown on less land over the period of rehabilitation. This may be done by deepening wells, extending rhettaras, or lengthening canal and pipeline systems. New drainage systems also may be needed to slow down soil salinization. Further exploration for groundwater and digging new wells also may often be necessary.

<u>Greenbelts and wood reserves</u>

Greenbelts can provide shade and some slight modification of local climate as well as furnishing wood, fruit, and some forage; they are both an alternative and

a base for future management. While they represent land at least temporarily out of
production, they are a long-run necessity. Eventually their productivity will more
than compensate any immediate loss. Greenbelts appear to be at present the one most
important physical measure in reversing regional desertification trends.

New crops and livestock, new products

In some areas, sheep can replace goats, cattle can replace both. Better breeds
introduced can produce more milk and cheese, more wool, more meat, better hides;
have lower mortality rates and grew faster. Both truck crops and orchards to handle
new food plants can be developed, either to feed local populations or for export, and
should be possible in many areas. Simple solar cookers can be made from oil cans,
and can reduce the use of wood and dried manures for fuel. Windmills could be used
in most desert regions to provide power;- they are not yet everywhere adopted.

Management

Pasture rotation according to seasons of forage development and volumes of forage
produced must be adopted, and closures strictly enforced. Much greater forage yields
from a given area are thus possible eventually, though during the initial rehabilitation period the land must be out of production.

Water management can be improved. The crops with highest yield should have priority for irrigation. Canals should be lined to avoid leakage losses, and have fringe
vegetation removed to avoid transpiration losses. Storage of water surpluses that
occur outside the growing season should also be provided. Present sources of water
can be exploited more efficiently, and new sources be developed in some instances.
Mulching and other soil-moisture-saving methods can be applied.

New and more effective farming methods can be introduced;- new crop rotations,
new tools, new systems of cultivation and irrigation. All can show considerable
benefits; but outside help will be needed in quantity.

EDUCATION AND TRAINING

Land use traditions and methods have been passed from father to son for hundreds
of generations, with some gradual improvements as time passed. However, within the
past century research has shown the way to better methods. These need to be taught
the the arid land farmer, to become part of his traditions. It is not enough to furnish new tools without showing how best they should be used, whether the tool is a
plow or an irrigation system or a crop rotation.

The several things suggested for flood and erosion control have proven workable
in many parts of the world, but they will not work unless maintained and continued.
It is often difficult for the farmer or grazier to understand future benefits to be
derived from long-term projects;- they must become parts of an accepted land use
tradition to succeed.

As new crops and new land use systems are introduced, their acceptance can be
speeded by marketing assistance. This may include establishment of farmer coopera-
tives, improvements in storage and transportation of farm products, and perhaps the
teaching of new crafts to make new uses of both the traditional and any new products.
Outside help will be most effective when it is shown that the new methods are more
productive and more profitable;- getting the cooperation of just one farmer or grazier
to let his land and his operation be used to demonstrate the new methods and crops,
will, as the land produces better yields and the livestock greater weight gains,
encourage his neighbors to do likewise. This in turn will make any immediate benefits foregone in the interest of flood and erosion control easier to accept.

SUMMARY

Floods and erosion are significant factors in degradation of land and desertification. There are corrective measures that can be applied; but none that will
have any lasting effect until the abusive treatment of the land is corrected. In
most of the arid and semi-arid regions where desertification is moving the fastest,
overgrazing is the principal factor. The resultant loss of the protective cover
vegetation and damage to the soil structure and infiltration capacity increases the
rates of erosion by both wind and water as well as the frequency and severity of
flooding and the occurrence of flood damages. Both soil and water are wasted, and
productivity lost.

Windbreaks, woodlot planting, range forage improvement, introduction of rotation grazing and systems for conservation farming, terracing, rehabilitation of
gullied areas, dune control, contour trenching, channel lining, checkdams, drainage
improvements; -- all are effective elements in reestablishing the quality and productivity of land that has been abused and degraded. However, these measures will be
of little use without changes in attitudes toward land use and stimuli to improve
land use methods. These can be developed by the introduction of new crops and
livestock and farming methods and grazing systems, which will be adopted when shown
to be more productive and more profitable.

SOME ELEMENTS OF THE DEVELOPMENT AND MANAGEMENT OF DESERTS

Recommendations of the Sacramento Conference

The Conference presented a selection of experiences and technologies from both developed and developing countries. To utilize the information presented and developed in the discussion, it will be necessary for the participants and their associates and their governments to select the strategies and techniques that will best meet the needs of their people.

This report has used the term "desert" in the broad sense of the Conference. It is recognized that both true deserts, with no vegetation whatever, and arid lands are included. Most of the recommendations apply to arid lands, some apply to both.

DESERTIFICATION

Many millions of people are presently being endangered by desertification processes in a region of about one million square kilometers bordering on desert areas. The factors leading to this desertification are fairly well known but not always well understood. These would include:

1. The rapid growth of population which has resulted in increased flocks, overgrazing and depletion of woody vegetation for fuel.
2. Incorrect land and water management and destructive techniques of cultivation and irrigation.
3. Climatic variation and long-range climatic change.
4. Social, political and cultural factors.

Millions of hectares of formerly productive soils are lost annually, some of this loss being irreversible. Obviously steps must be taken to contain and even to reverse these destructive processes. The technologies presented at this conference are partly well known and even if there is still a great deal to be learned about them they could theoretically be put into practice today.

Populations have been growing at a rapid rate so that even where development projects have been successfully implemented, per capita income has not substantially increased and at best, these projects have provided only a partial answer to the problems of desertification.

It seems clear that development programs will have to be accompanied by an intensive program of education and by far-reaching social and political reforms. Changes in custom and tradition where necessary have to be generated from within. The developed nations can help by:

    a. Providing capital and technical assistance for im-
        provement of educational facilities and services.
    b. Providing credits, subsidies and other incentives for
        measures leading to better land management and to
        reduction of population and livestock pressures.
    c. Establishing produce and marketing institutions.
    d. Aiding in the establishment and operation of local and
        regional research centers.
    e. Aiding in the development of local extention services
        to assist in the transfer of information developed
        through research to the farmers and other users.

DESERT DEVELOPMENT AND STRATEGIES

A number of the papers read at the Conference placed heavy emphasis on techniques and technology involving management of some basic components such as water, mineral, energy sources (including petroleum, solar, and geothermal) and the primary and secondary productivity of natural and agriculturally modified ecosystems.

Regional and national planners charged with strategy selection are often faced with simultaneously conflicting situations, especially when, as usual, the financial, labor, material, or other resource is limited. In such situations strategy decisions are often made with the hope of achieving maximal short term returns in order that other strategies may be activated later. However, in most cases two or more strategies are being carried out simultaneously. Even if local governmental agencies are emphasizing only a single strategy (e.g., a cash crop produced for export), local peoples are generally involved in food crop production, animal husbandry and harvesting of native fruits, berries and wildlife species as well as, perhaps, limited mining or home industry.

A number of different strategies are evident in the papers presented but most appear to be centered around the maximal acquisition and most efficient utilization of that basic resource in short supply -- WATER!

Further, most appear to involve the use of water for some agricultural purpose -- either field crops or domestic animals. Specific projects included in the reports that could be emphasized as strategies include the following:

    1. Development of coastal zones and regions, utilizing the
        extra food resources and cheap transportation supplied
        by the marine interface.
    2. Maximize storage of excess waters during wet seasons
        with inexpensive compartmented reservoirs and similar
        devices for use in dry seasons.

3. Where transportation to markets and technology exists,
    grow and export high value crops. In some cases this
    can include the use of controlled environment chambers
    (greenhouses).
4. By irrigation (from underground, imported surface
    waters or local impoundment in wet season) develop,
    where appropriate, local agriculture including field
    crops and animal husbandry. Although a desert area
    does not need to produce its own food, it must produce
    something resulting in an export surplus, so food can
    be purchased.
5. Utilize tourism (together with natural areas, game
    reserves and game farms) as high labor intensive source
    of foreign exchange.
6. Where possible, utilize extractive industries (mining).
7. In all possible situations (agriculture, mining)
    accomplish processing with local processing plants and
    labor, if water requirements are not excessive. The
    added value and reduced shipping charges will result in
    added value to local populations.
8. Creation of desert reserves in all areas is recommended
    to preserve a portion of the natural flora, fauna and
    land forms. Such reserves are valuable for study,
    research, and enjoyment and to preserve the indigenous
    germ plasm. If this is not done many species may
    become extinct.

WATER RESOURCES AND MANAGEMENT

Deserts and arid lands are by definition deficient in water supplies. Any
strategy for management or development, be it only the improvement of life for the
existing population, the extension of agriculture and grazing, or mining, petroleum development, industrial development or the encouragement of tourism, must
consider the most effective use of available water.

It is likely that a combination of strategies will be employed, depending on the
resources, needs, and political attitudes of the country. There sould be a
thorough survey of the water supplies available including surface supplies, ground
water, geothermal water, rainfall and the possibility of imports. The type and
extent of development must depend on the results of such a survey. It may be
necessary to adjust quality standards and use patterns to the quality of water
actually available. Maximum use must be made of water conservation measures and
the reclamation and reuse of waste waters.

Long range planning is essential. For example, in some areas there are large
bodies of fossil groundwater. Such water should be used benficially, even though
there is no recharge, but there must be a decision as to what rate of use will be
most beneficial to the people.

Some techniques to maximize the effective use of available water include improved
irrigation techniques, evaporation control, salinity control and drainage, and
water harvesting to maximize use of rainfall. Institutional and political solutions must be found, as well as technological answers. The economics of development must be realistic. The goal of long-term water planning should be a
safe and reliable supply of water designed to shield the local economy from
drought and depletion of available water.

AGRICULTURAL DEVELOPMENT

Cropping, Ranching, and Tree Plantations

Since some desert areas are being invaded by the plow, it seems important that this situation is given a close second look in many developing countries. This calls for a proper resources survey including soil, water and other investigations.

Lands which can sustain limited cultivation must be so identified and other lands should be considered for retirement from cultivation and diversion for other alternative land uses, e.g., ranching, tree plantation, widlife development, recreation, etc.

Grazing and harvesting of trees and other vegetation would have to be limited or the process of degradation would not be reversed.

Adequate means should be used to minimize losses to crops and native vegetation caused by pests.

Low water requirement and drought resistant crops and crop varieties including forage crops deserve special attention where cropping and grazing are considered desirable. Other practices, e.g., soil and water conservation practices, water harvesting, mulch farming, suitable conservation rotations, mixed cropping, etc., may be considered.

Scientists and extention workers should evaluate indigenous and traditional agricultural/farming practices in the region, for easy adoption and quick results. These methods should simultaneously be evaluated in the light of modern knowledge so that further improvement may be possible.

Mixed Farming

Recognizing the place of plant species of grazing value, shrubs and trees and by-products of some field crops as may be grown in the area, animal production in the area would deserve special attention in addition to cropping. Such an approach would provide for year-round work of a productive type to the rural people in many developing countries.

Where data on relative economics of cropping vs. animal husbandry vs. mixed farming is limited, the need to generate it is indicated. Wildlife studies need special consideration. Grazing behavior of mixed herds of domestic animals and also mixed herds having a suitable mix of wildlife with domestic animals calls for systematic studies. Wildlife may be an excellent source of meat and hides, but will not provide milk or wool. Upgrading of animals and appropriate health care have to be provided for.

Grazing and Livestock Management

In many countries, over-grazing of low rainfall desert areas is among the major factors responsible for desertification.

    1. Studies on carrying capacity of such areas should be made where such information is lacking so that regulation of grazing may be considered against this background.

2. Unrestricted grazing be discouraged as much as rapidly possible. This could be enforced easily on lands held by the Government. For private lands, suitable education and, if necessary, legislative action may be needed.

3. Instead of numbers of low quality animals, every effort has to be made to build the population of quality animals adapted to the area by reducing unproductive or less productive animals. Cross-breeding to a limited extent and introduction of new breeds/species of animals may be considered in this context.

4. Both immediate and long-range programs for installing watering points have to be formulated and implemented so as to reduce excessive grazing pressures on lands around the existing or newly developed watering points.

5. A suitable program for upgrading of the degraded areas should be developed to improve vegetative cover, reduce the incidence of erosion, and to improve the economy of the areas. Revegetation of degraded areas with natural vegetation of economic value must be considered as a first step. Further improvement through reseeding and management of better types of grasses and other vegetation, both locally available and introduced species/types should be scientifically evaluated.

6. Development and improvement of marketing systems would require close attention in areas where such arrangements are lacking.

Land Policy

Lands in many desert countries tend to be used beyond their capabilities, due to population pressures which leads to land deterioration. Land allotment policy in such areas has to be developed in keeping with the land capability. Lands which cannot sustain cultivation must be adequately safe-guarded by prescribing appropriate land use and needed practices.

The Wildlife Resources of Deserts

One of the most abused resources of the desert areas of the world is the indigenous animal resource. Its potential has, to date, been greatly underrated.

Sizable numbers of human desert dwellers have traditionally, and still do, depend on the wildlife resource as their major and often, only, source of protein. In Botswana, for example, an estimated 45% of the human population depends upon wild animals for protein.

Many desert dwellers produce various items derived from the indigenous animals. Often this may be the only source of cash income. This economic resource requires minimal infrastructural, no physical changes to the region, little or no capital investment, and necessitates no changes in life styles.

We highly recommend that the various wildlife species be investigated further wherever sizable populations currently or formerly existed. This investigation should be set up as a large-scale, long-term, on-the-ground practical trial with careful accounting of all variables.

Not until the results of these trials are available will we know whether, with low management resources and energy input, a higher or equal level of productivity (as compared to domestic stock) can be obtained.

In many cases even with lower productivity, optimal land use may prove wildlife raising as the only long-term sustained yield protein producing system possible. In many such places the domestic animals now present should be removed.

INDUSTRY

Industrial development can be an important factor in the transformation of an underdeveloped area into a modern society. The desert offers several advantages which could be utilized for industrial development such as: wide spaces, a dry climate and, in many cases, easy disposal of waste effluents without ecological deterioration.

The industrial potential of agriculture products could rapidly be tapped. The by-products of regular agricultural crops such as cotton seed and bulbs, molasses, bran and others could be the basis for the manufacture of many products. Desert grown industrial crops such as jojoba, guayule, acacias, etc., could be the basis for many sophisticated chemical products.

Governments can aid in establishing industry in desert areas by:

      a. Granting special incentives to investors who establish their manufacturing facilities in these areas.
      b. Training workers and technicians.
      c. Establishing the necessary infrastructure.
      d. Extending aid for marketing and export.
      e. Inventorying all available resources.

Countries planning mining of minerals including oil and gas should realize that these are depletable resources whereas desert industrial plants are renewable resources. Thus, careful planning is required to avoid ghost towns and economic disruptions.

The conference recommends that UNITAR and the United Nations Industrial Development Organization (UNIDO), in cooperation with FAO organize a conference to discuss:

      Arid and semi-arid zone plants as a source of industrial raw material and its processing possibilities.

      Minerals including petroleum and their utilization in deserts.

      Types of industries and industrial operations which most efficiently utilize water resources.

      Different forms of incentives to encourage the development of the above industries.

COASTAL RESOURCES

In addition to other natural resources in the hinterland of the desert coastal area, maritime oriented activities including off-shore resource exploration would seem to be decisive to the strategy for the development of the coastal desert and/or arid regions of the world.

Traditional human activities oriented towards exploitation of living marine resources, in the coastal desert and/or arid areas would be enhanced if successful mariculture were applied, modern fishing methods and fleets were introduced and rational fish processing and marketing schemes were developed.

Traditional maritime industries such as ship-building, drydock and ship repair and maintenance facilities should be regarded as compatible development schemes for coastal desert and/or arid region.

In the context of the emerging environmental constraints and other limitations to further industrial growth of the coastal areas in developed regions of the world, on one hand, and the need for continued and sustained economic growth for the benefit of mankind on the other, it would seem reasonable that institutional as well as economic policy guidelines for long term development of coastal desert and/or arid regions should be compatible with environmental consideration and consistent with yet to be defined policies for a new economic growth.

TOURISM

Tourism as a means of exploiting and developing the potential of arid and semi-arid areas should be carefully considered. Such developments are embryonic in most desert areas, except in some coastal areas.

The following aspects of the desert have tourism potential:

        a. Wildlife (viewing, photo safaris, big game hunting).
        b. The unique environment of the desert/observation,
            detailed examination (study tours) of certain aspects,
            e.g., wild foods/scenery.
        c. Health spas (tourists going to the dry region for
            health reasons).

The development of mineral spas or general spas is being viewed as a new potential strategy of fostering tourism in deserts.

A careful evaluation of the impact on desert residents, and the relative benefit to them (e.g., employment, investment opportunity) must be made before undertaking tourism development, to assure that there will be substantial benefit to the local people. A substantial infrastructure is needed for tourist development including transportation, food services, health services, and water supply.

ENERGY AND MINERALS

Energy

In desert countries energy is as vital a supply as in non-desert areas. The use
of firewood for cooking still is wide-spread in desert areas and is continuing to
destroy the remaining trees and bushes, and unless a country can organize on a
regular basis the planting of trees or bushes for firewood supply, countries
should study carefully the possibilities of replacing the use of firewood with
other energy sources.

The introduction of electricity in desert areas in addition to its energy application is vital for communications and many other uses.

Wind power units should be considered in those areas where wind conditions are
favorable. Unfortunately precise local wind condition surveys have not been
carried out in most areas, and until such surveys are available, the application
of wind power cannot be judged.

Solar energy at the present stage of development is useful for water warming, but
for electricity generation the costs are too high at present. Geothermal energy
is known to exist in many desert areas, and when geothermal energy can be developed it could also be used among many other applications for electricity generation.

If a country possesses sedimentary areas it should, as a matter of course, be a
primary task for the country to explore its underground potential, and in such
areas there is a possibility of finding fuel of many types from conventional oil,
oil shale, tar sands, and gas to coal, lignite, and peat. It is equally obvious
that the discovery in a desert country of large deposits of oil and gas or coal
cannot only contribute to the local energy supply, but could also provide an
important source of income for the country, in addition to local employment. In
general, in the energy policy of a desert country the destruction of bushes and
remaining forests through use as firewood should gradually be eliminated and any
available source of energy discussed above should be introduced and electricity
should be made available as soon as possible to the whole population.

MINERALS

It is likely that all desert areas contain mineral resources of one type or the
other. There are many mineral resources which are specific to desert areas
because they are formed in the deserts, such as saline deposits, including a
variety of salts such as potash, as well as other minerals, such as lithium.
Depending on the geology of a desert area one can expect to find materials from
gold to silver, lead, zinc, and many others, including bauxite, the raw material
for aluminum. Unfortunately, mineral exploration requires high skill and risk
capital; moreover, many of the desert areas have limited geological information,
frequently no detailed geological mapping, and modern geochemical and geophysical
surveys covering only very limited areas. It is recommended, therefore, that
desert areas consider mining exploration as one of their primary tasks, including
training of their own geologists and geochemists and geophysicists so that the
government at a fairly early stage will be able to evaluate its own underground
resources, and on the basis of this, can make policy decisions on which resources
to develop first.

Modern mineral development in desert areas need not to be water intensive, and the present situation where mineral deposits are left undeveloped because of water shortages could in many cases be overcome, by application of dry processing methods. In addition, the reuse and reprocessing of mining water can sharply reduce the total quantity of fresh water which may be required. Furthermore, most mining processes allow the use of low quality water, and so on; and thus, if the mine is within proximity of agricultural areas or urban centers, the mining development, except for the need of the personnel, could be covered by waste water.

The same policy should be applied to the oil industry in desert areas which still use in some countries large quantities of fresh water. Thus, the development of mineral resources in desert areas need not be a burden on the limited water resources of a desert country, but could make a considerable contribution to the development of such a country through the income effect, the employment effect, and supply of energy, and other material also needed in the country.

HEALTH

Introduction

This topic falls into two categories:

1. Maintaining the health, or improving the health, of the resident and immigrant desert population; and
2. Health care as a focus for tourists potential. The inherent dry nature of deserts, it should be noted, means health problems are probably generally of a less chronic nature than in more humid areas.

Recommendations

a. Examination of the traditional desert-peoples diet could reveal foods and means of keeping a desert population healthy;
b. Strategies for bringing health-care to resident desert populations should be further investigated and different country experiences collated. Mobile health services (ranging from flying doctors to barefoot doctor type) is particularly stressed.
c. Tackling nutritional needs and desert-specific health problems and diseases is critical to other developments.

INFRASTRUCTURE

In some desert areas there may be large supplies of ground water or other resources that have not been developed because there is no infrastructure. To make development possible there must be transportation systems (roads, railroads, airports) and communications (telephone or radio). There must be energy resources, a marketing structure and community services. The latter includes health services, education and vocational training, and government functions.

The infrastructure should be self-supporting eventually, but it is probable that outside capital will be needed to provide the initial services so development can begin.

PARTICIPANT LIST

- A -

Dr. Mansur M. Aba-Husayn
Assistant Deputy Minister for
   Research and Development
Ministry of Agriculture and Water
Riyadh, Saudi Arabia

Adly Abdel-Meguid
Senior Industrial Development Officer
United Nations Industrial Development
   Organization
United Nations, Room 2766
New York, New York 10017

Dr. Ahmed Gamal Abdel Samie
Vice President
Academy of Scientific Research and
   Technology
101 Kasr El Eini Street
Cairo, Egypt

Adolfo Aguilar
Center for Third World Economic
   and Social Studies
Cor. Profirio Diaz 80
Mexico 20, D.F.

Rhousmane Ahmoud
Chef de Service Planification
   Sectorielle, Ministere du Plan
B. P. 862
Niamey, Niger

Shahid Akbar
Chief (Corporate Planning)
Oil and Gas Development Corporation
Shafi Chambers, Club Road
Karachi-4, Pakistan

Kola Aladejana
Acting Director of Forestry
Federal Department of Forestry
P.M.B. 5011
Ibadan, Nigeria

Omar A. Mughram Al-Ghamdi
Mechanical Engineer, Saline Water
   Conversion Corporation
P.O. Box 4931
Jeddah, Saudi Arabia

Dr. Abdullah Al-Jahmi
Director General of Plant Production
Ministry of Agriculture
Sidi Misri
Tripoli, Libya

Laika Khidir Al-Jibury
Scientific Research Foundation
Jadiriya
Baghdad, Iraq

Ahmad A. Al-Mansour
P.O. Box 2
Alkhobar, Saudi Arabia

Nasser Mohamed Amer
Director, Irrigation Department
Ministry of Agriculture
P. O. Box No. 1161, Khormksar
Aden, P.D.R. Yemen

Abdullah Abdir-Razik Arar
Regional Land and Water Development
   Officer
Food and Agriculture Organization
P. O. Box 2223
Cairo, Egypt

Mazhar Aslam
Manager, Groundwater Development
Resources Development Corporation
Dawood Centre
Karachi, Pakistan

Mohammad Osman Atif
Officer in Charge of Personnel Training,
    Forestry Department
Ministry of Agriculture
Kabul, Afghanistan

El Hag Makki Awouda
Head, Management and Services Division,
    GUM Arabic, Sudan Forest Service
c/o Forestry Department
Khartoum, Sudan

Mesfin Aytenffisu
Regional Manager,
    Harar Regional Office
Ethiopian Water Resources Authority
P. O. Box 2
Harar, Ethiopia

Eduardo Azuara Salas
Center for Third World Economic
    and Social Studies
Cor. Profirio Diaz 80
Mexico 20, D.F.

            - B -

Dr. Omar Ba
General Director
Sahalien Institute, CILSS
c/o I.E.R.
Bamako, Mali

Dr. Amanclyte M. Babayev
Desert Institute
744012 Ashkhabad
U.S.S.R.

E. Bakutina
c/o State Committee for Science &
    Technology
11 Gorkey Street
Moscow
U.S.S.R.

Dr. Joseph Bamea
Co-Director of Conference
United Nations Institute of Training
    and Research
801 United Nations Plaza
New York, New York   10017

Roshan B. Bhappu
Vice President and General Manager,
    Mountain States Research and
    Development Co.
P.O. Box 17960
Tucson, Arizona   85731

Michael Bloome
United Nations Institute of Training
    and Research
801 United Nations Plaza
New York, New York   10017

Sergio E. Bonilla
Vice President, INIA
Santiago, Chile

Boualem Bousseloub
Resources Planner, California
Department of Water Resources
    P.O. Box 388
Sacramento, California   95802

Jakov Bradanovic
Centre for National Resources, Energy
    and Transport
United Nations
P.O. Box 20 G.C.P.O.
New York, New York   10017

Dr. Michael Bradley
Associate Professor
Department of Hydrology and Water
    Resources
University of Arizona
Tucson, Arizona   85721

Dr. Donat B. Brice
Staff Chemical Engineer
California Department of Water
    Resources
P. O. Box 388
Sacramento, California   95802

John E. Bryson
Chairman, California State Water
    Resources Control Board
1416 Ninth Street
Sacramento, California   95814

William E. Bullard, Jr.
4931 N.E. Roselawn
Portland, Oregon 97218

Joseph I. Burns
President, World Affairs Council of
    Sacramento
600 Forum Building
Sacramento, California 95814

William E. Bye
Environmental Engineer
U.S. Environmental Protection Agency
401 "M" Street, S.W.
Washington, D.C. 20460

- C -

Professor Chang Sen-dou
Department of Geography
University of Hawaii
Honolulu, Hawaii 96822

Bertrand H. Chatel
Chief, Technology Applications Section,
    Office of Science and Technology
United Nations
New York, New York 10017

Dr. C. Brent Cluff
Water Resources Research
University of Arizona
Tucson, Arizona 85721

Dr. E. Lendell Cockrum
Head, Department of Ecology and
    Evolutionary Biology
University of Arizona
Tucson, Arizona 85721

A. Gene Collins
Project Leader, Petroleum Production
    Research
U.S. Energy Research and Development
    Administration
P.O. Box 1398
Bartlesville, Oklahoma 94003

Keith R. Cooley
United States Water Conservation
    Laboratory
4331 East Broadway
Phoenix, Arizona 85040

Carlos Correa Sanfuentes
Regional Secretariat for Planning and
    Coordination
Prat 350 La Serena
P.O. Box 387
La Serena, Chile

Kit Cullen
California Office of Economic
    Opportunity
555 Capitol Mall
Sacramento, California 95814

- D -

Sami R. Daghistani
Institute for Research on Natural
    Resources
Jederiah
Baghdad, Iraq

Dr. B. C. Dando
Epidemiologist
World Health Organization
Box 54
Gaborone, Botswana

Ali Darag Ali
Ministry of Agriculture and Natural
    Resources
P.O. Box 199
Khartoum, Sudan

Christian DeClerq
Economist, U.N. Economic Commission for
    Western Asia
P. B. 4656
Beirut, Lebanon

Dr. K. James DeCook
Associate Hydrologist, Water Resources
    Research Center
University of Arizona
Tucson, Arizona 85721

Claire T. Dedrick
Secretary for Resources
California Resources Agency
1416 Ninth Street
Sacramento, California 95814

Georges Cohen deLara
Chief, Department of Water Quality,
    Sogreah, Inc.
6 rue de Lorraine
Grenoble, France

Philippe de Seynes
Director, Programme of Future Studies
United Nations Institute of Training
    and Research
801 United Nations Plaza
New York, New York 10017

Aliou M. Diallo
Programme Management Office
Sahelian Unit, Office of Technical
    Co-operation
United Nations
New York, New York 10017

Efren Diaz
Pergolas 192
Xochimilco, Mexico D.F.

John R. Donnell
Chief, Oil Shale Section
Chemical Research Branch
U.S. Geological Survey
P. O. Box 25046; Stop 939
Denver, Colorado 80225

                - E -

Lic. Rodolfo Echeverria Zuno
Center for Third World Economic and
    Social Studies
Cor. Profirio Diaz 80
Mexico 20, D.F.

Stahrl W. Edmunds
Dean, Graduate School of Administration,
    University of California Riverside
Riverside, California 92521

Dr. William C. Ellet
Director, Marketing Research
Northrop Corporation
1800 Century Park East
Los Angeles, California 90067

Azouz Ennifar
Director, Tunisian National Tourist
    Office
630 Fifth Avenue
New York, New York 10020

Dr. Mohamed Aly Ezzat
Undersecretary for Desert Irrigation
Ministry for Irrigation
70 Gumhuriya Street
Cairo, Egypt

                - F -

Michael W. Fall
Fish and Wildlife Service
U.S. Department of the Interior
Denver Wildlife Research Center
Lakewood, Colorado 80255

Egidio Feliu Acevedo
SERPLAC
Edificio Gobierno Regional
Iquique, Chile

Dr. Elias Fereres-Castiel
Area Soil and Water Specialist
LAWR: Water Science and Engineering
University of California, Davis
Davis, California 95616

Robert F. Fingado
Chief of Implementation, Delta Branch,
    Central District
California Department of Water
    Resources
P. O. Box 160088
Sacramento, California 95816

Walter W. Fisher
    Metallurgist
Arizona Bureau of Mines
University of Arizona
Tucson, Arizona 85721

Carlton E. Forbes
Chief Engineer, California Department
of Transportation
1120 N Street
Sacramento, California 95814

                - G -

Kay Gary
University of California Davis
Davis, California 95616

James J. Geraghty
President, Geraghty and Miller, Inc.
P. O. Box 17174
Tampa, Florida 33682

Professor McGuire Gibson
Oriental Institute
University of Chicago
Chicago, Illinois 60637

Herbert W. Greydanus
Chief, Division of Planning
California Department of Water Resources
P. O. Box 388
Sacramento, California 95802

Dr. Joseph G. Gringof
Director, Middle Asian Regional
   Hydrometeorological Research
   Institute
Uzbekistan, Tashkent
U.S.S.R.

              - H -

Dr. Robert M. Hagan
Professor of Water Science
University of California Davis
Davis, California 95616

Dr. Hassan H. Hajrah
Head of Biology Department
Faculty of Science
King Abdulaziz University
P. O. Box 1540
Jeddah, Saudi Arabia

Walter G. Halset
United Nations International Labour
   Organization
P. O. Box 1966
Lusaka, Zambia

Dr. Adnan Hardan
Agricultural Consultant
Ministry of Planning
P. O. Box 2018
Baghdad, Iraq

Edward L. Hastey
State Director
U.S. Bureau of Land Management
2800 Cottage Way
Sacramento, California 95825

Thomas J. Henderson
President, Atmospherics, Inc.
5652 E. Dayton
Fresno, California 93727

Lee F. Hermsmier
Agricultural Engineer
Imperial Valley Conservation Research
   Center
4151 Highway 86
Brawley, California 92227

Lonnie Hiebert
Saline Processors
Box 21
Claremont, California 91711

Diane Hilden
University of California Davis
Davis, California 95616

Dr. Colin W. Holloway
International Union for Conservation
   of Nature and Natural Resources
   (IUCN)
1110 Morges
Switzerland

Dr. David Hopcraft
Director, Wildlife Ranching and
   Research
Athi River, Box 47272
Nairobi, Kenya

H. E. Horton
Assistant Regional Director
U. S. Bureau of Reclamation
2800 Cottage Way
Sacramento, California 95825

Abdoulgader Houdairi
Hydrogeologist
P.O. Box 4411
Tripoli, Libya

Dr. William L. Hughes
Director, Engineering Energy Laboratory
Oklahoma State University
Stillwater, Oklahoma 74074

Dr. John L. Hult
President, Application Concepts and
   Technology Association
P. O. Box 1731
Santa Monica, California 90406

              - I -

El Mahdi H. Ismail
Director, Agricultural and Industrial
   Section
Ministry of Planning
P. O. Box 600
Tripoli, Libya

Dr. Shawkat Abdel-kader Ismail
Advisory to H.R.H. Governor
Saline Water Conversion Corporation
P. O. Box 4931
Jeddah, Kingdom of Saudi Arabia

- J -

Dr. Merle H. Jensen
Environmental Research Laboratory
University of Arizona
Tucson International Airport
Tucson, Arizona 85706

E. Mostafa Jermoumi
Chef de Service de la Mise en Valeur
    Agricole de Marrakech
Ministère de l'Agriculture et de la
    Réforme Agraire
Direction Provinciale de l'Agriculture
B. P. 21
Marrakech, Morocco

Darshan Johal
Centre for Housing, Building and
    Planning
United Nations
New York, New York 10017

Dr. Jack D. Johnson
Director, Arid Lands Studies
University of Arizona
845 N. Park Avenue
Tucson, Arizona 85719

Professor Arthur W. Jokela
Department of Landscape Architecture
California State Polytechnic University
3801 West Temple Avenue
Pomona, California 91768

Professor Paul H. Jones
Department of Geology
Louisiana State University
Baton Rouge, Louisiana 70803

- K -

Dr. Michel Keita
Institute de Recherche en Sciences
    Humaines
B. P. 318
Niamey, Niger

Roy J. Kelly
Planning Economist
Metropolitan Water District of
    Southern California
P. O. Box 54153
Los Angeles, California 90054

Ngare Kessely
Conseiller
Ministère des Affaires Etrangères
N'djamena, Tchad

M. Said Kettanah
Scientific Research Foundation
Jadiriya
Baghdad, Iraq

Dr. Nikolai G. Kharin
Biologist, Desert Institute
Academy of Sciences of Turkmen S.S.R.
GSP 24, 744012 Ashkhabad
U.S.S.R.

Ali Kholdbarin
Expert, Soil Conservation Bureau
Forest and Range Organization
Ministry of Agriculture and Natural
    Resources
Farahg Onobi-Mansour
Farahnal 25
Tehran, Iran

Robert E. King
Consultant
580 Shore Acres Drive
Mamaroneck, New York 10543

Ira E. Klein
Consulting Geologist
1521 El Nido Way
Sacramento, California 95825

- L -

Frank L. Lambrecht, D.T.M.L.H.
International Health Program, Health
    Sciences
University of Arizona
Tucson, Arizona 85724

Everard Mervyn Lofting
Engineering-Economics Associates
1700 Solano Avenue
Berkeley, California 94707

Hussain A. K. Lolo
Undersecretary, Ministry of Agriculture
Sanaa, Yemen Arab Republic

Janece Long
Administrative Assistant
California Secretary of State
925 "L" Street, Suite 605
Sacramento, California 95814

Manuel Lopez, Jr.
Regional Director, U.S. Bureau of
   Reclamation
P. O. Box 427
Boulder City, Nevada 89005

- M -

F. G. "Phil" Macias
Pacific Southwest Planning Officer,
   U.S. Department of the Interior
450 Golden Gate Avenue
San Francisco, California 94102

Dr. Joel Maltos Romo
Secretaria de Agricultura Zonas Aridas
Apartado Postal 426
Saltillo, Coahuila, Mexico

Billy E. Martin
Regional Director, U. S. Bureau of
   Reclamation
2800 Cottage Way
Sacramento, California 95825

John W. Masier
Chief, Planning Investigations Branch,
   San Joaquin District
California Department of Water Resources
P. O. Box 5710
Fresno, California 93755

Dr. W. Gerald Matlock
International Agriculture
University of Arizona
Tucson, Arizona 95721

Melvin Mattson
General Engineer
Membrane Processes Division
Office of Water Research and
   Technology, U.S. Department of
   the Interior
Washington, D.C. 20240

Richard L. Maullin
Chairman, California Energy Resources
   Conservation and Development
   Commission
1111 Howe Avenue
Sacramento, California 95825

Dr. Fernando Medellin-Leal
Director, Instituto de Investigacion de
   Zonas Deserticas
Universidad Autonoma de San Luis Potosi
Apartado Postal 458
San Luis Potosi, S.L.P., Mexico

Ing. Salvador Mejia-Acevedo
Center for Third World Economic and
   Social Studies
Cor. Profirio Diaz 80
Mexico 20, D.F.

Professor Aden B. Meinel
Optical Sciences Center
University of Arizona
Tucson, Arizona 85721

Dominick Mendola
Vice President and Director of
   Operations
Solar Aqua Systems, Inc.
P. O. Box 88
Encinitas, California 92024

Bashkar P. Menon
Center for Economic and Social
   Information
United Nations
New York, New York 10017

S. W. O. Menzel
Chief Executive
Reed Irrigation Systems
8400 K Magnolia
Santee, California 92071

Jewell L. Meyer
Area Soil and Water Specialist
University of California
Cooperative Extension
433 County Center, III Court
Modesto, California 95350

Robert W. Miller
Chief, Interstate Planning
California Department of Water
   Resources
P. O. Box 388
Sacramento, California 95802

Mohamed Khattar Mohamed
General Director for Planning
Secretariat of Municipalities
P. O. Box 2739
Tripoli, Libya

Lucien Monition
Scientific Advisor, Bureau de recherches
  geologiques et minieres
B. P. 6009
45018 Orleans Cedex, France

Virginia S. Mueller
Vice President, World Affairs Council
  of Sacramento
Stanford Brothers Warehouse
106 L Street
Sacramento, California 95814

- N -

Stephen J. Nicola
Senior Fishery Biologist
California Department of Fish and Game
1416 Ninth Street
Sacramento, California 95814

F. Niknam
Director General, Soil Conservation
  Bureau, Forestry and Range
  Organization
Ministry of Agriculture and Natural
  Resources
Tehran, Iran

Rose M. Nonini
Special Assistant
California Department of Water Resources
P. O. Box 388
Sacramento, California 95802

- O -

Mohamed J. Omar
Planning Director, Settlement Development
  Programme
P. O. Box 1407
Mogadishu, Somalia

Dr. James E. Osborn
Chairman, Department of Agricultural
  Economics
Texas Tech University
P.O. Box 4169
Lubbock, Texas 79409

Dr. B. F. Osorio-Tafall
Director General
Center for Third World Economic and
  Social Studies
Cor. Profirio Diaz 80
Mexico 20, D.F.

Abdullah Ouderdan
Forest Director
Forestry Department
Kabul, Afghanistan

John Dalmas Owvor Onyango
Deputy Chief Conservator of Forests,
  Forest Department
P.O. Box 30513
Nairobi, Kenya

- P -

Patricia Paylore
Office of Arid Lands Studies
University of Arizona
845 N. Park
Tucson, Arizona 85719

Professor Zvi Pelah
Head, R. and D. Authority
Ben-Gurion University of the Negev
P. O. Box 1025
Beer-Sheva, Israel

Dr. Ishwar Prakash
Principal Animal Ecologist
Central Arid Zone Research Institute
Jodhpur, India

Dr. Robert M. Pratt
Special Assistant
California Department of Food and
  Agriculture
1220 N. Street
Sacramento, California 95814

Edgar P. Price
Special Assistant to the Regional
    Director
U.S. Bureau of Reclamation
2800 Cottage Way
Sacramento, California 95825

- Q -

Dr. William J. Quirk
Deputy Chief Scientist for the Global
    Atmospheric Research Project
NASA Goddard Institute for Space Studies
2880 Broadway
New York, New York 10027

- R -

Joseph Ramamonjisoa
Directeur du Project de Development de
    l'Androy
Ministere du Development Rural et de la
    Reforme Agraire
Anosy
Tananarive, Madagascar

Robin R. Reynolds
Co-Director of Conference and
Deputy Director, California Department
    of Water Resources
P. O. Box 388
Sacramento, California 95802

Robin R. Reynolds III
California State Water Resources Control
    Board
1416 Ninth Street
Sacramento, California 95814

Ronald B. Robie
Director, California Department of
    Water Resources
P. O. Box 388
Sacramento, California 95802

Frank E. Robinson
Water Scientist, Imperial Valley Field
    Station
University of California
1004 E. Holton Road
El Centro, California 92243

Bruce C. Rogers
Special Assistant
California Secretary of State
925 L Street, Suite 605
Sacramento, California 95814

Dr. Virgilio Roig
Director, Instituto de Investigaciones
    de Zonas Aridas
Universidad Nacional de Cuyo
Gregorio Torres 2085
Godoycruz
Mendoza, Argentina

Richard E. Rominger
Director, California Department of
    Food and Agriculture
1220 N Street
Sacramento, California 95814

Dr. Adam Rose
Assistant Professor,
Department of Economics
University of California Riverside
Riverside, California 92521

Professor Boris G. Rozanov
Assistant to the Secretary-General,
    United Nations Conference on
    Desertification
U. N. Environment Programme
P. O. Box 30552
Nairobi, Kenya

- S -

Antonio Sainz Unzueta
Jefe Dpto. de Suelos
Ministerio de Asuntos Campesinos y
    Agropecuarios
Carlos Blanco No 841
Casilla de Correo No 5780
La Paz, Bolivia

Nampaa N. Sanogho
Ingenieur Principal, Direction
    Nationale des Eaux et Forets
B. P. 275
Bamako, Mali

Dr. Albert Sasson
Division of Ecological Sciences
U.N.E.S.C.O.
Paris 75007, France

Margaret W. Savage
Thermo. Mist Company
9528 Lemoran Avenue
Downey, California   90240

Glenn B. Sawyer
Supvr. Land and Water Use Analyst,
   California Department of Water
   Resources
P. O. Box 388
Sacramento, California   95802

Daniel P. Schadle
Supervisor, Lands, Development and
   Maintenance
Arizona Game and Fish Department
2222 W. Greenway Road
Phoenix, Arizona   85023

Dr. Joel Schechter
Director, Applied Research Institute
Ben-Gurion University
P. O. Box 1025
Beer-Sheva, Israel

Frank J. Schober, Jr.
Commanding General, California Military
   Department
P. O. Box 214405
Sacramento, California   95821

Abdul Gasim Seif El Din
Senior Research Officer
GUM Research Division
P. O. Box 302
El Obeid, Sudan

Gonsalo Sevilla Rodriguez
Instituto de Reforma Agraria y
   Colonizacion
Carrio 1040, Casilla 146-A
Quito, Ecuador

Richard P. Sheldon
Chief Geologist
U. S. Geological Survey
Reston, Virginia   22092

Dr. Marilyn L. Shelton
Assistant Professor of Geography
University of California Davis
Davis, California   95616

Clark W. Shraeder
Sierra Club
Box 621
Broderick, California   95605

Dr. Ousmane Silla
C.I.L.S.S.
B. P. 7049
Ouagadougou, Upper Volta

Dr. Ranbir Singh
Director of Agriculture
Government of Rajasthan
Jaipur, India

Dr. Subrata Sinha
Director, Environmental Geology (W.R.)
Geological Survey of India
C-33, Lajpat Marg
Jaipur 302001, India

George I. Smith
Geologist, U.S. Geological Survey
345 Middlefield Road
Meno Park, California   94025

Professor Norman S. Smith
Arizona Cooperative Wildlife Research
   Unit
219 Biological Science Building
University of Arizona
Tucson, Arizona   85721

Dr. J. Herbert Synder
Director, Water Resources Center
University of California Davis
Davis, California   95616

Alberto A. Sojit
Project Director
United Nations Development Program/
   Inter-American Development Bank
UNDP
P. B. O. 197D
Santiago, Chile

William Spaulding, Jr.
Geothermal Advisor, U.S. Fish and
   Wildlife Service
U.S. Department of the Interior
1730 K Street
Washington, D.C.   20240

Professor Brian Spooner
Department of Anthropology
University of Pennsylvania
Philadelphia, Pennsylvania   19174

Terry G. Spragg
A.C.T.A.
221 Third Place
Manhattan Beach, California   90266

Dr. Preston Stegenga
Director, International Center
California State University, Sacramento
6000 J Street
Sacramento, California 95819

Wesley E. Steiner
Executive Director
Arizona Water Commission
222 North Central Avenue, Suite 800
Phoenix, Arizona 85004

C. L. Stetson
Chief, San Joaquin District
California Department of Water Resources
P. O. Box 5710
Fresno, California 93755

Professor H. H. Stoevener
Department of Agriculture and Resource
    Economics
Oregon State University
Corvallis, Oregon 97331

                    - T -

Michael Takla
Project Engineer, Bechtel Inc.
P. O. Box 3965
San Francisco, California 94119

Harold E. Thomas
1339 Portola Road
Woodside, California 94062

A. Toure
Directeur de la Protection et
    Amélioration Efficace
Espèce Agro.-Pastorale
B. P. 170
Nouackchott, Mauritania,
West Africa

P.A. Towner
Chief Counsel, California Department of
    Water Resources
P. O. Box 388
Sacramento, California 95802

Dr. Thaddeus C. Trzyna
President, Center for California Public
    Affairs and
Chairman, Sierra Club International
    Committee
P. O. Box 30
Claremont, California 91711

George L. Turcott
Associate Director
Bureau of Land Management
U.S. Department of the Interior
Washington, D.C. 20240

                    - U -

Zafar Uddin
Deputy Director of Research
Agricultural Research Council
F-7/2 Islamabad, Pakistan

                    - V -

Frank P. Vandemaele
Senior Technical Adviser
United Nations Development Programme
Room DC-2062
New York, New York 10017

                    - W -

Kenneth Wahl
Water Resources Division
U. S. Geological Survey
855 Oak Grove Avenue
Menlo Park, California 94025

William E. Warner
Consultant
2090 Eighth Avenue
Sacramento, California 95818

Robert O. Watkins
District Director, California
    Department of Transportation
P. O. Box 847, 500 South Main
Bishop, California 93574

Gary D. Weatherford
Deputy Secretary for Resources
California Resources Agency
1416 Ninth Street
Sacramento, California 95814

Richard A. Weaver
Wildlife Manager, California
    Department of Fish and Game
1416 Ninth Street
Sacramento, California 95814

Ernest M. Weber
Engineering Geologist
Colorado River Board of California
107 South Broadway, Room 8103
Los Angeles, California 90012

James L. Welsh
Chief, Statewide Planning Branch,
   California Department of Water
   Resources
P. O. Box 388
Sacramento, California 95802

Ruth Welsh
California Department of Water Resources
P. O. Box 388
Sacramento, California 95802

Elizabeth Wily
Ministry of Local Government and Lands
P. Bag 006
Gaborone, Botswana

Dr. Vicente M. Witt
Chief, Environmental Program Support
World Health Organization
Pan American Health Organization
525 Twenty-Third Street, N.W.
Washington, D.C. 20037

                    - X Y Z -

Dr. James A. Young
Range Scientist
Agricultural Research Service
U.S. Department of Agriculture
Reno, Nevada 89512

INDEX